3rd International Conference on Engineering Geophysics 2015

AA005398

Al Ain, United Arab Emirates
15-18 November 2015

ISBN: 978-1-5108-9926-1

Printed from e-media with permission by:

Curran Associates, Inc.
57 Morehouse Lane
Red Hook, NY 12571

Some format issues inherent in the e-media version may also appear in this print version.

Printed with permission by Curran Associates, Inc. (2019)

For permission requests, please contact Society of Exploration Geophysicists
at the address below.

Society of Exploration Geophysicists
8801 South Yale, Suite 500
Tulsa, Oklahoma 74137-3575
USA

Phone: (918) 497-5500
Fax: (918) 497-5557

www.seg.org

Additional copies of this publication are available from:

Curran Associates, Inc.
57 Morehouse Lane
Red Hook, NY 12571 USA
Phone: 845-758-0400
Fax: 845-758-2633
Email: curran@proceedings.com
Web: www.proceedings.com

TABLE OF CONTENTS

ENGINEERING GEOPHYSICS: ADVANCES IN MULTI-METHOD INTEGRATION

NEAR-SURFACE GEOPHYSICS: NEAR-SURFACE GEOPHYSICS FOR FORENSIC APPLICATIONS

NEAR-SURFACE GEOPHYSICS: ARCHAEOGEOPHYSICS

ENGINEERING GEOPHYSICS AND NATURAL HAZARDS

ENGINEERING GEOPHYSICS AND NATURAL HAZARDS: ENVIRONMENTAL GEOPHYSICS APPLICATIONS AND NATURAL HAZARDS

ENGINEERING GEOPHYSICS: APPLICATION FOR THE ENERGY INDUSTRY

ENGINEERING GEOLOGY FROM GEOPHYSICS: GROUND STABILITY, MATERIAL PROPERTIES, SITE RESPONSE, COASTAL/SEDIMENTATION, STRATIGRAPHY, ETC.

ENGINEERING GEOPHYSICS: NEW APPROACHES AND DATA PROCESSING

NEAR-SURFACE GEOPHYSICS: GRAVITY

NEAR-SURFACE GEOPHYSICS

HYDROGEOPHYSICS: GROUNDWATER AND SHALLOW WATER STRUCTURES AND CHARACTERIZATION

BARINETERS OF GEOPHYSICS METHODS: STATE OF THE PRACTICE, STATE OF THE ART; THINKING OUTSIDE THE "OUTSIDE THE BOX" TOOL DEVELOPMENT

POSTERS

المؤتمر العالمي الثالث
للجيوفيزياء الهندسية
Third International Conference
on Engineering Geophysics

Adapting the oil exploration model to contaminated site characterization: A cost benefit perspective

John H Bradford, Boise State University, Boise, ID, USA

For decades, the oil industry has employed a working model for hydrocarbon exploration in which large-scale geophysical surveys are undertaken prior to a second phase of intensive, targeted drilling. This latter phase may be conducted in conjunction with further focused geophysical studies. The geophysical surveys provide lateral coverage and continuity that are used to drive placement of drilling locations. The reason for this approach is simple: wells are expensive relative to geophysical surveys. Also, practical limits on lateral coverage preclude optimization of exploration targets based on well information alone.

In concept, the problem of contaminated-site characterization is analogous to oil exploration; yet in practice, managers have been slow to adopt the use of geophysics at the front end of a characterization project. A more common scenario is that drilling serves as the primary exploration tool. If geophysics is used, it is often at a later stage of characterization and often only after well data are demonstrated inadequate. The reason for the reluctance to deploy geophysics first is not entirely clear. Certainly the economic drivers are similar–the lateral coverage afforded by geophysical methods in the near surface is continuous and relatively inexpensive compared to drilling or other direct sampling methods such as a cone penetrometer.

We can make some key observations that may provide some insight into the reluctance to apply geophysics: (1) site managers are more likely to have an engineering background and may have little or no exposure to geophysics in their training–lack of experience leads to suspicion of "soft" geophysical data, and (2) the economy of contaminated-site characterization is driven by punitive interests, or rather fear of punitive action, whereas oil exploration is profit-driven. The latter may lead to a greater willingness to take risk. Taken together, these observations may reveal why a culture has evolved that does not utilize geophysics as part of its best-practices working model. However, given the substantial economic and technical advantages resulting from the use of geophysics, we as geophysicists must continue to push education of site managers with the goal of bringing the full benefit of geophysical methods to bear on keeping our water supply safe and clean.

An example from Hill Air Force Base (AFB), Utah illustrates the problem. This site is typical of many U.S. Department of Defense and Department of Energy installations that dumped large quantities of liquid waste into open pits in the decades between 1940 and 1980. This misguided practice has left a legacy of clean-up problems, many of which remain to be solved. Common contaminates include light nonaqueous-phase liquids (LNAPL), typically light hydrocarbons (fuels), and dense nonaqueous-phase liquids (DNAPL)–chlorinated solvents are common examples.

Geophysical and Geotechnical Techniques: Complementary Tools in Studying Subsurface Features

Haydar Baker[1], Amir Gabr[1] & Mohamed Djeddi[2]*
[1]UAE University, P.O.Box 15551, Al-Ain, UAE
[2]USTHB, Laboratoire de Géophysique, Alger, Algeria

Summary

During site characterization for any construction project, certain hesitation and uncertainty attitude from the Geotechnical Engineers always existed towards the role of geophysical surveys; and the question which is always being asked on how the geophysical techniques can help and be useful to the geotechnical studies and to site selection procedures.

In this paper the important role of the geophysical methods is demonstrated through case studies, and how the geotechnical and geophysical techniques should be considered as complementary techniques. The results of geophysical survey can serve as guidelines for placing the geotechnical boreholes and the geotechnical data can be used to ground truth geophysical results.

Examples from the United Arab Emirates and different parts of the world are given. A reference is given to the initiative of Abu Dhabi Emirate to make the geophysical survey mandatory for obtaining the building permit for any type of construction.

Introduction

Geotechnical data is one important tool among others considered by engineers in their designs. However, it is known that the geotechnical data can in some cases be misleading and/or leads to various damages in the buildings' structures or in certain cases leads to catastrophes if there are no complementary systems of control. In other cases, geotechnical investigation could encounter complex subsurface conditions which require the use of geophysical technique to enhance the borehole data (Earley and Rudenko, 1988; Baker & Gabr 2011, Khaldaoui et al 2011). Engineers can build on any type of ground even on water without any problem by making special designs of basement and supports which are adapted to the "ground" conditions. Therefore, when a failure occurs in a building's structure, whose fault is it? Is it because of the design or is it because of the wrong ground truth information given to the engineers? To answer such questions and to avoid such discrepancy the geophysical techniques should be considered by the geotechnical engineers as their primary tools.

The geophysical methods are non-invasive and non-destructive. The use of the existing different geophysical techniques such as Shallow Seismic Refraction, Multichannel Analysis of Surface Waves (MASW), Seismic Tomography, Electrical Resistivity Tomography (ERT), Electromagnetic (EM), Ground Penetrating Radar (GPR) and others have been proven to be very effective in studying the near surface features and give the answers to various civil engineering problems encountered during and after the construction. These techniques can provide the physical parameters of the site and the subsurface model. However, the geophysical techniques remain dependant on cost-effectiveness and site conditions, i.e. the required depth of investigation, the site accessibility and the area, the nature of soil cover, the lithology, the groundwater, etc…

As a fact when utilizing the geophysical techniques in order to study the subsurface conditions of any construction site it is strongly recommended or necessary to use two or more geophysical techniques to provide more detailed information about the subsurface structures and lithology, in addition verification/calibration boreholes are also required. The depth of the boreholes should at least be equal to the investigation depth of the employed geophysical technique.

Many papers and case studies have been published on the role of geophysics in solving geotechnical problems but the doubt on the technical ability is being repeated by some to the present day. Indeed, geotechnical drilling gives detailed lithological and physical parameters of the site but it remains limited because of its small fingerprint and cannot be generalized to the whole area of the site even when a number of boreholes were drilled in the site and correlation was performed. The reasons are mainly due to the distance between the wells and the drilling depth. Positioning three to four boreholes in triangular or square settings would give excellent information on the lithology but it can easily miss a cavity or other geologic structure, e.g. fault, at the center of the setting. The advantage of the geophysical survey is that it gives a high resolution spatial coverage.

Since 2009 Al Ain City Municipality has requested the geophysical survey for all construction projects in the region. In 2011, the Municipality in collaboration with the UAE University conducted phase 1 of the Geozoning Project during which a reconnaissance geological, geotechnical, hydrogeological and geophysical surveys were conducted.

The results were presented as a map which divides Al Ain region into zones of possible risk where the implementation of the geophysical techniques and the number of methods to be used in each zone are compulsory.

In 2014, the Department of Municipality Affairs of Abu Dhabi Emirate issued a decree making the geophysical survey mandatory along with the geotechnical drilling to obtain the building permit for any type of construction in the Emirate.

Case Studies - The role of geophysical techniques:

1. Residential and Commercial Compound in Oran, Algeria:

The site was divided into many sectors. Each sector represents one to three building blocks. While preparing for the foundation platforms in sectors 4 and 6, a considerable sized cavity was accidently discovered just below the foundation limit (Fig. 1) in spite of the fact that a number of geotechnical boreholes were placed in the area. This discovery, which was not observed in any of the geotechnical boreholes, required a comprehensive subsurface investigation for any further geohazard sources that might exist in the site.

A detailed geophysical survey was conducted using different Electrical Resistivity arrays (Wenner, Dipole-Dipole and L-Shape/Baker), Slingram Electromagnetic technique (Horizontal and Vertical Dipoles) and microgravity method.

The survey measurement grid was set to be 5m x 5m and then reduced to 2.5m x 2.5m.

The combined investigation results revealed a honeycomb type of subsurface cavity network in the area (figure 2). The subsurface geophysical models showed that the dimensions of the cavities varied from few tens of centimeters to few meters.

Figure 1, Cavity location which was not seen by geotechniical borehole.

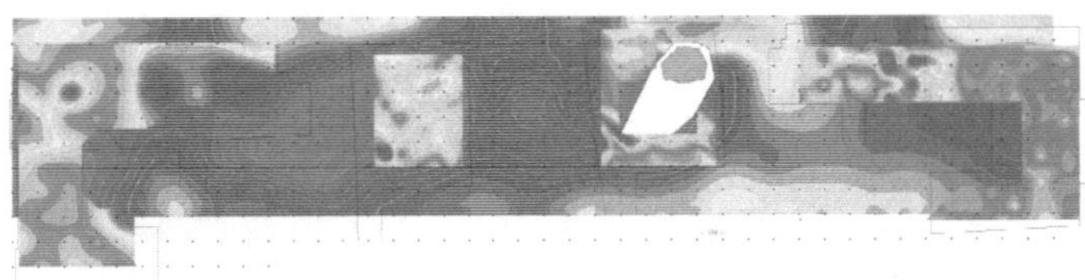

Figure 2, Site characterisation by different geophysical techniques.

2. Projects in UAE:

a) Residential Area in Al Ain:

Part of a residential area was built on an old and unmarked landfill site. The villas in the area which were constructed by taking into consideration the geotechnical results in a number of wells drilled in the region.

After a number of years, because of the nature of the landfill, ground settlement and irrigation water, cracks and wall deviations and inclinations started to appear in certain villas, road subsidence, etc.. (fig. 3). Detailed geophysical and in a later stage environmental and geotechnical surveys were conducted in the area. The results showed the extension and depths of the landfill (fig. 4).

Figure 3, Landfill effectives on building & roads.

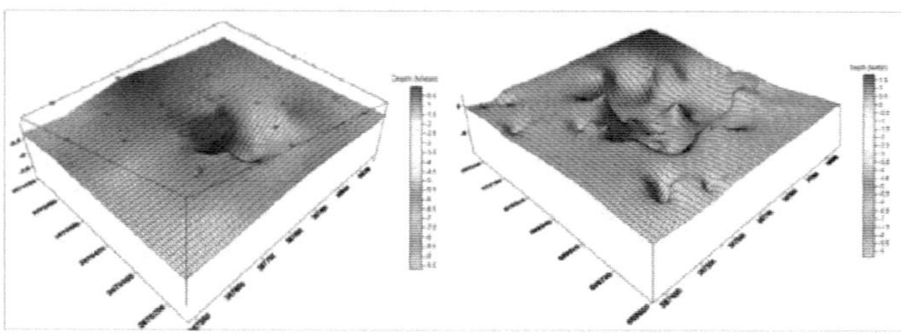

Figure 4, Geotechnical & Geophysical 3D results presentation.

Figure 4 shows a comparison between the geophysical (right) and the geotechnical (left) results. A clear advantage of the geophysical survey is indicated in the figure where more details of the landfill morphology can be seen.

Base on the result of the geophysical investigation as phase one and the environmental impact study as phase two, a number of villas had to be demolished in the affect site.

b) Sheikha Salam Mosque in Al Ain:

Because of the repeated problems in the Mosque's structure and before its third time reconstruction, it was decided to conduct a geophysical cross-hole electrical resistivity tomography to a depth of 30 meters for subsurface conditions investigation and to specify the source of hazards in the site.

Most of the geotechnical boreholes in the site which were drilled to a depth of 10 meters did not indicated the presence of a cavity or source of hazard. However, when a borehole was requested to a depth of 30 meters; there was a free-fall of 9.8 meters starting at -14.5 meters.

The results of the cross-hole electrical resistivity tomography showed the dimensions and extension of the cavity beneath the site.

3. Other examples in the Gulf region and worldwide demonstrated that in certain cases planning and designing building based only on geotechnical results and not conducting geophysical survey in the sites have created at later stages numerous geohazards problems (Styles 2011 and Miller 2013).

Which geophysical technique should be used?

Geophysical techniques are different and various and there is at least one technique which will be suitable to providing the answers on the site characterization.

Therefore, a number of questions should be asked prior to the selection of the technique. These questions can be summarized as follow:
What are we targeting in the site?
What are the possible physical properties of the site?
What is the required survey depth and with which resolution?
Which technique is more adapted to the site condition and survey requirements?
Finally and adding to the above, it is recommended to consider the cost effectiveness of the chosen geophysical technique.

Conclusions

For the past few years it has been demonstrated that whenever geophysical data were properly acquired, processed and interpreted, it can give detailed informative subsurface geologic model.

The site geophysical models can be used as guidelines to geotechnical boreholes implantation and detailed correlation between the boreholes sites.

It has been demonstrated that by conducting geophysical survey prior to geotechnical drilling can be a cost effective approach.

References

Baker H & Gabr A., 2011, Using Geophysical and Geotechnical Techniques to Delineate a Landfill in a Residential Area. First International conference of Engineering Geophysics. EAGE Conference Proceedings, EG38.

Earley K.H. and Rudenko D., 1988. Use of Geophysical Methods in a Geotechnical Investigation. Proceedings: Second International Conference on Case Histories in Geotechnical Engineering, June 1-5, 1988, St. Louis, Mo., Paper No. 2.64.

Khaldaoui F, Djeddi M, Abtout A & Baker H., 2011, Geophysical Investigation to Characterize Geological Risks – Detecting Zones of Dissolution in Karst Areas of Algeria. Second International conference of Engineering Geophysics. EAGE Conference Proceedings, EG37.

Styles P, 2011, Environmental Geophysics: Its role in the planning of 21st Century Cities. First International conference of Engineering Geophysics. EAGE Conference Proceedings, Keynote Presentation.

Miller R, 2013, Near-Surface Applied Geophysics: Unraveling the mysteries and identifying the hazards beneath the ground we walk on. Second International conference of Engineering Geophysics. EAGE Conference Proceedings, Keynote Presentation.

High Resolution Seismic Reflection in Geotechnical and Hydrogeological Applications

Marcellus Gregor Schreilechner (Joanneum Research) & Christoph Georg Eichkitz (Joanneum Research)*

Summary

Commonly, seismic reflection is used for hydrocarbon detection and monitoring of hydrocarbon reservoirs. The advantage of an active geophysical methodology with an acoustic wave field is that it can be performed according to needed target depths. For geotechnical and hydrogeological projects the targets are mostly in depths from few meters to several hundreds of meters. The vertical and lateral resolution depends strongly on the used frequencies and the velocities of the near surface layers. As high frequencies can be obtained with explosives, we are using for geotechnical applications mostly low charges of explosives or electrical detonators exclusively instead of a sledge hammer. In comparison with classical environmental geophysical methods such as geoelectrics, the application of seismic reflection has one big disadvantage – it needs huge financial and technical efforts. In this abstract we present different examples of seismic reflection for near surface applications. In the first example seismic reflection application for ground water detection is shown using sequence stratigraphic interpretation. The second example is a seismic pre-investigation for the underground railway of Vienna with ultra-shallow unexpected gas deposits and in the third example an ultra-shallow railway tunnel investigation in hard rock environment is presented.

Introduction

Seismic reflection imaging is a common tool for the interpretation of hydrocarbon reservoirs. For these purposes the target of investigation is often in depths of 1000 to 3000 m. Another field of application of seismic reflection is for engineering purposes. In this field commonly geoelectric, georadar, or seismic refraction profiling is applied. These methods have limitations in their resolution and their depth of investigation. Seismic reflection profiling offers the possibility to perform geotechnical investigations in depths from a few meters to several hundred meters. In this work three case studies for the application of seismic reflection profiling for engineering purposes are shown. In these case studies the target depth varied from approximately 20 m to 300 m.

Example 1 – High resolution sequence stratigraphy in the Styrian Basin, Austria

The Neogene Styrian Basin is located at the eastern margin of the Alps and is the westernmost subbasin of the Pannonian Basin System (Horváth and Tari, 1999). It has an elongate shape of about 100 km in length and 50 km in width. Figure 1 shows a typical seismic section through the south-eastern part of the Basin. The purpose to acquire this seismic section was to detect deep groundwater with good quality (not influenced by agricultural pollution) and sufficient quantity. Generally, groundwater wells in this region have productivities of about 5 liter per second. With the seismic reflection method combined with a sequence stratigraphic interpretation new wells were proposed. One of these proposed wells was drilled (see Figure 1, well KD) into an area with deeply inclined reflectors in a depth of around 220 m below the ground level. These deeply inclined reflectors are correlated with coarse grained sediments.

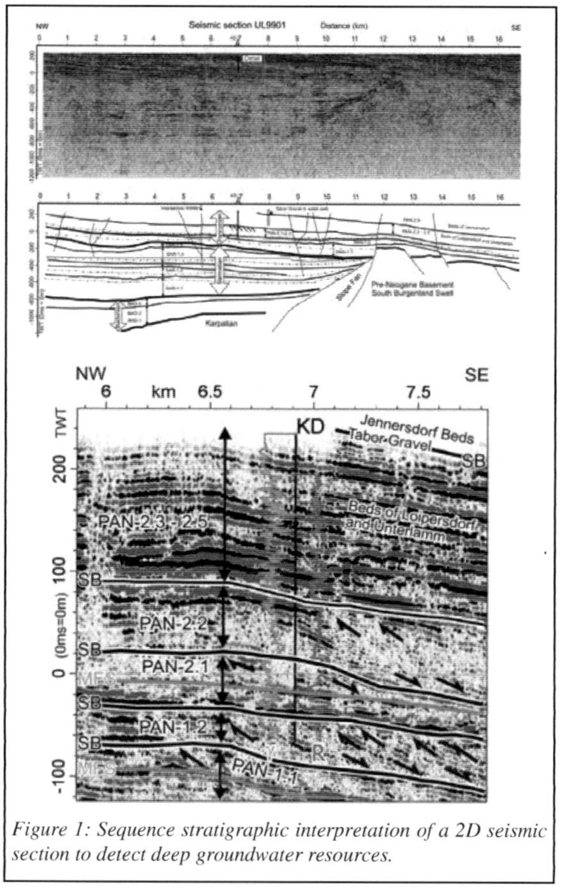

Figure 1: Sequence stratigraphic interpretation of a 2D seismic section to detect deep groundwater resources.

The new well KD has a high productivity of around 20 liters per second. It is the best productive well for the water supplier of that big area. With the help of sequence stratigraphic interpretation of seismic reflection data a very good overview of the sedimentary environment can be given. The methodology of sequence stratigraphic interpretation of seismic data (seismic stratigraphy) is based on the concepts of Mitchium et al. (1977). Vail et al. (1988), Posamentier and Allen (1999), Cantuneau (2002) and Coe (2003).

Example 2 – Deep foundations for elevated railway

The current subway network in Vienna should be expanded in southern direction. In an initial planning phase several exploratory wells were drilled along possible subway corridors. In one of these wells at a depth of 38 m below ground gas was encountered in a 20 cm thick sand layer. This gas bearing layer consists mainly of sand and gravel layers and belongs stratigraphically to Pannon D (Friedl, 1973 and Papp, 1951 and 1953). Bernhard (1993) described similar features in the southern Vienna Basin, which he interpreted as braided stream deposits. In the vicinity of this well the subway should come to the surface and continue on elevated tracks. To avoid hazards in the construction of the subway it was necessary to map the gas bearing horizon laterally. As shallow seismic reflection imaging has proven to be a valuable tool for identifying small targets at this shallow depths (e.g. Bachrach and Mukerji, 2001; Büker et al., 1998; Eichkitz et al., 2009; Khattri et al., 1979; Miller et al., 1989; Schmelzbach et al., 2005; Shtivwlman et al., 1998; Sloan et al., 2010), we decided to use this geophysical method for delineating the lateral extension of the gas bearing horizon. The aim of this project was to test the principal applicability of shallow reflection seismic for the given problem in an urban environment. Furthermore we wanted to test the effect of different geophone types, varying geophone and source distances, and different type of sources on the quality of the seismic reflection profile. The area of seismic acquisition is in urban Vienna, next to a road with heavy traffic and with a tramway line. The traffic is a big problem for the seismic acquisition as it causes a lot of random noise. To avoid this random noise it was necessary to do all measurements at night, when traffic is low and the tramway line is not operating.

In the seismic reflection profile a few structural elements are visible, which are interpreted as lateral faults with an additional vertical component. The gas-bearing sand horizon is only 20 cm thick. Thus, it is not possible to identify the top bottom of the horizon as this is below the seismic resolution. Nevertheless, it is possible to illuminate this zone as it is within the seismic visibility. Additionally, we can see high amplitude reflectors (bright spot, Khattri et al., 1979) in the data caused by the gas-bearing horizon. The gas-bearing horizon terminates toward the south at the south-dipping lateral fault (Figure 2).

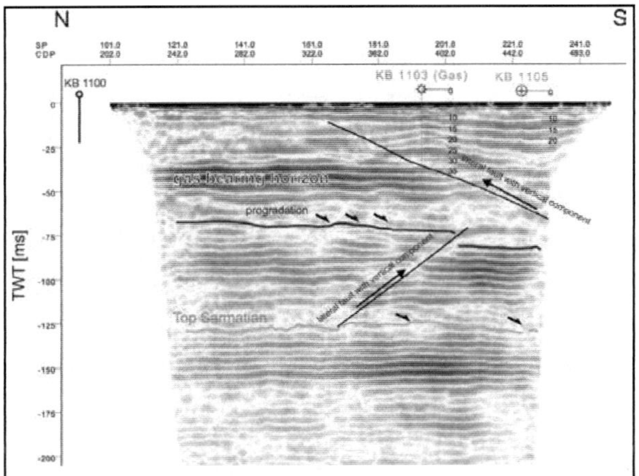

Figure 2: Interpreted seismic section in the time domain. The gas-bearing horizon has a thickness of 20 cm; therefore it is not possible to resolve the top and bottom of this horizon. Nevertheless, the gas-bearing horizons appears as a "bright spot."

Example 3 – Railway tunnel

In the course of preinvestigations for the Semmering rail tunnel it was necessary to determine the thickness of valley infill sediments of the Fröschnitztal and to identify possible fault zones in the vicinity of the planned tunnel. For this purpose a seismic reflection, a seismic refraction, and a geoelectric profile should be acquired.

The acquired seismic reflection profile has a total length of approximately 1250 m and a geophone spacing of 5 m. The source for the acquisition was small amounts of explosives. Processing of the raw data followed a standard seismic processing scheme. Additionally, we applied depth conversion of the seismic section. For this purpose a velocity model based on stacking velocities, checkshot measurements, and velocities derived from seismic refraction. In the seismic reflection profile multiple faults could be identified (Figure 3). Furthermore, it was possible to identify the transition from carbonatic rocks to crystalline rocks. Based on the seismic reflection data it was not really possible to determine the thickness of the valley infill sediments. To overcome this problem, the thickness of the valley infill sediments was determined from the seismic refraction measurements.

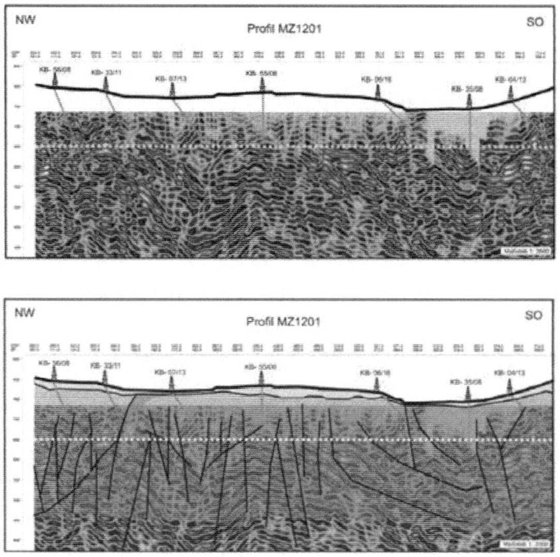

Figure 3: Uninterpreted and interpreted seismic section MZ1201. The yellow dotted line indicates the depth of the planned rail tunnel.

Conclusion

Sequence stratigraphic interpretation of a high resolution 2D seismic section is a good basis for groundwater detection. Steeply inclined clinoforms can be correlated to coarse grained sediments and therefore groundwater reservoirs can be detected very precisely. In areas with water scarcity seismic reflection investigation is highly recommended.

With the test profile WL1001 it was demonstrated that it was possible to image a gas bearing horizon at a depth of around 38 m. The best results in an urban environment with high noise level can be achieved by acquiring the data at night and using a high fold seismic section. Using geophones with a resonance frequency of 40 Hz was additionally very helpful. With this experience it is possible to plan new seismic surveys using similar acquisition parameters to confidently identify similar geological feature conditions.

In the case of the seismic reflection profile for preinvestigation of the rail tunnel it was possible to interpret several fault zones in the seismic data. Additionally, we could determine the boundary between carbonatic and crystalline rocks. Due to small thicknesses of the valley infill sediments it was not possible to determine their thickness based on seismic reflection data.

References

Bachrach, R., and T. Mukerji, 2001, Fast 3D ultra shallow seismic reflection imaging using portable geophone mount: Geophysical Research Letters: **28**, no. 1, p. 45-48.

Bernhard, M., 1993, Geophysikalisch-hydrogeologische Untersuchungen Pannoner Tiefensüßwässer im nordöstlichen Wien. Ph.D. Thesis, University Leoben.

Büker, F., A.G. Green, and H. Horstmeyer, 1998, Shallow seismic reflection study of a glaciated valley: Geophysics **63**, no. 4, p. 1395-1407.

Catuneanu, O., 2002, Sequence stratigraphy of calstic systems: concepts, merits and pitfalls. Journal of African Earth Sciences, 35, 1 – 43.

Coe, A., 2003, The Sedimentary Record of Sea-Level Change. Cambridge University Press, Cambridge.

Eichkitz, C.G., M.G. Schreilechner, J. Amtmann, and C. Schmid, 2009, Shallow seismic reflection study of the Gschliefgraben landslide deposition area – interpretation and three dimensional modeling: Austrian Journal of Earth Sciences, **102**, no. 2, p. 52-60.

Friedl, K., 1931, Über die Gliederung der pannonischen Sedimente des Wiener Beckens: Mitteilungen der Geologischen Gesellschaft in Wien, **24**, p. 1-27.

Horváth, F. and G. Tari, 1995, IBS Pannonian Basin Project: a review of the main results and their bearings on hydrocarbon exploration. In: B. Durand, L. Jolivet, F. Horváth and M. Séranne (eds.), Thr Mediterranean Basins: Tertiary Extension within the Alpine Orogen. Geological Society, London, Special Publication, 156, 195-213.

Khattri, K., R. Mithal, and V. Gaur, 1979, Pattern space of seismic anomalies associated with hydrocarbon deposits: Geophysical Prospecting, **27**, p. 1528-1534.

Miller, R.D., D.W. Steeples, and M. Brannan, 1989, Mapping a bedrock surface under dry alluvium with shallow seismic reflections: Geophysics, **54**, p. 1528-1534.

Mitchum, R. M. Jr., Vail, P.R. und S. Thompson, 1977, Seismic Stratigraphy and Global Changes of Sea Level, part 2: The depositional sequence as a basic unit for stratigraphic analysis. In: C.E. Payton (ed.), Seismic Stratigraphy – Applications to Hydrocarbon Exploration. Memoir of the American Association of Petroleum Geologists 26, pp. 53 – 63.

Papp, A., 1951, Das Pannon des Wiener Beckens: Mitteilungen der Geologischen Gesellschaft in Wien, **39-41**, p. 99-193.

Papp, A., 1953, Die Molluskenfauna des Pannon im Wiener Becken: Mitteilungen der Geologischen Gesellschaft in Wien, **44**.

Posamentier, H.W. und G.P. Allen, 1999, Siliciclastic sequence stratigraphy: concepts and applications. SEPM Concepts in Sedimentology and Paleontology no. 7, 210 pp.

Posamentier, H.W. und P.R. Vail, 1988, Eustatic controls on clastic deposition II-sequence and systems tract models. In: C.K. Wilgus, B.S. Hastings, C.G.St.C. Kendall, H.W. Posamentier, C.A. Ross und J.C. Van Wagoner (eds.), Sea-level Changes: An Integrated Approach. Special Publication, Society of Economic Paleontologists and Mineralogists, 42, pp. 19 – 30.

Schmelzbach, C., A.G. Green, and H. Horstmeyer, 2005, Ultra-shallow seismic reflection imaging in a region characterized by high source-generated noise: Near Surface Geophysics, **3**, p. 33-46.

Shtivelman, V., U. Frieslander, E. Zilberman, and R. Amit, 1998, Mapping shallow faults at the Evrona playa site using high-resolution reflection method: Geophysics, **63**, no. 4, p. 1257-1264.

Sloan, D.S., and J. Harris, 2010, Shallow seismic investigation of surface deformation associated with the Kilmichael dome, Montgomery County, Mississippi, USA: Near Surface Geophysics, **8**, p. 451-457.

Assessment of Dynamic Characteristics of Contemporary Buildings under Earthquake Loading

Abdelrahman Ashri and Aman Mwafy*
Civil and Environmental Engineering Department, United Arab Emirates University, UAE.

Summary

This paper aims at evaluating the dynamic characteristics of modern multi-story buildings with different structural systems and diverse configurations under earthquake loading. Ten reference structures of low, medium and high-rise buildings are selected, fully designed and detailed according to the international building codes and construction practice adopted in the UAE and surrounding region. The selected structures vary in height from 2 to 100 stories, with three different lateral force resisting systems, namely flat slab-columns (FSC), shear walls (SW) and tube in tube (TIT) systems. The elastic and elongated dynamic cherectristics are estimated through eigenvalue analysis (EVA) and dynamic time history analysis (THA) using the fast Fourier transformation (FFT) algorithm. It is concluded that the average period elongation at the life safety limit state is 2.2, 1.6 and 1.1 for the FSC, SW and TIT systems, respectively. At the collapse prevention performance level, the period elongation for the above-mentioned systems is 2.8, 2.0 and 1.3, respectively. An effective stiffness of 0.5EI, 0.8EI and 1.0EI is proposed for design based on this study for the vertical members of the FSC, SW and TIT structural systems, respectively.

Introduction

The dynamic characteristics of structures, such as the periods of vibration, are essential for the structural design of buildings under dynamic loading. Reinforced concrete (RC) structures experience strength deterioration and stiffness degradation during earthquake shaking due to concrete cracking and steel yielding of different structural members. The structure becomes soft and responds with longer natural periods as a result of this degradation in strength (e.g. Elnashai and Mwafy 2002; Trifunac *et al.* 2001). Therefore, it is important to accurately estimate the dynamic cherectristics of structures during strong earthquakes.

The elongated fundamental period, which represents an important indicator of the degree of inelasticity of a structure under earthquake loading, is a key factor to evaluate the structural performance. One of the fundamental concepts learnt from the response of structures under strong earthquakes is that the level of damage and the dynamic characteristics of structures (i.e., periods of vibration and mode shapes) are closely correlated (Elnashai *et al.* 2010). A number of damage models based on period elongation were suggested to estimate the global damage of RC structures (e.g. Ghobarah *et al.* 1999; Williams and Sexsmith 1995).

Due to the rapid developments and large investments in the construction industry, the Abu Dhabi government has recently enforced unified building codes, namely the Abu Dhabi international building code (DMA 2013). This building code is based on the international building code with minimum amendments in the current implementation phase (ICC 2012). Hence, there is a pressing need for calibrating the seismic design factors of the new code, such as the periods of vibration, for buildings with different characteristics. The main objectives of this study are thus twofold: (i) to evaluate the dynamic characteristics of a wide range of building with different configurations and structural systems at various performance levels; and (ii) to propose effective stiffness values for analysis and design in order to arrive at a reliable estimation of earthquake forces.

Methodology and Results

Ten reference structures of low, medium and high-rise buildings are selected based on a collected inventory to represent the contemporary buildings in the UAE (Mwafy 2012). The selected structures are RC buildings of 2, 8, 18, 26, 40, 50, 56, 66, 80 and 100 stories. The ten buildings have six different layouts and three different structural systems, as show in Table 1.

Table 1: Selected structures to represent modern buildings.

No.	System	Ref.	No of stories	Structural system
1	i	2 St	2-story	flat slab-columns (FSC)
2		8 St	8-story	
3	ii	18 St	18-story	shear walls (SW)
4		26 St	26-story	
5		40 St	40-story	
6		50 St	50-story	
7		56 St	56-story	
8		66 St	66-story	
9	iii	80 St	80-story	tube in tube (TIT)
10		100 St	100-story	

The reference buildings are fully designed and detailed as per the latest international building codes and construction practice adopted in the UAE in order to arrive at the optimum concrete cross-sections and reinforcing steel for all structural members (ACI-318 2011; ICC 2012). Detailed three-dimensional (3D) finite element (FE) models are developed for the design of the reference buildings using the structural analysis and design software ETABS (CSI

2011). Figure 1 depicts the developed 3D FE models of the selected buildings along with corresponding structural system.

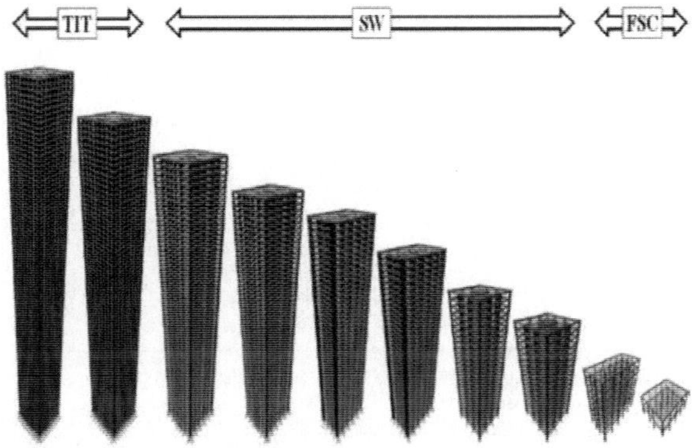

Figure 1: Design models of the reference buildings with corresponding structural system.

Moreover, detailed fiber-based analytical models are developed for the ten reference buildings using the modern analysis platform ZEUS-NL (Elnashai *et al.* 2012). The developed models are used to conduct EVA and THA. All buildings are effectively idealized to enable monitoring the stresses and strains during the inelastic dynamic simulations. For the SW buildings, a 2D idealization is adopted due to the symmetry of the buildings layouts and the negligible torsional effect. This modeling approach enabled reducing the number of elements and the analysis time significantly. On the other hand, for the FSC and TIT buildings, detailed 3D idealization approaches are adopted to accurately idealize the internal and the external structural systems and their intricate seismic response.

Two Performance levels are adopted in the present study to assess the period elongation, namely life safety (LS) and collapse prevention (CP) limit states. The interstory drift ratio (IDR) is considered as the primary performance criterion to estimate the damage states of the reference buildings. For the CP limit state, IDRs of 4.0%, 2.27% and 1.80% are adopted for the FSC, SW and TIT structural systems, respectively. The LS limit state was considered equal to 50% of the CP limit state (ASCE/SEI-41 2007).

The elastic and elongated periods are estimated through EVA and THA using the fiber-based and the design models of the reference structures. Figure 2 shows a comparison between the elastic fundamental periods obtained from the design models and fiber-based models of the ten reference structures, which represent different structural systems, buildings layouts and heights. The comparison shows that the elastic periods obtained from the design models are slightly longer than those obtained from the fiber-based models. This is attributable to the steel reinforcement considered in the fiber-based models unlike the design models in which the stiffness is only based on the concrete cross-sections. Moreover, the actual material strength values considered in the fiber-based models increase stiffness and reduce periods. The presented results validate the fiber-based numerical models of the reference structures, which are used to estimate the elongated periods.

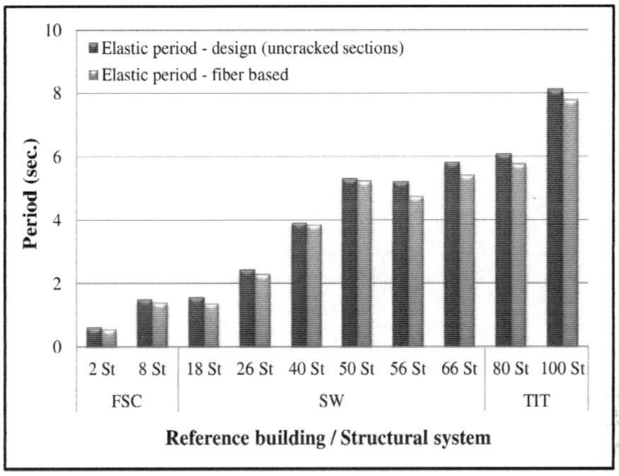

Figure 2: Comparison between elastic fundamental periods obtained from the design and the fiber-based models of the ten reference structures.

The elongated fundamental periods of the reference buildings are calculated in the present study at the adopted LS and CP performance criteria for different structural system. Figure 3 depicts a comparison between the elastic fundamental periods obtained from EVA and the inelastic periods estimated from THA at the LS and CP limit states. The period elongations at the above-mentioned performance limit states are illustrated in Figure 4.

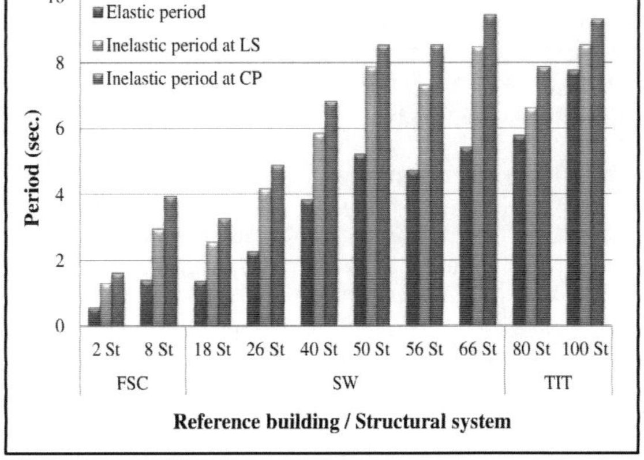

Figure 3: Comparison between the elastic fundamental periods and inelastic periods estimated at the LS and CP performance limit states.

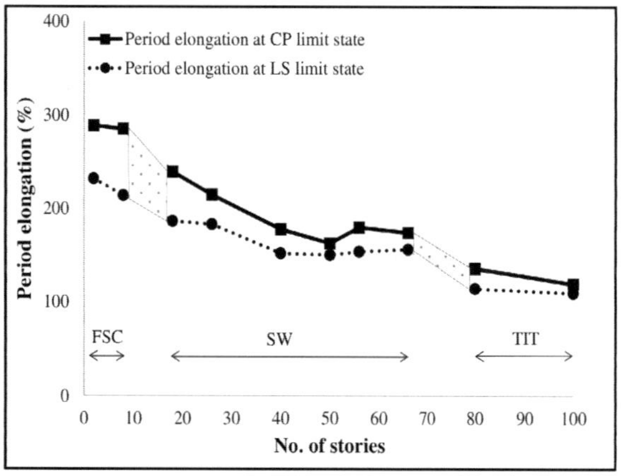

Figure 4: Comparison between the inelastic period elongation at the LS and CP limit states obtained from THA for ten reference buildings.

The results show that the average period elongation is 2.2, 1.6 and 1.1 for the FSC, SW and TIT structural systems, respectively. At the CP performance level, the period elongation is 2.8, 2.0 and 1.3 for the above-mentioned three systems, respectively. These levels of elongation clearly reflect the damage states of the structure under seismic loads. The observed high period elongation of the FSC system is attributed to the spread of inelasticity and damage in all structural members throughout the building height. The low period elongation of the TIT system is due to the limited inelasticity and concentration of damage in the very rigid internal core walls that represent the inner tube. The period elongation of the SW system falls between the two above-mentioned structural systems.

Effective stiffness values are typically used in elastic seismic analysis and design procedures. The building code for the design of concrete structures ACI-318 (2011) recommends an effective stiffness of 0.7EI for uncracked walls, 0.7EI for columns and 0.35EI for beams, where EI represents the stiffness of the uncracked section. The building standards for rehabilitation of existing buildings ASCE/SEI-41 (2007) recommends 0.8EI, 0.7EI and 0.5EI for uncracked walls, columns and beams, respectively. Other values were also suggested in the literature. For instance, Paulay and Priestley (1992) suggested 0.8EI for uncracked walls and columns and 0.4EI for rectangular beams. In order to evaluate the effective stiffness values used in design procedures, comparisons are conducted between the values recommended by the code provisions and previous studies with the results of the current work, as illustrated in Figures 5 and 6.

The results presented in Figures 5 and 6 show that the actual elongated periods of the FSC system at the LS limit state are higher than those obtained using the effective stiffness values recommended by the code provisions and previous studies. On the contrary, the elongated periods of the TIT system are notably lower than those obtained using the effective stiffness values recommended by the code provisions and previous studies. The elongated periods of the SW system are higher than those obtained using the recommended effective stiffness for the 18- to 50-story buildings, while the 56- and 66-story buildings show comparable periods from inelastic analysis and the recommended effective stiffness.

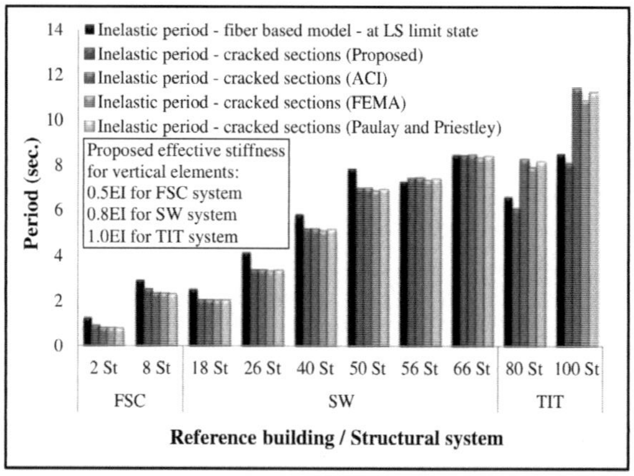

Figure 5: Comparison between the inelastic fundamental periods using different effective stiffness values along with those calculated from THA at the LS limit state.

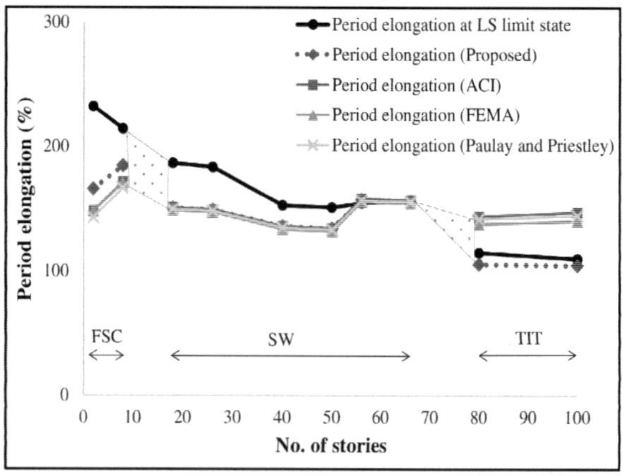

Figure 6: Comparison between the period elongations obtained using different effective stiffness values along with those calculated at the LS limit state.

Based on the present results, it is proposed in this study to reduce the effective stiffness of the FSC columns to 0.5EI. The suggested reduction in the column stiffness slightly increases the period of the 2- and 8-story buildings without exceeding the actual inelastic periods at the LS limit state. For the SW system, it is clear that the period elongation is related to the building height and layout. To avoid suggesting non-conservative stiffness values, it is recommended to use the values adopted by ASCE/SEI-41 (2007) and Paulay and Priestley (1992) for walls (i.e. 0.80EI).

For the TIT system, the results clearly reflect that the effective stiffness values recommended by design provisions and previous studies are highly non-conservative since they result in much longer periods than the actual inelastic periods at the LS limit state. This is expected to endanger the structure since the non-conservative periods may result in underestimating design forces. It is therefore recommended to use the full stiffness of structural members to arrive at conservative inelastic periods for the design of the TIT system.

Conclusions

It is concluded based on the results obtained from this comprehensive study that the period elongation at the LS limit state was 2.2, 1.6 and 1.1 for the FSC, SW and TIT structural systems, respectively. At the CP performance level, the average period elongation was 2.8, 2.0 and 1.3 for the above-mentioned systems, respectively. It was also suggested in this study to employ an effective stiffness of 0.5EI for the columns of the FSC structural system. For the SW system, the ASCE-41 recommended effective stiffness for walls (0.8EI) is adequately conservative. For the TIT system, the effective stiffness values recommended by the design provisions and previous studies were highly non-conservative. Hence it was suggested to employ fully non-cracked sections to arrive at conservative inelastic periods of vibration for the structural design of the TIT buildings.

Acknowledgements

This work was supported by the United Arab Emirates University under research grants nos. 31N132 and 31N070. This support is highly appreciated.

References

ACI-318, 2011, Building Code Requirements for Structural Concrete and Commentary, American Concrete Institute, Detroit, Michigan.

ASCE/SEI-41, 2007, Seismic rehabilitation of existing buildings, Standard ASCE/SEI 41-06, American Society of Civil Engineers, Reston VA.

CSI, 2011, ETABS - Integrated building design software, Computers and Structures, Inc., Berkeley, California.

DMA, 2013, Abu Dhabi International building code, Department of Municipal Affairs, Abu Dhabi, UAE.

Elnashai, A. S., Gencturk, B., Kwon, O.-S., Al-Qadi, I. L., Hashash, Y., Roesler, J. R., Kim, S. J., Jeong, S.-H., Dukes, J., and Valdivia, A., 2010, The Maule (Chile) Earthquake of February 27, 2010: consequence assessment and case studies, Mid-America Earthquake Center, University of Illinois at Urbana-Champaign Urbana, IL.

Elnashai, A. S., and Mwafy, A. M., 2002, Overstrength and force reduction factors of multistorey reinforced-concrete buildings: The Structural Design of Tall Buildings, 11, no. 5, 329-351.

Elnashai, A. S., Papanikolaou, V., and Lee, D., 2012, Zeus-NL - A System for Inelastic Analysis of Structures - User Manual, Mid-America Earthquake Center, University of Illinois at Urbana-Champaign, Urbana, IL.

Ghobarah, A., Abou-Elfath, H., and Biddah, A., 1999, Response-based damage assessment of structures: Earthquake Engineering & Structural Dynamics, 28, no. 1, 79-104.

ICC, 2012, International building code, International Code Council, Country Club Hills, IL.

Mwafy, A. M., 2012, Classification and idealization of the building stock in the UAE for earthquake loss estimation: Proceedings of 15th World Conference on Earthquake Engineering, International Association for Earthquake Engineering, Lisbon, Portugal.

Paulay, T., and Priestley, M. J. N., 1992, Seismic design of reinforced concrete and masonry buildings, John Wiley & Sons, Inc., NY.

Trifunac, M. D., Ivanovic, S. S., and Todorovska, M. I., 2001, Apparent periods of a building. I, Fourier analysis: Journal of Structural Engineering, 127, no. 5, 517-526.

Williams, M. S., and Sexsmith, R. G., 1995, Seismic damage indices for concrete structures, a state-of-the-art review: Earthquake Spectra, 11, no., 319.

Seismic pavement testing

Nils Ryden, Engineering Geology, Lund University, Sweden*

Summary

Seismic surface wave testing can provide valuable data for material characterization of pavement layer properties. Measurements can be performed as quality control during or after pavement construction. The thickness and stiffness of the top layer in a pavement construction can be evaluated in a simple manner based on Lamb wave propagation in a free plate. However, due to the complex interaction of higher modes of Lamb type of waves in the embedded layers the evaluation of deeper pavement layers is more challenging. A method for full-wavefield inversion is described and demonstrated on a real pavement construction. Recent developments towards non-contact scanning seismic testing of pavements using microphones is also presented.

Introduction

Traditional pavement design is based on empirical models, with trial and error as basis for the decisions. With this approach it is difficult to optimize natural and financial resources in each project. Therefore, modern pavement design is moving towards mechanistic and analytical models. A prerequisite for using these models is that material properties, such as Young's modulus (E-modulus), can be measured and validated in the field. Stiffness properties obtained from seismic velocities are low strain parameters that can be used in analytical pavement design (Ekdahl et al., 2004; Gudmarsson et al., 2014).

Since the early studies by Jones (1962) and Vidale (1964) seismic surface wave methods have been recognized as a tempting approach for nondestructive evaluation (NDE) of pavements. The introduction of the Spectral Analysis of Surface Waves (SASW) method (Heisey et al., 1982; Nazarian, 1984) has resulted in a widespread use of the method (Goel and Animesh, 2008). In this paper we describe an alternative procedure to the SASW method, which differs in data acquisition, data processing, and evaluation. The proposed approach is based on a theoretical study of guided waves in pavement systems (Ryden and Lowe, 2004). Data collection and processing is based on the Multichannel Analysis of Surface Waves (MASW) method (Park et al., 1999). The evaluation of pavement layer properties can be divided into two different methods; one quick and simple method for the stiff top layer and a second more advanced full wavefield inversion for all pavement layers.

Theory

A typical pavement structure consists of a very stiff asphalt or concrete layer, relatively stiff base layer (compacted granular material), and the subgrade material (natural soil). The stiffness ratio between the layers is usually large. Wave propagation along the pavement surface is similar to Lamb waves in free plates. The fundamental anti-symmetric mode (A0) is the dominating type of Lamb wave that can be excited in the stiff top layer from a vertical source at the surface (Ryden and Lowe, 2004). The lower velocity base layer is forced (tuned) to vibrate with the same frequency and horizontal wave number as the asphalt layer. However, since the stiffness of the base layer is much lower than the asphalt layer, higher modes of Lamb wave type are generated in the base layer. All these higher modes are leaky and energy is radiating down into the lowest velocity layer (subgrade).

A theoretical layer model (Table 1) is used to exemplify the main features of guided waves in pavements. In Figure 1a the first 120 modes from the theoretical layer model have been traced, in the complex wave number domain (Ryden and Lowe, 2004). Figure 1b shows the predicted phase velocity spectrum at the surface, calculated from the same layer model but by using the forward model from the proposed inversion algorithm (Ryden and Park, 2006).

Table 1: Theoretical pavement profile

Layer	h (m)	V_S (m/s)	ρ (kg/m^3)	ν
1	0.20	1400	2000	0.35
2	0.60	500	2000	0.35
3	∞	100	2000	0.35

In Figure 1a all dispersion curves are grey-scale coded with respect to the attenuation due to leakage. Black parts represent low leakage (efficiently propagating guided waves) and grey parts represent higher leakage (strongly attenuating waves). The two lowest dark phase velocity asymptotes (A and B) in Figure 1a correspond to interface waves with almost zero displacement at the surface. In Figure 1b the excitability of different modes is also taken into account and only propagating modes with dominating excitability at the surface are visible. Lamb wave dispersion curves, A0 and S0, calculated from the top layer as a free plate are superposed in grey in Figure 1b. At low frequencies these dark branches are separated but at higher frequencies all branches merge together and form the trend of the A0 dispersion curve corresponding to the top layer as a free plate (Ryden et al., 2004). This A0 trend can be used for a simplified inversion of the stiffness and thickness of the top layer (Jones, 1962). To evaluate the properties of embedded deeper layers the separated branches at lower frequencies need to be resolved. By comparing Figure 1a and Figure 1b it is evident that the measurable small branches of dispersion curves (Figure 1b) cannot be easily numbered.

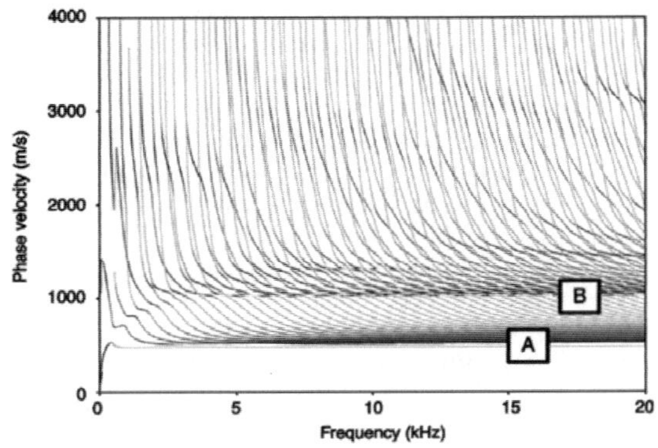

(a)

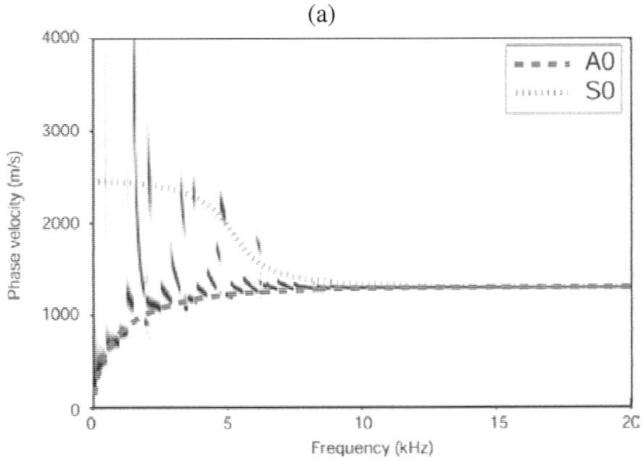

(b)

Figure 1: (a) Predicted theoretical dispersion curves from the example pavement profile (Table 1). (b) Predicted phase velocity spectrum at the pavement surface with superposed Lamb wave dispersion curves corresponding to the top layer as a free plate.

Evaluation of Top Layer

A field example is presented to illustrate how the stiff top pavement layer can be evaluated using Lamb waves in a free plate. Data was collected using a fixed accelerometer and multiple impacts (dx=0.05 m) at different distances from the receiver (Ryden et al., 2004). At this pavement structure in Malmoe, Sweden, the asphalt thickness (h) was known to be 0.13 m. All recorded traces have been compiled to a multichannel record, plotted in Figure 2a. Figure 2b shows the resulting MASW phase velocity spectrum (Park et al. 1999). Lamb wave dispersion curves corresponding to a free plate with V_S=2162 m/s, ν=0.30 and h=0.13 m are superposed on top of the phase velocity spectrum (Figure 2b). Most of the visible energy in the full wavefield image matches quite well with the A0 Lamb wave dispersion curve with the given thickness and a realistic shear wave velocity, representative for the asphalt stiffness at high frequencies (Gudmarsson et al., 2014).

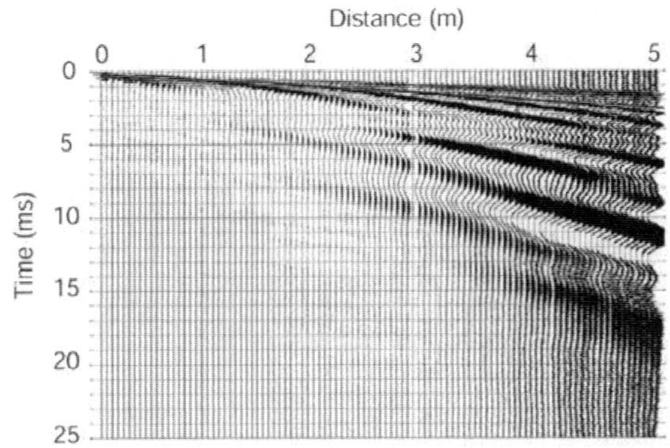

(a)

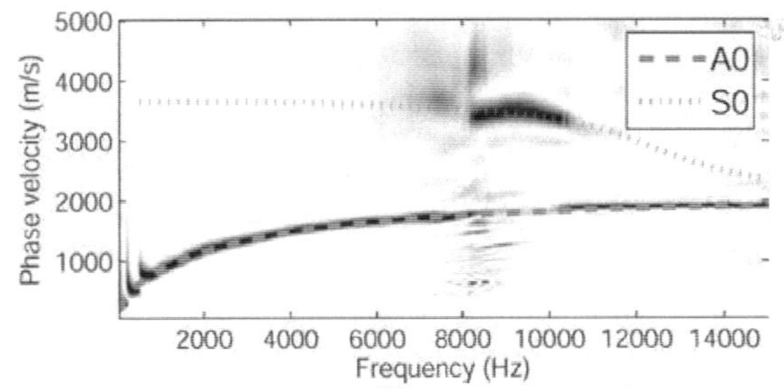

(b)

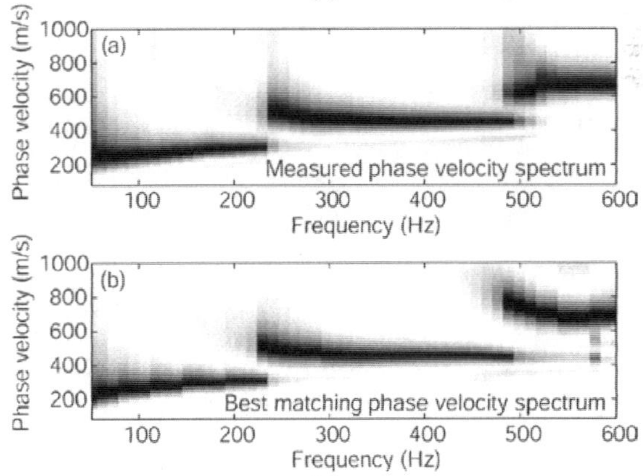

(c)

Figure 2: (a) Measured data at the pavement surface. (b) Corresponding high frequency phase velocity spectrum. (c) Low frequency phase veolocity spectrum and best matching theoretical phase velocity spectrum from a multilayered inversion of deeper embedded layers.

Evaluation of Multiple Embedded Layers

As discussed earlier in the theoretical section it is the measurable branches at lower frequencies that reveal the presence of the embedded deeper layers. In general it is not possible to identify the exact mode number of these branches of higher mode dispersion curves (Ryden and Park, 2006). To evaluate the shear wave velocity and thickness of the deeper embedded layers it is therefore important to utilize the full-wavefield spectrum. The proposed method to evaluate a multilayered model is based on a global search algorithm, fast simulated annealing, which is used to find the best possible match between the measured and the modelled complete phase velocity spectrum of surface waves (Ryden and Park, 2006). It should be noted that this inversion scheme can also be used in conventional surface wave applications (Socco et al., 2010; Dou and Ajo-Franklin, 2014).

The theoretical phase velocity spectrum (forward model) is calculated from a known layer model (m) and the field set-up geometry. Each layer is assumed to be homogeneous and isotropic with material properties defined by V_S, Poisson's ratio (ν), and the bulk density (ρ). The forward model should ideally mimic the field set up and measured response as closely as possible to generate a phase velocity spectrum that is directly compatible with the measured replica. Further details on the forward model are given in (Ryden and Park, 2006).

The predicted phase velocity spectrum (S_{pred}) is compared to the observed phase velocity spectrum (S_{obs}) through the objective function

$$M(\mathbf{m}) = \left(1 - \frac{\sum_{i=1}^{K}\sum_{j=1}^{N} \left|\mathbf{S}_{ij\,obs}\right| * \left|\mathbf{S}_{ij\,pred}\right|}{\sum_{i=1}^{K}\sum_{j=1}^{N} \left|\mathbf{S}_{ij\,obs}\right|^{2}} \right) \quad (1)$$

where M represents the mismatch between the two phase velocity spectra. The goal of the inversion procedure is to find a layer model whose corresponding phase velocity spectrum matches the measured spectrum best. This case represents the global minimum of the objective function, Equation (1).

The inversion procedure is here exemplified on the measured data presented in Figure 2a (Ryden and Park, 2006). The given layering at the test site is presented in Table 2. The given thickness of each layer is accurately measured and controlled during the construction of the pavement and the values in Table 2 can be considered as the true reference values. In an attempt to get the best possible reference seismic wave velocities, surface and compression wave measurements were taken on top of each layer during the construction of the pavement. The asphalt material is viscoelastic and this property by itself makes the seismic wave propagation become dispersive (Gudmarsson et al., 2014). To account for the viscoelastic properties, V_S of the asphalt layer is expressed as a power-law function ($V_S = b1\, f^{\,b2}$) of frequency (f) during the inversion. Here b1 and b2 are coefficients to be determined through the inversion procedure. Core samples of the asphalt layer were taken in the same section and tested in the laboratory. These measurements were carried out to get the best possible reference data for comparison with the results obtained from the inversion of all layers. However, it should be noted that V_S of compacted pavement materials is sensitive to moisture content, ageing effects, and the in-situ state of stress, and may consequently change with time and climate during construction of the pavement.

Figure 2c shows the measured and best matching phase velocity spectrum from the inversion. This is the low frequency part of the spectrum presented in Figure 2b. The corresponding best matching layer model is presented in Table 2 where the parameters from the inversion are given in parentheses. Inverted parameters, corresponding to a mismatch value M of 9.1%, agree quite well with the reference data also given in Table 2. The inverted thickness of the top layer (0.129 m) also agrees well with the evaluated thickness from the approximate Lamb wave analysis (0.130 m) (Figure 1b).

The inverted viscoelastic function for V_S ($880f^{\,0.070}$) of the asphalt layer gives a comparatively low velocity (1725 m/s) at high frequencies (15 kHz) when compared to the constant value obtained from the Lamb wave analysis (2162 m/s). This can be explained from the simplified viscoelastic function used in the inversion. More viscoelastic coefficients are necessary to express V_S over a wide frequency range (Gudmarsson et al., 2014).

Table 2: Field pavement construction with results from the inversion of all layers in parenthesis.

Layer	h (m)	V_S (m/s)	ν	ρ (kg/m3)
1	0.130 (0.129)	$880\,f^{\,0.070*}$ ($888\,f^{\,0.059}$)	0.30	2400
2	0.300 (0.308)	350 (371)	0.11	2300
3	0.200 (0.203)	320 (368)	0.20	2100
4	>5.000	213 (192)	0.41	2100

*Obtained from laboratory test on core samples and representative for 13°C.

Towards Non-Contact Seismic Testing

In the example above seismic measurements were taken using one fixed accelerometer and multiple impacts at different distances from the receiver (Figure 2a). This approach is simple and flexible but very slow for practical implementation in modern pavement management systems.

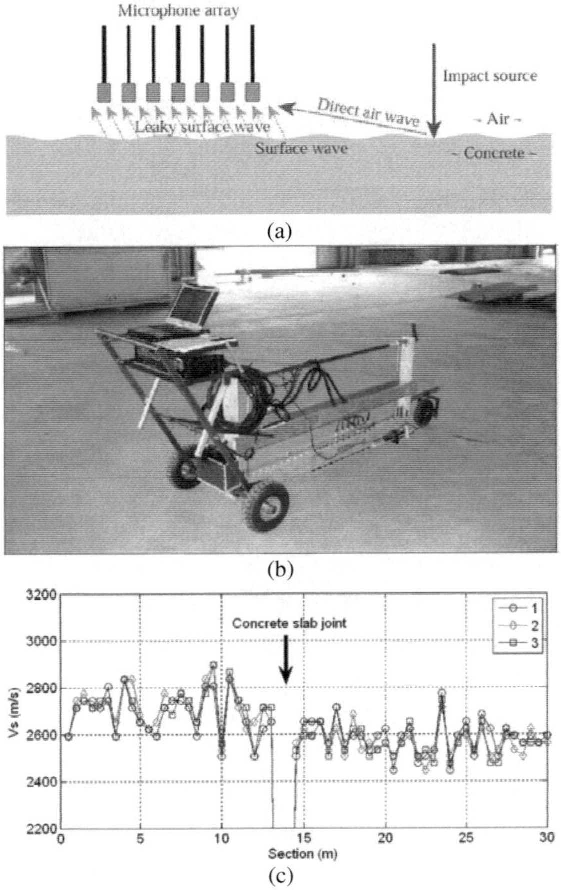

Figure 3: (a) Schematic illustration of a non-contact microphone array. (b) Prototype trolley with impact source and non-contact microphone array. (c) Results from a repeatability test along a 30 m long line over 2 concrete slabs.

To move towards faster and rolling (scanning) measurements non-contact sensors is an attractive alternative.

Zhu and Popovics (2001) showed that a directional microphone can be used as an air-coupled sensor to pick up the out-of-plane motion created by supersonic surface waves propagating over a concrete surface. In recent years this approach has been studied further focused on air-coupled non-destructive characterization of concrete and pavements (Ryden et al. 2009). Figure 3 shows an example where a multichannel array of microphones and an automatic source are attached on a small trolley so that measurements can be taken while moving. Results in Figure 3c demonstrate good repeatability in V_S measurements along a concrete slab using the prototype trolley in Figure 3b. However, an uneven surface where the distance between all microphones and the pavement surface is not constant can generate errors in the measurements (Bjurström and Ryden, 2014).

In the above example the surface wave velocity in the concrete is faster compared to the acoustic velocity in air. However, the same approach can also be applied to softer granular pavement layers utilizing the evanescent acoustic field in the air close to the soil surface (Ryden, 2010).

Conclusions

Results show that the measurable wavefield at the surface of a pavement structure is dominated by leaky quasi-Lamb waves in the first and second layer. The fundamental anti-symmetric mode of vibration is the dominating mode generated in the stiff top layer. This mode drives the complete system and continuity across the boundaries generates higher order modes in the embedded second layer. The interaction of leaky Lamb waves in the first two layers results in large variations in the excitability and the attenuation, so that only the waves corresponding to certain portions of the dispersion curves are measurable at the pavement surface.

The top layer thickness, and stiffness properties can be obtained automatically in the field by a Lamb wave analysis of the measured phase velocity spectrum. However, the inversion of deeper embedded layers is based on a more computationally demanding inversion of the full-wavefield spectrum.

Non-contact measurements using ordinary audio microphones as receivers have shown promising results for faster and more economically pavement testing in the future.

Acknowledgements

The financial support from the Swedish Transport Administration is greatly appreciated.

References

Bjurström, H., and N. Ryden, 2014, Effect of surface unevenness on non-contact surface wave measurements using a rolling microphone array: 41st Annual Review of Progress in Quantitative Nondestructive Evaluation, AIP Conference Proceedings, **1650**.

Dou, S., and J. B. Ajo-Franklin, 2014, Full-wavefield inversion of surface waves for mapping embedded low-velocity zones in permafrost: GEOPHYSICS, **79**, no. 6, EN107–EN124.

Ekdahl, U., P. E. Bengtsson, and N. Ryden, 2004, A new framework for analytical pavement design based on systematic control during construction work: Proceedings of the 14th Nordic Geotechnical Meeting, E55–E80.

Goel, A., and D. Animesh, 2008, Nondestructive testing of asphalt pavements for structural condition evaluation: a state of the art: Nondestructive Testing and Evaluation, **23**,121–140.

Gudmarsson, A., N., Ryden, H. Di Benedetto, C. Sauzéat, N. Tapsoba, B. Birgisson, 2014, Comparing Linear Viscoelastic Properties of Asphalt Concrete Measured by Laboratory Seismic and Tension–Compression Tests: Journal of Nondestructive Evaluation, **33**, 571-582.

Heisey, J. S., K. H. Stokoe, and A. H. Meyer, 1982, Moduli of pavement systems from spectral analysis of surface waves: Transportation Research Record, **852**, 22–31.

Jones, R., 1962, Surface wave technique for measuring the elastic properties and thickness of roads: Theoretical development: British Journal of Applied Physics, **13**, 21–29.

Nazarian, S., 1984, In situ determination of soil deposits and pavement systems by spectral analysis of surface waves method: Ph.D. thesis, University ofTexas.

Park, C., R. D. Miller, and J. Xia, 1999, Multichannel analysis of surface waves: GEOPHYSICS, **64**, 800–808.

Ryden, N., and M. J. S. Lowe, 2004, Guided wave propagation in three-layer pavement Structures: Journal of the Acoustical Society of America, **116**, 2902–2913.

Ryden, N., C. B. Park, P. Ulriksen, and R. D. Miller, 2004, Multimodal approach to seismic pavement testing: Journal of Geotechnical and Geoenvironmental Engineering, **130**, 636–645.

Ryden, N., and C. B. Park, 2006, Fast Simulated Annealing Inversion of Surface Waves on Pavements using Phase Velocity Spectra: GEOPHYSICS, **71**, no. 4, R49–R58.

Ryden, N., M. J. S. Lowe, P. Cawley, and C. B. Park, 2006, Non-contact surface wave measurements using a microphone: Proceedings of the Symposium on the Application of Geophysics to Engineering and Environmental Problems (SAGEEP), 1110-1115.

Ryden, N., M. J. S. Lowe, P. Cawley, 2009, Non-contact surface wave testing of pavements using a rolling microphone array: Proc. of the 7th Int. Symp. on Non-Destructive Testing in Civil Engineering (NDTCE'09).

Ryden, N., 2010, Towards non-contact surface wave testing of subsonic soil layers using microphones: GeoFlorida 2010: Advances in Analysis, Modeling, & Design, ASCE Conference Proceedings **365**, 139.

Socco, L. V., S. Foti, and D. Boiero, 2010, Surface-wave analysis for building near-surface velocity models Established approaches and new perspectives: GEOPHYSICS, **75**, no. 5, A83–A102.

Vidale, R. F., 1964, The dispersion of stress waves in layered media overlaying a half space of lesser acoustic rigidity: Ph.D. thesis, University of Wisconsin.

Zhu, J. Y., and J. S. Popovics, (2001), Non-contact detection of surface waves in concrete using an air-coupled sensor: in Review of Progress in QNDE, Vol. 20B, edited by D. O. Thompson and D. E. Chimenti, AIP Conference Proceedings, American Institute of Physics, 1261-1268.

Reliability maps for seismic tomography

Thomas Fechner, Geotomographie GmbH, Neuwied, Germany*
Solomon Ifeanyi Ehosioke, University of Goettingen, Germany
Sonja Mackens, Geotomographie GmbH, Neuwied, Germany
Lutz Karl, Geotomographie GmbH, Neuwied, Germany
Daryl Tweeton, GeoTom LLC, Apple Valley, Minnesota

Summary

Seismic borehole tomography has become a standard method and is routinely used for the detection of karstic phenomena and the delineation of geological structures. Seismic tomography is believed to be the seismic method promising highest accuracy and reliability. Remaining uncertainties due to a non-zero residual travel time fit during data inversion are often neglected. However, data quality has a significant influence on the accuracy of a travel time pick. We present tomographic inversion results where the signal-to-noise ratio of the first arrival times is considered as a data quality measure during tomographic inversion. This new data quality weighting scheme is supposed to provide more reliable inversion results. Information about the reliability of the tomogram provided along with the seismic tomogram may support the geophysicist interpretation. The effect of the data quality weighted inversion is studied on a field data set.

Introduction

The data quality of seismic signals varies significantly not only from site to site but also within a single data set. Signal quality can range from excellent to poor signal quality. In the case of excellent data quality first arrival time picking is more accurate compared to less reliable arrival time picks for poor signal quality. In general, the accuracy of a first arrival time pick associated with the seismic signal is not considered in the tomographic inversion procedure. Thus, the present tomographic inversion procedures mix travel times of different data quality and reliability and treat them equally weighted. Several authors have described weighting procedures for the well-known tomographic SIRT inversion (Krajewski, C. et al., 1989, Tinti, S. and Ugolini, S., 1990, Lehmann, B., 1992). The proposed weighting procedures consider geometrical properties of the tomographic inversion such as ray density or ray angle azimuth but not signal quality itself. Mackens et al. (2014) proposed the implementation of the data quality weighting using the signal-to-noise ratio of the seismic data and presented first results. This paper presents results obtained during a field campaign in 2014 (Ehosioke, 2014).

Data Quality Weighting

The signal-to-noise ratio influences the accuracy of a first arrival time pick, i.e. the larger the noise the lower the related reliability of the time pick. The signal-to-noise ratio is calculated by the ratio of the maximum first arrival amplitude and the average noise amplitude before the first arrival. This factor is called quality factor (QF) by Mackens et al. (2014) and implemented within the SIRT inversion routine (see equation 1):

$$\Delta s_n = \frac{1}{\sum_k QF_k} \cdot \sum_k \frac{QF_k \cdot \Delta t_k \cdot r_{n,k}}{\sum_m r_{m,k}^2} \qquad \text{eq. 1}$$

with

- Δt_k → traveltime residual of ray "k"
- $r_{n,k}$ → length of raypath segment (ray "k") in voxel "n"
- m → index of raypath segments belonging to ray "k"
- QF_k → quality factor of ray "k"
- Δs_n → slowness residual in voxel "n"

During inversion differences between measured and calculated travel times are weighted with their associated quality factors and finally summed up as weighted slowness corrections. If all quality factors are equal, equation 1 reduces to the well-known SIRT slowness correction for a cell. As a consequence, seismic velocities of rays with higher quality factor will have greater influence in a model cell than those rays with lower quality factor, i.e., the inverted velocity gets closer to the seismic velocities associated with higher weighted seismic rays. After the final iteration step a mean quality factor in each cell is calculated and can be plotted along with the seismic tomogram velocities. The map of the data quality factors is called reliability map and ranges from "0" to "1". A reliability map shows areas with low and high quality factors, thus with lower and higher reliability of the seismic velocities. It is recommended to interpret seismic velocities and their associated structures together with their reliability map. Whereas the interpretation of geological structures imaged using the standard inversion procedure cannot differentiate between well resolved and less well resolved structures the new approach allows a more sophisticated interpretation of a tomogram.

As the reliability map is calculated based on physical measurable values like the signal-to–noise ratio there is a resilient data quality value available and no subjective probability. Thus, the result is reproducible and the interpretation can be well documented. In principle, the data quality weighting used to calculate a reliability map allows any individual data quality criteria to be used during the inversion procedure.

Field Measurement and Data Processing

A seismic cross-hole tomography test was carried out in February 2014 at a test site in Hannover to study the heterogeneity of an aquifer. Two boreholes were available to a depth of about 100 m having a distance of approximately 90 m. The geology at the site is characterized by Quaternary fine sand and coarse gravel, underlain by marlstone. Collection of seismic data was carried out using a seismic sparker source type SBS42 from Geotomographie and a hydrophone string type BHC2 with passive sensors type MP25 (Geospace). Sampling interval of the seismograph type DMT Summit was set to 1/16 kHz and shot and receiver spacing was set to one meter.

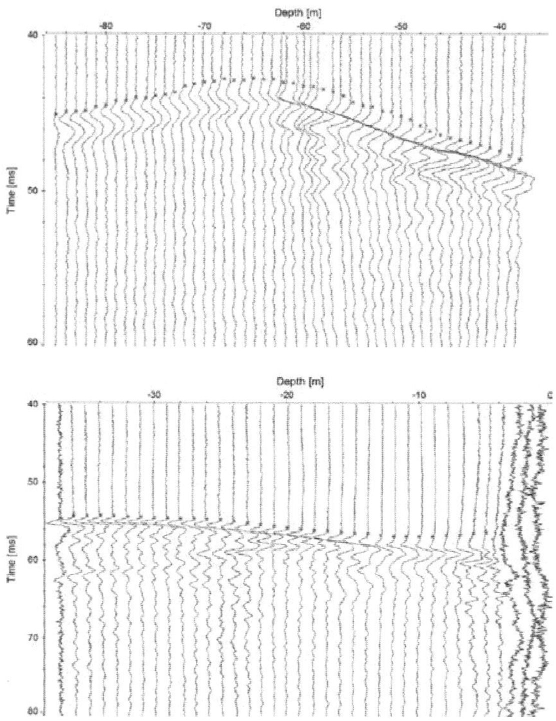

Figure 1: Seismic example records (shot depth at 75 m for upper record and 16 m for lower record)

Example raw data are shown for two different shot depths at 75 m and 16 m (see figure1). About 3350 first arrival times were picked and the associated quality factors were determined. Statistical analysis on the picking accuracy in respect to data quality showed that a signal-to-noise ratio of 16 can be regarded as sufficient to allow an accurate travel time pick. Thus, an uppermost quality factor threshold of 16 was selected. Values exceeding this threshold were set to 16 and normalized to 1. Inversion of the travel times was performed using the software GeotomCG (GeoTom LLC). A vertical cell size of 1 m and a horizontal cell size of 2 m were used.

Results and Interpretation

Velocity tomograms were calculated without and with data weighting. Figure 2 shows the seismic velocity tomogram without data weighting.

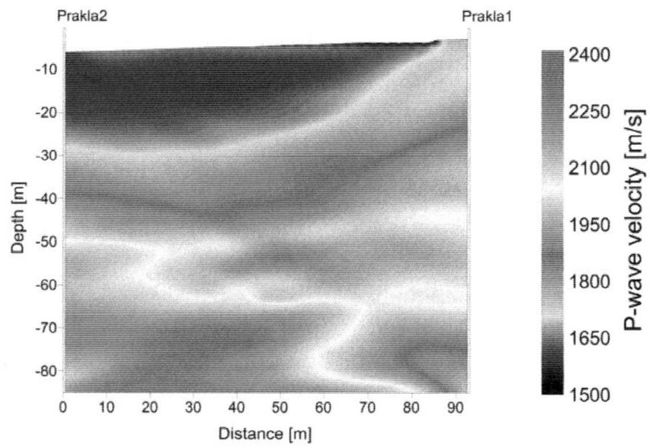

Figure 2: Seismic velocity tomogram (without data weighting)

According to the drilling results seismic velocities below about 1700 m/s belong to unconsolidated aquifer sediments. Velocities above 1700 m/s can be related to weathered to compact marlstone. Figure 3 shows the ray coverage plot for the calculated tomogram. It clearly shows highest ray coverage in the middle of the tomogram and lower ray coverage at top and bottom.

The ray coverage map as shown in figure 3 is typically used to conclude that highest confidence can be given to the center of the seismic tomogram as there is the highest density of ray paths. Anyhow, by introducing the data weighting into the inversion a somewhat different conclusion can be drawn.

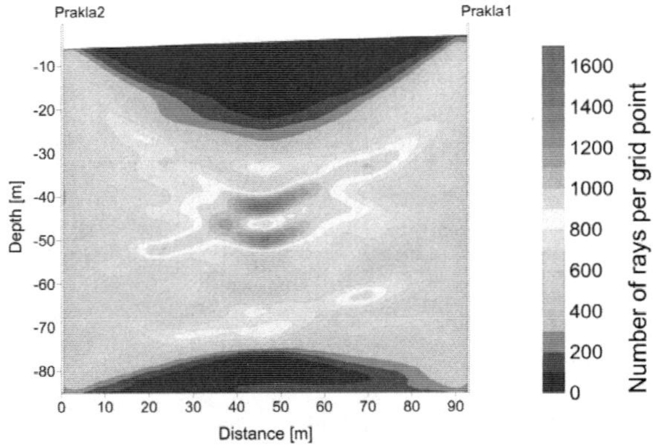

Figure 3: Ray coverage for seismic tomogram (without data weighting)

Figure 4 shows the map of the data quality obtained after inversion of the same data set introducing quality factors. This map is called reliability map and shows the distribution of the averaged data quality factors for each tomogram cell. Data quality values close to one can be related to an excellent reliability of the seismic velocity estimation whereas low data quality values are associated with a poor velocity estimate.

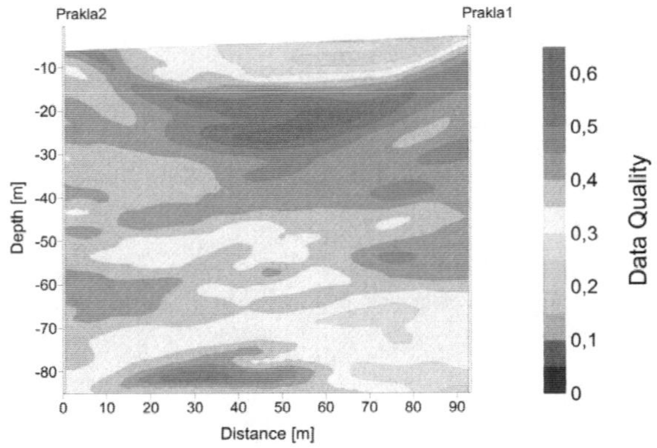

Figure 4: Reliability map of data quality factors (with data weighting)

The reliability map presented in figure 4 shows higher quality factors in the upper part of the map. The lower and the center part show medium data quality values. This result allows a different interpretation of the seismic tomogram obtained using weighted data quality factors shown in figure 5.

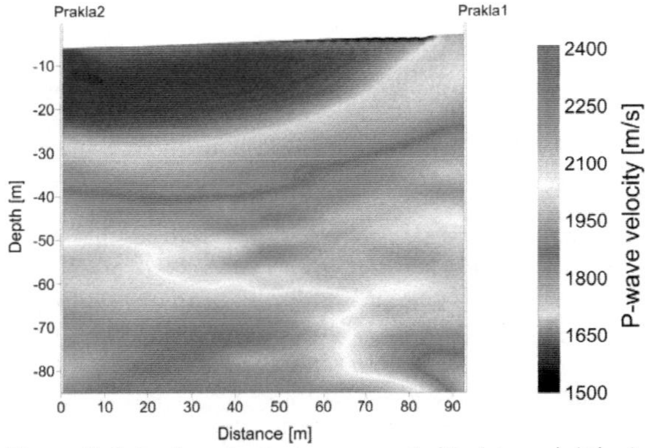

Figure 5: Seismic velocity tomogram (with data weighting)

After this the seismic velocities within the unconsolidated sediments are well determined thus the basin like structure of lower seismic velocities is well resolved and can be trusted. An interpretation based on ray coverage only would most likely lead to a conclusion that this structure might be related to the low ray coverage itself shown in figure 3. Lower data qualities are imaged in the lower right corner of the seismic tomogram and associated with the low seismic velocities of the marlstone. Obviously, the marlstone is strongly weathered in this area. Anyhow, according to the reliability map the velocity contours are not as reliable as in other parts of the tomogram. This is also in agreement with the map of the slowness residual error left after the final iteration due to a non-perfect travel time fit between measured and calculated travel times (see figure 6). This map shows the slowness residual Δs normalized by the slowness s of a cell. Areas with relative slowness residual errors close to 0 indicate good estimates of the local seismic velocity. Negative or positive relative slowness residual errors are a sign of underestimated or overestimated velocities, respectively.

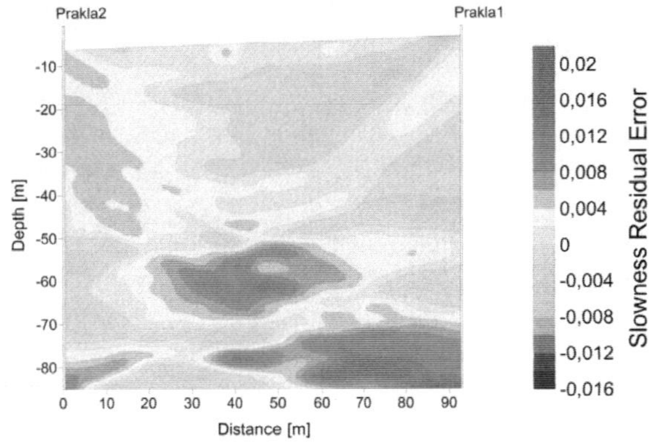

Figure 6: Map of the relative slowness residual error (with data weighting)

Conclusions

Reliability map calculation based on a data quality weighted seismic travel time inversion scheme has been presented. The reliability maps are meant to support the interpretation by geophysical engineers. Up to now velocity structures visible in tomograms are only supported by the interpretation of the ray coverage. Using the reliability map and quality factor values areas of higher or lower reliability of the seismic velocity can be documented. Additionally, the plot of the remaining relative slowness residual errors calculated after the final iteration provides helpful information about the reliability of a tomogram.

References

Ehosioke, S., 2014, Application of Cross-Hole Seismic Tomography in Characterization of Heterogeneous Aquifers, University of Goettingen, Germany, Master Thesis.

Krajewski, C., Dresen, L., Gelbke, Ch., and Rueter, H., 1989, Iterative Tomographic Methods to Locate Seismic Low-Velocity Anomalies: A Model Study, Geophysical Prospecting 37, Issue 7, P. 717-751.

Lehmann, B., 1992, Modellseismische Untersuchungen zur Transmissions- und Reflexionstomographie unter Verwendung der Gauß-Beam Methode, Phd-Thesis, Ruhr-University Bochum, Reihe A, Nr. 33 (in German).

Lehmann, B., 2007, Seismic Traveltime Tomography for Engineering and Exploration Applications, EAGE publications bv, DB Houten, The Netherlands, P. 37-57

Mackens, S., Karl, L., Fechner, Th. and Tweeton, D., 2014, Interpretation of Seismic Tomography Results Using Data Quality and Residual Error Maps, FastTimes, June 2014, Volume 19, Number 2.

Tinti, S. and Ugolini, S., 1990, Pre-Selection of Seismic Rays As a Possible Method to Improve the Inverse Problem Solution, Geophysical Journal International 102, Issue 1, P. 45-61.

Use of electrical resistivity tomography for mapping the extensions of an existing cavity, a case study from Doha, Qatar

Mahmoud K. Harb, Emad Sharif and Tanzeel Ur Rehman Sabir-Arab Center for Engineering Studies, Dubai, UAE*
h.mahmoud@aces-dubai.ae

Summary

The presence of natural voids and cavities in subsurface karstic limestones causes severe problem for civil engineering and environmental management. The presence of such features is one of the major concerns for the construction sector in Qatar. These cavities have led to ground subsidence eventually leading to damages to infrastructures such as depression and cracking of the buildings and roads. The main objective of this study was to indicate the extension of large cavity located in Doha. Only one main technique; Electrical Resistivity Tomography (ERT-2D), was proposed/specified and used for subsurface cavity detection. The geophysical survey was calibrated and verified by direct drilling of Fourteen (14) boreholes. Analysis and interpretation of the findings have identified existence of several subsurface open and sediment in-filled cavities. Correlation between ERT results and ground materials was established. Accordingly; risk map was established which identified the likely significance of the encountered risks to the proposed development/construction.

Introduction

Near surface geophysical methods have been widely used in geotechnical applications over the last two decades to detect underground cavities, cellars, caves etc. (Anderson et al., 2007). This important information is provided for geotechnical concerns like the detection of boundaries between geological formations, differences in the rock/soil water saturation, the presence of weak features and the depth to bed rock-etc. (Psomiadis et al., 2009). These cavities can be empty or partly filled-in with water or soil material or both. Different geophysical techniques are deployed to detect these cavity features to characterize the variations in the physical parameters of underground formations. To observe the physical parameters, geophysical techniques produce a set of data. Each of these parameters is related to one or more physical properties of the subsurface and to their spatial distribution (Psomiadis et al., 2009).The use of electrical resistivity technique in identification of cavities in buried walls and subsurface rocks at shallow depth has long been a common practice (Sarris and Jones 2000; Drahor et al., 2008; Berge et al., 2011, Leucci and Greco, 2012).The investigated site is located in Doha city, Qatar. Figure 1 shows the investigated area along with the geophysical survey lines. Initially thirty four (34) ERT profiles were conducted around the excavated area and twenty five (25) ERT profiles were conducted inside the excavated area. Based on the findings further twenty one (21) ERT profiles were conducted with high resolution up to 10m depth around the excavated area in order to have more confidence and to map the extent of the cavity more precisely. The study demonstrates the efficiency of the electrical methods for mapping subsurface conductive zones. As part of the investigation program and based on initial ERT results Eight (8) verification boreholes were drilled at selected locations within the site in order to verify the geophysical finding. Furthermore Six (6) additional verification boreholes were drilled within the vicinity of additional surveyed area for calibration and verification purposes. The boreholes were drilled to various depths up to 15m below existing ground surface.

The objectives of the geophysical study to locate and define the shallow/deep cavities, unconsolidated weak zones within the depth range of investigation to understand the nature and composition of the soil as well as the geological features under the site. Such sub-surface geologic hazards and their assessment lead to decide if the site is acceptable for such construction or there is a need for additional investigations.

Figure 1: Pictures of Existing Cavity

Geology:

The regional geology of the Arabian Peninsula area, the Qatar Peninsula geologically represents a part of the Arabian Gulf Basin. This basin is mainly composed of extensive carbonate sediments with different ages overlying the basement rocks and may reach up to 10 km in thickness. The outcropping rocks in Qatar are mainly of the Quaternary and Tertiary Ages, except Halul and Shara'wah Islands.

The geology of the study area is mainly of the recent and Tertiary rocks, which were covered by top soil; these are composed mainly of light brown silty sand with gravel of limestone. These materials are directly overlying the Fill Materials, Residual Soil, Simsima Member of the Eocene (upper Dammam Sub-Formation), Midra Shale, FahahilVelates Member and Rus Formation.

The Simsima Limestone is known to be upto 30m thick, the thickest development being in the extreme north of the peninsula, particularly away from the central arch. It is believed that a restricted thickness of Simsima Limestone has been deposited over the Central, Simsima Arches and Dukhan Anticline. Although the Simsima Limestone forms most of the land surface, its detailed stratigraphy and lithology are not well known, since exposures giving a complete cross-section do not exist. A general description of the Simsima as described by Cavalier (1970) and others would be a fine to medium grained off-white to pale brown and buff, poorly bedded, chalky crystalline calcareous limestone and dolomitic limestone with numerous vuggs and irregular joints often filled with weaker siltstone. Thin layers of pale green or red brown attapulgitic clays are occasionally present.

Over the southern half of Qatar the lowermost member of the Dammam Formation – comprises the attapulgitic shells of the Midra Shale and FahahilVelates Member. A general transgression at the beginning of the Midra Eocene results in a return to shallow marine deposition across the whole of Qatar and the formation of the Simsima limestone.

The rocks of Rus Formation in Qatar is generally composed of whitish or yellowish chalky dolomitic limestone occasionally intercalated with thin beds of greenish to brownish attapulgitic clay. In addition there are several narrow beds of whitish to greyish compact fossiliferous dolomitic limestone, the uppermost of which forms the top of the Rus Formation and is particularly well developed at Al Khor, where it has been used in the past for construction purposes. The occurrence of numerous and extensive beds of evaporates, mainly gypsum and anhydrite is a particular feature of the Rus formation. The gypsum beds, which individually can be upto 4m thick, are present beneath ground level throughout southern Qatar and extend northwards to a belt between the Dukhan Anticline the central arch and south of Al Khor. However, it should be mentioned that at the north of this belt, evaporates are absent from the Rus Formation. This absence may result from the arch activity, rising during the Lower Eocene, and thus, preventing deposition of evaporates rather than their deposition and subsequent dissolution. Where present however, the upper beds of gypsum have subsequently been extensively dissolved by circulating ground-waters causing the overlying strata to slump in place giving rise to the many surface collapse structures common in southern and central Qatar.

Methodology

Electrical methods are used to determine the location, thickness of the unconsolidated materials, nature, and extent of the geological materials like clay beds and bedrock and depth of the water table ((Robinson and Coruh, 1988). Electrical resistivity methods have been applied to many groundwater situations where the resistance of the fluid in the ground varies, for example this happens when there is a plume of saline groundwater that may be the result of salt water intrusion or leachate from a landfill (Fetter, 2001). In electrical resistivity methods, one can measure electric current, electric potential and electromagnetic field induced either naturally or artificially. These are classified on the basis of type of energy source, artificially or naturally, these include Self Potential (SP), Electrical Resistivity (ERS), Electromagnetic (EM) and Induced Polarization (IP) method and Electrical Resistivity Tomography (ERT).

A direct investigation utilizing geophysical ERT technique was used to achieve the objectives of the investigation. This method was chosen for this investigation because of its known efficiency for detecting subsurface voids and weak zones in similar site conditions.ERT uses a multi-core cable, that can be connected with a number of electrodes, pre-planted to the ground at fixed distance knowing as electrode spacing (dx), one of the ends of the multi-core is plugged into controller box (switch box), which in turn is connected with the main control unit (resistivity meter). To start the measurement, the current (I) is injected through transmitting electrodes and the potential difference (voltage V) between two receiving electrodes is measured. After the completion of the first reading, the system will switch off some electrodes toward the start of the section and activate other electrodes toward the upper side of the section for the next reading. The system continues this process until using the largest possibility for using the electrodes as potential and current with the first electrode spacing ($dx=x$), which is called Set No.1 ($n=1$). After completing the sequence of measurements with electrode spacing ($dx=x$), the next sequence of measurements with ($dx=2x$) electrode spacing proceeds. Through these steps the depth of the exploration will increase. This process is continued and repeated for measurements until use of the largest possible distance between the electrodes. At the end of the field measurement, there will be hundreds to thousands of resistivity readings which can be used to generate 2D (two dimensional) resistivity section. After completion of the field work and the measurements passed the quality verifications, the acquired data can be converted for 2D - electrical section utilizing appropriate software which is closely correlated with subsurface geology.

Data Acquisition and Processing:

The aim of the ERT survey was to detect and map the lateral and vertical extension of existing cavity within the site. This required designing a systematic grid of measurement lines to obtain 2D ERT data. The ERT survey was, accordingly performed with 1.0 - 1.5 m electrode spacing respectively in order to achieve high resolution data. Measurements were obtained using the dipole –dipole array chosen due to its known for good lateral resolution which is important for locating isolated targets (Voids).

Following field data collection, the measured apparent resistivity values were then used to build a pseudo section.

Table 2: Summary of ERT Survey Design Parameters.

Survey Purposes	ERT (2D)
Configuration	Dipole –Dipole (Max n =8 & Max Dipole =6)
Electrode Spacing	1 m
Profile Spacing	2 m
No. of Electrodes	112

Following field data collection, the measured apparent resistivity values are then used to build a pseudosection. The pseudosection gives a very approximate picture of the true subsurface resistivity distribution, however it is a distorted picture because the apparent resistivity values depend on the type of array used as well as the true subsurface resistivity. Nevertheless, the apparent resistivity pseudosection is the base to estimate the true resistivity section. This can be achieved using iterative least square inversion methods. The procedure for processing the apparent resistivity values to achieve the true resistivity values can be summarized into 5 basic steps:

1) Check the quality of the raw data (measured apparent resistivity) and filter the raw data to be processed.
2) Run the least square inversion procedure with the appropriate inversion parameters.
3) Remove bad data points (if exist) to decrease the inversion RMS error.
4) Run the least inversion procedure again and repeat until the desired RMS error is achieved.
5) Display the resulting true resistivity sections in an appropriate color scale to delineate anomalies if exist.

Results and Discussions:

The electrical resistivity tomography survey provided a two dimensional image of the subsurface along the proposed site at specific traverse line locations. The ERT results indicated that the study area of the project site exhibited distinct resistivity conditions and contrasts. Analysis and interpretation of the findings has identified existence of several small isolated subsurface anomalies of geotechnical concern in the investigated area. The detection of features in an ERT survey is dependent on the presence of a significant contrast between the feature of interest and the surrounding material. A cavity can be identified in the inverted section of ERT (2D) with high reliability, if a small area (high resistivity) is embedded within a zone of low resistivity above or below the water table. There is abrupt behavior of water table in this area, much variations encountered few boreholes are dry however water table ranges from 9.0m to 13.2m. ERT survey results do not immediately identify the composition of anomalies which may be caused by variations in geology and/or moisture content (or other factors). Final data interpretation is greatly enhanced by calibrating or benchmarking the electrical resistivity images against existing site data and/or follow-up confirmation boring data. The project area was divided into two main areas; around (outside) the excavated area and within (inside) the excavated area.

Around the Excavated Area: The ERT survey within this area was conducted in five different directions around the excavated area. The finding of each part is described below: Part 01 includes nine lines. Verification Boreholes (VBH-01, VBH-02 and VBH-08) were drilled within this part to confirm the nature of the obtained anomalies, as a result; VBH-01 did not encountered any cavity which means the type of such anomalies around this borehole is not related to existing cavity while VBH-01, VBH-02, VBH-08 and VBH-09 confirmed existing open and in-filled cavities, which means the similar anomalies around these boreholes are related to the same type of cavities. Part 02 includes five lines. Verification boreholes (VBH-01, VBH-02, VBH-04, VBH-05 and VBH-10) were drilled to confirm the nature of the obtained anomalies, as a result of these boreholes the marked anomalies within this part are related with open and in-filled cavities, particularly VBH-05 confirms that open cavity was located at depth 14.1 up to 14.45m followed by in-filled cavity for depth up to 18m bgl as shown in Figure 2. However, the lateral extension of this cavity probably extends under the existing building, whereas, geophysical survey profiles could not continue further due to existing building. Part 03 includes five lines. Verification Boreholes (VBH-03 and VBH-04) were drilled to confirm the nature of the obtained anomalies, as a result of these boreholes the marked anomalies within this part are related with in-filled cavity and open cavity. Part 04 includes nine lines. One Verification Borehole (VBH-06) was drilled to confirm the nature of the obtained anomaly, as a result of this borehole the marked anomalies within this part is related with 2m in-filled cavity as shown in Figure 3. Part 05 includes six lines. Verification Borehole (VBH-07 and VBH-11) was drilled to confirm the nature of the obtained anomaly, as a result of this borehole the marked anomalies within this part is related with 0.4m open cavity.

Within the Excavated Area: This part includes 25 lines, the obtained results indicated one particular area at southern-west of the excavated area specified by very high resistivity values followed by very low resistivity values. This area presented in Figure 5 below. Based on the correlation between the resistivity values that encountered within the excavated area and the resistivity values that encountered outside of the excavated area which was calibrated by verification boreholes, the high resistivity values would reflected from big/strong boulders and very strong Simsima limestone layer. However, the high resistivity was followed with very low resistivity, this behavior could indicate existing open and/or in-filled cavity.

Additional ERT Survey around the Excavated Area: Based on above findings additional 21 ERT profiles were conducted around the excavated area in order to have more confidence and to map the boundary of the cavity more precisely up to 10m depth, additional survey area was divided in two parts explained below:

Part No. 06 lies in the east of the project area and includes eleven ERT lines. Part 07 lies in the west of the project area and includes ten ERT lines. Verification boreholes (VBH-12, VBH-13 and VBH-14) were drilled to confirm the nature of the obtained anomalies and to confirm the limitation of the existing cavity, as a result of these boreholes only VBH-14 confirms that open cavity was located at depth 7.3 up to 8.5m bgl as shown in Figure 4 However, the lateral extension of this cavity probably extend under the existing building, whereas, geophysical survey profiles could not continue further due to existing building.

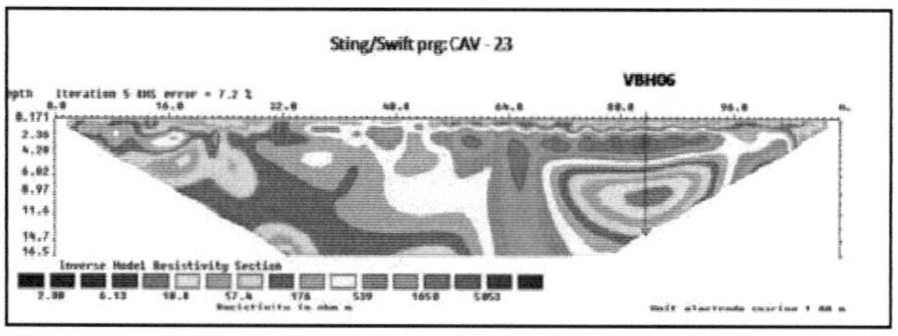

Figure 2: ERT profile with verification borehole no. 6

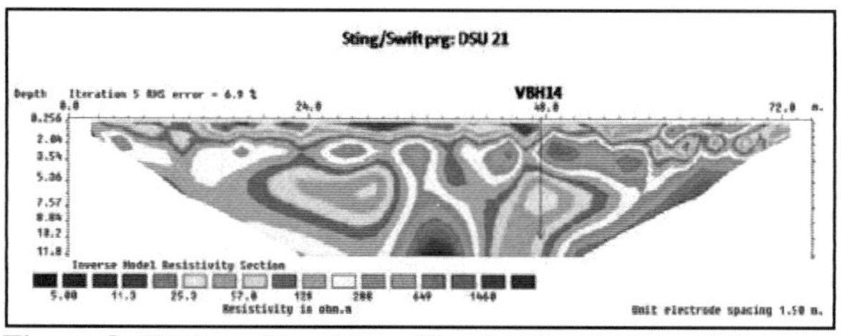

Figure 3: ERT profile with verification borehole no. 14

Conclusions:

Based on the comprehensive analyses of the geophysical results and verification borehole results, the following conclusions can be drawn from the present survey:

1. Only one geophysical technique was used in this survey (ERT); however it is noted that the survey was conducted along closely spaced profiles (at 2 - 4m spacing) with high resolution (electrode spacing at 1–1.5 m) and high quality data was obtained. Further, fourteen (14) numbers of verification boreholes were employed to confirm and verify the geophysical anomalies at different locations within the site. Furthermore, three diagraphy boreholes were also considered in the final interpretation.
2. Correlation between ERT results and geologic materials showed relatively large range of resistivity values for the different material types. This is mostly attributed to differential weathering and fractured nature of the rocks.
3. Subsurface anomalies with different levels of severity have been identified within the site at different depths. Nine out of fourteen verification boreholes confirms existing open and/or in-filled cavities.
4. VBH-05 and VBH-14 (very near to existing building) confirm that open cavity was located at depth 14.1 up to 14.45 followed by in-filled cavity for depth up to 18m bgl in VHB-05 and open cavity was encountered between 7.3 up to 8.5m bgl in VBH-14, However, the lateral extension of this cavity probably extend under the existing building, whereas, geophysical survey profiles could not continue further due to that building.
5. Based on the correlation between the borehole logs and the geophysical anomalies, a risk maps were generated. However, in terms of shallow cavity detection, the risk areas can either be high or medium risk zone. High risk areas are associated with strong anomalies which were frequently repeated with parallel and perpendicular ERT profiles as well as confirmed cavity within the range of 10m by direct drilling. While Medium risk areas are associated with strong anomalies which were randomly distributed along ERT profiles but no confirmation of shallow cavity while relatively deep cavity can be located within this zone.
6. The combination results confirmed that the site is critical and risky; small scale of open and in- filled cavities could be present in the vicinity of risk zones which need to be considered prior to any construction works.

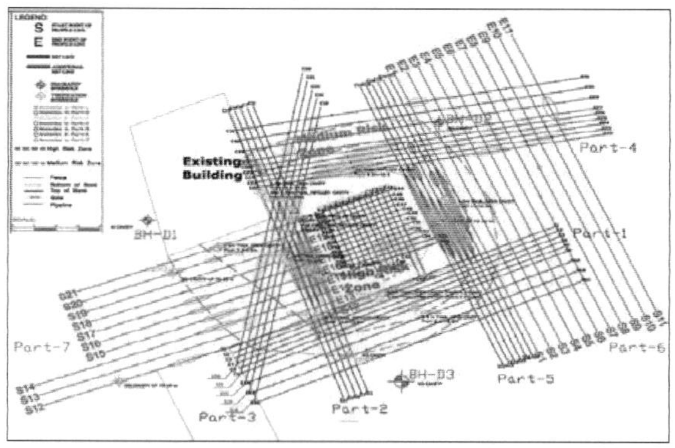

Figure 4: Risk assessment map

Imaging of Landslide Deposits around Uzungöl Lake (Trabzon-Turkey) by Refraction Tomography and MASW

G. Vanli Senkaya[1,], H. Karsli[1], M. Senkaya[1], R. Güney[1]*
[1]Karadeniz Technical University Geophysical Engineering Trabzon Turkey, [*]gulseda-vanli@hotmail.com

Summary

Uzungöl Lake and around are located on the landslide deposits which occurred due to an historical landslide. As constructing and commercial, social and cultural development are increasingly rising in the area, seismic risk analyses are need to be accurately done due to the peculiarities of the place where it is located. However, there is not enough geoscience study or no geophysical survey around Uzungöl Lake to reveal sub-surface conditions. Refraction tomography and active surface wave analysis (MASW) were used to image the thickness of land slide deposits, determine P and S-wave velocities of sub-surface. The analyses from 16 profiles demonstrate that the thickness can be seen around 15-20 meters on the west side of the study area, it reaches 30 m and more on the south side. Also, the general appearance of deposit thickness is that it increases towards the lake and maximum thicknesses obviously occur on coastal of the lake.

Introduction

Uzungöl Lake is one of the most important touristic places in Trabzon and Turkey with its natural beauties and flora. The area is in need of more commercial, social and cultural developments due to increasing recognizability in last decades. However, there is no geophysical study about the sub-surface conditions to answer the growing construction demands. Therefore, constructions are being continued in uncontrolled situation with conventional experiences. At the same time, it is well known that Uzungöl Lake is a result of a historical landslide (Alkan, 1996 and Altundağ, 2008) also many new landslides can be seen on area cause of excessive raining and snowing events and high slopes. So, the first meters of the sub-surface that is used for constructions have many irregularities in terms of deposit thicknesses and wave velocities. In this condition, detailed imaging of land slide deposits and determination P and S-wave velocity is becoming key role for stable and reliable constructions.

In this study, it is aimed to image the sub-surface, reveal the land slide deposits and determine P and S-wave velocities of geologic units by seismic methods. To achieve this, 16 profiles were assigned on 4 regions around the Uzungöl Lake and. While seismic refraction data were acquired for tomographic solutions, active and passive multichannel surface wave data were used to obtain 1D and 2D S-wave profiles by MASW. As the result of combining of all conclusions together; land slide deposits thicknesses, lateral and vertical changes of geologic units, velocity structures of deposits and base rock are produced to the maximum depth being able to be imaged.

The Location and Geology of Study Area

Uzungöl, 99km from Trabzon (Fig. 1) and on the altitude of 1100m, takes place in a valley surrounded by high mountains (Fig. 2) in the constitution of Soğanlı and Kaçkar mountain ranges in the East Pontid Tectonic Belt, where there are the rainforests of Turkey. While basalt has a widespread area as the bedrock, sedimentary masses, basaltic lava, tuff and agglomerate units can be seen over the bedrock (Altundağ, 2008). Also, the gravelly and silty interlayers can be seen as an intermediate layers. The uppermost arable unit is generally characterized by the gravelly-silty clay. The simplified geologic column section is presented in Fig 3.

The drillings around the lake by state hydraulic works (DSI) in 1982 (Fig. 4) demonstrate that the thickness of top soil over the bedrock and unit characterization is very changeable around the study area, especially south side of the lake. This is because, the Uzungöl Lake was formed as a result of the historical landslide also many new landslides has been occurring around the lake due to excessive raining and snowing activities and high slopes.

Figure 1: The study area

Figure 2: Appearance of Uzungöl Lake between high mountains

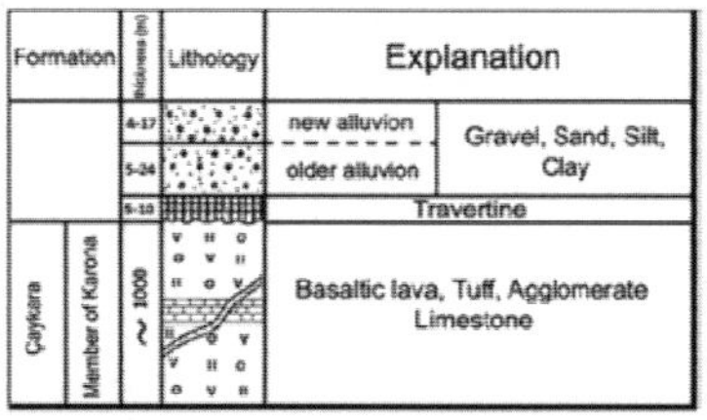

Figure 3: Simplified geologic column section for study area

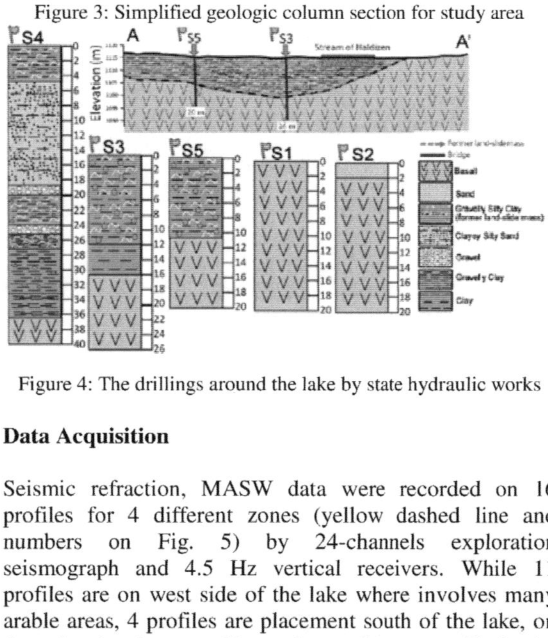

Figure 4: The drillings around the lake by state hydraulic works

Data Acquisition

Seismic refraction, MASW data were recorded on 16 profiles for 4 different zones (yellow dashed line and numbers on Fig. 5) by 24-channels exploration seismograph and 4.5 Hz vertical receivers. While 11 profiles are on west side of the lake where involves many arable areas, 4 profiles are placement south of the lake, on the other hand we could get data only one profile in the north of the lake cause of high slope of field.

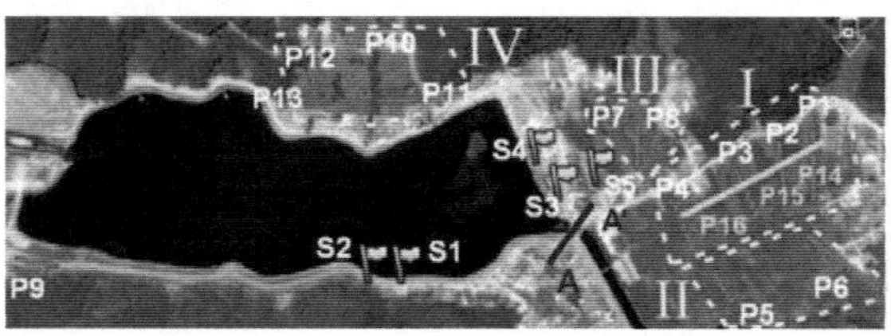

Figure 5: Appearance of profiles and drillings in the study area

The data were obtained by 10 kg sledge hammer with minimum 3 vertical stacks for refraction and MASW. The recording time is 0.5s, 1s and minimum 32s for refraction, MASW, respectively. Receivers sampling changed between 2 or 3 m according to available spread length. Since we aim to use refraction data for tomographic solution, we try to sample the spreads as possible as much shots and so each profile has minimum 8 shots. Also, the UTM coordinates saved on critical points of profiles and interpolated for all shots and receivers due to avoid topographic effects.

Data Evaluation

While the first-arrival travel times from refraction data was inverted by using a simulated-annealing and non-linear optimization (SeisOpt by Optim, LLC), MASW data inverted by KriSIS code which has many local search options (Kritikakis and Vafidis, 2011).

Zone 1 includes 7 profiles and is the most detailed zone when compared others (Fig. 6). The tomographic sections show that thickness of land slide deposit changes between 15-20m for higher part of zone, also base rock depth and irregularity increase with Profile 15 (as seen P15 on Fig. 6c).

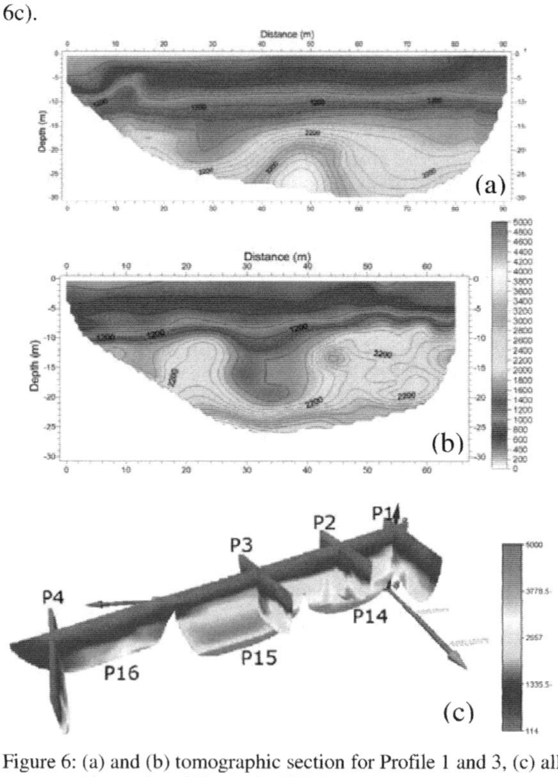

Figure 6: (a) and (b) tomographic section for Profile 1 and 3, (c) all tomographic results of Zone 1 in UTM coordinate.

While 1D S-wave velocity sections (Fig. 7) are compatible with tomographic sections, 2D S-wave section, which is generated by combining all 1D S-wave depth profiles along the slope, demonstrates increasing of land slide deposits.

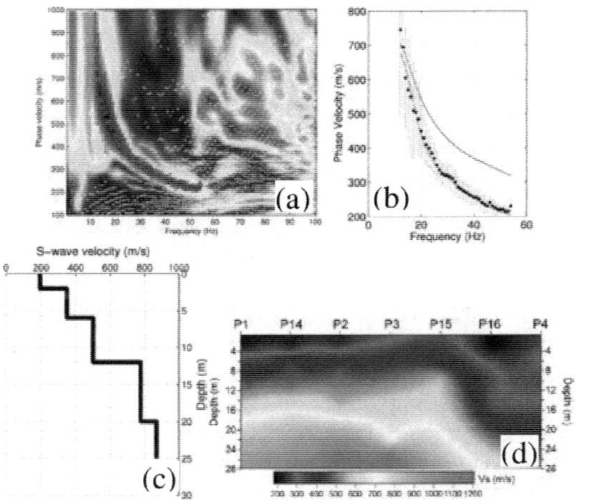

Figure 7: (a) Rayleigh dispersion curve for Profile 1, (b) dispersion data (black dot), initial (blue line) and final (red) dispersion curves, (c) 1D S-wave velocity profile, (d) 2D S-wave velocity section for Zone 1

Zone 2 is located left side of Zone 1 and has Profile 5 and 6 (Fig. 8). Tomographic section and 1D S-wave velocity profiles show that the top soil, which is arable area for local people, thicker on Profile 5 than Profile 6 (all 1D S-wave velocity profiles give in Fig. 12). Also, base rock is deeper approximately 2 m for Profile 5.

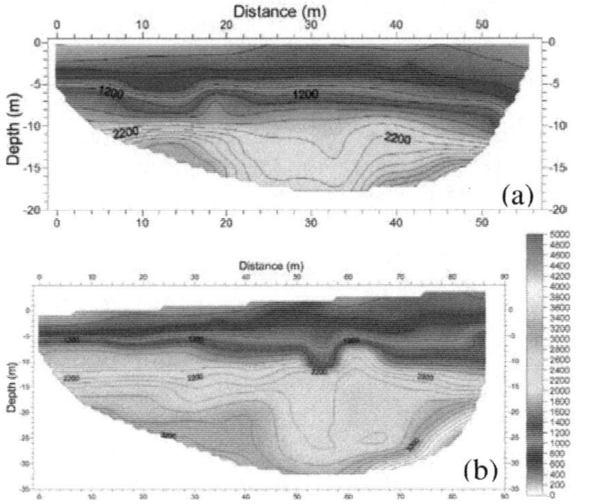

Figure 8: Tomographic section for (a) Profile 5 and (b) Profile 6

Zone 3 includes Profile 7 and 8 (Fig. 9). Even so, both profiles are very close to each other, the land slide deposit thickness is very different. While the base rock interface can be seen on Profile 7, it is not possible for Profile 8. This variation can be seen clearly on S-wave velocity depth profile, too.

Zone 4 has the same situation Zone 3. If we compare P10-P11 and P12-P13 twins, P11 and P13, which are closer the lake, have thicker land slide deposit than P10 and P12, respectively (Fig. 10 and Fig. 11).

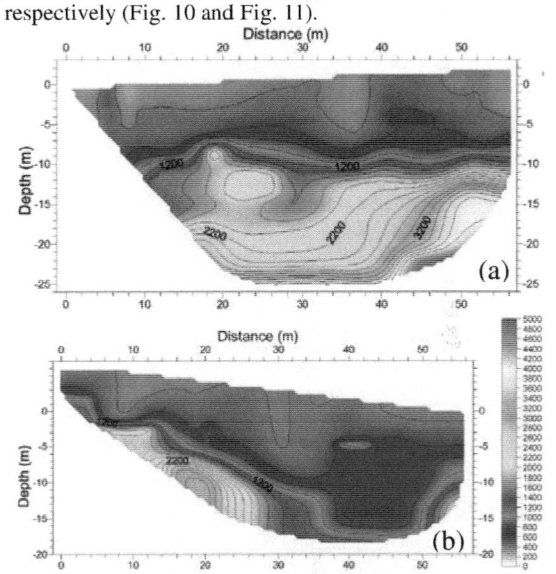

Figure 9: Tomographic section for (a) Profile 7 and (b) Profile 8

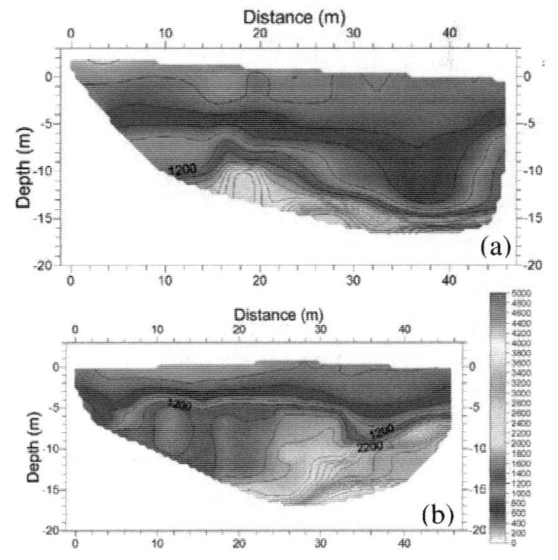

Figure 10: Tomographic section for (a) Profile10 and (b) Profile 11

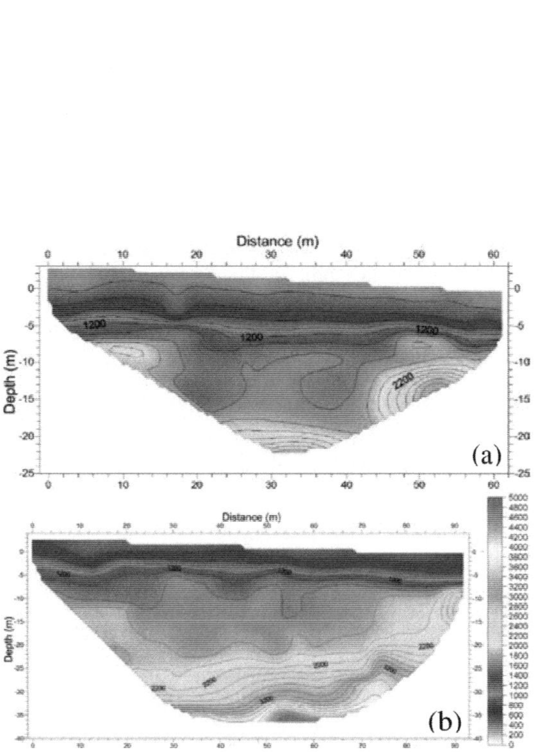

Figure 11: Tomographic section for (a) Profile 12 and (b) Profile13

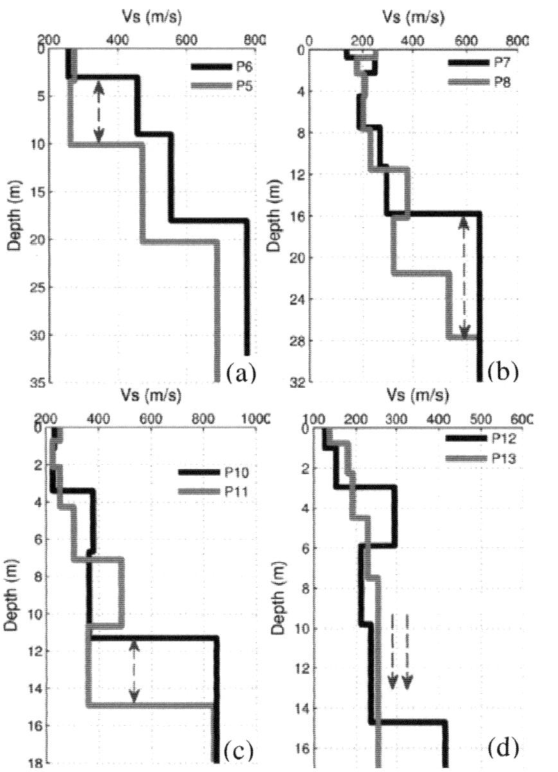

Figure 12: 1D S-wave velocity profiles for (a) Profile 5 and 6, (b) Profile 7 and 8, (c) Profile 10 and 11, (d) Profile 12 and 13

Conclusions

The landslide deposit thickness until the base rock has very irregular character in the study area. While the thickness can be seen around 15-20 meters on the west side of the study area, it reaches 30 m and more on the south side. The thickness of deposits increases towards the lake and maximum thicknesses obviously occur on coastal of the lake. It can be seen three main geologic units on the study area; the first unit represents arable soil and characterized with gravely silty clay. While P-wave velocity changes 300-480 m/s, S-wave velocity is around 240 m/s. The second unit refers transition units between upper layer and base rock. P and S-wave velocity changes 600-1300 m/s and 350-550 m/s, respectively, according to its character (compact arable soil or gravel and silty sand). The base rock generally is characterized with basalts. The first meters of base rock shows weathered character and P-wave velocity progressively decreases till 1900 m/s but deeper parts are more compact and P-wave velocity is seen around 3500-3900 m/s. On the other hand, S-wave velocity changes between 870-930 m/s.

Acknowledges

We would like to thank Dear Satish Pullammanappallil, Ph.D., Chief Scientist, Optim for his contribution for tomographic solutions and George Kritikakis from Technical University of Crete for his kindness for usage of his code.

References

Alkan, S., 1996, Carried agglomerate material determined to Uzungöl lake and A research on measures for preventing filling lake, Master thesis, KTÜ Institute of Science and Technology.

Altundağ, D., 2008, Environmental changes in Holocene period in Uzungöl and its catchment (in Turkish), Msc. Thesis, Ankara University, Turkey.

Bulut, F., 1989, Site investigation of the Çambaşı (Trabzon-Çaykara) dam and Uzungöl hydroelectric power plant in terms of engineering geology (in Turkish). Ph.D. Thesis, Karadeniz Technical University, Trabzon, Turkey

DSI, 1982, The engineering geology report of Of-Solaklı project Uzungöl-1 unit (planning stage), Trabzon, Turkey

Kritakis, G.S. and Vafidis, A, 2011, Different Alternatives of surface waves data inversion. Applications on environmental and geotechnical case studies, 6th Congress of Balkan Geophysical Society, P05, Budapest, Hungary

EVALUATING OF THE VISIBLE CRACKS AND RELATED DAMAGE IN THE PAVEMENT ROAD (MARIA ISLAND BRIDGE-ABU DHABI CITY) UTILIZING SEISMIC MULTI ANALYSIS SURFACE WAVES (MASW) AND GROUND PENETRATION RADAR (GPR)

Dr. Mohammed Hassouneh-Technical Manager-Tatweer for geophysical studies and consulting

Summary

This paper describes the objectives, the methods used, and the results of the investigation. It also will provide recommendations. The objective as stated in this paper is to evaluate the visible cracks and related damage on the new constructed road in Abu Dhabi city utilizing two geophysical techniques; seismic Multi Analysis Surface Wave (MASW) and Ground Penetration Radar (GPR). MASW technique is a recently developed seismic method that deals with relatively lower frequencies and shallower investigation depth ranges than do conventional high-resolution seismic methods. It provides shear-wave velocity (VS) information of near-surface materials in a highly cost-effectively manner. Because of the relatively significant value of this (VS) information in most geotechnical engineering projects and also because of the relatively simple in-field operation and data processing, it is gaining popularity among engineering communities.

Ground Penetrating Radar (GPR) is a non-destructive ground survey method that can be used in assessing roads, railways, bridges, airports, tunnels and environmental objects. Its main advantage is the continuous profile it provides of the road structure and subgrade soil. The GPR technique is becoming an increasingly important tool especially for structural evaluation of roads in the design phase but recently it has also been used more frequently in quality control and quality assurance of paving and construction projects.

<u>Analysis and interpretation of the findings from MASW and GPR scanning surveys has confirmed that there are loose materials existing at a depth between 2.5 and 4 m under the cracked road. This very week layer of loose unconsolidated materials influenced by the strong vibrations from the vicinity new tower construction. The vibrations generated seismic waves (ground shaking) that caused rapid compaction, rearrangement of soil particles and</u>

settling of subsurface materials (particularly in the layer of loose, non-compacted, and variable sandy sediments). The visible and invisible cracks on the road have been developed due to the *amplification* of the ground shaking effects by the *unconsolidated materials* under the asphalt pavement road.

Introduction

The Geophysical Multi Analysis of Surface Waves (MASW) and Ground Penetration Radar (GPR) method's (fast and non-destructive surveys) have been provided a fine alternative to the visual pavement conditions evaluation such as destructive methods (drilling, coring and pit excavations). Geophysical MASW and GPR investigations were conducted on Al Maria island Bridge in Abu Dhabi City to assess/evaluate the cause of the major visible cracks (defects in the new road). Cracking in road structures cause severe problems if not corrected early enough, the development of cracking may result in wholes or partial collapses, causing major operational risks and potentially life-threatening hazards.

Multi-channel Analysis of Surface Waves (MASW):

First introduced in Geophysics (1999), the multichannel analysis of surface waves (MASW) method is one of the seismic survey methods evaluating the elastic condition (stiffness) of the ground for geotechnical engineering purposes. MASW first measures seismic surface waves generated from various types of seismic sources—such as sledge hammer—analyzes the propagation velocities of those surface waves, and then finally deduces shear-wave velocity (Vs) variations below the surveyed area that is most responsible for the analyzed propagation velocity pattern of surface waves. Shear-wave velocity (Vs) is one of the elastic constants and closely related to Young's modulus. Under most circumstances, Vs is a direct indicator of the ground strength (stiffness) and therefore commonly used to derive load-bearing capacity. After a relatively simple procedure, final Vs information is provided in 1-D and 2-D formats.

MASW data has been utilized to provide shear wave maps of the subsurface for design site classification (Site class E, D, or C) per the 2009 International Building Code and to map voids, top of bedrock, and soft or weaker bedrock zones.

This information is important in both geotechnical and environmental applications.

Ground Penetrating Radar (GPR):

GPR is a remote sensing technique, it is non-invasive, non-destructive, relatively quick and efficient and highly accurate when used in appropriate situations. GPR can be used to map any object capable of being penetrated by radio waves, just as X-rays image the internal structure of the body. A sensitive detector is used to record weak radio waves reflected from objects and other underground 'contact points', such as differences between soil layers. Measurements were made at many points to ensure an accurate representation of the area being examined. GPR data are transformed into an image on a computer. Hyperbola (arches) displayed on this image, indicate the placement and depth of objects (e.g. cracks) in the survey area.

Figure 1: View of Crack over which MASW & GPR Profiles were collected at 300 and 800 MHz. Left side Picture Shows the MASW Profile over the Flow Path of the Highway.

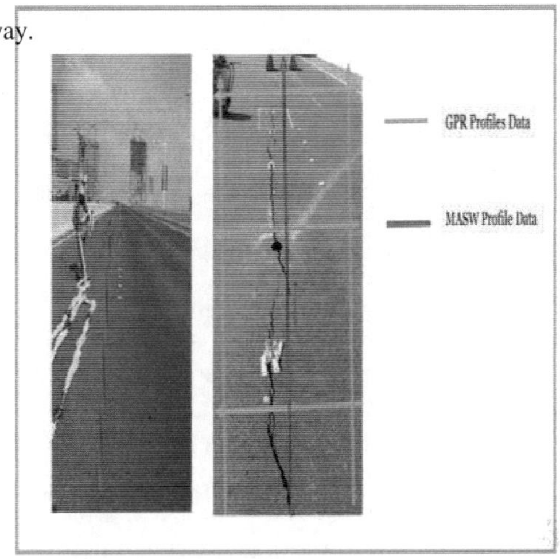

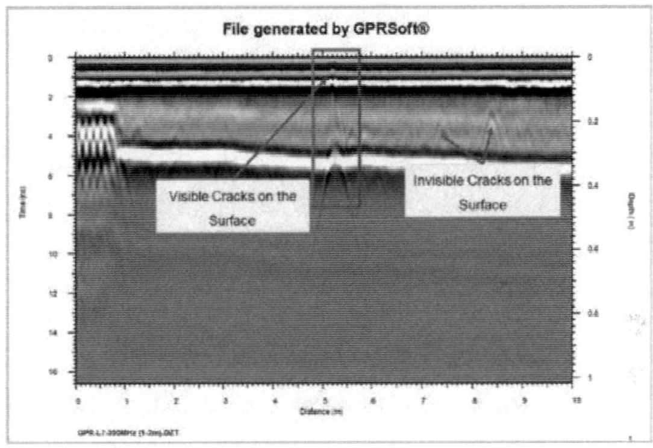

Conclusions

1. Analysis and interpretation of the findings from MASW and GPR scanning surveys has confirmed that there are loose materials existing at a depth between 2.5 and 4 m under the cracked road. This very week layer of loose unconsolidated materials influenced by the strong vibrations from the vicinity new tower construction. The vibrations generated seismic waves (ground shaking) that caused rapid compaction, rearrangement of soil particles and settling of subsurface materials (particularly in the layer of loose, non-compacted, and variable sandy sediments). The visible and invisible cracks on the road have been developed due to the *amplification* of the ground shaking effects by the *unconsolidated materials* under the asphalt pavement road.
2. The likely risk of open surface sinkhole in the cracked road is low. The wave velocity is increasing in the compacted and decreasing in the loose unconsolidated materials.
3. Shallow isolated feature of extremely low velocity values are presented on the S-Wave velocity profile crossing the site. These locations are correlating very well with highly unconsolidated less competent materials.
4. S-wave velocity showed a good relation with solidity and density, the low velocities are associated with soft, loose materials and high shear wave velocity associated with strong top compacted road materials.

Technology and Geoforensics

Ruffell, Alastair B.Sc HONS, PhD, FGS, School of Geography, Archaeology & Palaeoecology, Queen's University, Belfast, N.Ireland, BT7 1NN a.ruffell@qub.ac.uk

Summary

From the landscape-scale of searching for buried objects using Global Navigation Satellite Systems, through the challenge of finding sunken objects in water, to the non-destructive analysis of trace evidence, technology is fundamental to recent advances in Geoforensics (the use of Earth Science techniques in legal enquiries). At the macro-scale, wearable devices that allow both the remote direction of personnel, as well as recording their location and tracks, can be used to plan, carry out and review searches in featureless areas (moorland, forest, deserts). At the medium (tens of metres) scale, the search and inspection of water bodies is a major challenge for both serious crime as well as engineering and archaeology. Combined side-scan/forward-facing sonar and boat-operated ground penetrating radar (more correctly, water penetrating radar) has proven very effective in the search for victims of drowning, suicide and homicide, as well as bridge and quay/jetty inspection: both use the same technology. When recovered (sometimes using the above technology), trace (less than 2g of material) geological evidence is often critical in establishing a case, in the absence of DNA, blood, fibres or fingerprints. A major issue is what to do with such samples? Incorrectly deployed destructive analysis will preclude other investigations. A suite of new devices, from digital colormeters, Fourier Transform Infrared Spectroscopy, Portable XRF and the in situ analysis by x-ray diffraction are now available to the forensic investigator, to better inform what the trace sample is, and thus the most appropriate method of further, probably destructive, analysis.

Macro-scale

A feature-rich or object-rich environment is a challenge to manage a forensic-based search; so is an area that is featureless. An area with what we may consider 'too many' features/objects could be a forest (especially with non-standard planting, or a wild forest), a rock or boulder field or an area with abundant sand dunes. Such a plethora of features can be disorienting, may block line of sight and communications and can lead to areas that are not searched or are over-searched (inefficient use of human and canine resources). Featureless environments are as challenging to search effectively: water bodies, salt flats, peat bogs are all good examples. How do the search team know a) where they are and b) where they have already searched? They may leave flags, paint marks, footprints and have line of sight, all of which are good, but ultimately vanish. However, both the feature/object-rich and featureless areas present difficulties in planning, managing and recording a search that can be solved using GNSS (Global Navigation

System) technologies. Harrison & Donnelly (2009) suggest that a search be split into pre- syn- an post-search phases. The pre-search should comprise a full desktop study of solid and drift geology, soils, historic and recent land-use, past and recent orthoimagery, with any other available information such as local knowledge. The most recently-acquired imagery (satellite, aerial or map-based) should be geo-located and used to plan the search using GNSS technology, which can guide the direction search personnel take, but also record where deviation to suggested routes took place, and why. This geolocated integration of the pre- and syn-search allows for comprehensive planning as well as thorough analysis of what was searched and where. This type of geolocated planning, real-time direction and recording is not done often enough, yet from experience it is obvious this needs to be carried out, as it is simple, cost-effective and reliable in a court of law.

Meso-scale

Should the macro-scale search be effective at defining smaller areas of interest, or indeed the crime scene itself, then technology can likewise assist in both conducting the search as well as recording what was done at the scene. Two examples are provided, one in the effective sampling of a scene, such as provided by McKinley & Ruffell (2009), or in the difficult process of searching small (tens of metres diameter) bodies of water: a particular challenge. Few forensic scientists state publicly what samples they took, their context or their distribution. Pye (2009) is honest in his approach, in showing some theoretical sampling strategies for domestic gardens, parks, pathways and other locations. This common-sense attitude needs to be combined with taking a statistically-valid number of samples (such as in McKinley & Ruffell, 2007), in a digitally-referenced framework such as a scan of the scene by LiDar, which is commonly used in accident investigations, so why not as standard in serious crime scenes as well? Aside from the above issue of sampling for geoforensics, crime scene science is generally well researched and well-managed for the more common types of homicide, rape, assault and terrorist-related offences. These include domestic, urban, suburban, industrial or abandoned industrial locations, as well as some rural scenes. The search of unusual locations is less-well documented, and these include bricked-up, concrete immolation or burial, and body or object disposal in water. Both pose particular challenges for searching and sampling of a scene or potential scene of crime. Objects hidden behind walls and beneath concrete are of interest and statistically shown to be found by confession, innocent witness, and then cadaver dog and GPR indication (Ruffell *et al.* 2014). Deposition of objects, especially cadavers, in water can be split into marine and non-marine. The marine environment is challenging for both perpetrator and investigator alike, it being open, large, unpredictable and thus uncontrollable: to summarise it is open to chance. For instance, a body disposal at sea may result in the body sinking, floating, sinking again to the thermocline or halocline, then sinking again to remain submerged, or moved by currents or fishing trawlers, or discovered by divers. The non-marine environment comprises rivers, lakes, ponds and ditches. Fast flowing rivers are unusual body or object deposition sites, whereas ponds and ditches in temperate areas of the Earth are often accessible, controllable and thus common locations for the placing of bodies and other crime-related items (weapons, explosives). Such small water bodies pose distinct challenges for the search coordinator, drainage can be a logistical problem, with environmental consequences. Geolocation when operating from small boats is a challenge; divers may have very limited visibility. The use of side-scan or forward-looking sonar is particularly effective for mapping the underwater scene in such locations (Schulz *et al.,* 2013). Examples of diver-operated, forward-imaging sonar for the imaging of piers, jetties, bridges and river or lake banks show how for unburied objects and scene mapping this technique is very powerful. Objects such as cadavers, weapons, contraband etc. submerged in sediment in freshwater areas are very difficult to locate, as only CHIRPS (high resolution seismic) and water-based GPR (WPR) will be able to assist in location.

Micro-scale

Trace (sub-gram) soil, rock and other crystalline debris evidence has to be analysed qualitatively, non-destructively and rapidly to inform both searches by crime investigators and what subsequent analyses should take place. The x-ray diffraction analysis of *in situ* materials can achieve this (see Krill, 2003) but has yet to gain widespread use or be tested in a research project or scientific publication. This presentation will show how rigorous, non-destructive analysis can be conducted on adhering soils or rock dust.

Context. 1. Murray & Tedrow (1992, following the works of Sugita & Marumo and Palenik) have developed a comprehensive flow chart of best practice for the sequence and range of analyses that can and should be undertaken when comparing crystalline materials for forensic comparison. 2. This best practice is tried and tested, follows other scientific protocols and should be adhered to where possible. 3. Current murder enquiries in N.Ireland demonstrate two reasons why this best practice could not be followed, forcing a review of methods that was untested in a full scientific analysis such as this. These were: a) where only very small amounts of possible clay-rich soil and rock debris (>1mg) were recovered as mud spatter on clothing, footwear and vehicle interiors and b) where large numbers (hundreds) of seized clothing and footwear items, each with the small spatter marks described above, had to be screened for more expensive, time-consuming but ultimately more accurate palynological, isotopic, QemScan or DNA analysis. In each case, there was insufficient sample material to be sequentially analysed, as in standard practice. Similarly, it was desirable for screening to be as non-destructive as possible, so that subsequent analysis could be performed. 4. These scenarios will become common for two reasons. First, as criminals become forensically aware, so their clean-up operations become more efficient, leaving only the trace evidence described above. This is especially relevant to organized gangs be they terrorist or not, who have the finances and facilities for efficient clean-up operations. Second, the need for sophisticated forensic analysis that require pristine samples will increase as defence or prosecution use whatever techniques are not available to their opposition to establish suspect to scene comparisons. The maintenance of sample integrity by non-destructive analysis will be of major benefit in this latter case. 5. Two problems arise with Murray & Tedrow's best practice protocol when applied to the above scenarios. First, although such small amounts of material as described could all be picked from fabric, examined, split by grain size with coarse material mounted for optical microscopy and fine material analysed by SEM and XRD, this sequence takes time that could be critical in the continued prosecution of the case (Murray & Solebello, 2002). Second, in the studies embarked upon by the proposer, it was found that the reason these materials were mud specks in the first place was due to their fine-grained nature, thus the traditional, rigorous sequence of analyses provided only limited information on the materials. 6. Recent work by Keaney et al. (2009) has demonstrated the use of x-ray diffraction analysis on fine-grained dust-covered or mud-spattered sections of the original suspect material (clothing, footwear) as a qualitative mineral determinant for comparative purposes. These studies are reported either in a manufacturers magazine (Krill, 2003) or are unpublished: furthermore they use only selected substrate materials. The quality of X-ray diffraction analysis data is dependent on substrate type, sample preparation and analytical software. Otherwise, the technique is well-established and trustworthy. Thus, Kugler's work establishes the efficiency of the technique. This work suggests using XRD as a non-destructive screening method in forensic analysis. The potential value in the fight against crime can be shown with an example. A number of mud-stained fabric items are seized. The stains are mud- or dust spatter of a few millimeter-diameters that the suspect may not have noticed in clean-up. This mud has to be compared to a scene or indeed excluded from other alibi locations. The investigating officer and forensic scientist have a choice: do they embark on traditional analysis, of sample extraction, petrography, particle size analysis leading to specialist analyses of suitable material, now limited in quantity and with a higher possibility of compromise? Or do they decide that because of the very small amounts of fine-grained material, they should guess at the most suitable single destructive analytical technique. Visual examination of colour, grain-size, possibly grain shape may allow discrimination and the elimination of some items but a more sophisticated, rapid and non-destructive technique is required in informing this decision, thus allowing rapid decisions on what further scientific investigations need be undertaken, limiting the number of time-consuming and costly subsequent analyses, aiding the preliminary screening and providing comparative data for conjunctive analyses. The rapid acquisition of results can be important in sampling a potential scene as close to the time of the event as possible, before changes (such as building works, erosion and weathering) have compromised scene integrity. The above rationale demonstrates that rapid (hours, days), non-destructive, qualitative information on crystalline trace evidence (mud spots, rock dust coatings, residual soil/rock in washed items) that can be used at the time conjunctively with visual description and later with other analyses would aid the time- and increase cost-effectiveness.

Recent advances in high-frequency surface-wave methods

Jianghai Xia[1,2], *Lingli Gao*[1], *Yudi Pan*[1], *and Chao Shen*[1]
1. Subsurface Imaging and Sensing Laboratory, Institute of Geophysics and Geomatics, The China University of Geosciences, Wuhan, China
2. Hubei Subsurface Multi-scale Imaging Key Laboratory, Institute of Geophysics and Geomatics, The China University of Geosciences, Wuhan, China

Abstract

Multi-channel Analysis of Surface Waves (MASW) analyzes high-frequency Rayleigh waves to determine near-surface shear (S)-wave velocities. The method is getting increasingly attention in the near-surface geophysics and geotechnique community in the past 20 years because of its non-invasive, non-destructive, efficient, and low-cost advantages. They are viewed by near-surface geophysics community as one of most promise techniques in the future. However, they face unique problems related to extremely irregular velocity variations in near-surface geology or man-made constructions, for example, highway, foundation, dam, levee, jetty, etc., which are not solvable by techniques or algorithms widely used in earthquake seismology or oil/gas seismic exploration.

Calculation of dispersion curves by existing algorithms may fail for some special velocity models due to velocity inverse (a high-velocity layer on the top of a low-velocity layer). Two velocity models are most common in near-surface applications. One is a low-velocity half space model and the other a high-velocity surface layer. The former model results in a complex matrix that no roots can be found in the real number domain, which implies that no phase velocities can be calculated in certain frequency ranges based on current existing algorithms. A solution is to use the real part of the root of the complex number. It is well-known that phase velocities approach about 91% of the S-wave velocity of the surface layer when wavelengths are much shorter than the thickness of the surface layer. The later model, however, results in that phase velocities in a high-frequency range, calculated using the current algorithms, approach a velocity associated with the S-wave velocity of the low-velocity layer NOT the surface layer. These results conflict with the feature of surface waves and results in an incorrect inversion model. We proposed a direct and essential solution to correctly calculate surface-wave phase velocities due to this model. The algorithm that we newly developed can handle any arbitrary velocity models, which is the foundation of high-frequency Rayleigh-wave methods and is critical to near-surface applications.

Non-geometric wave exists uniquely in near-surface materials, especially in unconsolidated sediments. It occurs in near-offsets in high-frequency Rayleigh data. It is valuable for a quick and accurate estimation of S-wave velocity of the surface layer. Our study shows that non-geometric waves are leaky waves and dispersive. Leaky surface wave could cause misidentification when treating the leaky-wave energy as fundamental or higher modes. Such misidentification will result in wrong inversion results.

Multichannel Analysis of Love Waves (MALW) analyzes high-frequency Love waves to estimate near-surface S-wave velocities. Based on 2D geometry spread in which source and receivers are placed along a same line, current MALW fails to work in three-dimensional (3D) seismic acquisition system. This is because Love-wave particle-motion direction is perpendicular to its propagating direction, which makes it difficult to record

Love-wave signal in 3D geometries. We proposed to perform 3D MALW via recording two orthogonal horizontal components (in-line and off-line components) at each receiver point at the same time, and then rotating and composing those two components to recover Love-wave data. A synthetic model and two real-world examples verified the validity as well as the high effectiveness of our proposed method.

Geophysical Data Fusions for Resolution Improvement

Gang Tian, Shahid Ali Shaikh, Abdullah Lizan, Ran Chen, Yiming Wang
Department of Earth Science, Zhejiang University, China

Abstract

Nowadays, data fusion is quite a modern word for information technology. Actually, there has been a steady increase for decades in the application of geophysical data fusion to improve resolution. We carried out the data fusion study mainly from geophysical sensors' stage.

At first, in order to solve the coupling problems occurred in different surface conditions, the special coupling geophone system for a certain condition is designed to match ordinary geophone data by site observation comparison. Then the coupling filter was applied to the ordinary geophones which were used at all surface cases. After that, the filtered data were treated as original acquired data. That means that the coupling problem can be solved by designing few special coupling geophone systems and filtering the data obtained by ordinary geophones. The example was given in a desert area.

Secondly, it is known that there are some geophones with different frequency responses for choose in field survey which is a dilemma for geophysicists. We considered combining the frequency advantages of different kinds of geophones to broaden the effective bandwidth of the acquired data by using a partial frequency match filtering method in data processing which helps to improve the resolution of seismic reflected. We collected data using two kinds of geophones, normal geophones with natural frequencies of 60Hz and 100Hz and towable cable geophones which can receive signal frequency up to several hundred Hz. We applied a partial frequency match filtering method, which has improved the bandwidth frequency by focusing the data from above geophones in data processing, and achieved results showing the existence of the target layer. The example was given in an archaeological geophysical investigation case.

Thirdly, we know that the penetration depth and resolution of energy of Ground Penetrating Radar (GPR) is purely relying upon the central frequency of a particular antenna. When the central frequency of transmitting and receiving antennas is not the same then the swap effect is different. We found that the impact of transmission antenna frequency is greater on the data. The major impact of receiving antenna frequency is on the resolution. Therefore it can be possible that the penetrating depth and the resolution with the optimum combination configure of antennas could be improved than the one with normal configurations of antennas as well as by applying a match filter afterwards. We will show an example in a test site for archaeological purpose.

And lastly, the data from Electrical Resistivity Tomography (ERT) methods were studied in observation point of view. It is know that the inversion data from different array types have different features, which could create various geo-electric image results and bring some difficulties to interpretation. The simulation and observation of 2-D ERT integrated data of Wenner and Schlumberger arrays in a test site were studied, since these two configurations have different characteristics for the sensitivity of resistivity variation. The observation of 3-D ERT from the site integrated data of pole-dipole and dipole-dipole arrays were also tested, because the inversion data have different depth values. The simulation and experimental results show that the integrated data method discussed here improves the precision of resistivity imaging.

Comparison Tests for Surface Electromagnetic Prospecting System(SEP)

Qingyun Di*, Zhongxing Wang, Changmin Fu, Zhiguo An
Key Laboratory of Shell Gas and Geoengineering, Institute of Geology and Geophysics, Chinese Academy of Sciences, Beijing, 100029
*qydi@mail.iggcas.ac.cn

Abstract

In recent years, induced electromagnetic methods applied in deep ore body prospecting, such as CSAMT and MT method have undergone rapid and great development. However, land EM exploration instruments were almost monopolized by several big companies from U.S., Canada and Germany. To develop Chinese own equipment and technology, we undertook the Sinoprobe project to research surface electromagnetic prospecting(SEP) system in 2010, andsome achievements have been obtained.SEP system is a land electromagnetic instrument developed by our research group; including high-temperature superconductor magnetic sensor(HTSC), fluxgate sensor, induced magnetic sensor and receiver box, which could be used for CSAMT, AMT and MT data acquisition.

In2012, a number of integration tests were carried out at Hebei, Liaoning, Gansu, et al. Based on the problems exposed during these tests, SEP research group conducted thorough analysis, and made some optimization and improvements on this system. In 2013, a test of overall ability of SEP system was conducted at Caosiyao molybdenum mine in Xinghe, Inner Mongolia, China in order to test the performance and reliability of SEP system in practical exploration. Caosiyao molybdenum mine is a super-large-scalemolybdenum mine. Even though detailed geological survey has been done in this area and geological structuresand mineral distribution revealed by boreholes are quite clear, minerals in this area are yet to be exploited. Hence, there is little interference, which is beneficial for data acquisition and instrument comparison in the field.

Using different schemes, several contrast tests were performed. Firstly, we compared the HTSC with commercial induced magnetic sensor using the same station. The comparison result showed that in the presence of interference, HTSC was inferior to the latter in high and low frequency band, while the performance of both sensors in quiet environment was consistent. Hence, there is still opportunity for improvement on the HTSC sensor. Secondly, using the same sounding point, we also compared fluxgate sensor with commercial induced magnetic sensor via different frequency range. The result showed fluxgate sensor was more suitable for low-frequency MT survey. Thirdly, we did transmitter comparison, and the result indicated that SEP transmitter performed better than GGT-30 and TXU-30 in the frequency band lower than 700Hz. Fourthly, we compared the observation data from SEP system with those fromV8 and GDP32 system at the same transmitting location and profile. The obtained sounding curves and inversion results were consistent. Finally, we compared SEP with V8 system in MT test, and got consistent result. The study showed that the SEP system can be used for a long-time MT observation, and its performance was comparable with V8 system. In conclusion, the comprehensive test showed that the self-developed SEP system's quality has reached or exceeded that of similar foreign commercial products, and it is competent for practical exploration.

Applications of 3-D modeling & inversion technology in AMT exploration with with varied topography

Ruan Shuai[1,2,], Wang Xuben[1], Zhou Jun[1], Tang Ji[2]*
1 State Key Laboratory of Oil and Gas Reservoir Geology and Exploitation, Chengdu University of Technology, Chengdu 610059, China
2 State Key Laboratory of Earthquake Dynamics, Institute of geology, China Earthquake Administration, Beijing, 100029, China

Summary

In a Kimberly exploration project, which is very hard to locate the real target, with prior information from 1-D inversion of typical sites, we build a 3-D background model to calculate AMT response influenced by topography and non-homogeneous near-surface, then use model response to correct measured data's impedance phase invariant. The corrected phase invariant terms to be a powerful tool to rule out misrepresentative abnormality on traditional 1-D and 2-D inversion. All these qualitative analysis give use better understanding of real resistivity distribution so that both the lower & upper bound even initial model of 3D AMT inversion can be sketchily determined. Finally, bound constrained 3-D inversion results lead to a more reliable resistivity model for geological interpretation.

Introduction

Our challenge is to locate Kimberly pipes from an extremely distorted AMT dataset by violently varied topography acquired from Southwest China (Figure 1). The vertical low resistivity belt-shaped area on pseudo-section or 1-D inversion, which normally treated as phase shift effect, could possibly be the target we need to find.

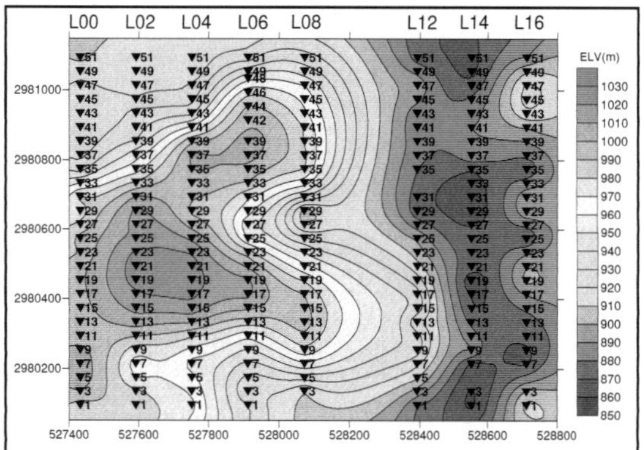

Figure 1: Surface topography map with observation sites

So we build two 3-D resistivity models based on 1-D Occam inversion (Constable C S, Parker R L. 1987) model from some typical AMT sites without too much phase shift phenomena on apparent resistively curves: one is a background layered 3-D model with the real topography and real sites' locations (MB), the other model named "MA" can be obtained by simply add two vertical low resistivity "pipes" on MB (Figure 2). The typical MT response can be calculated by 3-D staggered grid finite difference method (Mackie R L, etc. 1993).

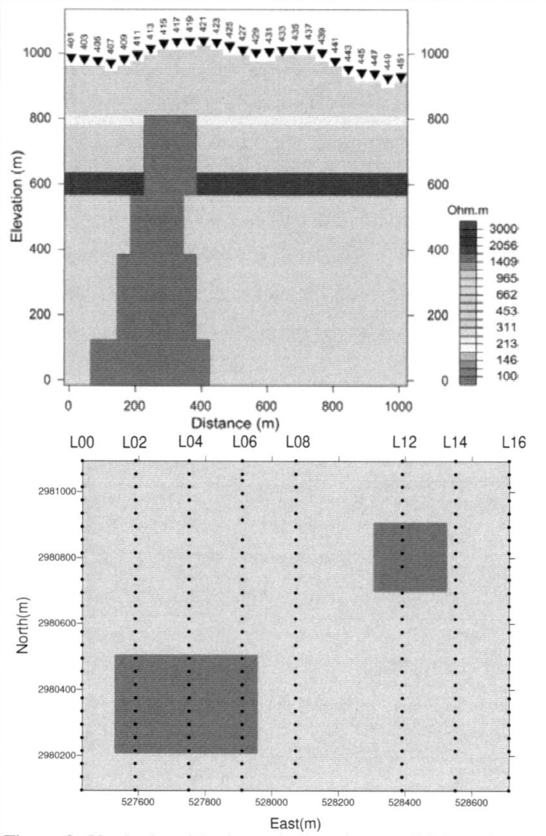

Figure 2: Vertical and horizontal slices image of 3-D resistivity model "MA" for analysis
up: slice on line L04;
down: slice on elevation as 0m.

The compared results shows there is almost no difference between MA's and MB's typical MT data response on two polarity's impedance phases and impedance phase invariant (Figure 3). The data seems to be completely distorted by the topography, which give us noting to locate the two pipes we set on model MA (other parameters have similar behaviors).This phenomena tells us impedance phase which supposed not to be influenced by topography need a precise correction, and implies that direct 1-D and 2-D inversion may lead to some misunderstanding toward this kind of resistivity model.

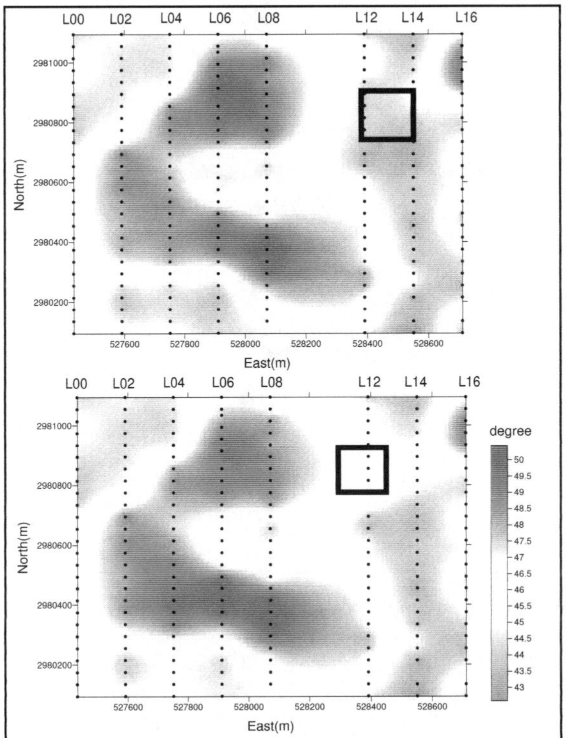

Figure 3: Impedance phase invariant slice maps on 500Hz frequency of two models (up: MA, down: MB, black square: differences can be recognized)

Impedance Phase Invariant Correction

Since the response of background model MB is calculated, the corrected impedance phase invariant (shorten as CIPI) can simply obtained by deducting MB's impedance phase invariant from MA's data, for similarity as normal impedance value behavior, deducted value then need to add 45 degrees (Figure 4). In real measured data case, whole process is almost the same, but replaces MA's data by smoothed observed impedance phase invariant (Figure 5). The CIPI can not only recover the true horizontal location of two low resistance pipes we set for model MA (Figure 4), but also help us recognize the wrong low resistivity area calculated by 1-D or 2-D direct inversion (Figure 5). In 1-D inversion result, we can see too much low resistance zones which is not coherent with CIPI calculate by real measured data (red circles), and the distribution of resistivity seems strongly disorganized (Figure 5 up). The 2-D inversion (Figure 5 middle) can give relatively better results but many belt-shaped structures were shown due to every line's inversion resistivity value level is different, which might let us miss the low resistance area in middle (red and black square on Figure 5 down).

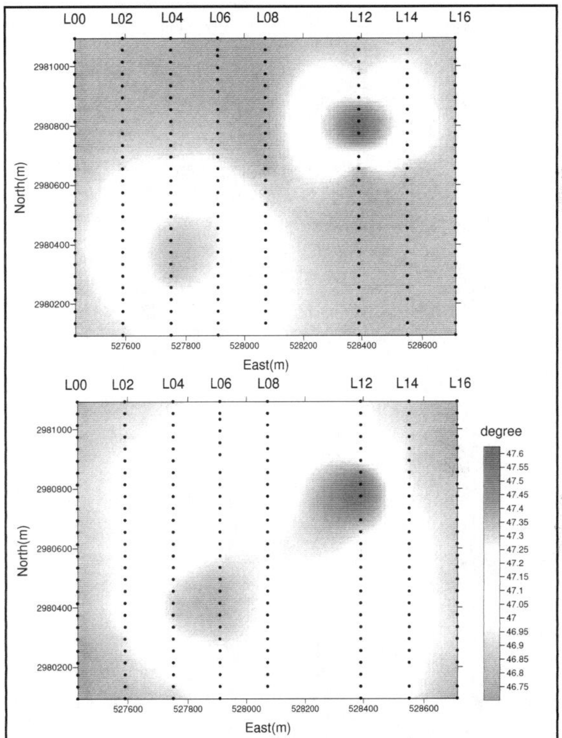

Figure 4: CIPI slice maps on two frequencies of model MA (up: 1000Hz, down: 10Hz)

Constrained 3-D Inversion

To avoid the shortage of 1-D and 2-D inversion, We developed a simply bound constrained Quasi-Newton 3-D MT inversion code based on LBFGS formula, which has the same kernel as LBFGS-B 3.0 Fortran code published by professor J. Nocedal on 1990s.

All noisy part and frequencies of measured data on death-band of 4 tensor impedance were removed. Then we use two homogeneous spaces (100Ω.m and 1000Ω.m) as initial

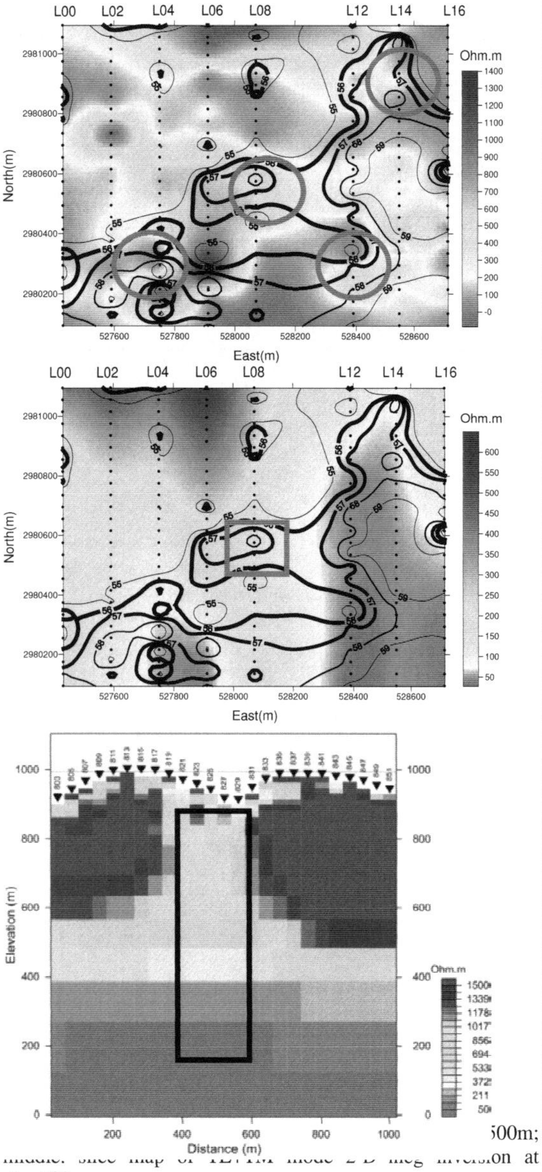

500m; middle: slice map of [illegible] inversion at elv=500m;
down: profile L08 of 2-D inversion)
contours: CIPI with value higher than 55 degrees;
red circle & square: possible low resistance area shown on 1-D CIPI;
black square: possible low resistance pipe shown on 2-D inversion profile.

model to do trail inversion, with every cell's resistivity varied from 1Ω.m to 100000Ω.m

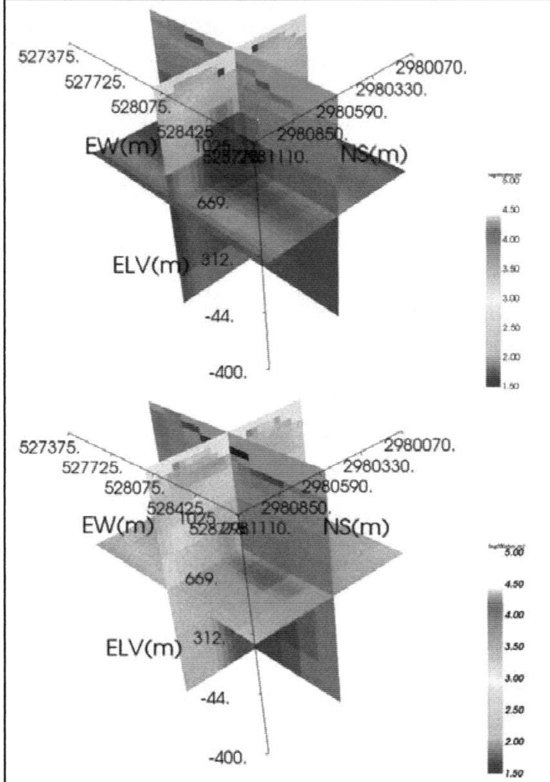

Figure 4: 3-D LBFGS-B trail inversion results of measured data
up: initial model as 100Ω.m;
down: initial model as 100Ω.m.

The trail inversion give use better choice of initial model, mean while, both the lower and upper bound of gird cells' resistivity can be sketchily determined as well. Finally, we use a two layered initial model with 1000Ω.m on elevation higher than 900m and 500Ω.m below, the first layer's resistivity bound is 500～100000Ω.m while second is 100 ～10000Ω.m.

The final constrained inversion results give us a clearly image of lower resistance zone in the middle of survey area, which is coherent with the middle trap of CIPI contours. All the analyzing and 3-D inversion procedure produces more reliable results at last.

Conclusions

In 3-D case, AMT impedance phase information can still be distorted by topography, meanwhile, 1-D and 2-D inversion can hardly recover the true resistivity model if our target is vertical which might cause similar image with phase shift effect on pseudo-section. By using the CIPI calculated based on 3-D modeling technology, we can avoid misinterpretations by 1-D and 2-D inversion, at the same time, qualitative analysis from CIPI and 3-D trail inversion can give us a better initial model and sketchy resistivity bounds for constrained 3-D AMT inversion to obtain better results.

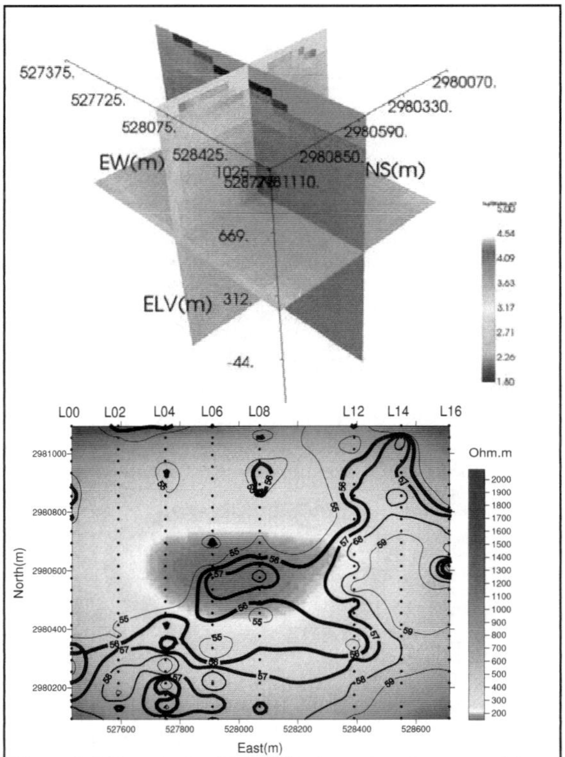

Figure 4: 3-D constrained LBFGS-B inversion of measured data
up: 3-D slices view;
down: slice map on elv.=500m;
contours: CIPI with value higher than 55 degrees.

Acknowledgments

This work is supported by National Science Instruments Developing Project (2011YQ05006007). Thanks to the kindly real measured data provider, No. 103 Geological prospecting party, Bureau of Geology and Mineral Exploration and Development of Guizhou Province.

References

Chen X B. 2010. Effects of topography to magnetotelluric measured data and two dimensional inversion correction with topography. China Geophysics 2010 - 26th annual of CGS & 13th academic congress of SSC. (in Chinese), 325.

Constable C S, Parker R L. 1987. Occam's inversion: a practical algorithm for generating smooth models from electromagnetic sounding data. Geophysics, 52:289–300.

Hu Z Z, Hu X Y, He Z X. 2006. Pseudo-three-dimensional magnetotelluric inversion using nonlinear conjugate gradients. Chinese J. Geophys.(in Chinese), 49(4):1126-1234.

Lu L Y, Zhang B X, Bao G X. 2003. Modeling of three-dimensional magnetotelluric response for a linear earth. Chinese J. Geophys.(in Chinese), 46(4):568-575.

Mackie R L, Madden T R, Wannamaker P E. 1993. Three-dimensional magnetotelluric modeling using difference equations: Theory and comparisons to integral equation solutions. Geophysics, 58:215–226.

Miao J C, Ruan S, Zhang Y. 2013. The application of the audio magnetotelluric sounding method to the precise interpretation of formal and reverse faults. Geophysical & Geochemical Exploration (in Chinese), 37(4):681-686.

Rodi W L, Mackie R L. 2001. Nonlinear conjugate gradients algorithm for 2D magnetotelluric inversion. Geophysics, 66:174–187.

Ruan S. 2013. Discussion of static shift effect in practical AMT data inversion and interpretation. China Geophysics 2013 - 29th annual of CGS. volume 24 (in Chinese), 631-632.

Shen J S. 2003. Modeling of 3-D electromagnetic response in frequency domain by using staggered-grid finite difference method. Chinese J. Geophs.(in Chinese), 46(2):281-288.

Siripunvaraporn W, Egbert G, Lenbury Y. 2002. Numerical accuracy of magnetotelluric modeling: acomparison of finite difference approximations. Earth Planets Space, 54(6):721–725.

Siripunvaraporn W, Egbert G. 2001. An Efficient data-subspace inversion method for 2D magnetotelluric data. Geophysics, 65:791–803.

The application of non-seismic data in near surface velocity modeling in surface volcanics area

Yang Zhanjun[1], Feng Wenyan[2], Lei Dong[2] and Jiang Wenbo[1]*
1. BGP Inc., China National Petroleum Corporation
2. China Southern Petroleum Exploration and Development Corporation

Summary

Near surface structure is extremely complex in surface volcanic rock areas. Uphole survey is a well used method to investigate surface volcanics since its high vertical resolution. Due to its higher costs and lower productivity, uphole survey is not performed with an enough density, which makes near surface velocity model unsatisfied, and leads to poor effects of seismic excitation and static correction. This paper presents an economic and effective solution, which take uphole survey and non-seismic survey to perform near surface structure investigation, and further establish near surface velocity model. A case in this paper shows the improvement of near surface velocity model and seismic static correction effect.

Introduction

It is well known that the near surface structure is extreme complex in surface volcanic rocks areas, where the volcanics distribution is irregular, their bottom interface and thickness vary quickly, lithology and velocity change fast in horizontal directions, and reversal velocity exists broadly. Especiously, the weathering action is speeded up in the south of China because of the rich rainfall, so it gets more difficult to well describe surface volcanics. Seismic exploration is strongly affected by the accuracy of surface velocity model, which determines the effect of seismic static correction. So the geophysicists is longing for a better solution to well define the spatial distribution of volcanics, establish a finer near surface velocity model, and further provide reliable data for static correction (Zhao, 2007).

Nowadays the shallow refraction survey and uphole survey are normally used in near surface structure investigation (Ding, 2007). The shallow refraction survey is a method to study low-depression velocity layers with first arrival of directive waves going through low-depression velocity layers or refractive waves going along near surface refraction interfaces, which can quickly measure the thickness of surface low-depression velocity layers and trace the high velocity layer top. But this method is useless when the refraction layer is thin, velocity contrast is too small or absent, or velocity reversal exists. Since the sand and clay layers is of lower velocity than the upper surface volcanics, the shallow refraction method is unsuitable for the structure investigation in near surface volcanic areas. Instead of that, uphole survey is a method to study low-depression velocity layers with first arrival of penetrated waves from borehole to surface, surface to borehole, or borehole to borehole, which can directly measure near surface velocity variation with higher vertical precision and is normally used for investigation of surface volcanic rocks. Since a large density of uphole shots are necessary in an exploration area to fight against the quick change of volcanic lithology, the cost and productivity normally are unacceptable. Seismic refraction wave first arrival is also used for investigation of near surface structure, which is picked up from seismic survey data for the inversion of near surface structure and velocity model. Since there is no stable refraction interface in near surface volcanic area and the quality of shot arrival is poor, it is always difficult to pick up satisfied first arrivals. Also the far offset first arrival is difficult to be picked up since the velocity of upper volcanics is higher than the lower sand and clay layers. So the static correction methods based on first arrival data are not applied in near surface volcanic areas. Then an effective and efficient solution to investigate the surface volcanics and build near surface velocity model is always expected in the complex surface area developed volcanic rocks.

Method and Techniques

1. Identifying surface volcanics with magnetics

Magnetics is an effective method to identify volcanics. Since different kinds of volcanic rocks are of different magnetic susceptibility while the susceptibility of sand and clay layers are closes to null, surface volcanics can be easily identified from sand and clay layers by using magnetic data. It is well known that the magnetic anomaly features of volcanics vary with their burial depths. Surface volcanics is characterized by high-frequency magnetic anomaly because of quick variation of volcanic lithology, even if the magnetic station spacing is small. On the contrary, volcanics located in depth is characterized by lower-frequency magnetic anomaly. So according to frequency characteristics, magnetic anomaly of near surface volcanics can be extracted to study its distribution (Yang, 2013). Figure 1a shows the application effect of magnetic anomaly to identify surface volcanic distribution, where high-frequency magnetic anomaly in the left part reflects the existence of near surface volcanics and low frequency magnetic anomaly in the right part reflects the absent of near surface volcanics, which well corresponds to the uphole lithology.

2. Mapping surface volcanic bottom with Transient Electromagnetic Sounding (TEM)

TEM is an artificial source time-domain EM method to detect 3D resistivity distribution, which transmits pulse magnetic field downwards with unearthed loop or earthed line source, and receives induced whirlpool field from subsurface medium with loop or electrodes at the period of pulse diapauses. TEM is of highest resolution in near surface depth range, which profits from the received signal of high frequency. The sand and clay layers are of very low resistivity while volcanics is of high resistivity. The resistivity of weathered volcanics is lower than tight volcanics and higher than sand and clay layers. So the burial depth to volcanics bottom can be easily defined with TEM data.

The task of TEM data processing is to transform time-domain EM raw data into depth-domain resistivity data with a higher precision. Figure 1b shows an inversion resistivity section, where an upper high resistivity layer overlaid on a lower resistivity layer in left part and a whole lower resistivity layer in the right part. The velocity curves of uphole located in the survey line confirm that there is an upper high velocity layer and a lower low velocity layer within the near surface depth range in the left part and a whole low velocity layer in the right part. It concludes that the apparent resistivity values of near surface rocks are related to the velocity value, and fatherly the velocity value can be transformed from the inversion resistivity data.

3. Jointly interpreting near surface structure and lothology

The joint interpretation of near surface structure is to perform comprehensive study making use of advantages of each geophysical method. In surface volcanic areas, taking uphole data as the control of lithology and burial depth to bottom at upohole locations, taking magnetic data to describe the distribution of near surface volcanics, taking TEM data to map the lothology and burial depth between uphole shots, a comprehensive study is carried out to infer the lothology distribution and conclude fine near surface structures. Uphole survey is charactered with high vertical resolution, and magnetic and TEM are of lower costs and higher productivity than uphole survey, so a near surface structure comprehensive investigation met seismic survey need at an acceptable cost can be obtained by controlling the depth with uphole data, and optimizing the burial depth in the whole area with magnetic and TEM data with denser station spacing.

Figure 1 shows a joint interpretation section with uphole and non-seismic data. Figure 1a is the residual magnetic anomaly reflecting the distribution of near surface volcanic rocks, where high-frequency anomaly on the left reflects the presence of surface volcanics and low-frequency anomaly on the right reflects the absence of surface volcanics. Figure 1b shows the TEM inversion resistivity section, where a high resistivity layer overlays on a low resistivity layer on the left, and a wholly low resistivity layer on the right instead, which characters fit well with uphole lithology logs and magnetic anomaly meanings. Especially, in the middle part of section, at the range from 7000m to 8000m, a high-frequency magnetic anomaly means the presence of near surface volcanic rocks while relative high resistivity reflects none of tight volcanics. Referring to the lower uphole velocity, a weathered volcanic zone is inferred. In geology, the volcanic rocks located at a higher topography suffered a long history of strongly weathering and denudation, and formed a weathered volcanic layer. The black line in Figure 1b represents the surface volcanic bottom, which plots out the interface between upper volcanics and lower sand and clay layers.

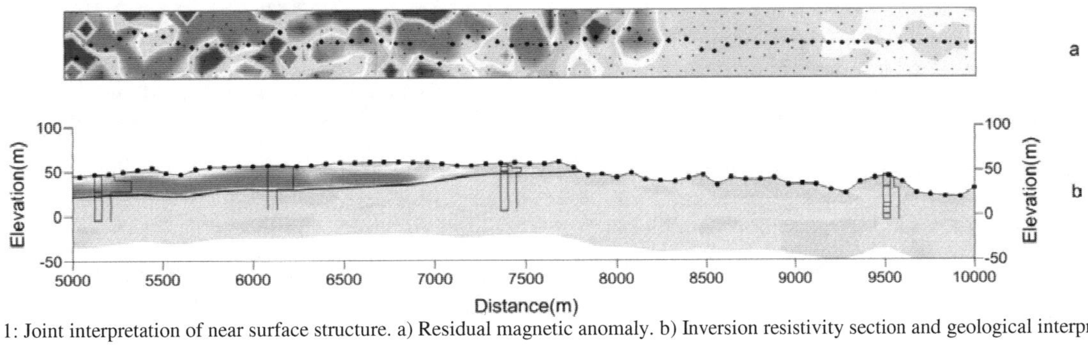

Figure 1: Joint interpretation of near surface structure. a) Residual magnetic anomaly. b) Inversion resistivity section and geological interpretation.

4. Establishing near surface velocity model

Uphole survey provides near surface vertical velocity data at uphole survey sites, but the velocity information between upholes is normally interpolated. Although the uphole velocity is of high accuracy, the near surface velocity model generally is of low accuracy for their larger spacing, which cannot meet the requirement of static correction.

Non-seismic data can provide variation information of underground medium lithology between upholes at reasonable cost and productivity. So we can take uphole velocity as control, study the horizontal variation of near surface lothology with non-seismic data, analyze the relationship between resistivity and velocity of each layer, and finally optimize the velocity model by utilizing resistivity data. Figure 2a shows the near surface velocity model derived from uphole survey. Though the uphole density is over 2 upholes per square kilometers, which is far higher than other survey areas, it is still found that the velocity model is of a low resolution which cannot meet the requirement of static correction. After magnetic and TEM surveys, a new velocity model, Figure 2b, is established jointly with uphole, magnetic and TEM data. Figure 2b shows that the resolution of the new model is improved. This model is more close to the real velocity model than Figure 2a.

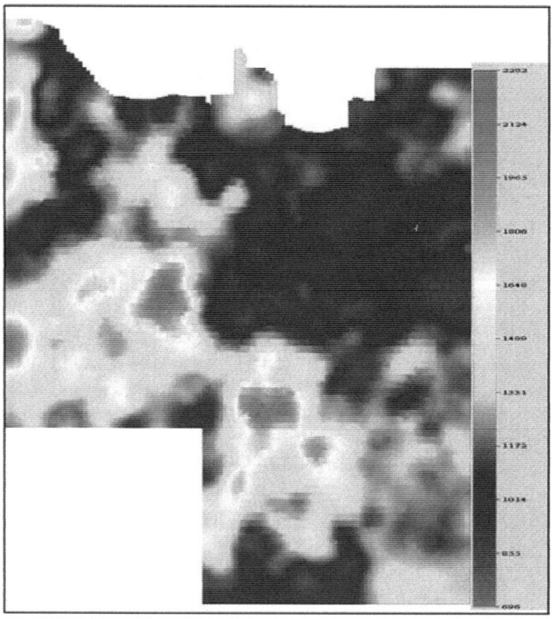
a

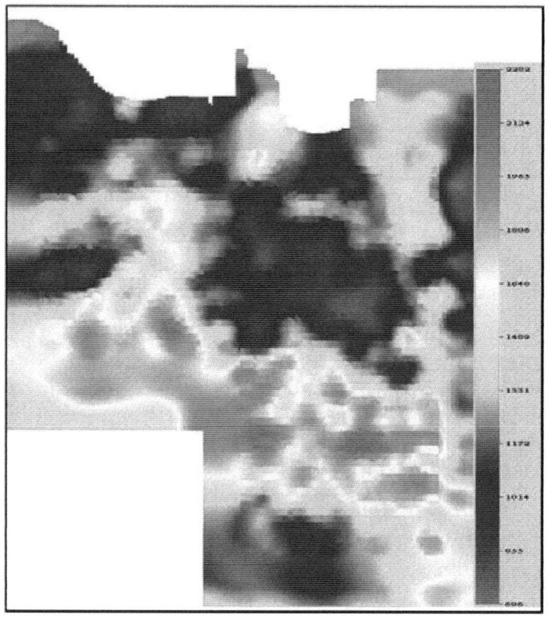
b

Figure 2: Comparison of uphole velocity model (a) and joint velocity model (b).

Cases

The study area locates in the north of Hainan Island, China, where the Quaternary volcanics widely develop, so the S/N ratio of seismic data acquired before is lower and static correction effect is unsatisfied. When second-time 3D seismic survey was carried out, magnetics and TEM surveys with denser station spacing accompanying denser uphole survey were performed for near surface structure joint investigation and modelling with the above-mentioned method.

Figure 3a shows a stack section after static correction with the velocity model derived from uphole survey, it was found that the events at a depth of 600ms is unsmooth, desultory, and difficult to be identified somewhere, and the event at a depth of 1000ms is more difficult to be identified. Figure 3b shows a stack section after static correction with the joint velocity model derived from both uphole and non-seismic data. It was found that the event at a depth of 600ms is more smooth, more continuous, and easy to be traced, and the event at the depth of 1000ms is improved and can be traced within a larger range. Generally speaking, the effect of static correction adopted the joint velocity model derived from uphole and non-seismic data is obviously improved.

Conclusions

In surface volcanic rock areas, the near surface velocity model established jointly with uphole and non-seismic data is better than the velocity model established with solo uphole data, and the static correction effect can be improved with the joint velocity model. Due to the lower costs and high productivity, the near surface velocity modeling method joint with uphole and non-seismic data is an economic and efficient solution to solve static correction problem in surface volcanic rock areas.

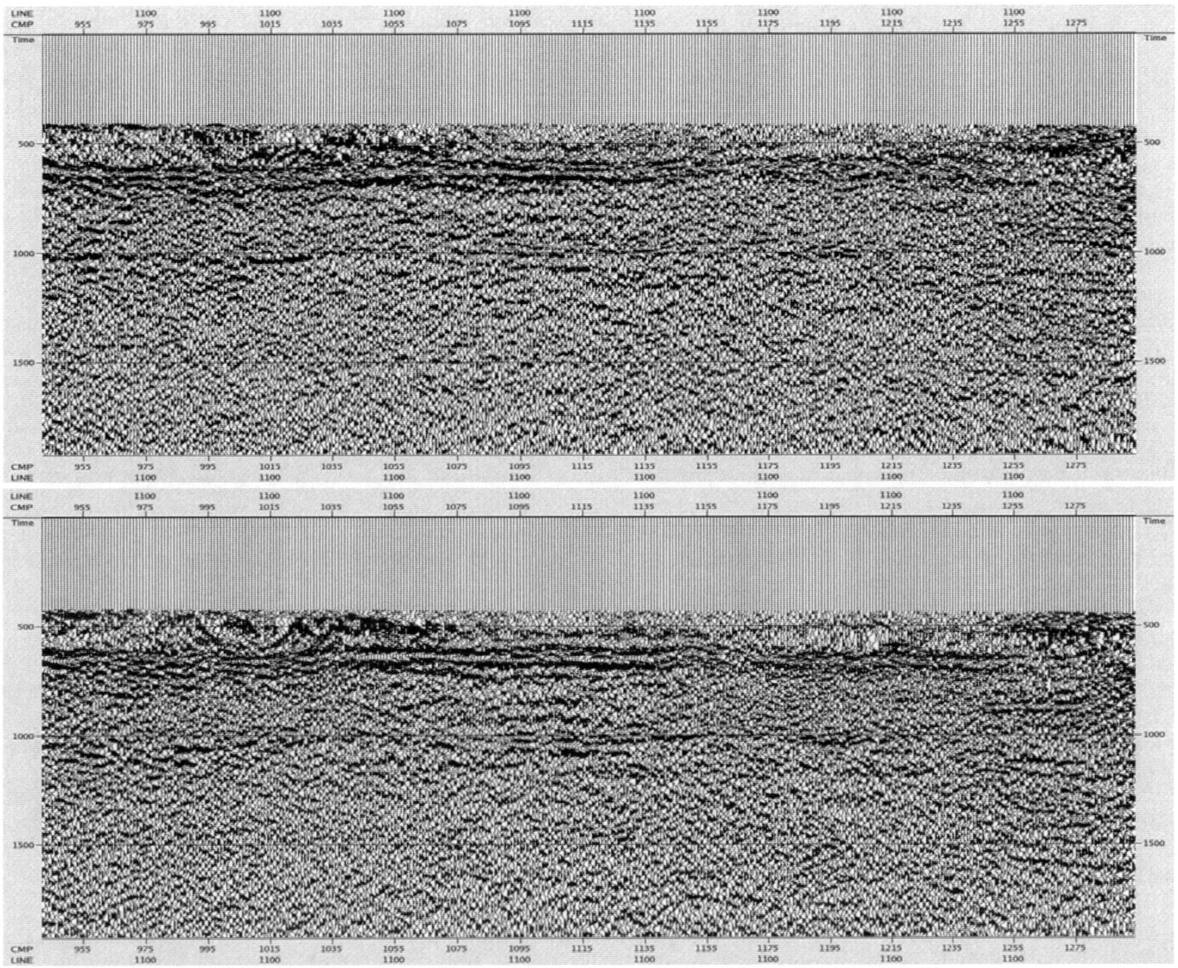

a

b

Figure 3: Comparison of stack sections with uphole velocity model (a) and joint velocity model (b).

Acknowledgement

This work was supported by the scientific study and technical development project of China National Petroleum Corporation (2013D-3808).

References

Ding J.R., Mao F.M., and Hao T.Y., 2007. Surface structure survey in the area where covered with the igneous rock near-surface, Progress in geophysics 22, 872-878.

Yang Z.J., Jiang W.B., and Lei D., 2013. The study method of magnetic and its application in surface volcanic distribution, CPS Geophysical workshop 2013, Expanded Abstracts, 710-713.

Zhao J.Z., Shi Y.M., Sun W.Z., Liu J., Zhang S.H. and Ma Q.L., 2007. Distribution of igneous rock and its effect on petroleum exploration in Fushan sag, Petroleum Geology 2007, 38-43.

Yang Z.J., Jiang W.B., Feng W.Y. and Lei D., 2015. The application of magnetics and TEM in seismic survey in volcanics-covered area, SEG/CPS GEM Chengdu 2015: International workshop on Gravity, Electrical & Magnetic Methods and their Applications.

GPR Signals Analysis of Post-tensioned Prestressed Concrete Girder Defects

Sixin Liu, Lei Fu, College of Geo-exploration Sci. & Tec., Jilin University, China*

Summary

Accurate inspection of duct condition in the post-tensioned prestressed concrete (PPC) is an essential part for GPR concrete inspection. The purpose is to inspect the grouting condition of the ducts where the strands are located, to find if there is void in the ducts, and if there exists water. In order to investigate the radar image characteristics of different PPC duct defects, a number of model girders were manufactured. Three major ducts are included in our study: 1) well grouted and no void (normal condition); 2) duct is half filled, and the void is filled by water or air; and 3) duct is not filled at all, and the duct is water or air filled. The data corresponding to seven different situations are acquired and processed. It is found that radar can detect the first interface in the duct, and the detailed structure inside the duct cannot be "seen" from the images directly. Characteristic curves help the interpretation very much. Complete void duct is the easiest to be differentiated from others. The signature for this situation is characterized with strong and clear reflection interface which becomes weaker as the void is water-filled. Normal condition shows weakest reflection interface. As for half void situation, front scan shows similar result to normal condition whether it is water or air filled, and back scan shows similar features to completely void situation. The experiment and analysis is helpful and instructive for practical engineering inspection.

Introduction

Regular inspection and maintenance of bridges with life span of more than 50 years is very necessary for their safe operation. The engineering inspection of the bridges usually utilizes the non-destructive methods such as electromagnetic method, direct current method, ultra-sonic method, seismic method (Liu and Guo, 2005) and thermal imaging, etc. Ground penetrating radar (GPR) is one of the most effective non-destructive geophysical methods, especially for shallow detection.

GPR survey can cover a wide area in a relatively short period of time for concrete evaluations. Due to recent hardware and software advances, real time cursory analysis can also be performed at the site. GPR has become an increasingly attractive method for the engineering community, in particular for shallow, high resolution applications such as concrete evaluation studies. GPR can be used to map rebar position, determine depth of concrete over rebar, identify air/water filled voids in or behind concrete, determine concrete thickness, identify rock pockets within concrete, identify leak grouted voids, determine asphalt thickness, determine rebar corrosion/concrete condition, map and define conduits and utility cables within concrete structures.

In order to investigate the radar image characteristics of different PPC duct defects, we have made model girders with three different ducts (figure 1 and 2): 1) well grouted and no void (good condition); 2) duct is half filled vertically, and void is filled by water or air; and 3) duct is not filled at all, and the duct is water or air filled. The data corresponding to seven different situations are acquired, and processed. The signals characteristics to different duct grouting conditions are analyzed.

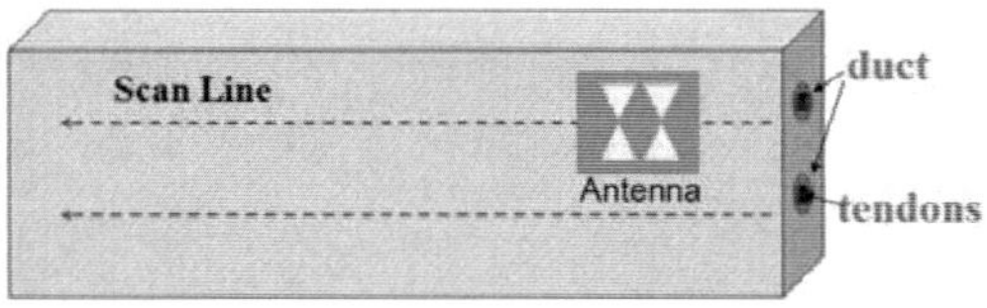

Figure 1: A 3D model of a PPC girder, which is 30cm wide, 50cm high and 200 cm long.

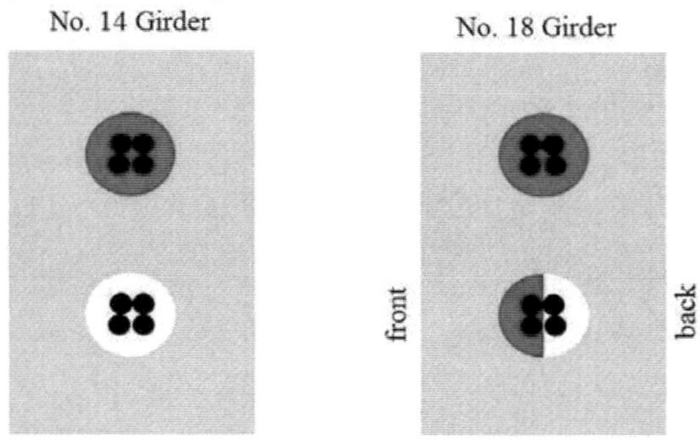

Figure 2. Cross-sections of No. 14 and 18 girders with different ducts grouting conditions. Top ducts of both girders are normally grouted. Down duct of girder 14 and 18 are void and vertically half grouted, respectively. The ducts are 11 cm in diameter.

Data collection and processing routine

RAMAC/GPR system with1.6GHz antenna made by MALA Geosciences was used as shown in figure 3 for the PPC inspection. Two bowtie antennas were enclosed in a small shielded box with a measuring wheel, antennas separation is 6cm. We consider two survey modes at the beginning: horizontal scan relative to the duct, vertical scan relative to the duct. The scan line is too short for the vertical scan in this situation and therefore the method is discarded. We use horizontal scan at last considering feasibility and efficiency. The antenna is at the same height as the duct during measurement to assure antenna is at the nearest position to the duct.

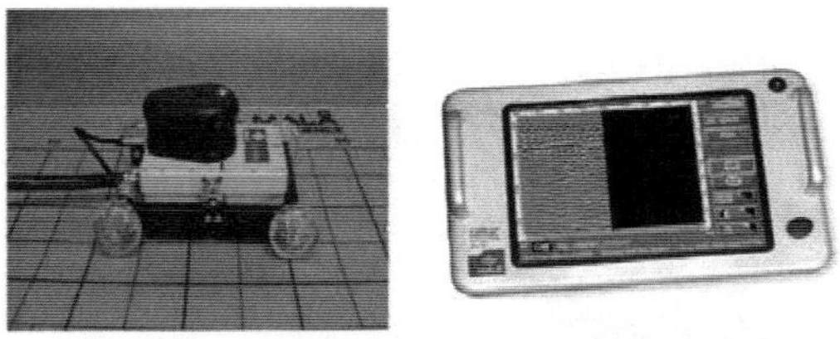

Figure 3. Ramac GPR 1.6 GHz antenna and control unit.

Data processing basically includes pre-processing and migration. The pre-processing has many options, we only choose direct current (DC) drift correction, time-zero correction, direct wave removal, and amplitude compensation. The processed results of each step are shown in figure 4.

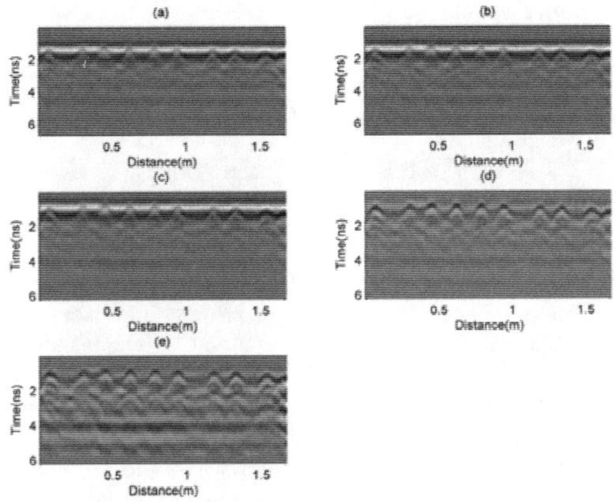

Figure 4. GPR profiles in different step. (a) raw data, (b) after DC drift removal, (c) after time-zero correction, (d) after direct wave removal, (e) after amplitude compensation. The data file is 723 acquired in a well grouted condition. The data processed effect is evident.

Radar images analysis and defects characteristics extraction

Radars images corresponded to seven situations are analyzed here, including normal condition and kinds of duct defects. Image characteristics corresponded to first interface, tendon, and other side of the girder are marked by number "1", "2", and "3".

To easy understand the results, the data acquisition parameters and the interpreted depth of tendons and different interfaces are summarized in Table 1. The depths of different targets are estimated manually.

1) Normal condition as the duct is well grouted. The radar images for this situation are shown in figure 5. The file name for each scan is marked on the top of each profile. These four scan are all carried out near the up duct of the girder No 14. The concrete/grout interface(1) show no (723 and1168) or small (773 and 977) reflections. The tendons(2) are all visible as marked in the figure. Top layer of rebar and the other side of the girder(3) are imaged clearly. We cannot see the rebar in the other side.

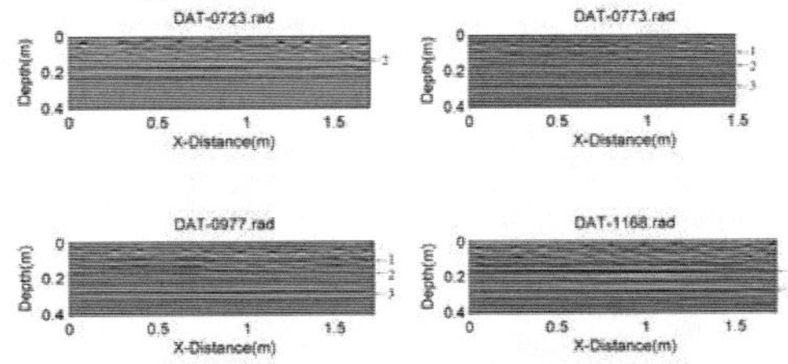

Figure 5. Radar images as the ducts is well grouted. The number above each image is the file name for each scan. The concrete/grout interfaces are visible sometimes(773 and 977) as marked by "1", the tendons are all visible as marked by "2", and the other side of the girder are visible as marked by "3", except when the data is absent.

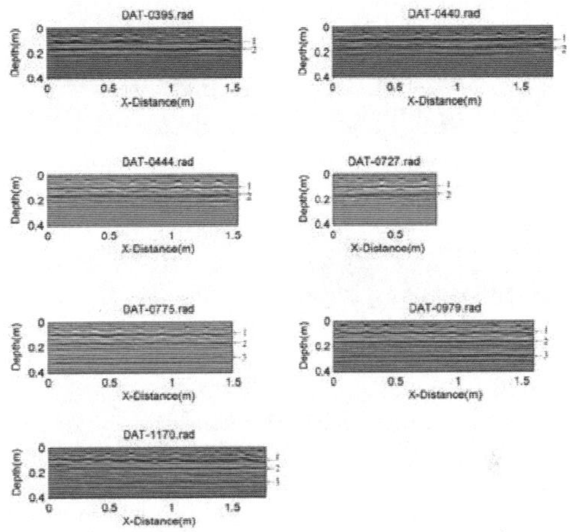

Figure 6. Radar image as the duct is completely void, and air-filled. The concrete/air interfaces, the tendons, and the other side of the girder are all visible as marked by "1" , "2", and "3", respectively, except when the data is absent for the other side of the girder.

2) The duct is completely void and filled with air. The data 395 were collected near the down duct of girder 18 before grouting. The data 440 were collected near the up duct of girder 14 before grouting. The data collected near the down duct of girder 14 are 444, 727, 775, 979, and 1170. The images are very typical and concrete/air interface(1) is very clear as shown in figure 6. The image characteristics differ from other situations clearly. Tendons(2) are visible in all case, especially in 395 and 979. Other side of the PPC girder(3) can be "seen" at about 0.3m sometimes..

3) The duct is completely void and filled with water. The data are collected near the up duct of girder 14. The concrete/water interface(1) is very clear as shown in figure 7. Other side of the PPC girder(3) can be "seen" at about

0.3m. Tendons(2) are visible in all case, but show change along the scan line. The two scans (985 and1172) show little difference.

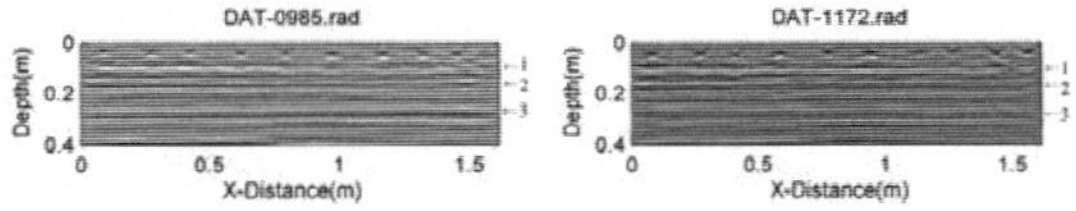

Figure 7. Radar image as the duct is completely void, and water-filled. The concrete/water interfaces, the tendons, and the other side of the girder are all visible as marked by "1", "2", and "3", respectively.

4) The duct is half void vertically and filled with air. The scan line is on the front side near the down duct of girder 18. The concrete/grout interface(1) is visible as shown in figure 8 sometimes. Other side of the PPC girder(3) can be "seen" at about 0.3m. Tendons(2) are visible sometimes, and show large lateral change in three scans (792, 1012 and 1162).

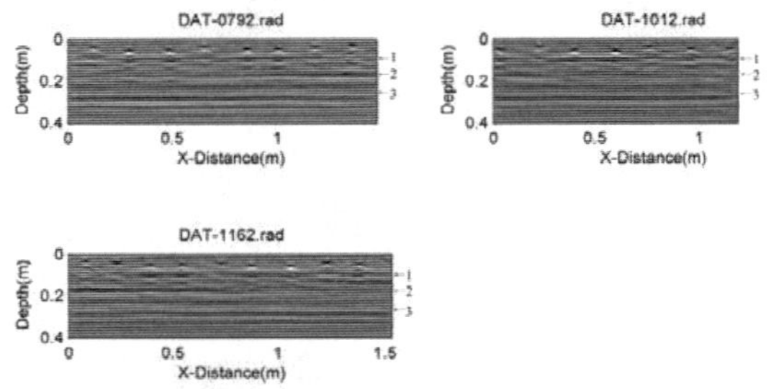

Figure 8. Radar image as the duct is half void, air-filled, and scanned from the front side. The concrete/grout interfaces, the tendons, and the other side of the girder are all visible as marked by "1", "2", and "3", respectively.

5) The duct is half void vertically and filled with water. The scan line is on the front side near the down duct of girder 18. The concrete/grout interface(1) is visible as shown in figure 9. Other side of the PPC girder(3) can be "seen" at about 0.3m. Tendons(2) are visible sometimes. Both the interface and the tendon show lateral change (1031 and 1164), sometimes invisible.

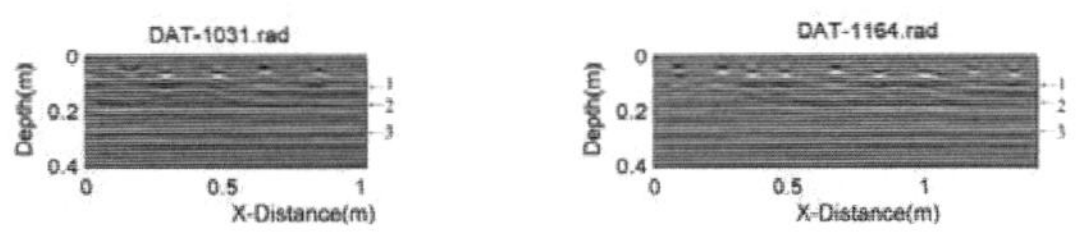

Figure 9. Radar image as the duct is half void, water-filled, and scanned from the front side. The concrete/grout interfaces, the tendons, and the other side of the girder are all visible as marked by "1", "2", and "3", respectively.

6) The duct is half void vertically and filled with air. The scan line is on the back side near the down duct of girder 18. The concrete/air interface(1) is very clear as shown in figure 10. Other side of the PPC girder(3) can be "seen" at about 0.3m. Tendons(2) are visible clearly. These scans (786, 1004, and 1159) show clearest results.

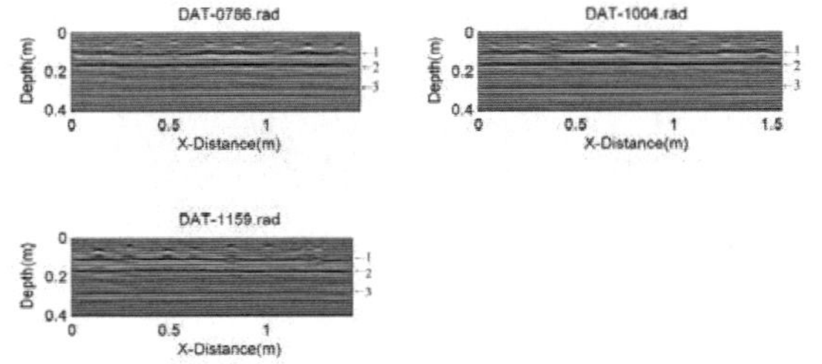

Figure 10. Radar image as the duct is half void, air-filled, and scanned from the back side. The concrete/air interfaces, the tendons, and the other side of the girder are all visible as marked by "1", "2", and "3", respectively.

7) The duct is half void vertically and filled with water. The scan line is on the back side near the down duct of girder 18. The concrete/water interface(1) is very clear as shown in figure 11. Other side of the PPC girder(3) can be "seen" at about 0.3m. Tendons(2) are visible sometimes, and shows lateral change (1025and 1166).

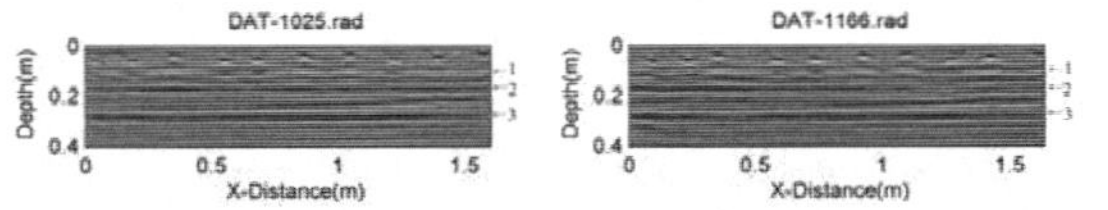

Figure 11. Radar image as the duct is half void, water-filled, and scanned from the back side. The concrete/water interfaces, the tendons, and the other side of the girder are all visible as marked by "1", "2", and "3", respectively.

The radar images are not so direct for interpretation and defects judgment. We tried to generalize the defects characteristics and make them concretely. We extract a characteristic curve for each radar image by averaging the data along the scan direction. Figure 12 show the seven situations listed in section 4.1.

Table 1 data acquisition parameters and interpreted depth of various targets

Figure no: Grouting condition	Filename	Scan position on a girder	1-First interface		2-tendon	3-otherside of the girder
Figure 5: The duct is well grouted	723	up duct/girder 14	concrete/grout	invisible	0.15m	invisible
	773			0.1m	0.17m	0.28m
	977			0.1m	0.17m	0.28m
	1168			0.1m	0.16m	0.28m
Figure 6: The duct is completely void (filled with air)	395	down duct/girder 18	concrete/air	0.1 m	0.16 m	invisible
	440	up duct/girder 14		0.1 m	0.16m	invisible
	444	down duct/ girder 14 as filled with air		0.1 m	0.16m	invisible
	727			0.1 m	0.15m	invisible
	775			0.09 m	0.16m	0.28m
	979			0.1 m	0.16 m	0.28m
	1170			0.1 m	0.16 m	0.28m
Figure 7: The duct is completely filled with water	985	down duct/girder 14 as filled with water	concrete/water	0.09 m	0.16 m	0.26 m
	1172			0.08 m	0.17 m	0.28 m
Figure 8: Half duct is grouted well(vertically), another half is filled with air	792	front side/down duct/girder 18	concrete/grout	0.09 m	0.17 m	0.28 m
	1012			0.09 m	0.17 m	0.28 m
	1162			0.09 m	0.17 m	0.28 m
Figure 9: Half duct is grouted well, another half is filled with water	1031			0.09 m	0.17 m	0.28 m
	1164			0.09 m	0.17 m	0.28 m
Figure 10: Half duct is grouted well, another half is filled with air	786	back side/down duct/girder 18	concrete/air	0.1 1m	0.16 m	0.28 m
	1004			0.11 m	0.16 m	0.28 m
	1159			0.11m	0.16 m	0.28 m
Figure 11: Half duct is grouted well, another half is filled with water	1025	back side/down duct/girder 18	concrete/water	0.09 m	0.17 m	0.28 m
	1166			0.09 m	0.17 m	0.28 m

For each kinds of defect situation, there are many curves. We observe the characteristic curves for certain case, and find they correlate very well. This means the repetitious scans for certain defect have consistency. The number in the legend refers to the file name for certain scan. The interested range for ducts defects analysis is from about 0.1m to 0.2m within which the signals reflect the different PPC ducts defects. We marked the positions of different targets, square for the position of concrete/air interface, circular square for the concrete/grout interface, ellipse for the concrete/water interface, circle for the tendons, and triangle for the other side of the girder. The positions of different targets are estimated manually.

Figure 12b and 12f show big similarity. The amplitude is large at 0.12m due to the large dielectric difference between the air and the concrete. These two situations are the easiest to differ from other situations. These two situations differ from each other little at 0.2 to 0.3m.

Figure 12c and 12g show some similarity. The amplitudes become weaker at 0.12m comparing to 12b and 12f situation, though the dielectric difference is extremely large between the water and the concrete theoretically. We contribute this to the fact that water invades the concrete and there is not abrupt interface. The phase change is opposite to 12b and 12f. These two situations can be differed from others. These two situations can be differed in the characteristic curves between 0.1 to 0.25m.

Figure 12d and 12e show very similar features, the amplitude are not so large, that means the dielectric difference between the grout and the concrete is small, and the radar cannot see void part in the duct whether it is filled with water or air. Figure 12a show similarity to 12d and 12e in their curves change characteristic. Figure 12a, 12d and 12e show too small difference to distinguish.

Mostly from the above analysis, the radar can detect the first interface in the duct. The detailed structure inside the duct cannot be seen in detail directly in the images, but they could be differed from each other in their characteristic curves. From figure 12, we classify (a), (d), (e) into first group, (b) and (f) into a second group, (c) and (g) into a third group. Three classes can be differed from each other. In the characteristic curves, (a), (d), (e) cannot be differed from each other, and (b), (c), (f), (g) can be differed from each other from the curves.

Conclusions

Accurate inspection of ducts condition in PPC is very important in concrete inspection. To identify and analyze GPR signal characteristics for different duct defects different model specimens have been made and radar data were collected as the scan lines along the duct, in contrast to conventional survey fashions vertical to the duct. Characteristic curves extracted from the images improve the interpretations greatly. It is found that the completely void duct is the easiest to differ from others. It is characterized with strong and clear reflection interface which becomes weaker as the void is water-filled. Well grouted condition shows weakest reflection interface. As for half void situation, front scan shows similar result to well grouted condition whether it is water or air filled, and back scan shows similar features to completely void situation. GPR can detect if the duct is grouted well, or void, and can distinguish if the void is water-filled or not using different scans.

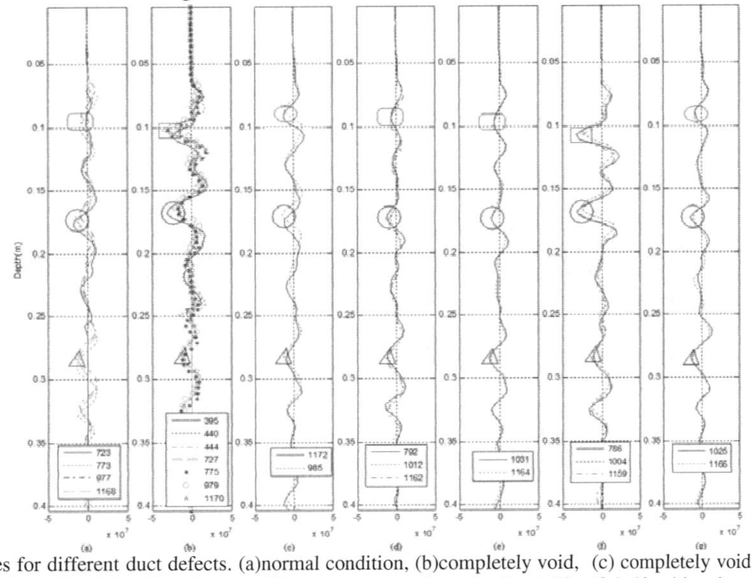

Figure 12. Extracted curves for different duct defects. (a)normal condition, (b)completely void, (c) completely void and fill with water, (d) half void and scanned from the front side, (e) half void, water-filled, and scanned from the front side, (f) half void and scanned from the back side, (g) half void water-filled, and scanned from the back side. We marked the positions of different targets with different signs, square for the concrete/air interface, circular square for the concrete/grout interface, ellipse for the concrete/water interface, circle for the position of tendons, and triangle for the other side of the girder.

CannyTEM – a high-resolution transient electromagnetic system based on mobile technology and artificial intelligence

Xing He and Rujun Chen, Central South University, Changsha, P.R.China;*
Hongchun Yao, Hong Wu, Jieting Qiu, Ruijie Shen, Qiang Ren, Honghua Shi, Hai Dong, and Shenglong Tan, Champion Geophysical Technology, Changsha, P.R.China

Summary

CannyTEM is a ground based high-resolution transient electromagnetic system, which is based on mobile technology and artificial intelligence for easy operation and good quality data acquisition. This paper presents the design, realization, and testing of CannyTEM. CannyTEM is composed of a transmitter, a receiver, two induction coils, and a smart phone or PDA. The transmitter is synced with GPS or cable from the receiver. Its turn-off time is 2.5 μs for 3 A injected current and 40 m × 40 m loop. The receiver owns low-noise amplifier, 24-bit ADC, CPLD, DSP, GPS and embedded computer. The bandwidth of high frequency coil with physical area as 6 m^2 is 1.2 MHz. The bandwidth of low frequency coil with physical area as 100 m^2 is 50 KHz. A smart phone or PDA offers friendly human-machine-interference for the operation of CannyTEM. A smart phone can control the receiver by WIFI and cabled network. The transmitter can send current waveform to the receiver by RS485. The receiver can acquire TEM data intelligently. The turn-off time, gain of each channel inside the receiver is computed and controlled intelligently. At last, we carried a test with PROTEM47HP. The testing result showed that our design is successful.

Introduction

Transient electromagnetic method (TEM) is a powerful in exploring target with high conductivity. It is widely used in the finding of metallic ores, underground water, polluted zone, salinity water, and so on (Danielsen et al, 2003). Traditionally, the human machine interface (HMI) of a TEM instrument system is designed by producer who produce TEM instrument, such as PROTEM and NanoTEM. The disadvantage of this approach is that it is difficult to maintain a state-of-art HMI (Karlson et al, 2008). In this paper, we use a smart phone as HMI for CannyTEM. The user can control the operation of CannyTEM with an arbitrary smart phone which supports Android. We also use mobile technology inside the receiver and transmitter to reduce the power consumption and add artificial intelligence. These approaches let CannyTEM easy to operate and acquire good quality data.

General design of CannyTEM

The CannyTEM intends to satisfy most requirement of TEM exploration on the ground. The compositions of CannyTEM are the following:

- 50 kW high power TEM transmitter system based on generator for mineral, geothermal, and hydrocarbon exploration..
- 1 kW fast turn-off TEM transmitter based on battery for engineering and environmental application.
- 1.2 MHz, 6 m^2 induction coil for engineering and environmental application.
- 50 KHz, 100 m^2 induction coil for general usage.
- 3 KHz, 3-axis fluxgate magnetic sensor for mineral, geothermal, and hydrocarbon exploration.
- 6 channels TEM receiver for general use.
- Smart phone (s) or PDA (s) for the control of TEM receiver and transmitter.

The idea is to use one or more smart phones control the operation of TEM transmitter and TEM receiver. For engineering and environmental usage, the TEM transmitter is controlled by a TEM receiver by a cable. And the TEM receiver is controlled a smart phone. Therefore, only one smart phone is needed. For large depth of investigation, the TEM transmitter and TEM receiver are usually synced by GPS. Therefore, two smart phones are needed to control transmitter and receiver independently.

There is an embedded computer inside high power TEM transmitter which can be controlled by a smart phone through WIFI. The high power TEM transmitter acted as a server. And a smart phone acted as a client. When a smart phone is powered on, it enables access point or WIFI spot. Then the TEM transmitter can find and joint access point. The TEM transmitter can be controlled by a PC through cabled network for debugging.

The idea is same for TEM receiver. The TEM receiver can be controlled by a smart phone through WIFI. The acquired data and setting are displayed on the screen of a smart phone. The user can control the TEM receiver easily by input from touch screen (Francone and Nigay, 2011).

The artificial intelligence is used widely in CannyTEM. And there are carefully designed hardware and software to support artificial intelligence. The transmitter can measure and analysis the waveform of injected current intelligently. It can use optimal setting for a given load and debug most bugs by analysis the recorded waveform of the injected current.

The receiver can remember user's setting even power off. Because the computing power of a smart phone is very strong, it can carry out advanced signal processing algorithm for excellent signal to noise enhancement. It can analysis the characteristic of interference and select best method to suppress interference. Furthermore, it adjusts gain of amplifier based on the analysis of interference and signal.

Hardware design of CannyTEM

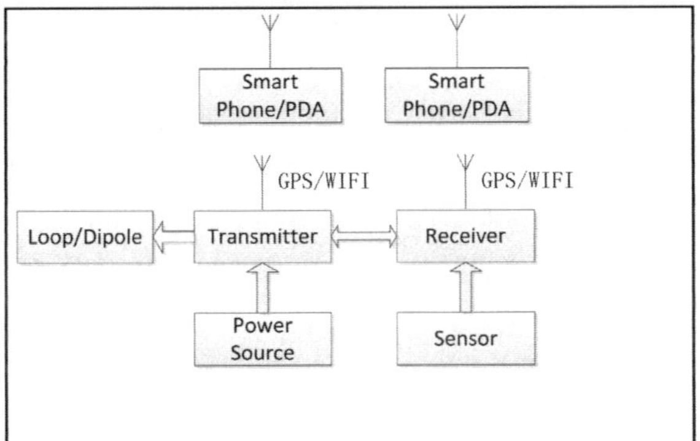

Figure 1: Simplified Diagram of CannyTEM

CannyTEM tries to offers a unified solution for ground TEM system. Therefore, figure 1 shows maximum components of CannyTEM. The transmitter and receiver are synced with GPS signal. However, the receiver can control the operation of transmitter by a cable. This function is needed when GPS signal is not available or the separation between the receiver and the transmitter is less than 30 m.

It is impossible to use a single transmitter to attack all problems in application (Fitterman and Anderson, 1987). We design two kinds of transmitter. The first one is targeted to engineering and environment application. It must be controlled by the receiver. The main parameters of transmitter controlled by receiver are following:

- Power sources: 12 ~ 36 V batteries.
- Maximum output current: 30 A
- Current acquisition: full waveform or turn-off waveform controllable
- Typical performance: 2.5 μs for 3 A injected current and 40 m × 40 m loop
- Sync: cable
- Communication: RS485

The second on is targeted to mineral, geothermal, and hydrocarbon exploration. It is optimized for maximum current output and voltage output (Figure 2). It can be controlled by a receiver and a smart phone. If it is controlled by a smart phone, it is synced to GPS signal and transfers acquired current data to a smart phone by WIFI. The main parameters of transmitter controlled by smart phone are following:

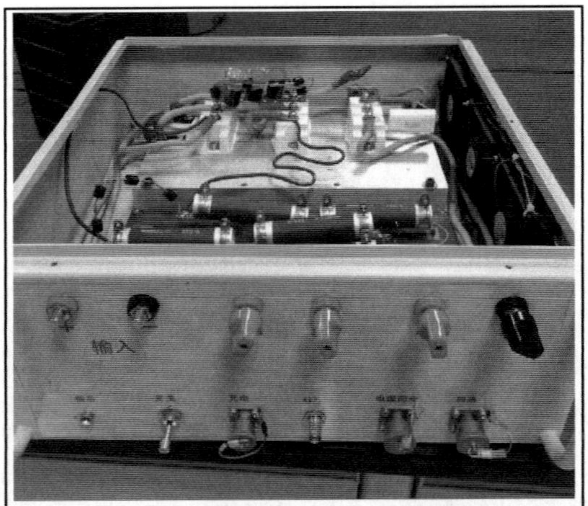
Figure 2: Prototype of 50 kW CannyTEM transmitter

- Power source: 70 kW generator.
- Maximum output current: 100 A.
- Maximum output voltage: 500 V.
- Rated output power: 50 kW.
- Current acquisition: full waveform or turn-off waveform controllable
- Typical performance: 500 μs for 100 A / 500 V and 1500 m × 1500 m loop
- Sync: cable, GPS
- Communication: WIFI, Ethernet

The receiver is designed as a general purpose TEM receiver satisfying all kind of TEM application. It adopts ultra-low noise amplifiers building two-stage instrument amplifier (McCracken et al, 1986). CPLD and DSP are used for real-time digital signal processing. The embedded computer system is also designed by us. It is an updated version embedded system based on AT91RM9200 + Linux (Chen et al, 2010; Jieting et al, 2010). It adopts 400MHz CPU AT91SAM9G45, 128M DDR 2 SDRAM and 1GB NAND flash. The main parameters of general purpose TEM receiver are following:

- Analog channels: 6, 3 for high/middle frequency band, 3 for middle/low frequency band.
- Bandwidth of analog channels: 3 MHz.
- Noise: < 1.3 nV/sqrt(Hz) @ 125 gain.
- Gain: 1, 5, 25, 125, 625 controllable.
- ADC converter: 24 bits, 1.2 MHz ~ 96 KHz controllable, one for each channel.

- DSP: 24 bit, 200 MHz DSP56311.
- Sync: cable, GPS, OCXO/ atomic clock.
- Communication: WIFI, Ethernet

The advantage of our general purpose TEM receiver is to adopt atomic clock. With the help of atomic clock, we can make TEM measurement at underground while the TEM transmitter works on the ground for a long time.

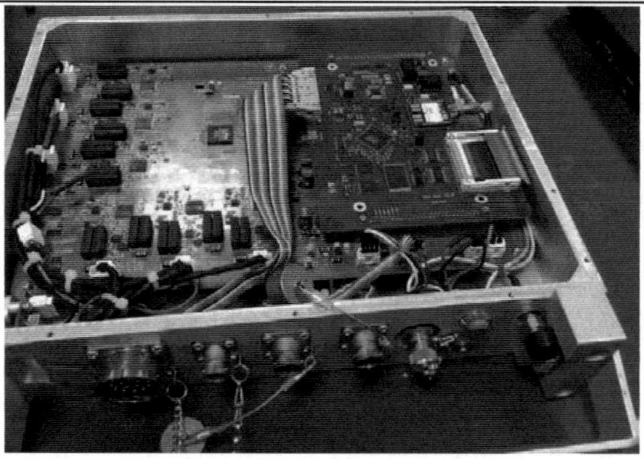

Figure 3: Prototype of 6-channel CannyTEM receiver

Figure 3 shows the prototype of 6-channel CannyTEM receiver. The PCB board in the bottom is 6-channel data acquisition board including analog part, ADC, CPLD, DSP, GPS, and atomic clock. The PCB board on the top is embedded system designed by us. The CPU AT91SAM9G45 receives data from DSP56311 by two SSCs.

It's necessary to design induction coil to receive TEM signal. There is a contradiction between the sensitivity and bandwidth of induction coil (Schamper et al, 2014). If we want to improve the sensitivity of an induction coil, the bandwidth of induction coil will decrease. If the bandwidth of induction coil is not enough, we can't get information form near surface. Therefore, we developed to kinds of induction coils to attack this problem. A high frequency induction coil is designed for engineering and environment application. And a middle frequency induction coil is developed for mineral exploration and other problem needing large depth of investigation. It's may be better to use fluxgate magnetic sensor for geothermal and hydrocarbon exploration which require the depth of exploration > 1000 m.

Figure 4: Prototype of 1.2MHz high frequency induction coil for TEM signal sensing, shielding included.

Here are the main parameters of high frequency induction coil (Figure 4).

- Bandwidth: 1.2 MHz.
- Physical area: 6 m^2, no pre-amplifier is used for maximum dynamic range.
- Diameter: 70 cm.
- Gain of buffer: 1.
- Battery of Buffer: +/- 12V.

The main parameters of middle frequency induction coil are following.

- Bandwidth: 50 kHz.
- Physical area: 100 m^2, no pre-amplifier is used for maximum dynamic range.
- Diameter: 70 cm.
- Gain of buffer: 1.
- Battery of Buffer: +/- 12V.

Software Design of CannyTEM

The software design of CannyTEM includes software running on smart phone, software running on embedded computer, program running on DSP, program running on CPLD, program running on the transmitter. The only software interfaces with a user is software running on a smart phone although there are a lot of software working for CannyTEM.

When CannyTEM is power on, the server program in embedded computer will check the hardware of receiver and transmitter. Alarm signal will alert user if the system's self-check fails. It will inform the user if everything is OK by LED.

The smart phone with android is preferred. The application inside a smart phone will enable access point or WIFI spot. Then the CannyTEM can find the access point and join it automatically. Then the application on a smart phone can control CannyTEM easily (Figure 5). The user can use both hands and touch screen to select different functions or input parameters needed in TEM acquisition.

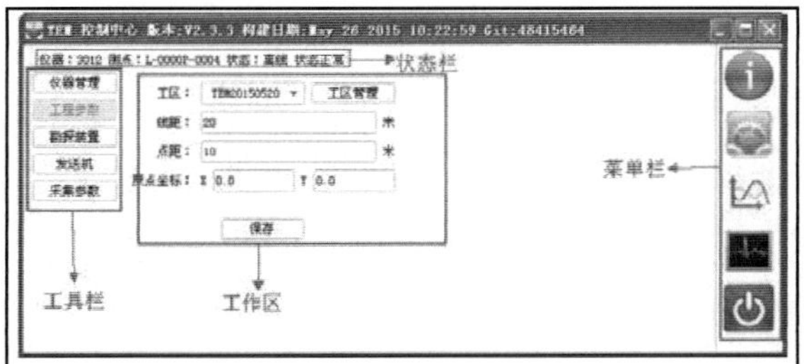

Figure 5: User interface of CannyTEM.

Test of CannyTEM

We carried a test with PROTEM47HP. The loop size is 40 m × 40 m. The injected current is 3 A, and the transmit frequency is 237.5 Hz. We found that the result got by CannyTEM is much better than that of PROTEM47HP (Figure 6). Artificial intelligence in CannyTEM may play an important role for good quality data.

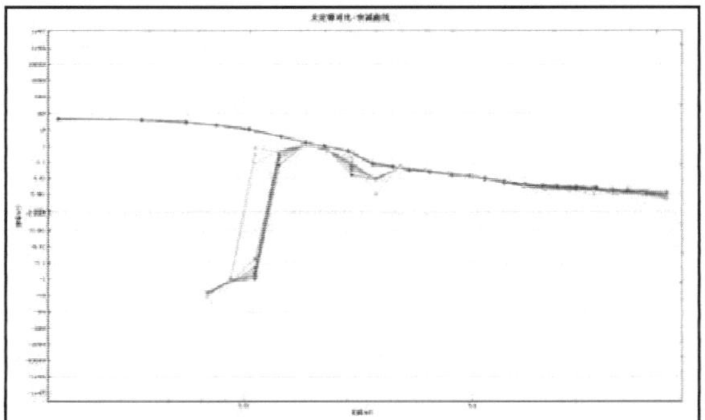

Figure 6: Comparision of result got by CannyTEM and PROTEM47HP. Curves with some oscillations at early stage are acquired by PROTEM47HP. Curves with no oscillation are acquired by CannyTEM.

Conclusions

Mobile technology and artificial intelligent may be the most important technology in 21st century. These technologies may bring revolutionary changes for geophysical instrumentation and geophysical application. We use these technologies in CannyTEM, an intelligent TEM system to satisfy the most requirements in geophysical application. We carried a testing with PROTEM47HP, a TEM system based traditional technology. And a better result is got by CannyTEM.

REFERENCE:

Chen, R., Zhangxiang, H., Jieting, Q., Lanfang, H., and C. Zixing, 2010, Distributed data acquisition unit based on GPS and ZigBee for electromagnetic exploration. In Instrumentation and Measurement Technology Conference (I2MTC), 2010 IEEE (pp. 981-985). IEEE.

Danielsen, J. E., Auken, E., Jørgensen, F., Søndergaard, V., and K. I. Sørensen, 2003, The application of the transient electromagnetic method in hydrogeophysical surveys. Journal of Applied Geophysics, 53(4), 181-198.

Fitterman, D. V., and W. L. Anderson, 1987, Effect of transmitter turn-off time on transient soundings. Geoexploration, 24(2), 131-146.

Francone, J., and L. Nigay, 2011, Using the user's point of view for interaction on mobile devices. In 23rd French Speaking Conference on Human-Computer Interaction (p. 4). ACM.

Jieting, Q., Rujun, C., Zhanxiang, H., and Q. Kailin, 2010, Development of control system in electromagnetic acquisition station based on ARM and Linux. In 2010 IEEE Instrumentation&Measurement Technology Conference Proceedings (pp. 1082-1086).

Karlson, A. K., Bederson, B. B., and J. Contreras-Vidal, 2008, Understanding one handed use of mobile devices. Handbook of research on user interface design and evaluation for mobile technology, 86-101.

McCracken, K. G., Oristaglio, M. L., and G. W. Hohmann, 1986, Minimization of noise in electromagnetic exploration systems. Geophysics, 51(3), 819-832.

Schamper, C., Auken, E., and K. Sørensen, 2014, Coil response inversion for very early time modelling of helicopter - borne time - domain electromagnetic data and mapping of near - surface geological layers. Geophysical Prospecting, 62(3), 658-674.

Acknowledgments

Thank Dr. Huaifeng Sun's help in arranging PROTEM47HP for testing.

Application of 2D Microtrmor Survey Method to Detect and Map Beijing Tugou-Gaoliying Ground Fissure

Peifen Xu, Key Laboratory of Shale Gas and Geoengineering,
Institute of Geology and Geophysics, Chinese Academy of Sciences
Suqun Ling, Geo-Analysis Institute Co. Ltd (Japan)

Ground fissure in addition to small faults in the plain areas of Beijing, and even Northern China, form the most common and destructive geological hazards. They have become key risky geological factors to the urban construction and development. Even though it may be avoidable to build major structures above the ground fissure, there is little choice to construct high-speed railways, subways, and underground pipelines without penetrating these hazardous geological features. It is, therefore, critically important to identify and map exact locations of the ground fissure and their spatial distributions. The map will provide valuable information for structural designing or planning to avoid these geological hazards as much as possible to ensure safe construction and operations of these major structures.

The Beijing Tugou–Gaoliying ground fissure is located near Gaoliying town in Changping District, Beijing. It was first developed in 1980s, stretching 6 km from the North Circular Ring at the northeast to the Beiqijia Hospital at the southwest. It crosses Beijing-Chengde Expressway and then passes by Xiwanglu village, Tangzitou village, and Tugou Village. It strikes roughly NE45°~60°, parallel to Huangzhuang--Gaoliying fault (Figure 1). It is responsible for wall cracks, surface deformation, and local ground subsidence. Recent geological study shows its close correlations with the activities of the major Huangzhuang-Gaoliying fault nearby.

The Beijing-Shenyang high-speed railway is the important part of the medium to long-term railway network plan, so called "four vertical and four horizontal" high-speed railway. The railway construction, starting from Beijing, passing by Chengde in Hebei province, Chaoyang city, Fuxin in Liaoning province, and finally ending at Shenyang North Station, has been officially started in late 2013, expecting to be completed in 2019. The Beijing-Shenyang high-speed rail, following the Beijing-Chengde Expressway in the study area, crosses the ground fissure obliquely at Gaoliying. We used a two-dimensional Microtremor Profiling Method to locate the ground fissure and its spatial distribution.

Two-dimensional Microtremor Profiling utilizes a 2D observation system to record and analyze microtremor signals (Okada, 2003) to detect and map geological structures, such as fracture systems. Based on the station array technic, it utilizes the vertical component of microtremors to estimate the phase velocity of Rayleigh waves (Aki, 1957; Okada, 2006; Ling, 1994). A generalized inversion method was then applied to obtain 2D apparent S-wave velocity sections that will be used to interpret or detect geological features. The method is recognized by far the most effective and convenient method to obtain S-wave velocity structures. It is especially useful in overly crowded areas with a lot of man-made noises and constructions that make conventional geophysical data acquisition almost impossible. To detect the Tugou-Gaoliying ground fissure, we used 10.4 m spacing between the observation points. The two-circular observation system has radii of 6 m and 12 m, respectively. The survey line was arranged perpendicular to the ground fissure. The obtained apparent S-wave velocity section is shown in Figure 2. Velocity variation is shown by the colors, as seen by the color bar located on the right. The figure clearly shows geological features such as fractures associated with S-wave velocity changes.

Microtremor 2D velocity section shows that the ground fissure (located between 8 and 9) extends to earth' s surface to the bottom left of the profile. At the depths of approximately 30 m, there is a small (1 m to 2 m) dislocation across the fracture surface. Rock formations in the vicinity of the fracture show an obvious drag and deformation. S-wave velocity above the fracture surface is clearly lower than that in the strata below the surface. The result indicates stronger cracking, fracturing, and deformation on the topside of the fracture, causing slower S-wave velocity. The presence of extensive fractures also significantly increases porosity and permeability on the upside of the fracture. Recent geological study on the trench (Wang et al., 2014) indicated the soil above the fracture is mechanically much weaker (more compressible) than the soils below the surface because of the presence of fractures and cracks. This difference is particularly evident within the 6~15 m below the surface, but the difference decreases with depth.

This is the first time for us to utilize the 2D Microtremor Profiling Method to detect and characterize the Beijing Tugou-Gaoliying ground fissure. The 2D profiling section clearly shows the shape and extent of the ground fissure and its relationship with the nearby Huangzhuang-Gaoliying fault. Based on our results, we recommend trying to avoid building major structures on the ground fissure. For high-speed railways, we recommend building large bridges across the ground fissure. Our data provided valuable information for railway planning and construction.

In short, 2D microtremor profiling method has become an efficient and powerful tool to detect and map geological features such as fractures in the over-populated urban areas with severe noises.

References

Aki K. Space and time spectra of stationary stochastic waves, with special reference to microtremors. Bull. Earthq. Res. Inst., 1957, 35: 415 – 456.

Okada H. "The Microtremor Survey Method", Society of Exploration Geophysicists, Geophysical Monographs Series Vol. 12, 2003, translated by Koya Suto, Society of Exploration geophysicists, Tulsa, 1-135.

Okada H. Theory of efficient array observations of microtremors with special reference to the SPAC method. Explor. Geophys., 2006, 37: 73-85.

Ling S Q. Research on the estimation of phase velocities of surface waves in microtremors [Ph. D. thesis] (in Japanese).Hokkaido University, 1994.

Xu P F, Li C J, Ling S Q et al. Mapping deeply-buried geothermal faults using microtremor array analysis. Geophys. J. Int., 2012, 188(1): 115-122.

Xu Peifen, Shi Wen, Ling Suqun et al., Mapping Spherically Weathered "Boulders" Using 2D Microtremor Profiling Method: A Case Study along Subway Line 7 in Shenzhen, Chinese J. Geophys. (in Chinese), 2012, 55(6): 2120-2128.

Wang Rong, Lu Yong, Yang Yan et al, Study on the difference of soil physical and mechanical in Gaoliying, Beijing., in "Peoceeding of the national ,symposium on earth fissures geological disaster prevention", 2014, 20-25.

Acknowledgments This research was supported by the National Natural Science Foundation of China (Grants No. 41274143 and 41330643).

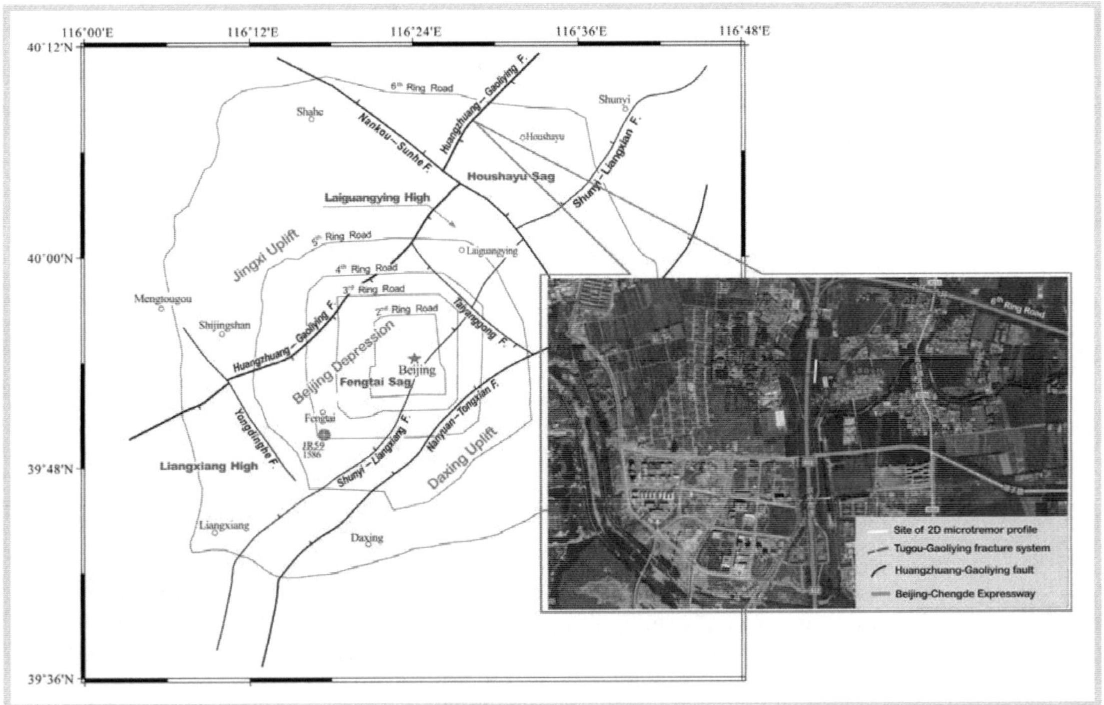

Figure 1: Map showing the major geological features of Beijing region, the locations of the Huangzhuang--Gaoliying fault and its ground fissure, and the site of the two-dimensional Microtremor Profiling (the lower right insert map).

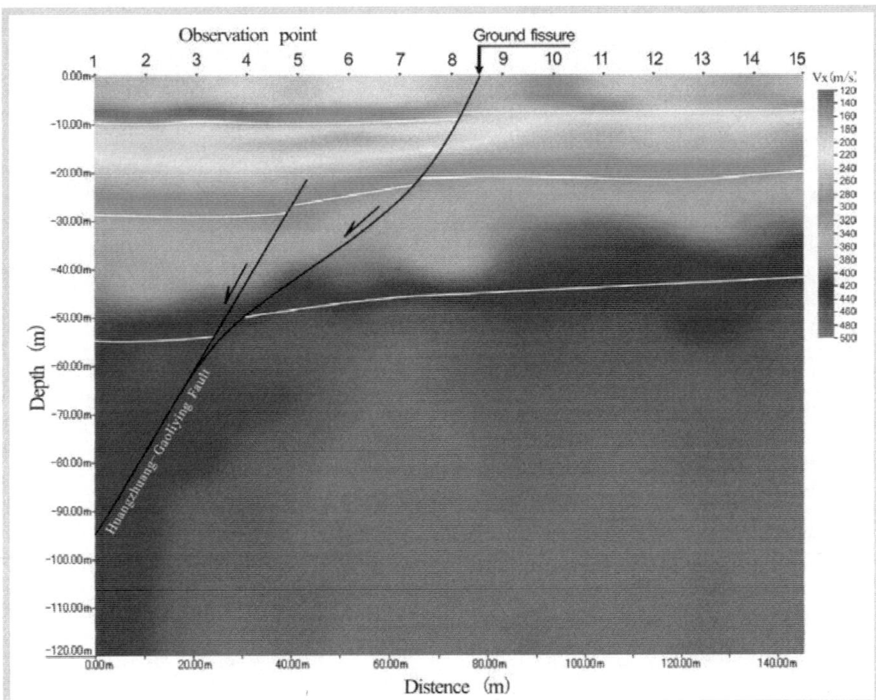

Figure 2: Two-dimensional apparent S-wave velocity section

Integrated Geophysical Survey to Investigate a Construction Site in Makkah City, KSA

Fathy Bahloul (DMT GmbH & Co.KG.), Dirk Orlowsky (DMT GmbH & Co.KG.)*

Summary

Geophysical survey methods used in combination with borehole data for verification provide the most integrated solution for the identification of subsurface conditions. An integrated geophysical survey using CMP-refraction seismic, seismic refraction tomography, electrical resistivity tomography and borehole logging was conducted at a construction site in Makkah, Saudi Arabia. The survey was performed mainly to determine the thickness of the saturated zone, the depth of the groundwater and the depth of the bedrock as well as to delineate near-surface faults and weak zones.

All geophysical results (seismic tomography, resistivity and borehole logs) are combined and jointly interpreted together with the boreholes data in order to calibrate the final interpretation and to generate an integrated underground model.

The chosen interpretation system ensures the generation of exact and consistent spatial images of the subsurface and underground conditions.

Introduction

The use of the geophysical methods in the engineering investigation is obviously increased. The geotechnical engineering issues and topics within a depth ranging from few meters to several hundred of meters are the main interest of shallow geophysical instigations.

The most of the geophysical methods are non-intrusive and can operate in different environmental conditions. They provide a detailed and continuous underground section in comparison to the point data collected by boreholes. Each geophysical method measures the contrast arising from variations in the corresponding physical property to reveal the subsurface conditions. Geophysical methods used in combination provide a most integrated underground model. The borehole data can be use to verify the geophysical results.

A construction site in Makkah, Saudi Arabia, was investigated using combination of geophysical methods. Objectives of the geophysical survey were mainly to determine and characterise the thickness of the saturated zone, the depth of the groundwater and the depth of the bedrock as well as to delineate near-surface faults and weak zones. The target depth was 50 m below ground surface.

The geophysical survey consisted of combined measurements of CMP-refraction seismic, refraction seismic tomography and electrical resistivity tomography and borehole logging. Once the geophysical data have been collected, processed and interpreted a model of the geological structure can be produced.

The geophysical measurements had been performed along numerous profiles with total lengths of nearly 7,000 m seismic and nearly 5,000 m resistivity surveys. Figure 1 illustrates the site conditions.

Figure 1: Site conditions

The results of the CMP-refraction seismic and refraction tomography methods are used for the description of the stratigraphy (weathering layer, ground water table, and bedrock) and for the identification of weak/shear zones in the near surface layers and in the upper bed rock.

The standard seismic refraction method assigns a constant velocity to each refracted layer. But layers with constant wave velocity are often far away from the real conditions in the underground. Usually the underground shows substantial variations of velocity in the individual layers due to the embedding of foreign materials. Particularly, cavities or erratic blocks show a strong reduction or increase of seismic velocities in comparison to their surroundings. A method for allocating the seismic velocity distribution within such layers and lateral velocity changes is the refraction tomography. For this, special requirements are demanded on data acquisition. This method is involving a considerably higher measuring effort than the standard refraction seismic. The shot point distances need to be selected very close to enable a dense description of the velocity distribution within the underground. In addition to the first arrivals used in refraction surveys the CMP-refraction method uses also amplitude and frequency characteristics in a full-trace processing to generate time sections similar to that of reflection seismic.

The hydrological features of the site are determined by electrical resistivity tomography. Using the resistivity tomography a high resolution image of the electrical conductivity distribution in the underground will be calculated, where the resistivity changes in the vertical direction, as well as in the horizontal direction along the survey line are measured. The resistivity results are combined with the seismic results and known geological information from boreholes so that a comprehensive model of the underground can be determined. In addition borehole measurements, such as optical and acoustic televiewer as well as full wave sonic log were performed in 4 boreholes.

Geophysical Methods

Seismic methods in its simplest form use a near surface source to generate seismic waves which travels through the subsurface. At interfaces that have an impedance contrast (related to elastic parameters, propagation velocity and density), a part of incident waves is refracted along the surface of the layer boundary. The portion of the energy refracted at a layer interface depends on the ratio between the seismic impedances on both sides of the interface. Analysis and recognition of seismic energy arrival patterns at different seismic sensors (geophone) allows estimation of depths to reflectors and average velocity between the reflectors and the earth's surface.

The seismic survey was conducted using a vacuum accelerated weight drop, vakimpak, and Summit acquiring system (DMT). The spread length was a tread-off between the need of long distances to observe head waves from deep layers and the need of short distance to ensure a high signal to noise ratio and to allow a reliable first arrival picking especially in this noisy environment. Maximum offset was 350 m with 72 active channels.

The Electrical Resistivity Tomography was conducted using the Resecs acquiring system. This multichannel system allows a continuous subsurface recovery over a long distance. The pseudosection is displayed on the screen during the survey to enable direct quality control in the field. Wenner-Schlumberger configuration with a maximum spread of 240 m and an electrode spacing of 5 m was used.

In addition borehole measurements, such as acoustic and Optic televiewer as well as Full Wave Sonic were performed in 4 boreholes.

The acoustical televiewer is a geophysical borehole imaging tool which provides an oriented 360° acoustic (density/velocity contrast) image of the borehole wall. Optical televiewer is a tool which generates a continuous, unwrapped oriented 360° image of the borehole wall using an optical imaging system. Full Wave Sonic tools emit a sonic/ultrasonic pulse which travels through the mud and the formation. The signal was received by four receivers array.

Data processing and interpretations

Seismic velocities are calculated with the iterative SIRT-Method (Simultaneous Iterative Reconstruction Technique) using a gradient model for starting the inversion. The start model parameters, which are needed for the SIRT-algorithm, were determined by standard refraction seismic

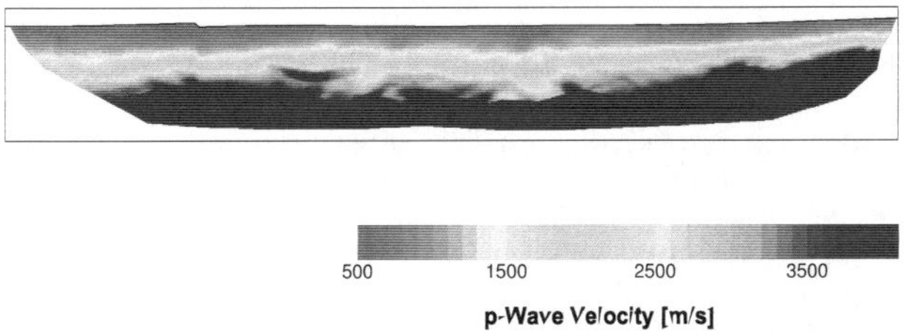

Figure 2: a) Refraction seismic velocity distribution obtained from refraction tomography. b) The CMP refraction results displayed as wiggle traces. The yellow bar marks those profile parts which do not show a clear refractor. In addition, ellipses denote a lateral change in the seismic sections. This might be caused by faults, week zones or similar structures. c) Inversion results of Electrical Resistivity Tomography

techniques. In this case a wave front inversion algorithm was used for travel time inversion. Figure 2a shows refraction seismic velocity distribution obtained from refraction tomography.

With the CMP-refraction seismic technique the shallow underground is described using the whole information (amplitude, frequency, phase characteristics) of the wavetrain, following the first break. The main process of CMP-refraction seismics is a partial RADON transformation of the refracted wave field which results in a display of the analyzed refractor which is comparable to the results of a stacking process in reflection seismics. The refracted amplitudes of the RADON stacked sections were investigated, normalized to the mean amplitudes of the first-break-phases and displayed on a map of the investigated area (Orlowsky et al., 1998). Figure 2b illustrates the CMP refraction results displayed as wiggle traces. The yellow bar marks those profile parts which do not show a clear refractor. In addition, ellipses denote a lateral change in the seismic sections which might be caused by faults or week zones.

The recorded apparent electric resistivities was analyzed and adjusted for the inversion process. A high resolution image of the electrical conductivity distribution in the underground was calculated with the use of corresponding inversion algorithms. During the calculation of theoretical resistivity values the underground model is changed iteratively so that a minimum between the theoretically calculated apparent resistivity values and the measured values is found. The result of this inversion procedure is a 2D resistivity model (depth profile) which displays the lateral resistivity distribution of the underground structures along the profiles. Figure 2c shows the inversion results of the electrical resistivity tomography.

The results of the borehole logging processing are displayed in Figure 3. The acoustic televiewer data mainly the amplitude log reveals the most fine structures. The optical televiewer data (BHTV-Image) shows the in-situ core. For structural orientation all visible, regular structures can be marked in the same way as with the acoustic televiewer. The full wave sonic log data are processed like seismograms. The data are collected using array tool consists of one transmitter and 4 receivers. This allows exact calculation of p- and s-wave velocities of the formations.

The first 3.7 m of the borehole are cased with a concrete pipe. That is clearly visible in the optic and acoustic televiewer logs. The Full wave sonic log shows in the depth between 3.2 m and 4.2 m a breakout zone. A few structures are also seen in a depth of 4.2 m to 6 m which correlate with the other logs. Below 6 m depth the optical televiewer (BHTV image) reveals very clear crystalline formation which matches with the Full wave sonic log and the obtained p-wave velocities as well as with the Rock Quality factor calculated from the amplitude of the acoustic televiewer log.

All geophysical results (CMP-refraction seismic, refraction seismic tomography, resistivity and borehole logs) are combined and joint interpreted together with boreholes data in order to estimate the final underground model. The interpretation has been performed with PETREL software (Schlumberger Information Solutions). The chosen interpretation system ensures the generation of exact and consistent spatial images of the underground conditions.

The integrated interpretation from the borehole investigation together with the depth section obtained from the refraction seismic tomography and the refractor depths interpreted from the CMP-refraction seismic method shows three different units. The first unit coincides mainly with seismic velocities about 1400 m/s. Thus, this boundary describes the changeover between different water saturated levels of sand and silty gravel layer. However, below this boundary the sand and gravel is completely water saturated with seismic velocities of about 2500 m/s. The third unit is identified as granodiorite bedrock with seismic velocity higher than 3500 m/s. The top of the bedrock is irregular and covered with weathered rocks, with seismic velocity

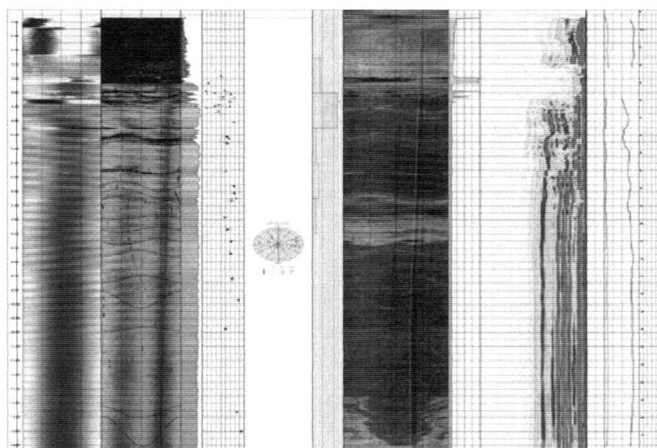

Figure 3: Processing results of the borehole logging. From left to right, acoustic televiewer (Travel Time and Amplitude), Rock quality (calculated from Amplitude), optic televiewer image (BHTV image), Full wave sonic (first channel data) and p- & s-wave velocity.

less than 3500 m/s.
Figure 4 shows the thicknesses of the water saturated and unsaturated layers. The lateral extend of the groundwater aquifer have been determined and its thickness have been also calculated.
The interpreted model and the information about the water saturated and unsaturated zone as well as the revealed structure features have been used to produce a general ground water model in this area.

Conclusions

The geophysical surveys are performed in an integrated frame work so that they complement one another to accommodate their varying levels of resolution and response to provide a comprehensive image of the subsurface conditions at the investigation area to meet the groundwater study objective.
The integrated interpretation of different geophysical methods created more adequate model of the subsurface. In our case the combination of seismic refraction tomography, CMP-refraction and electric resistivity tomography has given a lot of hydrogeological information which have been used to evaluate a detailed groundwater model.
The chosen interpretation system (for CMP refraction seismic and ERT) ensures the generation of exact and consistent spatial images of the underground conditions by utilizing overall visualization and modeling approaches.

Reference

Orlowsky, et al., 1998, Combination of Common Midpoint-(CMP-) Refraction Seismics with the Generalized Reciprocal Method (GRM). Journal of Applied Geophysics, 39, 221-235.

Acknowledgments

We thank Saudi Geological Survey (SGS), Dar al Handasah (shair and partners) and Saudi Binladen Group (SBG) for giving us the opportunity to be a part of this project. Special thank to Dr. Samar Shoman (SGS) and DR. Abdel Rahman Abdel Raouf (Dar al Handasah) for their support.

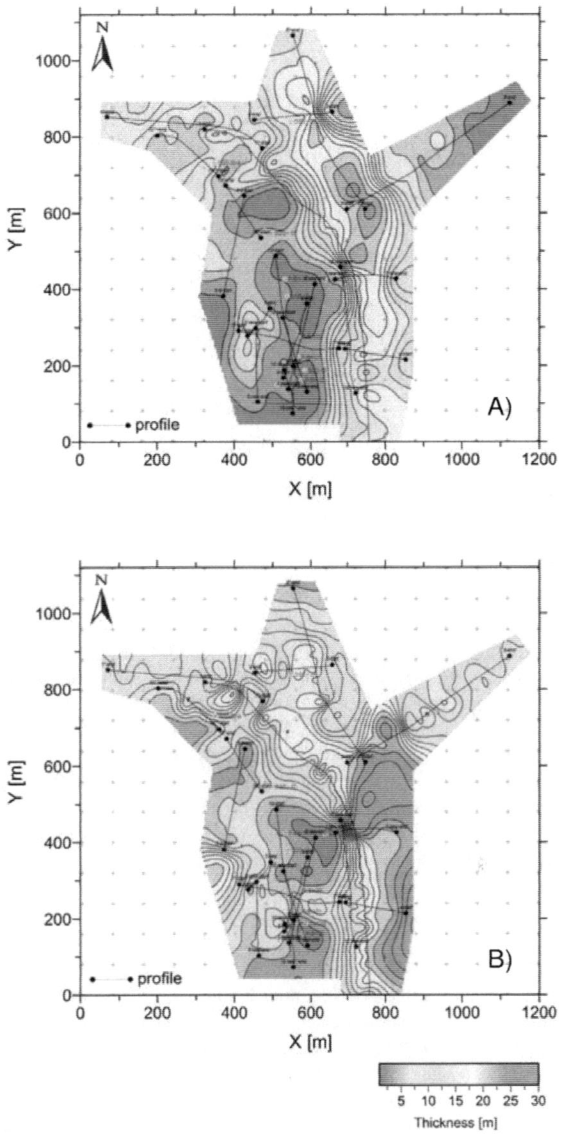

Figure 4: Map of the area under investigation with: a) Thickness of the unsaturated sand and silty gravel. b) Thickness of the saturated zone

Innovative use of Engineering Geophysical for Site Investigations

Dr. Sanjay Rana, Director, PARSAN Overseas Pvt. Limited, New Delhi, India; sanjay@parsan.biz

Summary

Engineering geophysics is an efficient means of subsurface investigations. The areas of application of engineering geophysics are only limited by the imagination of practicing professionals. The key to success is right choice of method and integration of various geophysical techniques to remove ambiguity to the extent possible. There are certain site conditions routinely considered as adverse to application of geophysical methods. It has, however, been noticed that little innovation in field data acquisition methodology and integration of techniques can overcome most of these challenges and provide valuable information in these 'unfavorable' site conditions as well.

The objective of paper is to demonstrate efficacy of various techniques with little innovation in field procedures.

Introduction

Any major construction work like power plants, high rise buildings, bridges etc., calls for a well-considered approach in view of the restricted timeframe and finance. In planning and development of project adequate knowledge of the geotechnical conditions at site is important besides the other factors that are involved. Application of tools and techniques that are helpful in enhancing efficiency of the geotechnical evaluation study is therefore preferable.

Engineering geophysics is an efficient means of subsurface investigation. The merit of application of this low cost aid lies in its ease of deployment and rapidity in providing a reliable knowledge of the underground over a large area, substantiating the requisite geotechnical evaluation studies thereby. Technological advancements and development of portable digital data acquisition instrument systems have increased the versatility in evaluating underground conditions and site characterization.

The state-of-the-art subsurface geophysical investigations are helpful towards minimizing involvement of the conventional direct exploration methods, aiding in accelerated and economical development of the construction projects.

It is an established fact that a single geophysical method cannot resolve all the problems associated with subsurface investigations. As an example, seismic refraction cannot 'see' low velocity zones under rock interface, which might be present under certain geological conditions. There is therefore need for integrated application of various geophysical techniques like seismic refraction, resistivity imaging, ReMi, Crosshole/ downhole/ uphole seismic and seismic reflection, to determine various properties of subsurface like bedrock quality and depth, low velocity zones (even under rock interface), fault/ fracture/ shear zones, water lenses, tunnel route geology etc.

Present paper discusses certain unique applications of these geophysical methods in conditions generally considered to be adverse to such applications. The discussed cases include:

1. Integrated Geophysical approach for investigation and monitoring of dams
2. Electrical Resistivity Tomography to detect cavity in sandstone, conducted on exposed sandstone, which is considered difficult due to absence of top soil to put electrodes.
3. Streaming Potential Survey (in combination with electrical tomography, GPR, ReMi and Seismic Refraction) on dam top having 1m thick concrete slab, posing problems for contact of porous pots.
4. Geophysical investigations across flowing river, having floating electrodes.
5. Geophysical surveys to detect leakages from water pipelines.
6. 3D GPR survey to model cavities behind concrete slabs formed due to seepage of water from slab joints, coupled with electrical resistivity imaging & streaming potential to map seepage paths.

Choice of Geophysical Method

The appropriate techniques for investigations should be chosen based on:

- Objective of investigations
- Resolution required
- Depth penetration required
- Physical property to be defined
- Geology of the area
- Nature of target & host material

It is a proven fact the in most of the cases, use of just a single geophysical technique will not reveal unique results, as there is a vast overlap of any physical property for various geological materials. It is therefore recommended to use an integrated approach to uniquely resolve the issues.

Each engineering geophysics site is unique, and a unique approach is required to be worked out to uniquely solve the problem.

Application for Dam Investigations

Dams represent a major investment and huge benefits to population in terms of irrigation, power and flood control. Most of the big dams are very old and regular monitoring and maintenance of these dams is of utmost importance for continuing benefits. Unlike soil investigations, critical nature of dams, does not permit traditional invasive inspections by means of drilling, and such inspections are best avoided unless extremely important, and are done only when problem is too grave.

Although long recognized that dams need periodic inspection and monitoring, it has only been recently recognized that geophysical surveys can supplement the results of standard inspection and monitoring techniques. Geophysical surveys have been performed on a number of dams around the world, including India, and have yielded extremely useful insight into dam conditions. Seismic refraction surveys have been performed to provide a cross-section of the dam embankment and foundation materials in terms of seismic characteristics. Streaming Potential (SP) surveys have been performed to investigate seepage conditions within dam embankment and foundation materials, and abutment materials. Electrical Resistivity Tomography is also routinely performed on dams to determine internal saturation conditions in dams. ReMi is effectively used to determine shear wave velocities (Vs) in dams.

Various geophysical methods are available to investigate the problems of earthen, masonry, concrete or composite dams:

- Leak path detection
- Internal Erosion
- Identification of zone of water accumulation
- Cavity/ sinkhole
- Concrete degradation

No single geophysical technique can uniquely solve the problem due to a large overlapping of physical properties in various subsurface materials. To date, the use of geophysical methods to investigate dams has produced mixed results, partly because the application of these methods is not well-understood and partly because false positives cannot be tolerated. That is the reason why it becomes important to use a combination of geophysical methods to uniquely resolve the problem.

This is an example of seismic tomography conducted across downstream and upstream face of a concrete dam, showing clearly the fracture zones in dam body,

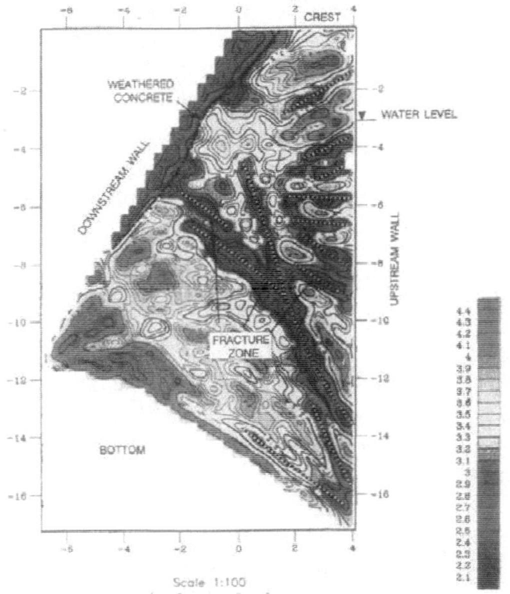

Figure 1: Dam Geophysics Example

Electrical Resistivity Imaging in Unfavorable Surface Conditions

Proper grounding of electrodes is critical for a successful electrical survey. It is therefore assumed that in conditions where top surface is exposed rock or concrete slab, these surveys cannot be conducted. It is, however, possible to carry out a successful electrical survey, even in such conditions, by using drill holes filled with brine to plant electrodes. The methodology was successfully used in various projects, including the example below, to detect a cavity in otherwise massive sandstone.

Appropriate array should be selected based on contact resistance achieved after planting of electrodes in brine filled drilled holes. Care also need to be taken to frequently check water content in these holes while survey is in progress, as dry holes might lead to high contact resistance and resultant spurious data.

This methodology can be successfully applied for investigations in all hard surfaces like hard rock, dam tops, concrete structures, tunnels etc.

This is an example of electrical resistivity imaging carried out to detect cavity in otherwise massive sandstone.

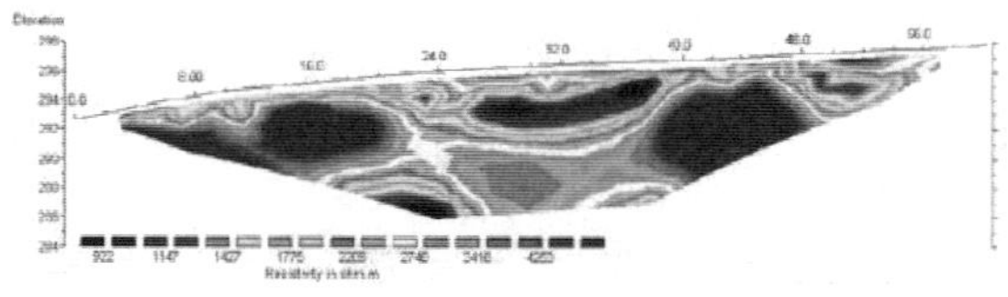

Figure 2: Electrical Resistivity Imaging in unfavourable site conditions (hard rock top surface)

Streaming Potential Survey in Unfavorable Surface Conditions

Streaming potential surveys require use of porous pots in contact with soil. It is therefore assumed that in conditions where top surface is exposed rock or concrete slab, these surveys cannot be conducted. It is, however, possible to carry out a successful SP survey, even in such conditions, by using copper sulphate saturated sponge on the surface, and then placing the porous pot over it. The methodology was successfully used in various projects, including the example below, to detect seepage path in a dam having 1m, thick concrete top.

This is an example of positioning porous pots on saturated sponge to ensure contact with ground

Geophysical Survey across Flowing Rivers

Subsurface information under the riverbed is of critical importance for project planning in various projects like selection of dam axis, pipeline crossing, bridge construction etc. As of now drilling is the only method used for the same, which is extremely time consuming and expensive in river conditions.

Various geophysical techniques like seismic refraction, electrical resistivity tomography etc., can quickly provide the information across river crossings. The data provides drilling contractors information of the thickness of overburden, the lithology of the overburden and the bedrock, and whether or not the bedrock is fractured, ensuring a successful project planning.

Data acquisition is carried out using portable equipment, boat and ropes. A river crossing of upto 300 meters can be easily investigated in less than a single day, providing information upto a depth of 50 meters. Various such projects have been carried out successfully, and results have been reconfirmed by drilling.

This is an example of data collection in a river with floating electrodes

Pipeline Leak Detection Using Geophysics in Difficult Conditions:

The pipeline leakage leads to waste of the material being transported, and also poses serious environmental risks. In certain cases the leaks leading to accidents can have catastrophic consequences.

Generally leak detection is done using noise loggers, noise correlators etc., which need direct contact with pipeline at few places. In many cases, however, pipeline is completely buried with no surface signatures. There are various geophysical tools available to detect leakages from pipelines, and localize such leakages. The techniques enable detection of leakages 'externally'. i.e. without physical access to pipelines, making them suitable for use in city and urban environment as well.

A typical engineering seismograph can be used to 'listen' to noise created by leaking fluid:

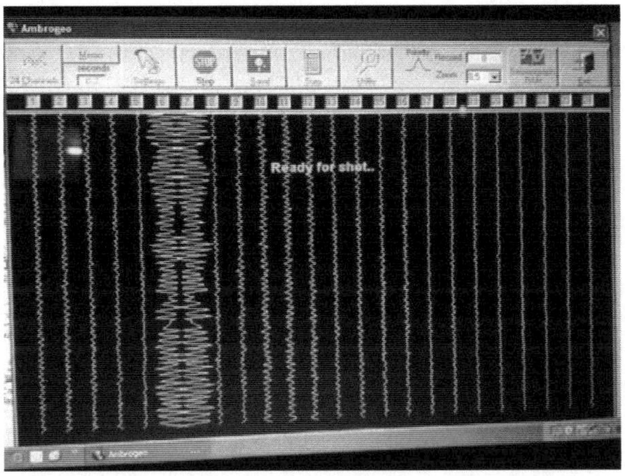

Ground penetrating radar is another excellent method to detect changes in soil permittivity due to leaking fluid:

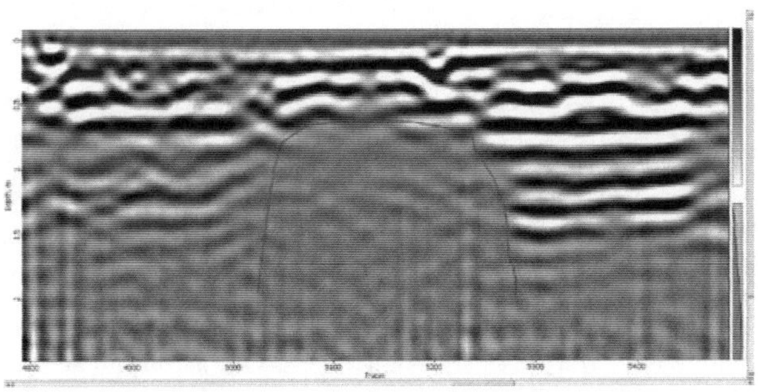

3D GPR for Cavity Detection Behind Concrete Slab:

Ground Penetrating Radar lines conducted at a close grid on concrete slab, and processed and modeled using advanced software like GPR-Slice, provide an in-depth information on extent and depth of various cavities and other features. The method can be coupled with other geophysical investigation techniques like electrical resistivity imaging to further map seepage paths and anomalous zones. The example below shows 3D cube based on GPR scanning of concrete slab.

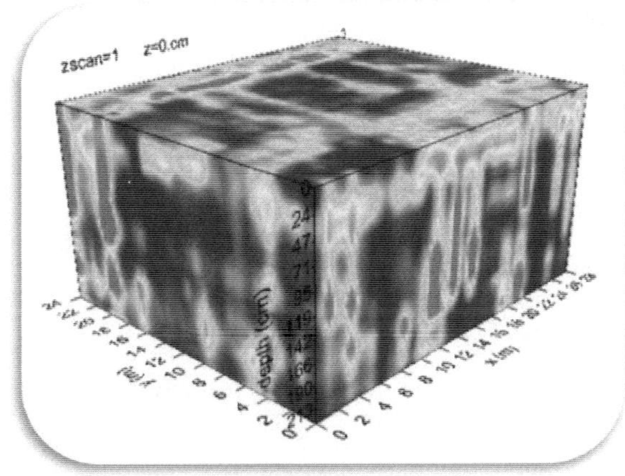

Conclusions:

Engineering geophysics is an efficient means of subsurface investigation. The merit of application of this low cost aid lies in its ease of deployment and rapidity in providing a reliable knowledge of the underground over a large area, substantiating the requisite geotechnical evaluation studies thereby. The state-of-the-art subsurface geophysical investigations are helpful towards minimizing involvement of the conventional direct exploration methods, aiding in accelerated and economical development of the projects.

Little innovation in field data acquisition techniques and integration of techniques can further enhance chances of successful implementation of geophysical approach in various engineering projects.

There is an urgent need to disseminate information on potential of geophysics in solving subsurface problems to the user groups.

Geothermal resource exploration in the region of Hmeïma (Central -Western Tunisia) using resistivity sounding and gravity data

Farhat Ben Mbarek [1], Wajdi Belkhiria[1], Mohamed Dhaoui [1]*

[1]Office National des Mines (ONM, Tunisia)

Summary

The Hmeima geothermal province, encompassing the structres of Boujabeur, Hmaima, Slata, Harraba and Guarn Halfaya, expresses promising geothermal anomalies in a number of wells. In this area, the potential geothermal aquifer is the Aptian reefal limestone.

The identification and exploitation of these geothermal resources require knowledge of the geometry of aquifers, their hydrodynamics and risks of contamination by Triassic salt deposits.

To attend this aims, a vertical electrical sounding (VES) survey containing 65 VES (AB/2= 1500 m) has been carried out to understand and evaluate the geothermal Aptian aquifer.

The interpretation of VES calibrated to the borehole data helped to image the geometry of the reefal Aptian aquifer. From a geological point of view, under the Mio-Plio Quaternary deposits, the borehole pass throw Albo-Aptian and Triassic strata. An isobath map at the top of Aptian interval shows a depth up to 1200m. This limestone is affected by a fault network with main faults oriented N70, N150 and N130 detected by gravity data analysis and in geoelectrical profiles. In particular, the major faults give a vertical drainage of the thermal reservoir in this area.

On the other hand, this study gave a detailed subsurface mapping of evaporates intrusions that can be a source of contamination of the reservoirs.

Key words:

Geothermal anomaly, reefal Aptian limestone, VES, Hmeima, north-west of Tunisia

Introduction

The Hmeima geothermal province encompassing the structres of Boujabeur, Hmaima, Slata, Harraba and Guarn Halfaya expresses promising geothermal anomalies in in a number of wells. In this area, the potential geothermal aquifer is the Aptian reefal limestone (Figure 1).

The identification and exploitation of these geothermal resources require knowledge of the geometry of aquifers, their hydrodynamics and risks of contamination by Triassic salt deposits.

Study area

The study area is located in Central-Western Tunisia, to the Algeria-Tunisian border. It is bounded on the north by Jebel Harraba, to the south by Jebel Boujabeur Slata, to the west by the Tunisian-Algerian border and on the east by Guarn el Halfaya and the Tajerouine city. It covers an area of about 800 km² (Figure 1).

Figure 1: Main geothermal provinces of Tunisia (adapted from Ben Dhia, 1992 and Boouri, 1994).

The study area is located in the Tunisian structural unit called "diapirs zone ."

Paleogeographically, the Hmeima region is located at the boundary between Cretaceous depositional areas: deep environment in Tunisian trough and essentially carbonated central platform.

In this area the Aptian outcrops widely and shows facies variations.

Lithologically, the main formations are represented by marlocarbonate alternations with the exception of the middle Albian limestone and of the reefal limestone of middle Aptian (Figure 2).

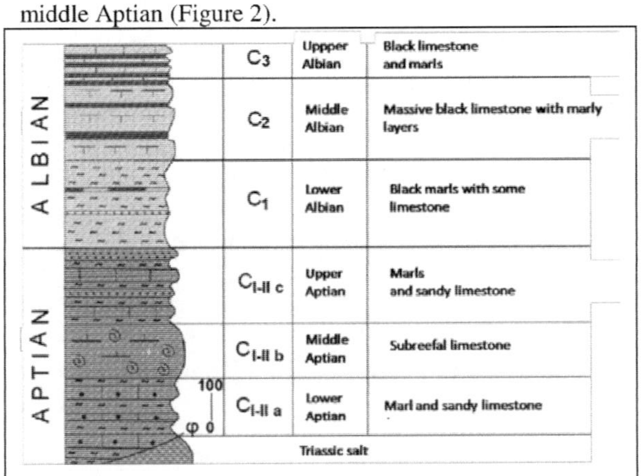

Figure 2: Simplified lithostratigraphical logs of the study zone.

Methodoloy

By geoelectrical methods that use DC, the resistivity of underground materials can be measured by injecting DC of known intensity through two electrodes (current electrodes) placed on the surface of the land. The electric potential difference (voltage) subsequently resulted from another two electrodes.

The electrical resistivity survey used in this study involves electrical sounding using schlumberger configuration with Georesistivity meter GRM 3000. The potential electrodes remain fixed and the current electrodes are expanded simultaneously about the center of the spread. When the distance between the electrodes gets too large, it then mandatory to increase the distance between the potential electrodes to have a measurable potential difference.

The vertical electrical sounding (VES) data are presented as depth sounding curve, which are obtained by plotting apparent resistivity values against electrode spacing on a log-log or bi-log graph paper. The depth sounding curves are classified based on layer resistivity combinations.
Finally cros sections was established based on the lithostratigraic information at boreholes and the correspondant resistivity.

Two geophysical methods was used in this study: Electrical and gravity surveying

Electrical survey

65 VES (Vertical electrical sounding) with an avergae equidistance of 3km distributed along 8 geo-electrical profiles between the structres of Slata, Hmeima, Harraba and Boujabeur :

- Hmeima-Guarn El Halfaya I, oriented NE-SW.
- Hmeima-Guarn El Halfaya II, oriented NE-SW.
- Hmeima-Guarn El Halfaya III, oriented NE-SW
- Slata-Harraba I, oriented NW-SE.
- Slata-Harraba II, oriented NW-SE.
- Hmeima-Harraba, oriented NNW-SSE.
- Boujabeur I, oriented N-S.
- Boujabeur II, oriented N-S.

Examples of geoelectrical profiles

- Geoelectrical cross-section Slata- Harraba I (Figure 3)

The cross-section Slata- Harraba I passes through 8 VES. It links Jebel Harraba to Jebel Slata. In the extreme NW of this profile, the Aptian limestone shows very important resistivity values of about 500 Ωm and lies at the depth of 100m.

Between Harraba and El Gara anticlines and at the Oued Mellegue, we note the presence of a syncline structure where the Aptian is very deep.

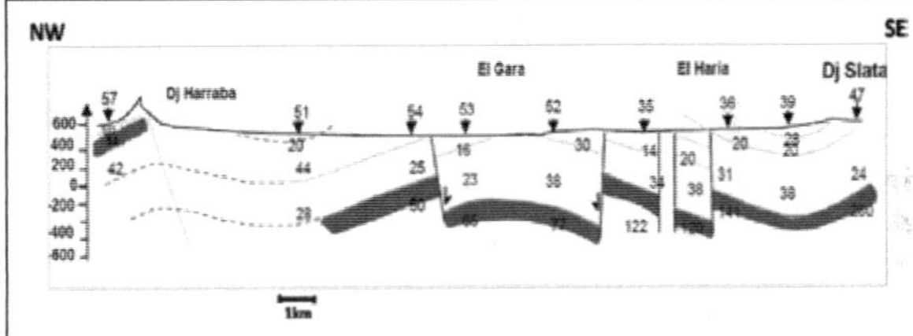

Figure 3: Geoelectrical cross section Slata – Harraba I.

- Geoelectrical cross-section Hmeima- Harraba II (Figure 4)

This cross-section passes through 6 VES.

A syncline structure is marked between Harraba and Bled Boughanem. The Aptian is not detected in this syncline. However to J.Hmeima, the Aptian shows high apparent resistivity (68-97 Ωm) with a thickness of 250m. It rests directly on a very conductive layer attributed to Barremian (facies clay dominance) or the Triassic deposit. We note the existence of a fault between VES 42 and 24 having a vertical throw that exceeds 200m.

The VES 22 is influenced by the vertical displacement of border faults that caused the collapse of the Aptian. The very low values of resistivity may be the responses of an essentially terrigenous fill (Bled Boughanem graben).

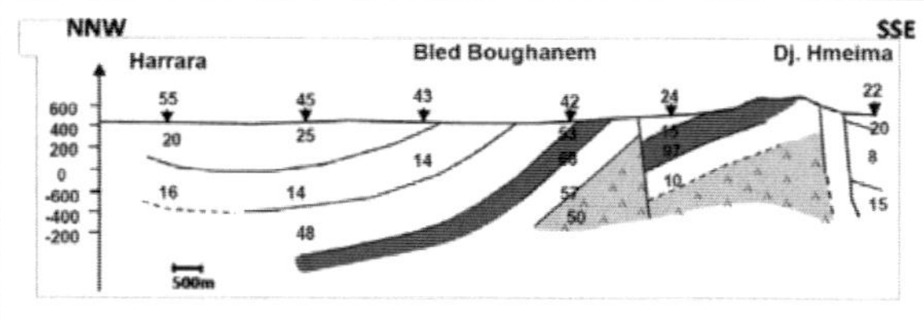

Figure 4: Geoelectrical cross section Hmeima – Harraba II.

Gravity data analysis

The Bouguer anomaly map (Figure 5) shows a negative anomaly over the Kalaa Khasba graben and a positive anomaly associated to Slata, Hmeima, Harraba and with a relative low amplitude over Guern el Halfaya.

The analysis of the Bouguer anomaly and the first vertical derivative maps (Figure 5 and 6) allowed the division of the region into three areas:

- A first collapsed area of Kalaa Khasba in the southeast oriented NWSE which has a Bouguer amplitude of -50 mgals limited by a strong gravity gradient marking the boundary faults with a throw exceeding 2000m filled by thick and light sediments.
- A second area collapsed in the center oriented EW and shallower than the graben previously described; it represents the Boughanem graben.
- A third area in the North with positive anomalies. They mark the presence at relatively shallow depths, dense structures that could be the response of the Aptian limestone.

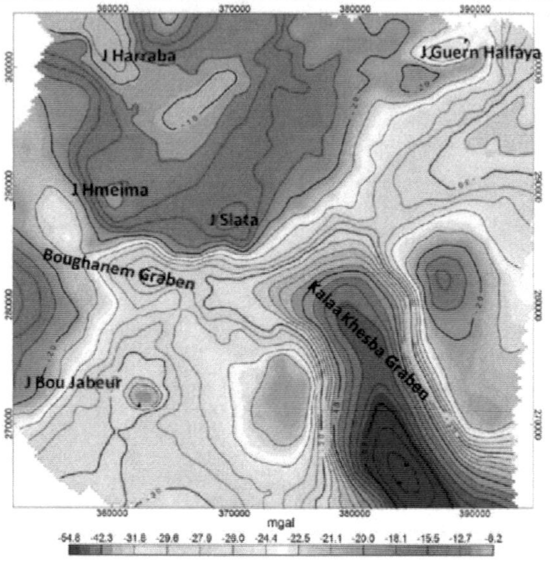

Figure 5: Bouguer anomaly map.

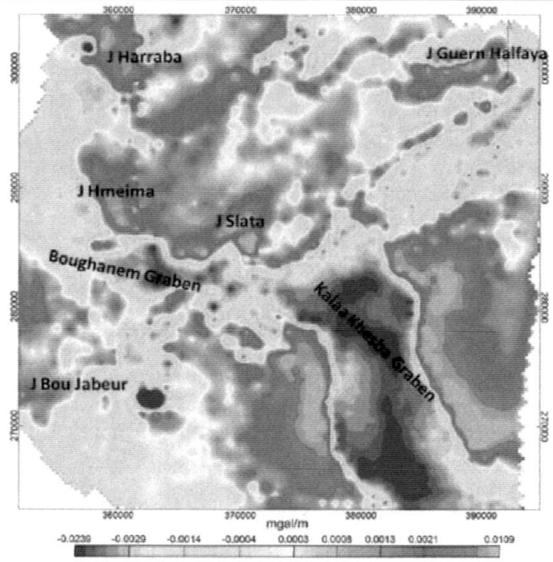

Figure 6: First vertical derivative map.

Results and discussions

On the basis of geophysical studies and taking into account surface geological and hydrogeological context of the region, we have found eight main structures oriented NE-SW.

- In this area, the main faults are oriented N70, N150 and N130. The border faults are responsible for the collapsing of the Boughanem graben and it is assumed to be related to an older extension movement.

- The Aptian shows a variable geometry and thickness.

- The Triassic sediments are largely represented in outcrops and in subsurface. It can be a source of contamination to the groundwater.
- We selected the areas where the depth of the electrical Aptian marker does not exceed 700m. In order of preference, we opt for a recognition operation on the electrical soundings sites where the risk of contamination by evaporates is low.

Conclusions

The study based on vertical electrical resistivity (VES) sounding and gravity measurements, undertaken in the region of Hmeïma, helped to understand the geometry and the extension of the Aptian potential geothermal aquifer. In addition, the studied region could be considered as a hydrogeological basin where the Aptian reefal limestones are likely to constitute a good aquifere for thermal water that could be exploited by a drill hole of 700 m depth.

Structurally, this geophysical survey helped to determine the faults network that played an important role in the hydrodynamism of the geothermal water. Furthermore, the area that can be contaminated by evaporates were limited.

References

Ben Dhia H., Bouri S., Meddeb N. and Zouari K., 1992, Hydrothermie de surface et bassins hydrogéothermiques en Tunisie, Coll. Hydr. Marrakech, 66 - 70.

Bouri S., 1994, Hydrothermie de surface et potentialités hydrothermales du nord tunisien. Identification des principales unités géothermiques : Thèse de 3ème Cycle, University de Sfax.

Geophysical application in preselecting a high-level radioactive waste site

Zhiguo An,Qingyun Di, Ruo Wang, Changmin Fu*
Key Lab of Shale gas and Geoengineering, Institute of Geology and Geophysics, ChineseAcademy of Sciences, PO. Box 9825, No. 19, Beituchengxilu Road, Chaoyang District, Beijing, 100029

Summary

This paper discussed comprehensively, the geophysical application in preselecting the high level radioactive waste (HLRW) site in northwest(NW) China. We analyzed 3D large-scale aeromagnetic and gravity results to interpret regional structures which controlled granite rock distribution, and evaluated the integrity of granite rock. Also, we conducted Controlled Source Audio-frequency Magnetotelluric method (CSAMT) survey at a potential site, which may be suitable for underground geological repository. The survey objective was to map out the fractured and crushed zones that resulted from geological structures like active fault, which would cause serious damage to the safety and stability of underground repository. The comparison between the interpreted results and geological information shows that both are in good agreement. The study proves that comprehensive geophysical exploration is helpful in detecting geological structures and evaluating the rock quality.It can also provide the reliable geophysical evidence for further study.

Introduction

In the world, the current solution for HLRW is deep geological disposal. An underground repository is constructed for safe storage or disposal of HLRW in intact surrounding rock between the depths of 500-1000mbeneath the ground surface. Via engineering materials and natural barrier, HLRW could be isolated and preserved for probably10000 years.

Since the mid-80s of last century, China has been conducting studies on preselected HLRW disposal sites (Wang *et al.*, 1999; Xu, 2002; Guo *et al.*, 2001; Wang, 2000)somewhere in the NW of China. The main lithology in the research area is granite. The study area is located in the Gobi desert, far away from residential district, and the annual rainfall of this area is less than 80 mm. Though the past studies of the site had a good foundation, the surface geological data are not enoughto understand the detailed underground information about the study area. Major geological informationsuch as the deep faults' extent which controls the mass distribution of rocks on sites, spatial shape and extension of rock mass, and bad geological structure in deep rock mass, have not yet been thoroughly studied. Therefore, the geological problem within the preselected area was the focus of the study. We used gravity and aeromagnetic data on a large scale to study the regional geological structures within and around the rock mass. Subsequently, in 2009, 2013 and 2014, CSAMT surveys were conducted to study the rock mass in more detail(An *et al*, 2013a,b; Di *et al*, 2015).

Tomography imaging of large-scale Gravity and Aeromagnetic data

To improve geological interpretation, tomography imaging (Cutler *et al.*, 1985), was applied to transform the gravity and magnetic field to relative density and magnetism data. Li and Douglas (1996, 1998, and 2003) presented a method and developed an algorithm for inverting surface magnetic and gravity data to recover 3-D susceptibility models and distribution of density contrast.

One of the study areas for Gravity and Aeromagnetic survey was shown in Figure 1. The gravity data were sorted out by China Geological Survey, while the aeromagnetic data were obtained by the China Aero Geophysical Survey and Remote Sensing Center. In Figure 1, the red blocks represent targeted rock mass. (An*et al*, 2013a,b).

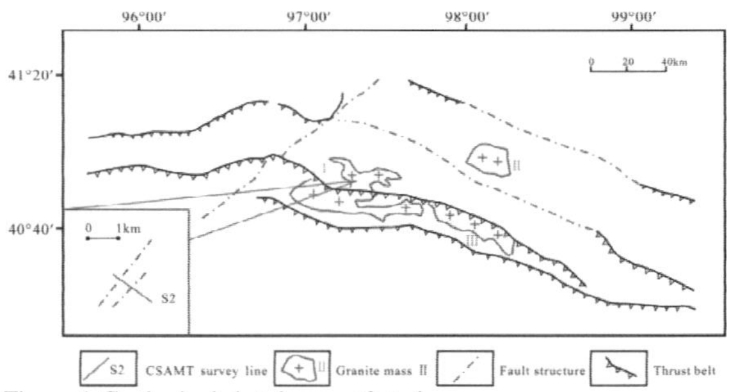

Figure 1:Geological sketch map of study area.

Based on previous study, we carried out a tomography imaging of gravity and aeromagnetic data to investigate the regional geological structures of a potential HLRW site in NW China. We demonstrated that Gravity and Aeromagnetic data allow us to delineate the deep structures from a depth of about 1 km to about 6 km. For the shallow depths, CSAMT data were able to provide much needed detailed information on possible weakened zones. To facilitate interpretation, we converted the Aeromagnetic and Gravity data into depth slices using tomography imaging. The results are presented as horizontal depth slices (Figures 2 and 3).

Results of Gravity data

The results of relative density data are shown in Figure 2.Observed at a depth of about2 km is some localized density variation, whose details could not be described. The distribution of the relative densityinthe uppercrustis still quiteuneven,but the overalldistributiondirection is northwest. Although the values of the relative density in the upper crust vary, there exist a large green area that indicates the rock mass is intact.In the lower crust,the relative density is even and the values of rock mass I, II, III and those in the NE part are higher, indicating the deep rock is relatively intact and in good condition.

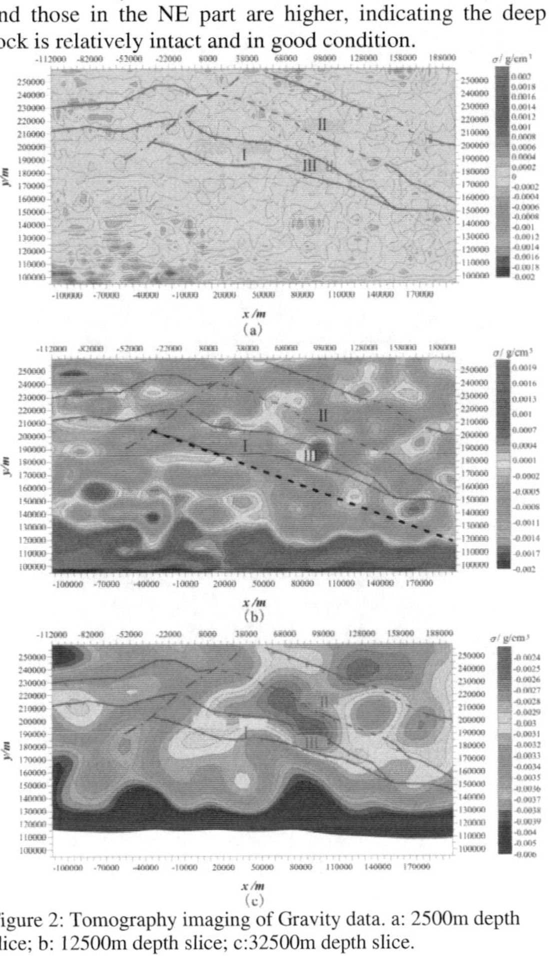

Figure 2: Tomography imaging of Gravity data. a: 2500m depth slice; b: 12500m depth slice; c:32500m depth slice.

Results of Aeromagnetic data

The results of Magnetic data can be seen in Figure 3. As shown in Fig.3a, rock mass I, II, and III with high magnetic values, located in themiddle of two geological structures, may be relatively uniform granite and metamorphic rocks. In Fig.3b, the high magnetic regions in Fig.3a becomes invisible. This shows that the granite and metamorphic rocks with high magnetic amplitude did not extend to this depth. Fig.3c shows a reversed high magnetic trend that extends to the depthof about 6km below the 1250 km depth slice. At this depth, the weak magnetic features of rock mass I and III are uniform and relatively regional. Overall, the magnetic data shows that the deep rock at1km depth with high magnetic amplitude(corresponding to granite and metamorphic rock)in the middle of two structures, may be quite stable.

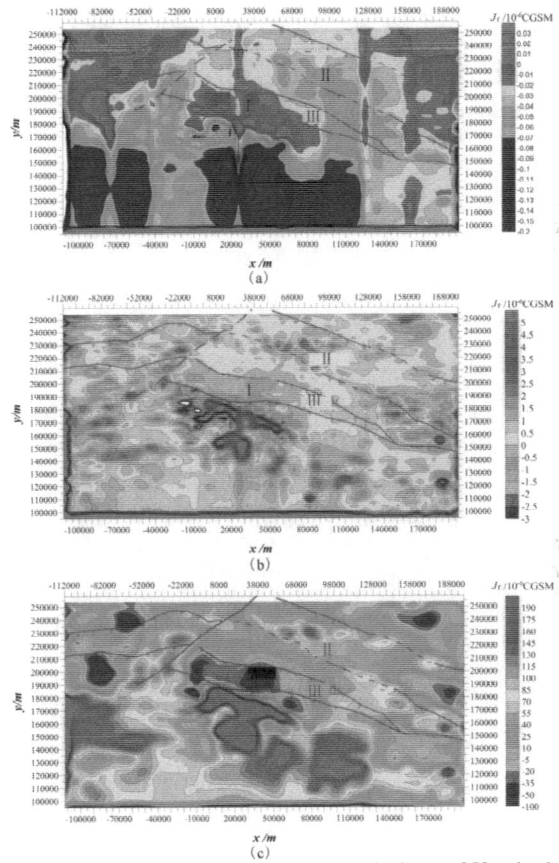

Figure 3: 3D tomography imaging of Magnetic data. a: 250m depth slice; b: 1250m depth slice; c:5750m depth slice.

After we've gotten the deep structural information and did a preliminary evaluation of rock mass, we still need to assess whether the preselected rock is intact or not. Therefore, we mapped out the weak geological structures within the rock mass using CSAMT method. We took two cases of CSAMT study as examples to present.

Data processing of CSAMT

Several processing steps were carried out on acquired CSAMT data. Firstly, based on physical property, we analyzed the original sounding curves to distinguish the electrical difference between intact rock and weak structure. Secondly, we marked the distortions. Thirdly, we applied EMAP filter to correct the static effect. The final step was to do 2D smooth-model inversion for far-field data. Furthermore, we performed geological interpretation according to the known geological information(An *et al*, 2013a).

Case one of CSAMT exploration

In fig.1, CSAMT survey was carried out along line S2. This was used to determine the location and extent of the two faults cutting across the rock mass I.

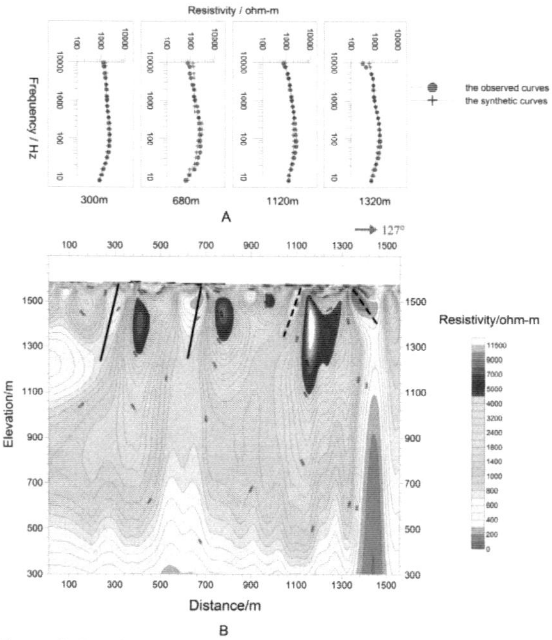

Figure 4: Result of CSAMT line S2. A: the fitness between observed data and synthetic data. B: inversion map.

The two conductive interfaces observed at 300 m and 680 m were interpreted as faults, and were verified by two nearby trenches. This shows that the CSAMT results are in good agreement with the known geological information. Conversely, no fault or fractured zone was observed at points 1140 and 1340, because the rock mass at these points are covered by Quaternary sand and gravel. However, the resistivity characteristics at these points are similar to that of the known faults mentioned above, that is, low-resistivity bands. Therefore, they are interpreted as faults.

Case two of CSAMT exploration

The second study area is still located in NW China. CSAMT data were acquired along 14 survey lines. Lines 1 and 14 were located in the north and the south region respectively. The line spacing used was 500m from line 1 to line 13 and 650m between lines 13 and 14. The lines were oriented in the N72°E direction asshown in Figure 5, having a uniform length of 3.24 km. The main bedrock exposed is medium-grained granodiorite.

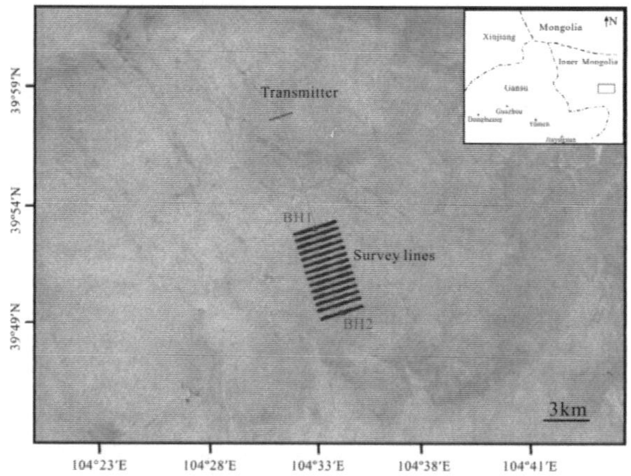

Figure 5:CSAMTsurvey sketch map of study area.

Result of line 14

2D inversion results of the 14 lines are shown in Figure 6. The resistivity characteristics along these lines are similar. To confirm the CSAMT result geologically, we compared the available borehole information(BH 2) with its respective survey line(line 14). In figure 7, the geological structure observed across this traverse is simple. From the surface to 1000mals, the resistivity characteristic shows mid-low resistant regions which are interconnected. Furthermore, below 1000mals, the resistivity value becomes very high. This shows that the shallow rock is fractured and/or weathered, with certain amount of water present within its pores.The deep rock mass however, is still intact. Borehole BH2 was sited at distance 1560m along line 14.The drilling result revealedthat the subsurface rock is broken, and there exist a lot of fractures and crushed zones within the subsurface. Hence, the rock mass quality of the cored region is low. This is consistent with the obtained CSAMT results. The resistivity logging result can be seen in figure6(Di, *et al*, 2015).

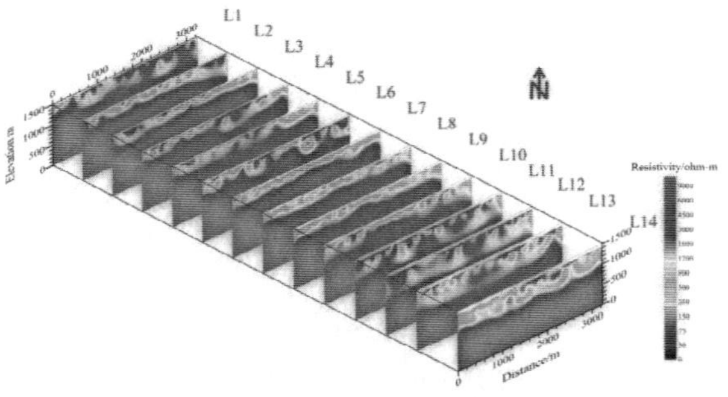

Figure 6: CSAMT 2D inversionresult of the 14 lines in 3D view.

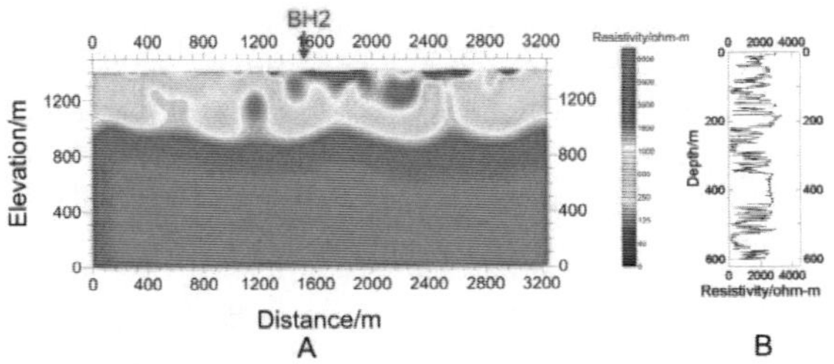

Figure 7: The CSAMT results of survey line 14.
A: The inversion map of line 14. B: The resistivity logging of BH2.

Conclusions

This paper discussed geophysical applications in preselecting HLRW site. In this study, we applied tomography imaging of large-scale gravity and aeromagnetic data to analyze and evaluate the integrity of rock mass within the study area. After the reconnaissance survey, we performed CSAMT survey to map out the geological structures within the targeted rock mass, and the obtained CSAMT results were in good agreement with the known geological information. In conclusion, the use of integrated geophysical methods in HLRW repository sitting is very effective, and could provide the necessary geophysical evidence for the design and construction of underground repository in future study.

Acknowledgments

The authors and their research group would like to acknowledge R&D of Key Instruments and Technologies for Deep Resources Prospecting (the National R&D Projects for Key Scientific Instruments), Grant No.ZDYZ2012-1-05, and thank the national defence science and industry project (No. 81261810).

References

An, Z.G., Q.Y. Di, C.M. Fu, C. Xu, B. Cheng, 2013, Geophysical Evidence through a CSAMT Survey of the Deep Geological Structureat a Potential Radioactive Waste Site at Beishan, Gansu, China, Journal of Environmental and Engineering Geophysics, 18(1), 43-54.

An, Z.G., Q.Y. Di, R. Wang, M.Y. Wang, 2013, Multi-geophysical Investigation of Geological Structures in a Pre-selected High-level Radioactive Waste Disposal Area in Northwestern China, Journal of Environmental and Engineering Geophysics, 18(2), 137-146.

Cutler, R.T., Langan, R., Love, P.L., and Bishop, T.N., 1985, A tomographic solution to the travel-time problem in General Inverse seismology: in Advances in Geophysical data processing, Vol. 2, Simaan, M. (ed.).

Di, Q.Y., Z.G. An, Z.X. Wang, 2015, CSAMT investigation for geological structures in a high-level radioactive waste preselected site, expanded abstracts, GEM2015, Chengdu.

Guo, Y. H.,T. X. Yang,S. F. Liu,2001, Hydrogeological characteristics of Beishan preselected area, Gansu province for China's high-level radioactive waste repository, Uranium Geology, 17(3), 184-189(in Chinese).

Li, Y., and D.W.Oldenburg, 1996, 3D inversion of magnetic data, Geophysics, 61, 394-408.

Li, Y., and D.W.Oldenburg, 1998, 3D inversion of gravity data, Geophysics, 63, 109-119.

Li, Y., and D.W.Oldenburg, 2003, Fast Inversion of large-scale magnetic data usingwavelet transforms and logarithmic barrier method, Geophys. J. Int., 152, 251-265.

Wang, J., G. Q. Xu, Y. X. Jin,et al.,2000, Study on regional crustal stability of Beishan, Gansu province, Beijing: Geology press(in Chinese).

Xu, G. Q., 2002, Preliminary determination for research on geological disposal of high-level radioactive waste in China in the period of 2000-2040,Uranium Geology, 18(3), 160-169(in Chinese).

Geophysical investigation of groundwater resources integrating seismic refraction tomography and resistivity data in Ibadan West, Nigeria

Olusegun K. Abass [1], Musa Bello [2]*
[1]Institute of Urban Environment, Chinese Academy of Sciences, Xiamen, 361021, PR China
[2]Department of Geosciences, University of Lagos, Akoka, Lagos, Nigeria

Summary

Secondary porosity formed within hard rock formations as a result of weathering, jointing, faulting or fracturing constitutes areas of special interest in groundwater prospecting. In this study, integrated electrical resistivity and seismic refraction surveys were carried out within the subsurface strata of Ibadan West, Nigeria, to delineate the groundwater potential and aquifer characteristics of the area. The integrated interpreted results revealed four layers overlying the resistive basement: the topsoil, dry sand, saturated sand and the weathered basement. The investigation also revealed the presence of a potential aquifer in the area, with resistivity values of 95 – 295 Ωm, and velocities range of 1650 – 2800 m/s while the depth of the aquifer ranged between 2.8 and 10.4 m. The forth lithological unit which has velocities ranging from 2500 – 3500 m/s and a resistivity value of 1124 – 1946 Ωm is considered the weathered basement. Comparing lithological logs from drilled boreholes with acquired geophysical data, the study area is considered suitable for groundwater development via shallow borehole construction.

Introduction

Hydrogeological and geophysical investigations are often conducted to assess groundwater potential and distributed information on subsurface hydrogeology (Kearey and Brooks, 1991). A number of geophysical means are commonly applied to image subsurface layers some of which include; seismic refraction tomography, electrical resistivity, ground penetrating radar, electromagnetic, geographic information system, remote sensing and gravity methods (Comair et al., 2012). Considering data accuracy, cost and location specificities (such as area geology), electrical resistivity method remains the best alternative and the most popularly used (Sirhan et al., 2011; Anomohanran, 2013).

However, reliable assessment of groundwater potentials and accurate identification of subsurface structure geometries and their corresponding hydrogeological parameters often requires the integration of proven geophysical techniques. Seismic refraction tomographic measurements are an alternative method of obtaining additional information between boreholes when uncertainties in the parameterization of other survey data exist.

In this paper, characterization of subsurface structures integrating information from different geophysical techniques to overcome some of the limits associated with subsurface structures interpretation is presented. Further, the groundwater potential and aquifer characteristics of the district are accessed to determine its suitability for groundwater development via shallow borehole construction.

Location geology and investigation technique

Field studies were undertaken in Ibadan West, situated in Southwestern area of Nigeria (Fig. 1). It lies at approximate latitude of 7° 24' and 7° 26' N and longitude 3° 29' and 3° 34' E. The area is characterized by undulating topography and the drainage pattern is generally dendritic where the streams flow over uniform bedrock. The average annual rainfall is about 1200 mm with a range of vales from 803.2 mm to 1800 mm and a standard deviation of 230.4 mm. Movement of groundwater follows the surface topography with seasonal variations in water levels.

The study area is underlain by Migmatite Gneiss complexes to late intrusive rocks (Fig. 1). Rock types present include migmatite gneiss (granite gneiss), porphyritic micro-granites, and pegmatite. These rocks are associated with brittle and ductile structures such as joints, fractures, faults and lineation.

Resistivity data acquisition was conducted using ABEM SAS 1000 Terrameter (Dahlin, 1996) to obtain the apparent resistivity values along three siesmic refraction traverses (Fig. 1). Siesmic refraction data were acquired on 15 profiles using a twenty four channel Seismograph with 10 Hz geophones (ABEM Instrument AB, 1994). The geophones were spaced at 2 meters intervals and planted for optimum physical contact with the ground, making a total of 50 meters profile length. The trigger mode was manual (geophone input). Information contained in the traveltime data obtained from the seismic refraction survey was used to determine the depth and configuration of refracting layers (Kilthy et al., 1986).

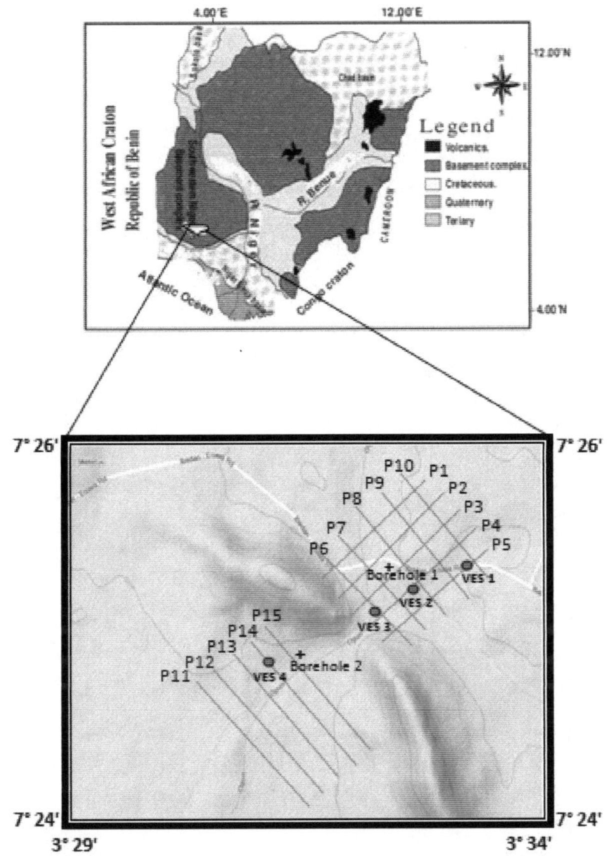

Figure 1: Location geology (adapted from Elueze, 1982,) and map of the study area showing seismic traverses, VES points and borehole locations

The aim had been to map fractures and weak formations (loose grounds) for further electrical soundings. Four vertical electrical sounding (VES) of the Schlumberger configuration was carried out with a maximum current electrode separation of 100 m. VES data were first interpreted by the conventional partial curve matching technique with two-layer master curves (Mooney et al., 1966) in conjunction with an auxiliary point diagram (Ebert, 1943) and afterward inputed for computer-assisted interpretation with RESIST software to obtain the true resistivity and thickness of the various layers.

Seismic refraction data were processed with tomographic inversion methods, using Pickwin and Plotrefa Software. In addition, two boreholes were drilled to determine the lithological characteristics of the subsurface zones.

Result and discussion

Quality siesmic tomography sections were generated and results of the interpretations are typically presented in color bar form, with different velocities shown in varied colors expressing the presence of either saturated, unsaturated, consolidated or unconsolidated layer(s). Fifteen tomographic model results are interpreted and sample profiles are shown in Fig. 2. Presence of low velocity zones in form of local depressions (Palmer, 1991 and Sheehan and Doll, 2003) and variable strata thicknesses within the overburden were detected. The raypath coverage plots of each model were generated to determine which part of the tomography sections are well defined. Table 1 shows the comparison of the tomographic model results with lithological log data obtained from the two drilled boreholes in the area, which was used as control for interpreting the tomography models. Data from the resistivity survey are also presented in Table 1.

The results of the analysis showed that the average depth to the weathered basement in the area is aprroxmately 10.4 m, with multiple interval velocities obtained from the surface to the basement. This implies that there are several strata within the overburden, which overlay the basement in the study area with variable thicknesses and velocities, presumably associated with dry, wet, and consolidated to unconsolidated sediments. In Fig. 2, profile 5, traverse 1, there are indications of depressions at two points, which shows a probable loose and unconsolidated ground characterized by sand and clayey materials. This low velocity zone be can be conveniently interpreted as a high permeable zone. By comparing the interpreted lithology of the profiles with the borehole report in Table 1, it is noticeable that the obtained seismic models clearly mapped most particularly, semi loose weathered formations in form of depressions, from which further resistivity sounding were conducted for detailed interpretation.

Apparent resistivity data obtained from the fitting curves and geo-electric section are shown in Fig. 3 & 4 respectively. Results from the fitting curves suggests that the resistivity model whose calculated apparent resistivity best fit the measurements, with a RMS-error of 2.7 and 3.0 respectively is characterized by; a top layer with a thickness of 1.0 and 1.4 m and apparent resistivities of 1606 and 1677.8 Ωm, a value typical of top soil. The second layer is approximately 1.3 and 1.4 m thick, with apparent resistivities of 935.8 and 1840 Ωm respectively, characteristics of dry sand. The third layer comprise a thickness of 2.8 and 10.4 m with apparent resistivities of 198.1 and 95.2 Ωm respectively. The range of apparent resistivity values and thickness of the third layer is

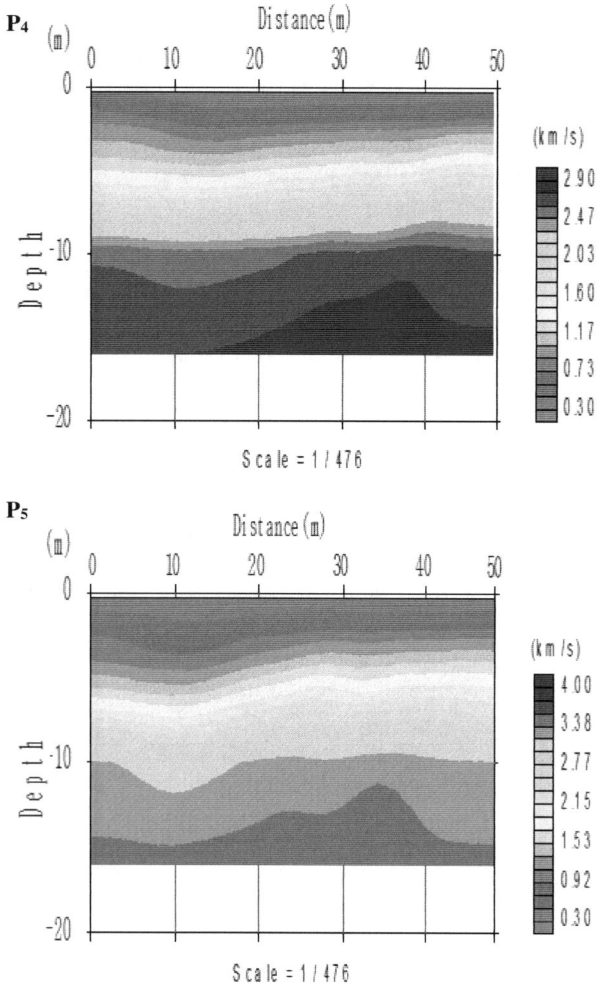

Figure 2: Seismic tomography imaging of profile 4 and 5.

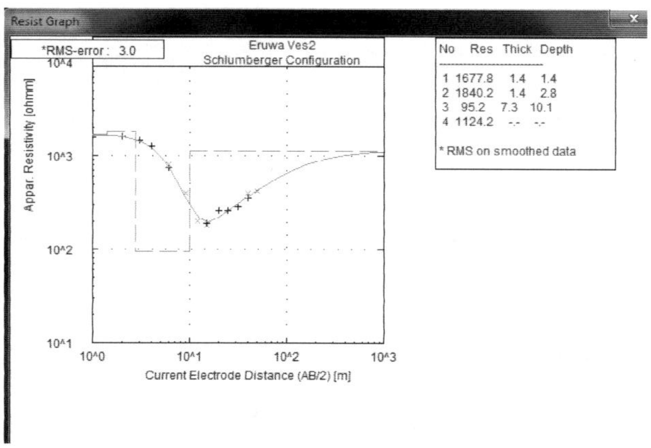

Figure 3: Vertical electric sounding curve obtained from the study area

Table 1: Results of seismic velocities, resistivities and depths from geophysical models correlated with two available Boreholes log data.

Depth (m)	P-wave velocities (ms^{-1})	Resistivity data (Ωm)	Geologic Interpretation	Borehole one		Borehole two	
				Depth (m)	Lithology	Depth (m)	Lithology
0 – 1.4	300 – 1700	1607 – 2992	Lateritic top soil	0 – 1.2	Lateritic clay	0 – 0.5	Lateritic clay
0.8 – 2.1	1200 – 1900	935 – 1840	Dry sand	1.2 – 2.3	Coarse sandy clay	0.5 – 1.4	Lateritic sandy clay
2.8 – 10.4	1650 - 2800	95 – 295	Saturated sand or weathered basement	2.3 – 4.6	Fine sand (brownish)	1.4 – 4.0	Medium to fine saturated sand
10.4 – down	2500 – 3500	1124 – 1946	Weathered basement/ Basement rock	4.6 – 7.0	Medium to fine saturated sand	-	-

interpreted as the weathered overburden saturated with pore water, as suggested by some authors (Ndlovu et al., 2010; Martinelli and Hubert, 1985). The forth layer have apparent resistivity values of 1683 and 1124.2 Ωm for VES 1 and 2 respectively and is presumed to be a weathered bedrock.. Other VES points show similar results as shown in Fig. 4. Comparison of the results of the seismic tomography and the geo-electrical sections with the borehole lithogic logs shows that different features from each of the models are incorporated to give a better interpretation of the subsurface characteristics of the study area.

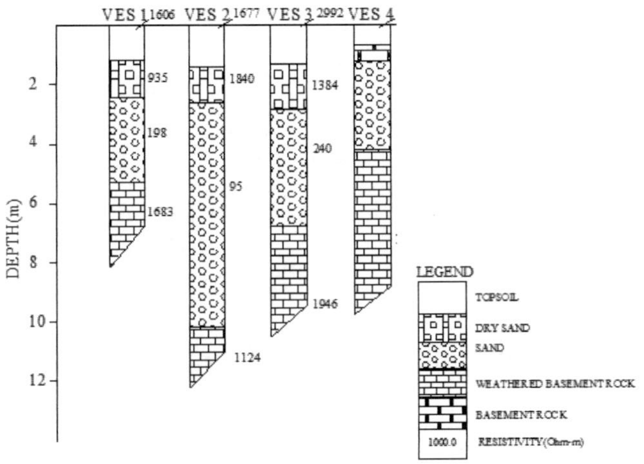

Figure 4: Schematic geoelectric section of the VES conducted in the study area

Conclusions

An integrated approach to describing subsurface heterogeneities by combining high resolution geophysical methods, such as seismic tomography and electrical resistivity technique, with borehole lithogical logs was presented. Tomographic and curve matching models were employed for the interpretation of the geophysical data. The application of the approach was consistent for identifying potential permeable zone which exhibits realistic heterogeneities. Hence, integrated geophysical approach appears to be capable of identifying subsurface structures relevant for the prediction of groundwater bearing zones.

Reference

ABEM Instrument AB, 1994, Equipment manual for TERRALOC MK6 software version 2.21: Bromma.

Anomohanran, O., 2013, Geophysical investigation of groundwater potential in Ukelegbe, Nigeria: Journal of Applied Science, **13**, 119–125.

Comair, G. F., D.C. McKinney, and D. Siegel, 2012, Hydrology of the Jordan River basin: watershed delineation, precipitation and evapotranspiration: Water Resource Management **26**, 4281–4293.

Dahlin, T., 1996, 2D resistivity surveying for environmental and engineering applications: First Break, **14**, 275–284.

Ebert, A., 1943, Grundlagen zur auswerkung geoelektrischer tiefenmr-ssulgon: garlands beitrage zur geopysik, BZ, **10**, 1–17.

Kearey, P., and M. Brooks, 1991, An introduction to geophysical exploration: second ed: Blackwell Scientific.

Kilthy K., R.A. Norris, R.W. Mclamore, K.P. Hennon, and K. Euge, 1986, Seismic refraction at Horse Mesa Dam: an application of the generalized reciprocal method: GEOPHYSICS, **51**, 266–275.

Martinelli, E., G.L. Hubert, 1985, A report to the ministry of energy and water resources and development: master plan for rural water supply and sanitation, Norad: Hydrogeology, **2**, 109–118.

Mooney, H. M., E. Orellana, H. Pickett, and L. Tornheim, 1966, A resistivity computation method for layered earth models: GEOPHYSICS, **3**, 199-203.

Ndlovu, S., V. Mpofu, D. Manatsa, E. Muchuweni, 2010, Mapping groundwater aquifers using dowsing, slingram electromagnetic survey method and vertical electrical sounding jointly in the granite rock formation: a case of Matshetshe rural area in Zimbabwe: Journal of Sustainable Development in Africa, **12**, 199–208.

Palmer, D., 1991, The resolution of narrow low velocity zones with the generalized reciprocal method: Geophysical Prospecting, **39**, 1031–1060.

Sheehan, J.R., W.E. Doll, W. Mandell, 2003, Evaluation of refraction tomography codes for near surface applications: 2003 Annual Meeting, SEG, Expanded Abstracts, 26-31.

Sirhan, A., M. Hamidi, and P. Andrieux, 2011, Electrical resistivity tomography, an assessment tool for water resource: case study of Al-Aroub Basin, West Bank, Palestine: Asian Journal of Earth Science, **4**, 38–45.

Integrated geotechnical and engineering geophysical approach to control groundwater influx: a case study from Abu Dhabi, UAE

Rasheed Jaradat,[1], Younes Al-Omari[1], Anmar Abdulaziz[2]*
[1]Gulf Laboratory for Site Investigation & Material Testing, Abu Dhabi, UAE. E-mail: rasheed.jaradat@gulflab.ae
[2]Spectra Geotechnologies and Foundations Contracting, Abu Dhabi, UAE.

Summary

Shallow groundwater conditions force construction projects and contractors to conduct dewatering and/or water control procedures that tend to affect the competency of foundation material and design of structures. Uncontrolled groundwater may result in a condition causing piping, heaving, reduced stability of excavated slopes and pits, or reduced foundation ability to support loadings and structures (DOD, 2004). Conditions of shallow groundwater are of prime concerns to engineering projects, especially when excavations are carried out within earth materials characterized by elevated hydraulic conductivities.

An integrated geotechnical and geophysical approach was adopted to effectively control ground water influx. Shallow groundwater and poor rock characteristics have led to sever flooding of construction excavation. Multichannel Analysis of Surface Waves (MASW) identified continuous zones of weakness bordering excavation from the southern and western borders of the pit. Interpretation indicated that flooding is taking place below the sheet piles, where drilling of five boreholes around the tank area revealed severe loss of drilling water. The recovered samples showed highly weathered and fractured rock conditions. The loss of drilling water was observed in two boreholes, BH-1 and BH-2, with two levels (referenced to the existing ground level) of water loss; at 5.6m to 15.0m and 13.5m to 18.5m respectively. An engineering solution was proposed aiming at isolating excavation by developing a grout curtain in order to successfully control groundwater influx. A grouting proposal was carefully designed and is under execution.

Introduction

Geotechnical and geophysical investigations were carried out in an attempt to characterize shallow subsurface conditions at western Samha, Abu Dhabi, UAE (Figure 1). The site is located at approximately 40km to the SE of City of Abu Dhabi. It is proposed to host Storm Water Pump Station. The proposed structural design will utilize conventional concrete frame.

Figure 1: Location map of the study area.

Severe flooding took place across the construction site following localized open-cut excavation and sheet piling (Figure 2). Sheet piles have a depth of 12m. The clients' engineer stated groundwater flooding was controlled across an adjacent location, however within the current location uncontrolled and severe flooding took place upon reaching 6-6.5m below current ground surface. Geotechnical drillings indicated that groundwater surface was only located at 0.60m below current ground surface (Figure 3).

This study presents the results of an integrated geotechnical and geophysical approach to control groundwater influx into construction excavation. The study was carried out in consequence of uncontrolled groundwater flooding of the construction site, aiming to obtain more conclusive information about the geology on the shallow section and to characterize subsurface geological conditions and hazardous features; such as cavities, dissolution features, and weak and fractured zones that can lead to future risks. Additionally, an attempt was made to propose an engineering solution based on obtained results.

Methods

A conventional geotechnical campaign encompassing five (no. 5) boreholes was carried out. Two boreholes (BH-01 and BH-03) encountered significant drilling water losses. BH-01 showed about 70% water loss at 5.60-7.10m; and a more drastic loss of 90% at 7.10-7.50m and 13.0m; and 100% at a depth of 15m below existing ground level, respectively. On the other hand, BH-03 encountered water losses of 30% at 13.0-13.20m, and a drastic loss of 100% at a depth of 18.0m.

Figure 2: Flooded excavation fensed by sheet piles (12m).

Water losses were encountered within moderately to highly weathered sandstone sequence that is intercalated with gravel and gypsum. However losses within the deeper section were attributed to moderately highly weathered mudstone and gypsum intercalations, which then shift into more highly weathered gypsum sequence interbedded with thin mudstone layers. These sequences tend to show low SPT-N, RQD and TRC values.

A geophysical investigation study was carried out adopting a detailed Multichannel Analysis of Surface Waves (MASW) approach to investigate subsurface conditions. Four MASW profiles were carried out in a closed loop pattern surrounding the excavated area, such that profiles were laid out at 1-1.5m from sheet piles (Figure 3). Geophone spacing was 1m and a shot interval of 2m. SurefSeis (v. 3.6) software was used. The results of the geophysical study was interpreted and calibrated against drilled geotechnical boreholes.

Results

MASW sections revealed a shallow sequence alluvium deposits as indicated from estimated seismic velocities (<300m/s). Drilling results described this material as medium dense to dense slightly gravely silty sand (i.e. described as Fill Material). This is underlain by a distinguished weak to very weak (mod. to highly weathered) sandstone-gypsum intercalated sequence (i.e. described as Gypsiferrous Sandstone) that was locally bedded with conglomerate. The depth of this sequence along BH-1 and BH-2 was 3-8m., and 4.5-8.0m, respectively. Total Core Recovery (TCR) indicated a general decrease from 90% to 33% for BH-1 across the lower part of this sequence. On the other hand, TCR of the same sequence along BH-3 did not show this behavior, but indicated a general increase from 65% to 80%, suggesting a more contact sequence. The estimated Vs values were within the range of 280-340m.

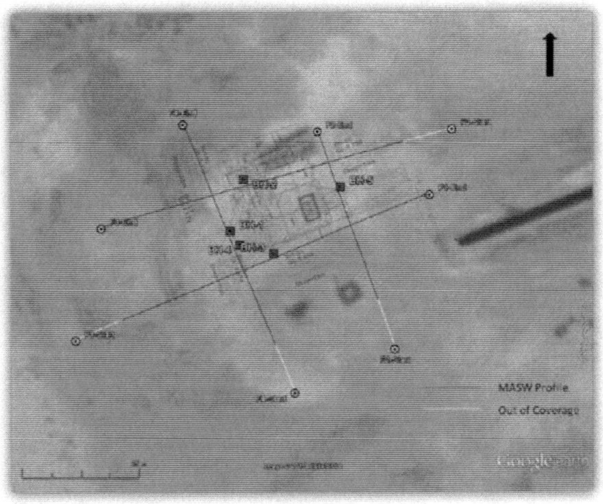

Figure 3: Field setup and borehole locations.

While the lines 1 and 2 showed correlated shear wave velocities for the weak Gypsiferrous Sandstone; profiles Line 3 and Line 4 showed a more drastic behavior, where modeled Vs values are relatively lower than what was seen across Line 1 and Line 2. However, a considerable change in thickness was noticed for the same sequence along BH-1 and BH-2 (5.0 and 1.5 m), respectively. The encountered Vs values across Line 4 and Line 3 were 250-260m/s suggesting a potential weak zone composed of gypsum intercalations of weak and weathered sandstone that was locally bedded with conglomerates. This was underlain by a sequence of highly weather calcareous mudstone interbedded with gypsum. The estimated shear wave velocities indicated relatively higher values than the upper sequence (420-600m/s).

Several localized low Vs anomalies (weak zones) were observed across a sequence of moderately weak to weak gypsum interbedded with mudstone as seen across lines 3 and 4. The observed loss of drilling water across boreholes BH-1 and BH-3 supported this conclusion. It was strongly believed that the shallow Gypsiferrous Sandstone sequence (i.e. down to a depth of 8m) was playing the major role in the mobility of groundwater by forming a conduit-like weak zone facilitating groundwater mobility towards the excavated hole from the west and northwestern directions (i.e. across the region containing the boreholes BH-1 and BH-2 (Figure 4).

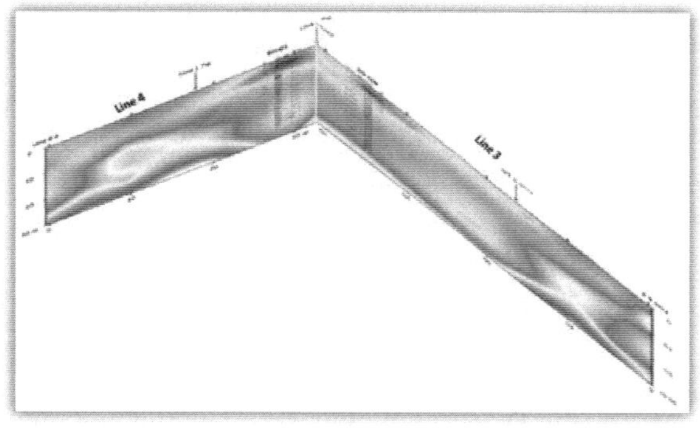

Figure 4: Composite presentation of the MASW lines 3 and 4 showing very close interpretation and correlated Vs values across the Tie line between the two lines located to the NW which is believed to be the direction for the groundwater movement towards the excavated hole.

As the groundwater table was relatively shallow (0.6m below existing ground level) across this regain, and located above the base of the excavated hole, excavation of the hole to a depth of 6.5m (i.e. 6m below the water table) created enough gradient to direct groundwater to move towards the hole. This is due to the presence of less competent weak and weathered sandstone sequence with gypsum intercalations. On the other hand, BH-4 which was located at a very close distance to BH-1, which had a thickness of 6.3m of the Gypsiferrous Sandstone sequence, did not show any drilling water losses and strongly supported this conclusion. The foreseen sequence across BH-4 suggests a more competent material where the TCR showed stable values ranging between 65-67%.

A grouting procedure was proposed aiming to eliminate or control water seepage below the pre-drilled cut-off wall sheet piles, such that the whole permeable section along the sheet pile will be grouted. A low pressure grouting approach will involve the injection of a cementitious grout into fractured rock layer to fill existing discontinuities and thus minimize the potential for water seepage (Figure 5).

A pre-grout tests will be carried out in 2 boreholes (20m deep), involving 5 numbers of packer permeability tests and 1 number of variable head permeability test in each hole. Following this a grout trial in each hole will be conducted to determine the hydraulic conductivity of the materials surrounding and below the excavation to decide on the depth and characteristics of the grouting required.

Once each side of the grouting is completed, 2 numbers of test holes at 3rd points down to the improved area will be drilled and 3 numbers of packer permeability tests in each (similar to pre-grout permeability test) and one number of variable permeability will be conducted below the soil section.

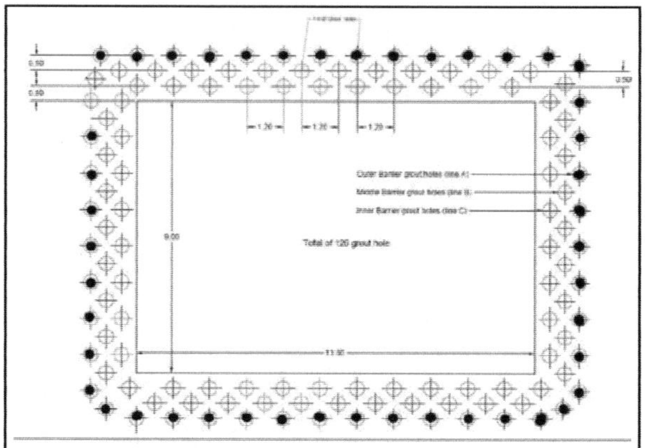

Figure 5: Grouting campaign plan composed of 2 rows anticipated to be executed, however depending on the results a third row can be needed.

Conclusions

A successful geotechnical and geophysical approach was proposed and carried out aiming to control groundwater influx into construction excavations.

Geophysical investigations revealed poor site conditions contributed to sever groundwater flooding incidence. Flooding of a sheet pile excavated hole can be attributed to the existence of less competent materials within the lower part of this sequence, as TCR declined sharply from 90% to 33% within BH-1 which showed 70-80% drilling water losses.

The integrated interpretation of borehole and MASW sections indicated that the region of intersection between the MASW lines 3 and 4, which were adjacent to BH-2 and BH-1, was mainly dominated by a main low velocity layer. Excavation across this area led to the formation of a conduit-like weak or fractured zone facilitating groundwater movement towards the hole (to the east and southeast direction). Excavation to a depth of 6.5m below ground surface created enough vertical gradient that accelerated the rate of groundwater mobility towards the hole, thus causing uncontrolled and severe flooding consequence. An engineering solution was proposed and being conducted aiming to isolate and control groundwater influx into the excavation pit.

Reference

Department of Defense, 2004, Dewatering and groundwater control, https://www.wbdg.org/ccb/DOD/UFC/ufc_3_220_05.pdf, accessed 20 May 2015.

Kanlı, A. I., P. Tildy, Z. Pronay, A. Pınar, and L. Hermann, 2006, VS30 mapping and soil classification for seismic site effect evaluation in Dinar region, SW Turkey: Geophysical Journal International, **165**, 223–235.

Park, C.B., Miller, R.D., and Xia, J., 1999, Multichannel analysis of surface waves: Geophysics, **64**, no. 3, 800-808.

Park, C.B., Miller, R.D., Xia, J., and Ivanov, J., 2007, Multichannel analysis of surface waves (MASW)-active and passive methods: The Leading Edge, 26, 60-64.

Rawlings, C. G., Hellawell, E. E., Kilkenny, W. M., 2000, Grouting for ground engineering: Construction Industry Research and Information Association.

Warner, J., 2004, Practical Handbook of Grouting: Soil, Rock and Structures: John Wiley & Sons.

A Collaborative Methodology for Ground Searches by a Forensic Geologist and Law Enforcement (Police) Officer: Detecting Evidence Related to Homicide, Terrorism and Organized Crime

Dr Laurance Donnelly BSc (Hons), PhD, CGeol, CSci, EurGeol, FGS, FGSA*
Chair, International Union of Geological Sciences (IUGS), Initiative on Forensic Geology (IFG), geologist@hotmail.co.uk

Commander Mark Harrison MBE
Australian Federal Police, IUGS-IFG & University of Canberra, Australia, Mark.Harrison@afp.gov.au

Summary

In the United Kingdom (UK), law enforcement (police) ground searches for burials for targets such as graves, firearms, drugs, items of value and evidence, have traditionally relied on a counter-terrorism search strategy. This developed in the aftermath of the failed attempts by the Provisional IRA to assassinate the former British Prime Minister, Margaret Thatcher in 1984.

In 1994, a geological search began for a missing person that was believed by the police to be buried in a shallow, unmarked grave, located in a remote upland range of hills in northern England. This search applied techniques and methods conventionally used in mineral exploration and engineering geology ground investigations. This included for example; analysis of air photos and satellite imagery, geological mapping and geomorphological observations, diggability assessments, the production of a Conceptual Geological Model (CGM), phased and systematic geophysical surveys and soil probing. The CGM enabled to most suitable suite of geophysical methods and associated search assets to be chosen, which were appropriate for the prevailing geological conditions and grave being sought.

In the early 2000s, this geological based search evolved to incorporate more conventional law enforcement, mainly counter-terrorism based search methods. The result, over the decade which followed, was the emergence of a new ground search strategy that combined both the geological and law enforcement approaches.

This new and innovative search strategy has raised awareness within law enforcement communities of the value in understanding the geology and ground conditions before the choice and deployment of geophysical instruments and other search assets. This collaborative methodology is based on the division of a search into three principal phases (or stages); known as Pre-search, Search and Post-Search. Each of these comprises series of sub-sub stages. This strategy enables the searcher to develop a methodology that best suits the prevailing geology and in context with the nature of the targets being sought. This method offers a high assurance, cost-effective, proportionate and pragmatic search to be deployed to negate or locate the presence of suspected target buried at a shallow depth in the ground. This strategy is now increasingly being used in the UK and internationally for open area searches for burials.

Search Strategy

Law enforcement (police) conduct ground searches to; locate buried or concealed objects/items, find missing persons and graves, deprive criminals of their resources and opportunities to commit crimes, protect vulnerable targets, gather intelligence and evidence. A geological search for a suspected grave began in northern England in the mid-1990s (still ongoing) with a review of available intelligence, data and information combined with reconnaissance (walk-over) surveys to observe the geology, geomorphological process, hydrogeology and to identify and logistical and operational constraints. This enabled the outer limits of the search are to be defined. The acquisition and analysis of aerial photography and satellite imagery provided information on potential ground disturbances associated with digging to be identified. The site visits and air photo analysis also assisted with the production of diggability maps and CGMs for the grave (Donnelly 2013). A control site was established to test the suitability, resolution and detection limits of the geophysical instruments. This also verified the soil augers could detect the desired items (clothing associated with a body).

In the early to md 2000s, law enforcement ground search strategies were applied to this ground search. These were techniques largely used in counter terrorism defensive or protective searches type (Harrison et al., 2006). This included the deployment of line searches, innovative geophysical instruments available from the police and military, Victim Detector Dogs (VDD) and the applications of behavioral profiling. This combined geological and police approach to a ground search resulted in a series of principal phases or stages being identified; known as the 'Pre-Search', 'Search' and 'Post-Search' stages/phases. This enables a ground search to be properly planned with appropriate, proportionate and cost-effective resources consistent with the nature of the target being sought, and it comprised 10 main tasks (Figure 1).

An open area ground search for a burial may be divided into three main phases/stages, as depicted in Figure 2 and managed by a Police or Law Enforcement officer (Figure 3). These may be further sub-divided into a series of sub-stages, as may be appropriate and relevant, and outlined below:

- Pre Search: This largely involves defining the search type, area, identification of the search assets and their method of deployment, planning and logistics. The authors recommend a ground search should extend from the macro to the micro scale and from the invasive to the non-invasive (Donnelly 2010; Donnelly & Harrison 2010) (Figure 4).

- Search: The implementation of the search should follow and agreed standard Operational Procedure (SOP). This should also be dynamic and flexible to be modified based on materially relevant field observations at the time the search is conducted (Figure 5).

- Post Search: Once a positive observational, geophysical, geochemical or VDD anomaly (also known as a 'hit' or 'indication') has been identified and verified, the recovery and recording stage of the search should take place. This may require additional support from, for example, a Crime Scene Manager (CSM), forensic scientist or forensic archaeologist with appropriate qualifications and experiences (Figure 6).

Applications of Geophysics as a Search Asset

In the recent past geophysics was used in law enforcement (police) ground searches for burials on the basis that the instruments chosen worked well at another geographical location where a 'similar' target was found. Relatively little regard, if any, was given to the geology and ground conditions. Alternatively, the choice of geophysical instrument used was pre-determined by the availability of the instruments within the search teams or the availability of a police search trained officer to use a particular geophysical instrument. Often these tended to be ground penetrating radar (GPR) with or without metal detection methods (ie. magnetic and/or electromagnetic techniques). In the UK, collaboration between geologists and police search advisors (PolSAs) has changed the way by which geophysical instruments are considered, tested and deployed in ground searches. The ground conditions and the physical properties of the detectables associated with the target must be carefully considered before the suitable suite of geophysical search assets can be chosen. Geomorphological, geological and hydrogeological processes that have been active in the time elapsed since burial took place must also be considered. No single geophysical method can locate graves or human remains in all different types if geological settings. The geophysical instruments must be considered following a detailed analysis of the geology of the burial site, the physical properties of the target and the presence of anthropogenic features that may inhibit the use of certain geophysical instruments, such as metal fences, pipelines and utilities (geophysical noise). If there is insufficient contrast between the target and the surrounding soils the target will not be detected. Furthermore, the target may also be undetected if it lies beyond the detection depth and resolution of the instruments deployed.

Typical geophysical techniques that could be considered to locate items desirable to be found by the Police include the following:

- Conventional metal detectors measure variations in magnetic susceptibility and may be used to locate weapons including Unexploded Ordnance (UXO) and Improvised Explosive Devises (IED), coins and firearms burial at shallow depths.
- Ground Penetrating Radar (GPR) measures high frequency electromagnetic radiation and has been used to locate, shallow, unmarked, clandestine graves related to homicide, grave in cemeteries that are not marked by headstones, weapons, and drugs. GPR may also be used to image the ground beneath areas of hard standing such as concrete, paved or tiled floors.
- Magnetic techniques measure spatial variations in magnet susceptibility and have been used to locate ferrous metal associated with unmarked graves, UXO, IED or containers concealing drugs, weapons.
- Electromagnetic conductivity methods measure variations in electromagnetic radiation and can could also be used to locate some graves and weapons, areas of waste or contamination, refuse, old mine shafts and underground utilities.
- Electrical resistivity methods measure variations in electrical resistance and may be used to locate burial sites in cemeteries or contaminant plumes.
- Gravity and microgravity techniques measure variations in the Earth's gravity field (density) and can be used to locate voids, mine workings, underground burial chambers and excavations in rock.
- Seismic techniques measure the travel time of reflected or refracted seismic waves and a such have been used to monitor explosions, plane crashes, impacts, nuclear weapons testing, military events, waste and contaminated ground.

Generally, 'real-time' analyses of the geophysical results will be desired for police and law enforcement searches. Some techniques a 'real time' indication of whether an anomaly may be present owing to the availability of field portable laptop computers, GPS tracking, GIS data management and software to permit the processing of geophysical data. Some techniques however, will require at least 12-24 hours, possibly longer to process the geophysical data (Figure 7). When possible, geophysical control sites should be established before the start of a ground search. These enables the operational suitability of the instruments to be determined, provides confirmation that the instruments have remained operational throughout a search, allows instrument calibration and provides the opportunity for new search team members to become familiar with their operational control and deployment. Furthermore control sites also provide verification of the search depths and resolution limits of the equipment. It is important that the items buried at the control site are similar to the items being searched for and the geological conditions are the same or similar to the defined search area. New intelligence or the availability of higher resolution search equipment may justify new searches taking place at a future date. Geophysical surveys have resolution and detection limits and do not offer a 100% assurance that a target will be fond or its absence confirmed, and this is important to convey to the investigating officers. Geophysics however, when aligned other search assets and deployed from a macro to micros scale and non-invasive to invasive methodology can provide a high assurance search strategy. Other techniques that may be considered, depending on the geology and context of the burial target, might include; remote sensing (Earth Observation), unmanned aerial vehicle (UAV, high resolution air photographs, soil and groundwater sampling to detect leachate plumes of volatile organic compounds that have emanated from the target, specially trained victim detector dogs and augering. A combined geological and law enforcement (police) approach may provide the greatest levels of assurance to help detect a target that has been buried associated with a criminal act, terrorism or organised crime. Or negate its presence so far as possible within the limits of the resources available. The geology, physical properties of the target and the detectability of associated items (such as metal on clothing, coinage or a spade/knife) should also be considered by the construction of a CGM (Figure 8). Further information on the applications of methods to assist with ground searches can be found in Fenning & Donnelly 2004, Ruffell and McKinley 2006, Pringle et al., 2013, Cheetham 2005; Ruffell & McKinley 2008; Pringle et al., 2012. Geophysical surveys may also be deployed on water, see for example, Ruffell 2006, 2014 and Parker et al., 2010.

Figure 7. Geophysical search for weapons, IUGS-IFG training, University of Messina, Sicily, 2015

References

Cheetham, P. 2005. Forensic geophysical survey. In: Hunter, J. & Cox, M. Forensic Archaeology: Advancers in Theory and Practice. Routledge, Taylor & Francis Group, London & New York, 62-95.

Donnelly, L. J. 2002. How forensic geology helps solve crime. All-Party Parliamentary Group for Earth Science. Westminster Palace, House of Commons, 12 March 2002.

Donnelly, L. J. 2010. The role of geoforensics in policing and law enforcement. Emergency Global Barclay media Limited, January 2010, 19-22.

Donnelly L. J. 2013. Use of Geology in Forensic Science: Search to Locate Burials. In: Elias S.A. (ed.)

Donnelly, L. J. & Harrison M. 2010. Development of Geoforensic Strategy & Methodology to Search the Ground for an Unmarked Burial or Concealed Object. Emergency Global Barclay media Limited, July 2010, 30-35.

Donnelly, L. J. & Harrison, M. 2015. A Guide to Forensic geology: Volume 1: Ground Searches for Burials (Draft 2.0). Report of the International Union of Geological Sciences, Initiative on Forensic Geology, (pending).

Donnelly, L. J. & Harrison, M. 2013. Geomorphological and Geoforensic Interpretation of Maps, Aerial Imagery, Conditions of Diggability and the Colour Coded RAG Prioritisation System in Searches for Criminal Burials. In: Pirrie, D., Ruffell, A. R. & Dawson, L. (2013). Environmental and Criminal Geoforensics. Geological Society of London, Special Publication, 384, 173-194.

Fenning, P. J. & Donnelly, L. J. 2004. Geophysical techniques for forensic investigations. In: Pye, K. & Croft, D. (eds) Forensic Geoscience: Principles, Techniques and Applications. Geological Society, London, Special Publications, 232, 11-20.

Harrison, M. & Donnelly, L. J. 2009. Locating concealed homicide victims; developing the role of Geoforensics. In: Ritz, K., Dawson, L. & Miller, D. (eds). Criminal and Environmental Soil Forensics. Soil Forensics Springer, 197-219.

Harrison, M., Hedges, C. & Sims, C.(Eds) 2006. Practical Advice on Search management and Procedures. National Policing Improvement Agency (NPIA), UK.

Parker, R., Ruffell, A., Hughes, D. & Pringle, J. 2010. Geophysics and the search of freshwater bodies: A review. Science & Justice Volume 46 No.50, 141-149.

Pringle, J.K., Ruffell, A., Jervis, J.R. Donnelly, L., J. McKinley, J., Hansen, J., Morgan, R., Pirrie, D. & Harrison, M. 2012. The use of geoscience methods for terrestrial forensic searches. Earth Science Reviews, 114(1-2), 108-123.

Ruffell, A. 2006. Under-water Scene Investigation Using Ground Penetrating Radar (GPR) in the Search

Ruffell, A. 2014. Lacustrine flow (divers, side scan sonar, hydrogeology, water penetrating radar) used to understand the location of a drowned person. Journal of Hydrology 513 (2014) 164–168.

Ruffell, A. R. & McKinley, J. 2008. Geoforensics. Wiley-Blackwell, Chichester, UK.

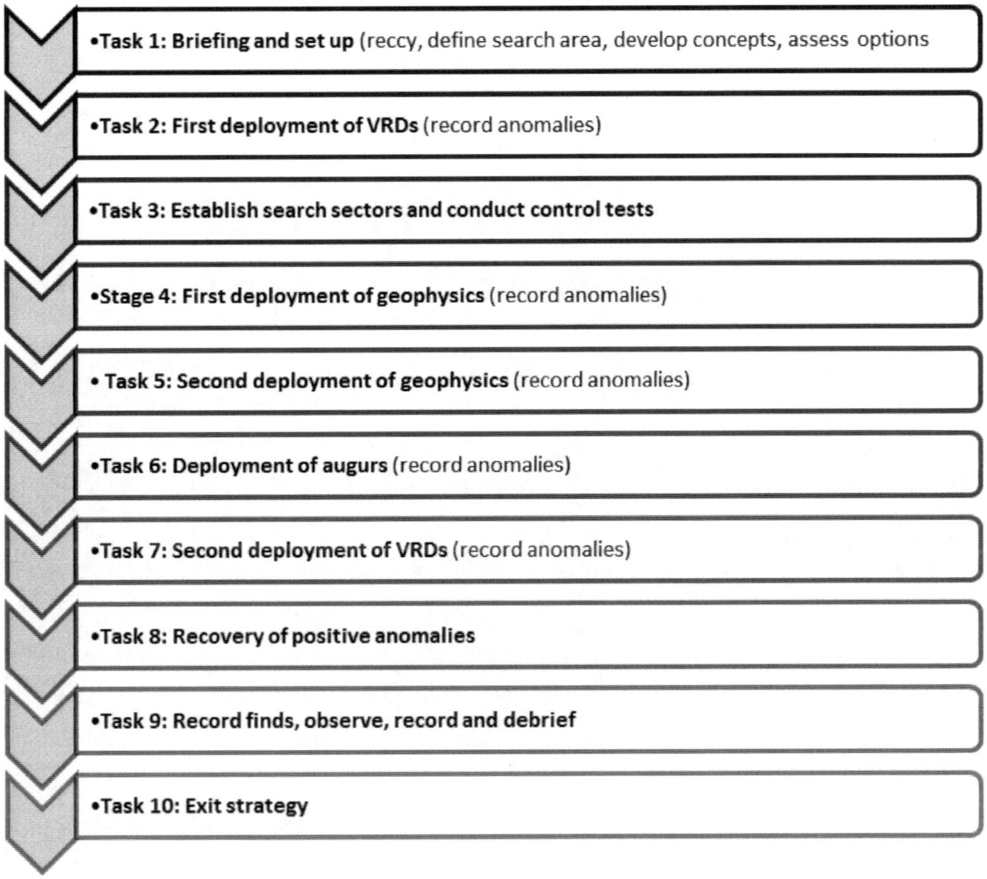

Figure 1. A ground search for a shallow,unmarked grave combining geological and law enforcement (police) stratgeies (Donnelly & Harrison 2015).

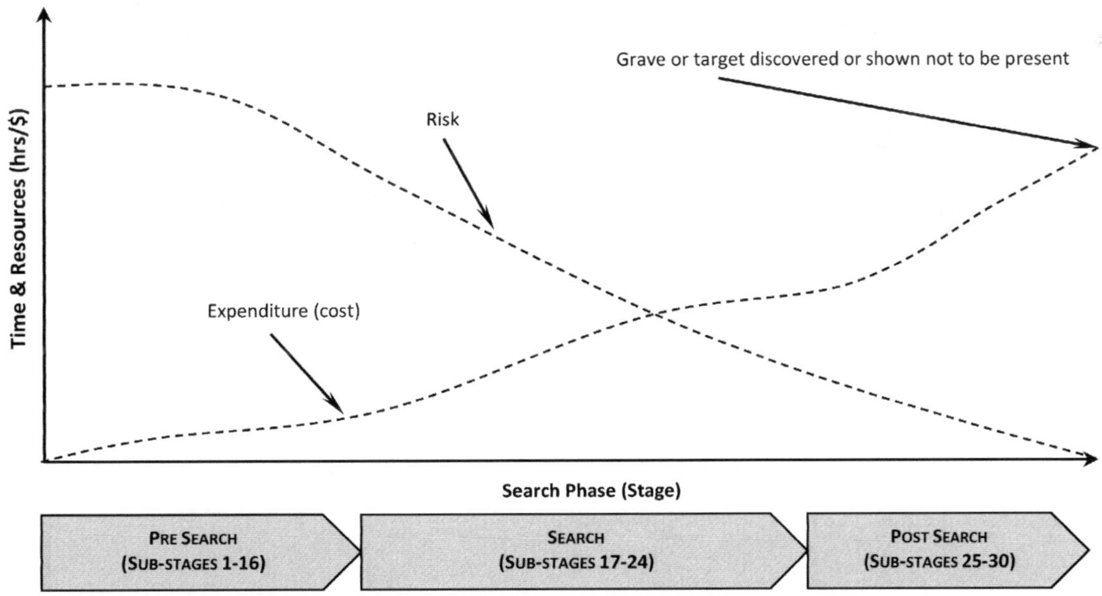

Figure 2. Idealised forensic geology search strategy for an open area search (modified after Donnelly 2002, Harrison and Donnelly 2008)

Figure 3. Example of the resources and personnel required for an open area ground search for a shallow, unmarked grave in the UK

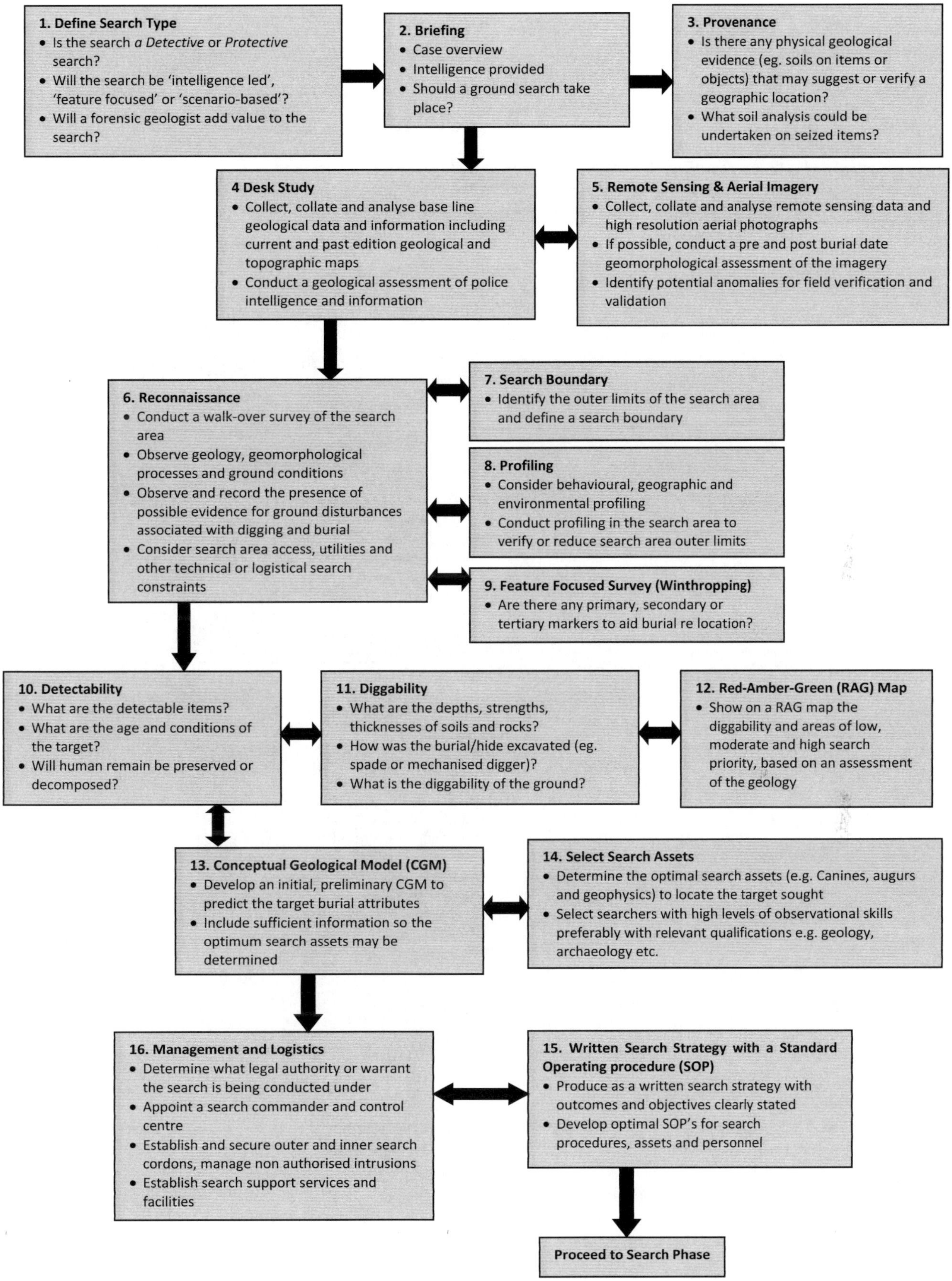

Figure 4. Pre Search

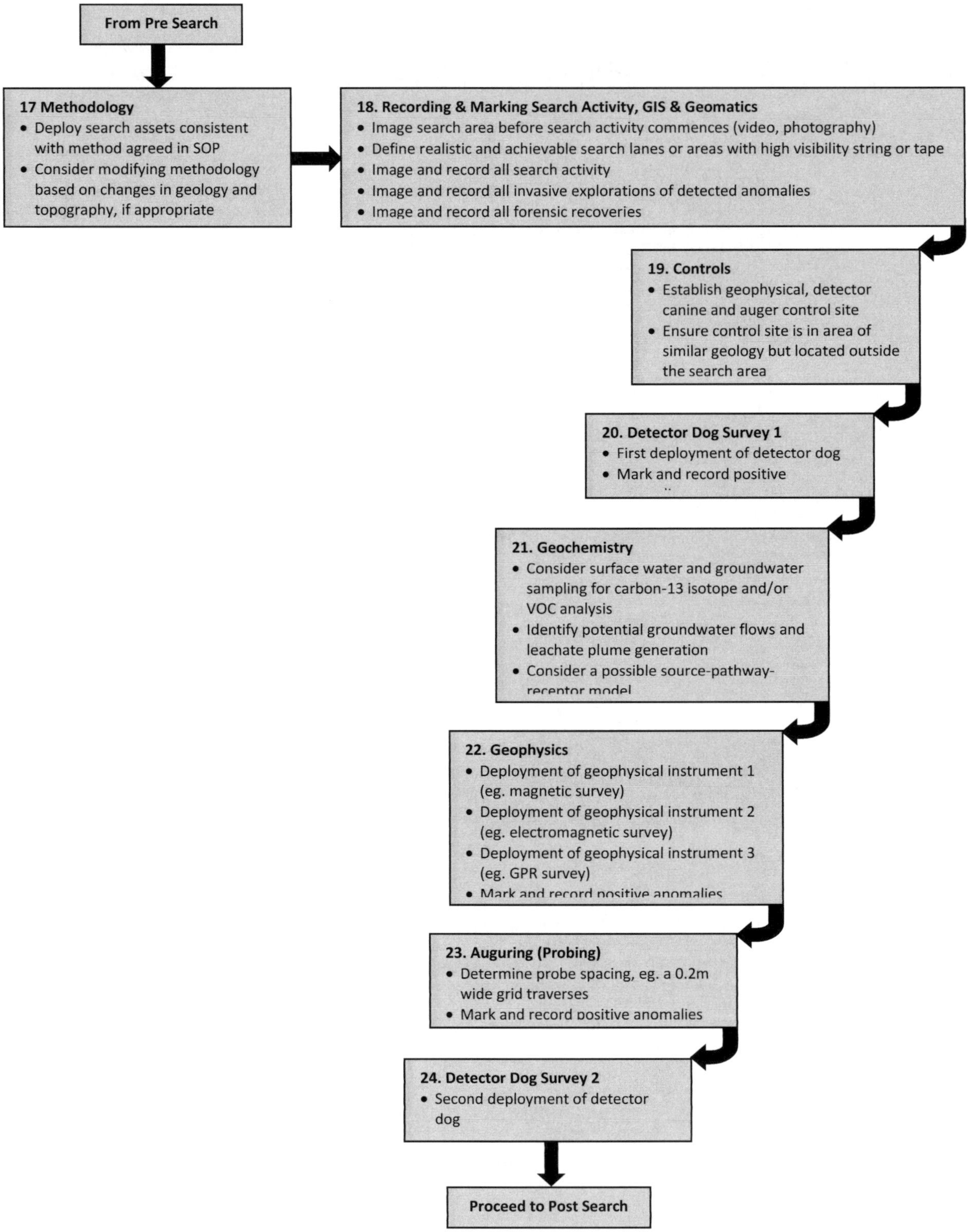

Figure 5. Search

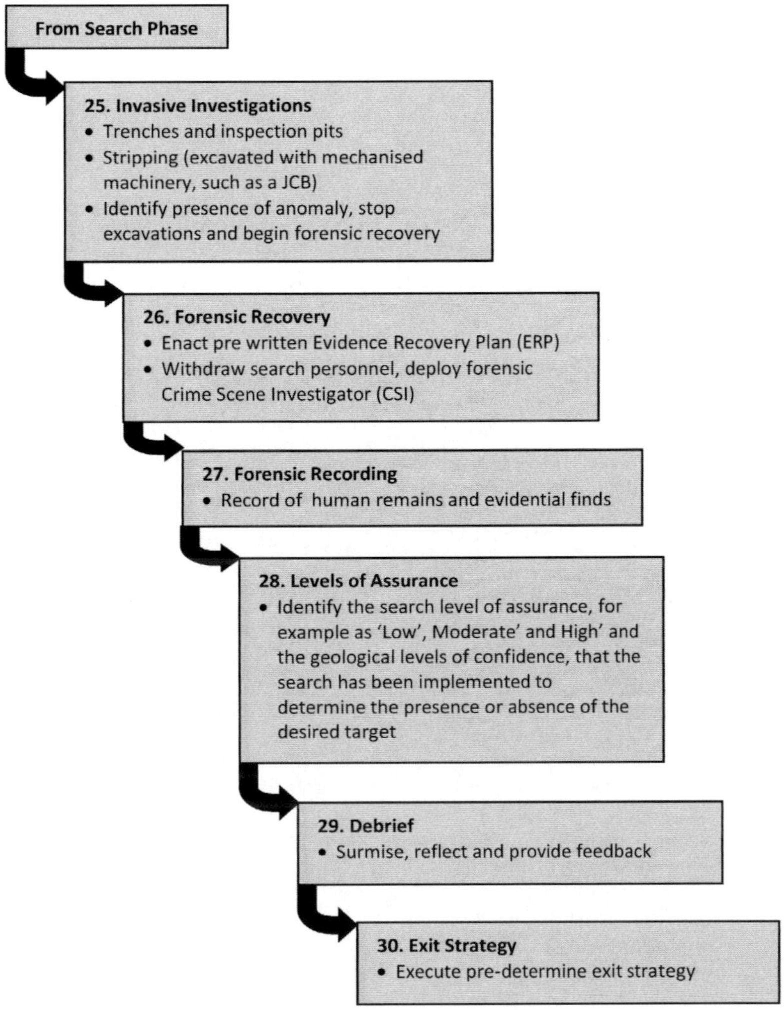

Figure 6. Post Search

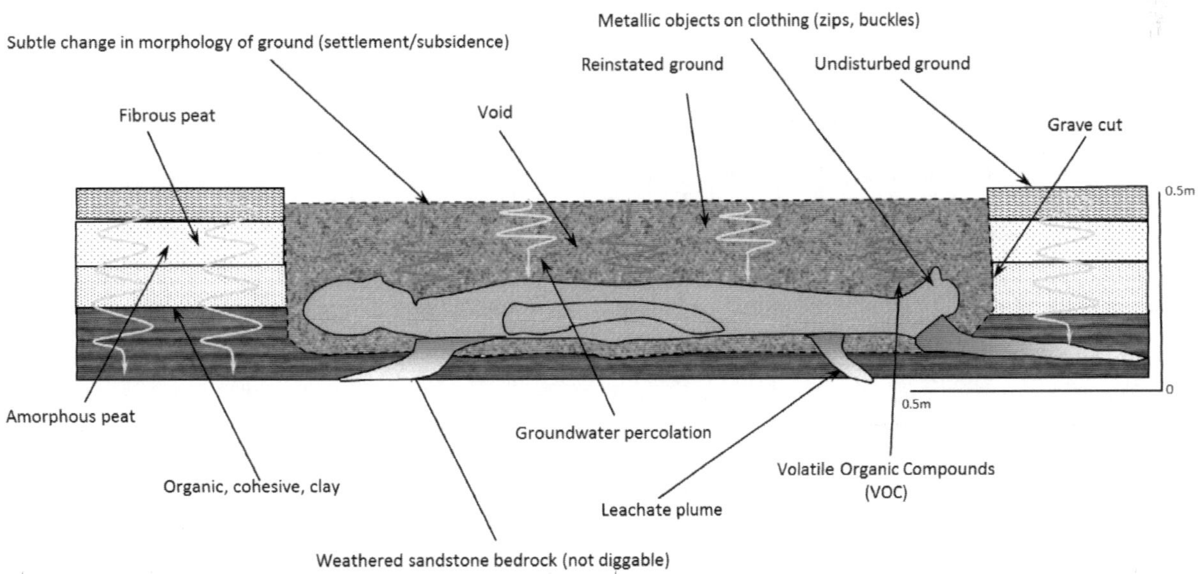

Figure 8. Conceptual Geological Model for a Geophysical Search for a Grave (after Donnelly 2002 in Donnelly & Harrison 2010)

Soil in Search and Evidence in the Case of Missing Person Pamela Jackson

Prof. Lorna Dawson BSc (Hons), PhD, CSci*
Lorna Dawson[1], Aelf Sampson[2], Adam McConochie[3], Karl Harrison[3], Gaille Mackinnon[3]
Ian Jefferies[4], Laurance Donnelly[5]

[1]The James Hutton Institute, Aberdeen, (lorna.dawson@hutton.ac.uk)
[2]Durham Constabulary, Durham Constabulary, Aykley Heads, Durham,
[3]Alecto Forensic Services Ltd, Preston,
[4]South Yorkshire Police, College Road, Doncaster, South Yorks, &
[5]Worley Parsons, Moorfield Court, 11a Alma Road, Headingley Leeds.

Summary

Soil can be a very useful technique in both determining provenance in a law enforcement/police search and also as physical evidence to determine if an offender or item was associated with a crime or location (Dawson and Hillier, 2010). This paper illustrates an example where soil information assisted in the search for a grave and also provided evidence that was presented in court. Operation Sorrento is also a good example of where there was effective team work, and illustrates interdisciplinary integration, which included police investigators, police search teams, a forensic geology ground search specialist, forensic service providers and forensic soil scientists, all with particular complementary expertise. Operation Sorrento was a large scale, high profile 'No-body Murder' investigation which lasted over several months, in the winter of 2013, stretching from County Durham to West Yorkshire, in England, UK.

Introduction

During the night of Thursday 7th March, 2013, the sons of Pamela Jackson called the Police reporting that they had not seen or heard of their mother since Saturday 2nd March 2013. Initial enquiries revealed that the last person to have seen her was Pamela's boyfriend, Adrian Muir. Due to some concerns from a comparison of Muir's initial account to the Police with the account he had given to Pamela's family, a Homicide investigation was quickly commenced, running alongside a Missing Person enquiry (MisPER). This developed through the phases of arrest, painstaking search, recovery of evidence and, in particular, the eventual recovery of the remains of the victim Pamela Jackson, analysis and interpretation, to full trial and ultimately conviction.

Search

Passive data from Muir's iPhone indicated a large area of hills which the suspect had visited in the times between Pamela's last sighting and the report of her being missing from home. The search and passive data strategies were intrinsically linked with other pieces of work ongoing, including for example Closed Circuit Television (CCTV), forensic, 'house to house' and witness interviews.

There was significant time pressure on the investigation team as early Crown Prosecution Service (CPS) charge authority had been sought and granted for a murder charge on Muir only 3 days after Pamela's disappearance. The search for Pamela Jackson was one of the largest search operations ever undertaken by Durham Constabulary, supported by specialist search resources from West Yorkshire, South Yorkshire Police, Greater Manchester Police, CAST (UK Home Office, Centre for Applied Science and Technology) and other underwater assets being among the specialist teams deployed. The objective of the search was to find Pamela's body and/or evidence relating to it. A preliminary conceptual appreciation of the geology was suggested for the suspected burial site located in the central Pennines of northern England. Here, the geology is dominated by coarse, strong, well-jointed and cross-bedded sandstones of Kinderscout Grit, of Namurian age. These were interbedded with weaker shales and mudrocks that have undergone weathering and erosion to form distinct escarpments, rolling hills in a moorland topography setting. These are covered by a veneer of peat, which has accumulated over the past c10,000 years. The peat may reach up to about 2-3m depth in places. In the context of a burial site, the distinct weathering of the slopes and rock exposures potentially enables navigation across the moorland landscape to facilitate the location of a grave. The peat is diggable with ease and a shallow grave may be dug and the peat/sub-soils and the displaced soils reinstated in less than five minutes, as determined in other search areas in this same range of hills (Donnelly and Harrison, 2013). Geophysical techniques potentially deployable in these type of geological settings were advised to include ground penetrating radar, magnetic and electromagnetic methods.

Operation Sorrento presented particular challenges with regard to the scope of the potential area of search. While mobile phone evidence suggested an upland location close to Cragg Vale, West Yorkshire, other intelligence suggested the possibility of a link with a reservoir site. Consequently a reconnaissance report was prepared by the forensic scientists that considered geological characteristics and geomorphological features across a number of target areas in the Halifax area. This also included control locations close to the suspect's address in Halifax, West Yorkshire, and the victim's address in Chester-le-Street, County Durham, including an area close to Turvin clough (Plate 1). In addition, the search report identified key exhibits taken from the suspect's car that might retain linking soil traces (i.e. gloves; Plate 2).

Production of the reconnaissance report was followed by targeted soil sampling at sites supported by both case-specific intelligence and generalised models of body deposition behaviour. Target landscapes were characterised by a variety of potential contact materials (as is found on the moorland around Turvin Clough, which featured highly organic upland peat bog overlying deposits of Kinderscout Grit).

Ultimately, intelligence-led targeting of soil samples helped provide a focus for searching efforts that led to the discovery of the remains of Pamela Jackson in a clandestine grave (Plate 1). Field observations by a skilled member of South Yorkshire Police's Victim Recovery Dog Team, near to the roadside, fine-tuned the geographic position of the grave. The location highlighted was a layby, which offered easy access to the moors via a wooden gate. On the right hand side of the gate was a natural steam flowing towards the roadside. The moorland was open and directly accessible on foot and had little natural visual protection from the roadside. The deposition site was dug into the banking of the stream further into the moor about 700m away from the parking area. The location of the deposition site on the stream bank gave concealment from the road side allowing Muir to cut into the banking and conceal the body without being compromised in his actions by any passers' by.

Using a dual soil analysis approach, by analysing both the inorganic (mineralogy) and the organic characteristics (hydrocarbon and alcohol fractions) of the soil, we were able to exclude certain areas as the source of the material found on the gloves from the boot of Muir's vehicle (home area, other areas identified as Muir having been parked during the initial period Pamela went missing). We were also able to home in on the area, seen in Plate 1, as being the location that most matched the characteristics of the material on Muir's gloves (Plate 2) found in the boot of his car. In parallel to this, field search teams identified anomalies in the edge of the road that indicated disturbance with unusually Kinderscout Grit sandstone near the soil surface. The Victim Recovery Dog (VRD) Handler noticed rocks, clay and soil, which had been deposited in the stream near to the site. On examining the distribution of the clay and boulders these were spread over a short distance, in an arc shaped pattern as if thrown from one location. He also noticed that an elongate rock appeared to have been moved from a patch of moss less than a metre away (it appeared to have different weathering patterns to others in the area and no lyceum was present). The grass and foliage around the boulder appeared to be relatively dry, compared to the surrounding foliage. The officer then took an auger sample near to the site (without interfering with the site) and noticed that the core was disrupted with subsoil located where the topsoil should have been (Plate 3). He then put his dog over the area and got a positive indication.

From the soil organic marker analysis it is clear that of the reference samples collected and analysed, sample KSH18 (300m west of collection point 3 Soyland Moor) was very similar to that sample collected from the glove (KSH42, Brushings from outer palmar surface of left glove LB/11). Furthermore, the organic marker characteristics (alkanes, alcohols, aldehydes, ketones) of the sample from the glove differed greatly from the other reference samples (KSH13, KSH24, KSH26 and KSH27). On the basis of soil mineralogy, sample locations KSH13 (Soyland Moor collection point 1) and KSH24 (Ogden Water collection point 8) were excluded as locations for the source of the material examined from glove LB/11. Also, there were also clear differences in mineralogy between both sample locations KSH26 (Booth Wood Reservoir collection point 9) and KSH27 (victim's house Chester-le-Street); these two locations were subsequently excluded as being the source of the material on the gloves found in the boot of Muir's vehicle and was eliminated from the search as being the source of soil on the gloves. Sample location KSH18 (Soyland Moor 'water' sample collection point 3) was the location most similar in terms of mineralogy and organic characteristics to the soil on glove LB/11. Pamela Jackson's body was found within 700m of this spot by police dogs once the area was further narrowed down through field observations. Soil samples were then collected at the grave, including contact point samples where the perpetrator would have stood and would have likely dug. The suspect vehicle was forensically sampled and traces of soil were found in the foot wells of Muir's car. These evidential soil samples were analysed in the same way as were the samples examined in the search phase of this operation.

Evidence

In court, under the UK adversarial system, with a judge and a jury, the strength of evidence that the soil like material found on the gloves originated from the grave site location was moderately strong for both organic and mineral soil characteristics. This evidence suggests that the person who was wearing the gloves (LB/11) (Adrian Muir) had been in contact with an upland moorland environment in northern England, as was characterised at the grave site location. In addition, the strength of the evidence that the soil like material found in some of the foot well mats (FNS and FOS) of the suspect's vehicle originated from both the organic and lower mineral horizons of the grave site location is moderately strong for organic characteristics. Signs of mottling and deciduous Molinia grass also provide additional common characteristics between the soil from the vehicle and the grave site location.

The ecological evidence suggested that there was transfer of soil to the suspect's vehicle from an upland moorland environment as was characterised at the grave site location. In addition, evidence was presented on Muir's fingerprint transferred to wrappings on flowers found within the grave, and CCTV imagery and phone records linked to the dates and times of Pamela's disappearance.

Muir was found guilty of manslaughter in Newcastle Crown Court on 30th September 2013, by the jury, and was sentenced to 18 years in prison. It is a crime that had to be proven, with the accused, Adrian Muir doing everything within his power to frustrate the investigation and discovery of Pamela's remains. He continues to deny that he is responsible for her killing and burial.

Conclusions

This case illustrates how soil analysis can be used effectively in both assisting in search for burials and in providing physical evidence in court. It also demonstrates the importance of team work and the benefit of cross disciplinary collaboration from the search phase to the evaluation of the evidence.

References

Dawson, L.A. & Hillier, S. 2009. Measurement of soil characteristics for forensic applications. SURFACE AND INTERFACE ANALYSIS, **42**, 363-377.

Donnelly, L. J. & Harrison, M. 2013. Geomorphological and Geoforensic Interpretation of Maps, Aerial Imagery, Conditions of Diggability and the Colour Coded RAG Prioritisation System in Searches for Criminal Burials. *In:* Pirrie, D., Ruffell, A. R. & Dawson, L. (2013). Environmental and Criminal Geoforensics. Geological Society of London, Special Publication, 384, 173-194.

Plate 1. Map of general search area focussing in on Tuvin Clough/Blackstone Edge Road.

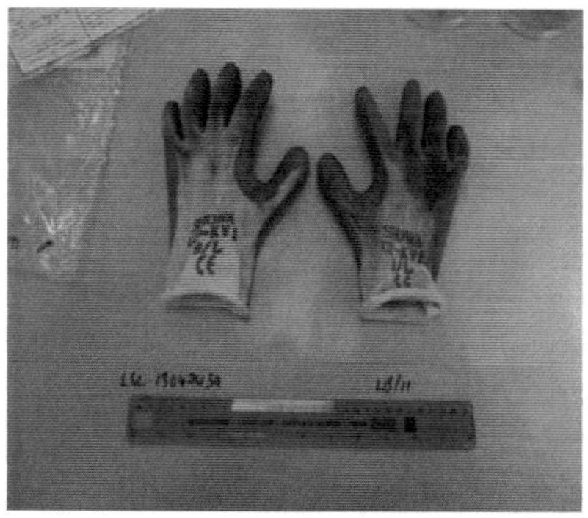

Plate 2. Gloves found in the boot of suspect Adrian Muir's vehicle. Back of right glove (LB/11).

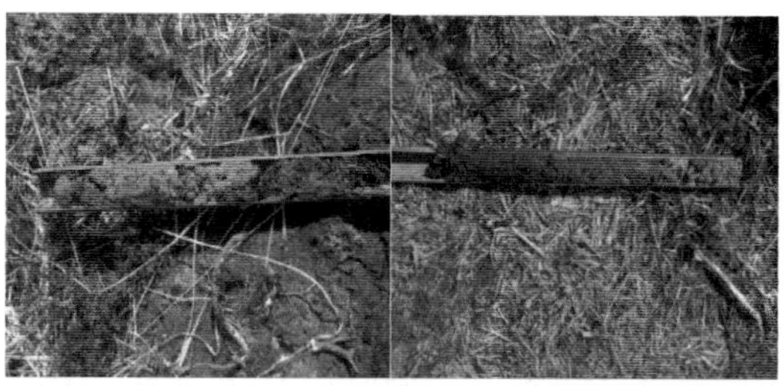

Plate 3. Soil cores showing: soil Sample taken from around 1 meter away from the deposition site. Soil Sample taken from the edge of the deposition site.

Site investigations using resistivity tomography: case studies from civil engineering and archeological examples

Farhat Ben M'Barek[1], Wajdi Belkhiria[1], Mohamed Dhaoui[1], Wissem Marzouki[1]*
[1]Office National des Mines (ONM, Tunisia)

Summary

The use of geophysical methods aims to determine the characteristics of the different rocks in order to understand their shape and their lithology. The fields of application of these methods are very extensive. This implies that the method we will choose for any study should be properly adapted to the problem. The team of Geophysics of the National Office of Mines carried out several studies in the areas of civil engineering and archeology using the electrical method.

Introduction

Geophysical prospecting using Electrical Tomography (ET) represents the most important methods used in site investigations applied to civil engineering and archeology (Barker, 1997). Indeed, this method yields significant insights in terms of the resistivity distribution for the subsurface studies.

The National Office of Mines plays an important role in integrated site investigations based on geological and geophysical studies.

Here we present three case studies from different locations in Tunisia. Each case has a particular objective and survey design.

Methodology

In this study, electrical tomography has been adopted using the dipole-dipole array and the suitable design: electrode spacing (a) and the investigation factor (n).

Siliana college site investigation using resistivity tomography

A resistivity surveying using electrical tomography was carried out in the site reserved to the construction of Siliana College (central-west of Tunisia).

This survey includes 7 parallel profiles with an inter-electrode spacing of 5 m and an investigation factor n=8.

This profile passes through the pressiometric test SP-3 (Fig). It shows the existence of a horizon characterized by large resistivities (40-150 Ω.m) reflecting the response of a heterogeneous and poorly compacted filling whith a thickness varying from 2 to 3.5 m. Under this horizon, we detect a medium resistivity layer (20-40 Ω.m) corresponding to the carbonate crust and tuff. These two levels seal a conductor substratum (resistivity <20 Ω.m) that corresponds to a clay formation. The morphology of the first layer shows that the extraction depth of the limestone crust is variable and followed by a filling.

Figure represents the resistivity profile with the corresponding geological interpretation

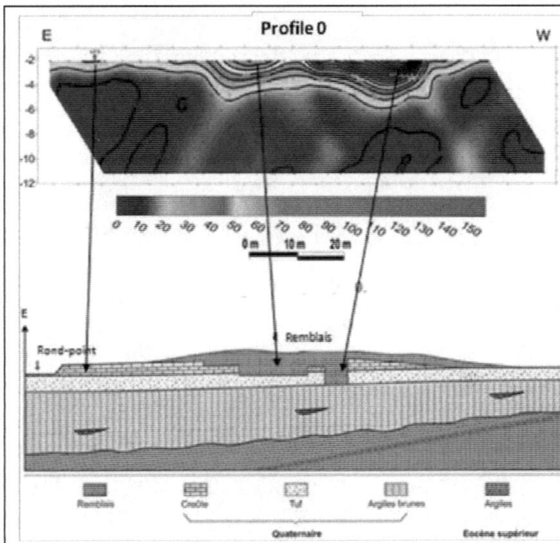

Figure 1: Profile 0 showing filling of the limestone exploitation.

Site investigation using resistivity tomography to understand construction problems: case of Al Farabi secondary school (Tataouine, south of Tunisia)

The second case consists of a resistivity survey conducted in the Al Farabi secondary school (south of Tunisia).

The prospecting of college buildings revealed a network of cracks affecting both the classes' building and the surrounding wall. Indeed, one can observe windows and doors do not close due to the deformation (Figure 9.3 and 9.4). The cracks to buildings and fence are oriented towards the toilets (which occupy the central part of the college) and have an increasingly high dip (Figure 9.1)

Figure 2: different problems affecting the building and the surrounding wall.

This profile has the same direction as the profile 1 is executed with a dipole a = 10m.

The resistivity contrast detected in this profile illustrates the heterogeneity of the bedrock on which the Al Farabi secondary school rests. Higher values of resistivity mark the response of a gypsum limestone whereas low resistivity values (7-18 Ω.m) are related to clay layers.

The bedrock is sealed with layers having low to medium resistivity (10-25 Ω.m) corresponding to red siltstones and conglomerates of Quaternary terraces and filling. The effect the water saturation is shallow and does not exceed 7 m.

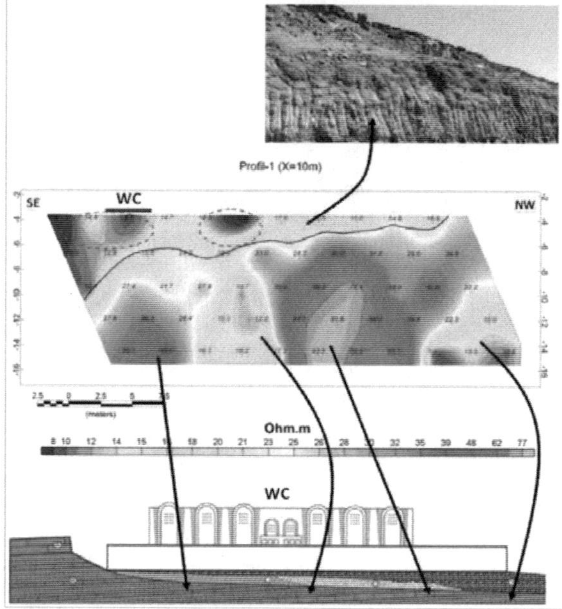

Figure 3: Profile1 and it geological interpretation showing clay limestone intercalations.

Role of geophysics when civil engineering projects meet great archeological sites: case of the tunnel project under the romaine aqueduct ELHNAIA

The fast development of infrastructure projects in a country of great history like Tunisia impose sometimes that civil engineering projects can encounter important archeological sites. The present project concerned the investigation of the passage of a tunnel under the romaine aqueduct ELHNAIA using the electrical tomography. The main objective of this investigation is to characterize the anchor depth of the ELHNAIA foundation and to understand if each foundation is isolated or not. Furthermore, we are interested by the lithology and the depth of the substratum.

This profile (Figure 4) is executed with a dipole length of 2.5m (Figure). At the dipoles 3 and 4 high resistivity values (95-100 Ωm) mark the clear presence of the anchor at a depth ranging between 1m and 1.8m. The underlying layer is composed essentially of a mixture of coarse sands and clays paste. Anchoring rests directly on bedrock very low resistivity (3-8 Ωm) related to the argillaceous substratum.

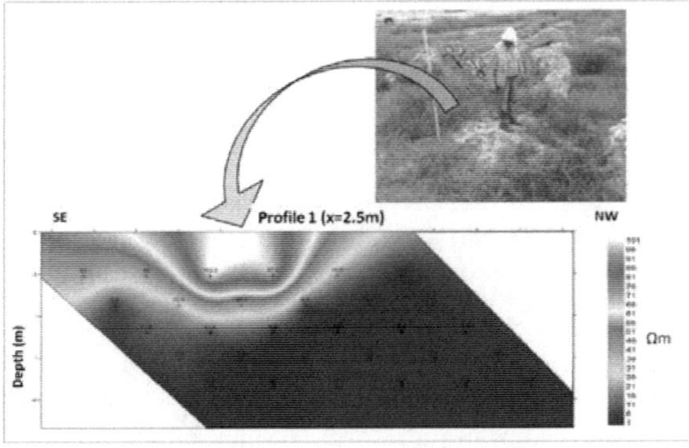

Figure 4: Resistivity profile showing the Depth of the Aqueduct foundation.

This profile (Figure 5) is parallel to the aqueduct in its eastern part and chose to guide the excavation of the tunnel. The investigation depth varies from 0 to 7.5m. characterized by the subdivision of resistivity values at two electric units. The first thin unit does not exceed 2m and exhibit average low resistivity values varying from 6 to 15 Ωm. It indicates a sandy clay facies in the uppermost part of the ground. This unit is based directly on homogeneous clays showing low resistivity (2 to 5 Ωm) confirmed by the geotechnical drilling and geological studies.

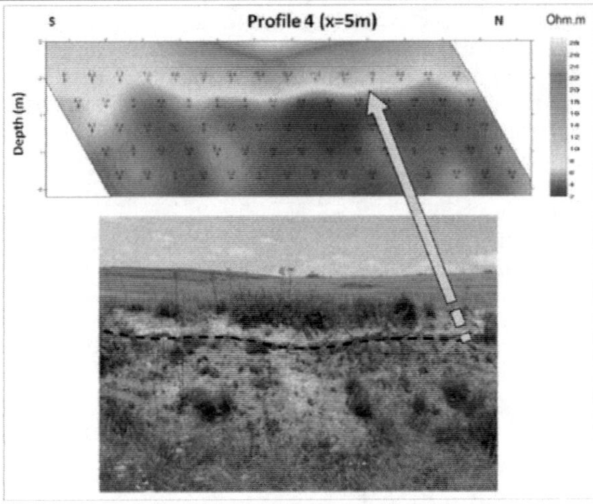

Figure 5: Lithology and depth of the bedrock.

Conclusions

The engineering site investigation using electrical tomography calibrated with the geological observation becomes an important step prior to construction. The analysis of a number of cases studies demonstrate that resistivity surveys are of great importance in site investigations.

As a non-destructive method, it helps to determine the depth to the bedrock for foundation purposes and tunnel and underground excavations. It also provides information on the degree of saturation of subsurface materials.

References

Barker, R.D. 1997. Electrical imaging and its application in engineering investigations. Pp. 37-44, in McCann, D.M., Eddleston, M., Fenning, P.J., and Reeves, G.M., (Editors), Modern Geophysics in Engineering GeologyThe Geological Society, London

Improving Target Identification and Characterization From GPR Data Using Isosurface Renders Technique

Mahmoud Gaballah[(1),(2)*]*; Deen Goodman*[(3)]*; Magdy Atya*[(1)]*; and Hany Salah*[(1)]
[(1)]*National Research Institute of Astronomy and Geophysics (NRIAG);11421 Helwan, Cairo, Egypt, Affiliation ID: 60030681.*
[(2)]*Now with College of Science and Humanities in Hawtet Sudair, Majmaah University, K.S.A.*
[(3)] *Geophysical Archaeometry Laboratory Inc., 20014 Gypsy, Woodland Hills, California, U.S.A.*

Summary

Ground–penetrating radar (GPR) is a high efficient and high–resolution geophysical technique used to investigate archaeological and engineering sites in a non-destructive manner. To effectively analyze GPR data sets, their combination with modern visualization techniques (e.g., isosurface rendering) has been acknowledged to facilitate interpretation beyond classical time slice analysis. We introduce two case study, in which we focus on the application of isosurface rendering to emphasize temporal and spatial variations within GPR archaeological data cubes and engineering problems. First case study main goal was to localize and potentially characterize cut-off tombs inside ferrous sandstone situated at Al-Nadura Temple, Al-Kharga City, New Valley. In the second case study we used GPR technique to evaluate the potentiality of this near surface geophysical method to image out the fracture system in a limestone plateau at Al-Mokattam site, Cairo, Egypt. The two case study GPR data were acquired with GSSI SIR 2000 system using 400 MHz shielded antenna and inline spacing 0.5m in both x and y direction. By comparing the visualized isosurface data to the conventionally processed time-slice, we demonstrate the superior interpretability of the isosurface technique for target identification and characterization. Based on these two case study, we illustrate the potential of such isosurface technique analyses towards a more comprehensive GPR data interpretation and target identification.

KEY WORDS: GPR, time-slice, isosurface renders, archaeology, Al-Mokattam, engeneering geophysics, fracture.

Introduction

GPR has become a well-known geophysical method for non-destructive archaeological exploration (e.g., Goodman and Nishimura, 1993; Gaballah et al, 2008). The GPR resolution capability (depending on frequency and soil properties), by far GPR resolution is greater than the resolution obtained by other geophysical methods, makes this technique suitable for high-resolution shallow studies like archaeological and near surface stratigraphy mapping (Nuzzo et al., 2002).

Typical 2D GPR profiles are characterized by discontinuity and complex reflection patterns with strong amplitude variations, which make the interpretation of this 2D profiles sometimes difficult. Therefore, 3D GPR data acquisition, and processing, seems to be easy technique to interpret and understand complex subsurface structures typically found at archaeological and engineering sites.

The use of 3D visualization technique is really important for different application in order to display the complex data in an easily understandable way. The most common technique to display 3D radar data is in time slice (or depth slice) maps (Conyers and Goodman, 1997). Horizontal slices may not be suitable visualization technique in the case of great subsurface complexity. For example, false amplitude anomalies can occur when the slicing planes cross dipping or undulating reflectors.

3D isosurface technique is a very powerful method in geological interpretation of 3D GPR data. The 3D amplitude isosurface rendering displays amplitudes of equal values in the GPR study volume. Shading is usually used to illuminate these surfaces, giving the appearance of real geological or archaeological structure in subsurface. The threshold amplitude setting is crucial for obtaining useful results from isosurface method. Values of about 60-75 of the maximum amplitude value usually provide the best results. However, we have particularly studied every case area (Al-Nadura site and Al-Mokattam site) separately.

Illumination in isosurface maps is produced by calculation to emphasize the features contained in the data volume. The color of an individual pixel on the screen is computed by combining contributions from each particle intersecting the ray. This allows the viewing of heterogeneity within objects with appropriate adjustment of the opacity.

In this paper, an application of 3D visualization method using isosurface renders technique are presented along with comparison to conventional time slice technique at two different sites ("Al-Nadura" archaeological site and "Al-Mokattam" geological site).

Site Description and Data Acquisition

The first case study was Al-Nadura temple site (Figure 1) which is one of the fortresses that had been built by the roman to secure the handle along Darb Al-Arbain route (Darb Al-Arbain is defined as the salver's caravan route for the handle between Assuit at the Nile River and Sudan). Al-Nadura site is located just to the north of El-Kharga city and was built by the Romans either during the rule of Antonius Pius or Hadrian. The temple may have been dedicated to the spouse of Amun, and it is in ruin. This temple is about 700 years younger than Hibis Temple, and belongs to the 2nd century A.D.. During the roman period, many fortresses had been constructed to secure this route (e.g. Um El-Dabadeb, Al-Nadura, and Al-Zayyan temples) (Michel, 1999).

The second case study was Al-Mokattam site in Cairo (Figure 2) which consists of three large limestone plateaus at the east edge of Cairo; Al-Mokattam plateau represents a serious dangerous problem especially at the borders, where at the southern part area there have been many landslides due to the fractured limestone (Moustafa et al., 1991). These landslides caused damage to many houses and hotels around the area.

GPR data were acquired using SIR 2000 from GSSI attached with 400MHz shielded antenna, providing a good compromise between resolution and depth penetration. Data were obtained using odometer survey wheel with 0.5m inline spacing in a zig-zag form direction. GPR scan traces were recorded at every 2 cm with 512 sample/scan using sampling interval 0.15625ns producing a two-way travel time with good reflection and penetration to about 80 ns. The scan area was about 52m x 42m for Al-Nadura site while the area for Al-Mokattam site was 22m x 8m; For these two case studies GPR data were acquired in both x and y-direction. A total of 190 profiles in x and y-direction were acquires at the site "Al-Nadura", while a total of 62 adjacent profiles were collected at the site " Al-Mokattam" with 0.5m inline spacing at the two sites.

GPR Standard Data Processing and Visualization.

All GPR dataset acquired during site surveying were processed in the same way using a standard processing sequence. These processing sequence include: (1) The zero time is adjusted by delaying the first break by the time it takes the direct airwave to travel from transmitter to receiver. (2) The ''dewow'' [Annan, 2003] step removes very low frequency components of the data associated with either inductive phenomena or analog/digital conversion. (3) Background Removal used to enhance the curved and dipping reflectors and to remove or attenuate the effect of horizontal reflectors. (4) Bandpass filter; amplitude spectrum were plotted and used to determine the signal frequency bandwidth for the 400MHz antenna. The frequency bands chosen for filtering were 120-220 and from 530-630. (5) Running trace averaging used to really smooth things up and make features more visible. (6) Velocity estimation were based on diffraction hyperbola analysis with ReflexW (Sandmeier Scientific Software).

Velocity of 0.1 m/ns was used in this paper to convert the two-way travel time section into depth information. Figure 3 shows typical 2D GPR vertical cross-section after data processing extracted from two different survey sites. For Al-Nadura archaeological site, high amplitude reflection in the 2D profiles probably correlated to the intercalation of ferrous material in sandstone. A lot of relatively large hyperbola anomalies cloud be observed in the data and we think that it might be related to the cut-off subsurface tombs. For Al-Mokattam limestone plateau, vertical fractures and irregular displacement features appearing as strong reflection patterns.

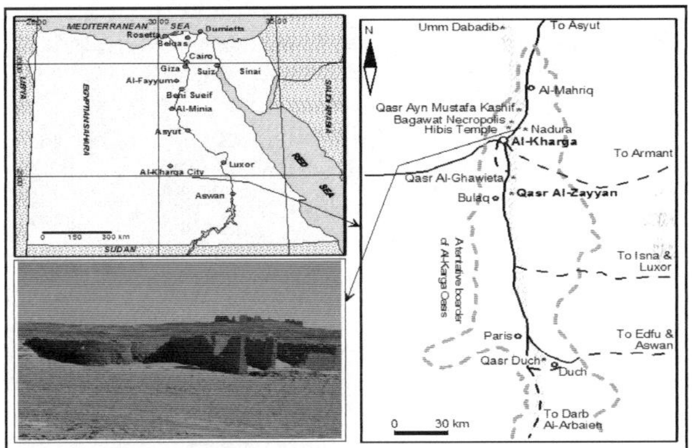

Figure 1. Location map for Al-Nadura temple survey site Al-Kharga Oasis, Egypt.

The GPR data were gridded on a regular grid using a natural neighbor routine and the 3D time-slices were plotted. GPR-SLICE (c) v7.0 (http://www.gpr-survey.com) was used for processing and visualization of the isosurface renders technique.

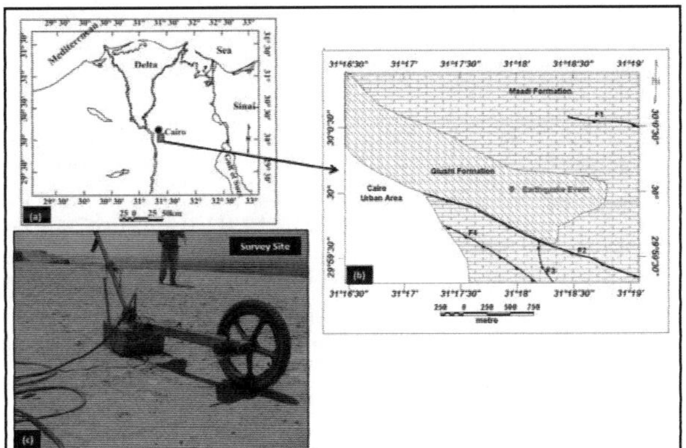

Figure 2. (a) Location map of Al-Mokattam survey site, Cairo, Egypt. (b) geologic map of the study area (modified after NARSS 1997 and Sultan 2010). (c) GPR surveying with 400 MHz antenna with odometer survey wheel.

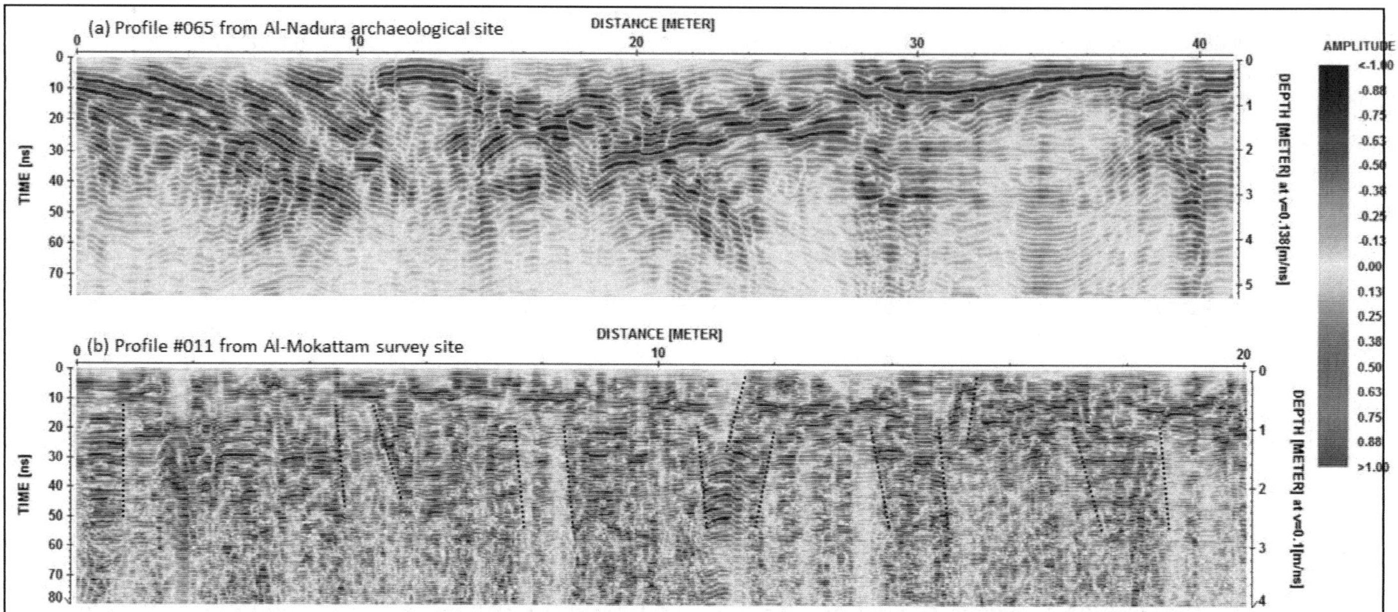

Figure 3. Example of 2D GPR profiles at two different sites (a) GPR data from Al-Nadura Site after signal processing and (b) Data from Al-Mokattam fractured plateau. basic signal processing includes: zero time-shift; background removal, bandpass filtering, and running average trace. (a) Strong reflection probably caused by the ferrous sandstone and relatively large hyperbola is related to cut-off tomb. (b) Vertical fractures and irregular displacement features appearing as strong reflection patterns.

Timeslices images are calculated by creating 2-D horizontal contour maps of the wave energy from a specified time value across parallel profiles, using GPR-SLICE (c) v7.0 Ground Penetrating Radar Imaging Software. Time slice data sets were generated by spatially averaging the squared wave amplitudes of radar reflections in the horizontal as well as the vertical. Data binning was chosen to generate averages that were 1-2 wavelenths in time, and about half the size of the profile spacing. The data were gridded using a Kriging algorithm that included a search of all data within a 0.75 m radius of the desired point to be interpolated on the grid. Using these spatial averages, interpolated and solid 3D volumes of reflections amplitudes were generated (Figure 4 and Figure 5)

Result and interpretation

The results are presented in 3D animation using real time Open GL graphic displays in which 3D time slice fence diagram are mixed together along with isosurface cubic volume from the 70% strongest reflector (Figure 4 & 5). Using spatially binned and interpolated volumes of radar data shows better continuity in mapping subsurface reflections, which are more useful and provide archaeologist and geologist an interpretable form of data presentation.

Figure 4 shows time-slice image at depths 2.8 including a some strong reflections form a prominent subsurface structures marked by focused arrows. After discussion with the archaeological expert in the site we believe that these anomalies are related to some new Roman tomb foundation. These tomb locations appeared in the 2D vertical cross-sections as a relatively large hyperbolic reflection from tomb edges (Figure 3a). It was accepted between archaeologists that there is a relation between the occurrences of the ferrous sandstone and the cut-off roman tombs.

Figure 5a shows the direction of the fractures in the x–y plane and the x–z plane, according to the isosurface 3D volume generated from the 70% strongest reflectors. Figure 5b interpretation overlay image shows how 3D isosurface can effectively characterized subsurface fracture planes. In addition, isosurface images not only shows the fracture direction but also it continuity. Geologist believes that most damages to old constructions and foundation may be due to swelling of the clay filling the fractured limestone and the existence of connected fracture in Al-Mokattam site.

Conclusions

Within the isosurface rendering technique (Figure 4& 5), the visibility of tombs location as well as subsurface fracture continuity, can be easily detected.

3D isosurface rendering procedure inherently enhanced the contribution of 3D features in comparison to that of conventional time-slice, so that the final image (Figures 4 & 5) is quite easily readable for the non-specialist to find target location and dimensions. Together with information from archaeology and geology of the site, the location, depth, size and vertical overlapping of the buried subsurface objects were effectively estimated from non-destructive GPR result

We have demonstrate that 3D visualization using advanced techniques (i.e. isosurface rendering technique), going beyond standard time-slice imaging flows, allowing for a comprehensive and more detailed interpretation for complex targets within GPR data sets. Visualizing such a subsurface target features helps to guide archaeologists and geologist for improving interpretation of a variety of GPR data sets and to locate the critical spots for excavation.

Acknowledgments

The authors would like to thank the Egyptian Supreme Council of Antiquities for giving us a permission to use Al-Nadura site for GPR survey. Thanks also extended to my staff at NRIAG for helping in field data measurments.

Refernce

Gaballah, M., Grasmueck, M. and Sato, M., 2008, High resolution 3D GPR applied to Archaeology for characterizing accurate Subsurface Structure", Proc. of the 119th of SEGJ conference, Tokyo, 157-160.

Goodman, D., and Nishimura, Y.,1993, Ground radar view of Japanese burial mounds. Antiquity 67 (255), 349-354.

Nuzzo, L., Leucci, G., Negri, S., Carrozzo, M.T., Quarta, T., 2002, Application of 3D visualization techniques in the analysis of GPR data for archaeology. Annals of Geophysics 45 (2), 321– 337.

GPR-SLICE (c) v7.0; http://www.gpr-survey.com/

Michel, R., 1999, Site militaries romaines de l'oasis de Kharga. Bulletin de l'Institut Fransais d'Archeologie Orientale 99, 1999, p. 377-396.

Annan, A. P., 2003, Ground Penetrating Radar Principles, Procedures and Applications. Mississauga, ON, Canada: Sensors and Software, 2003, p. 278.

Conyers, L., B., and Goodman, D., 1997, Ground-Penetrating Radar – An Introduction for Archaeologists (AltaMira Press, A Division of Sage Publications, Inc.), pp. 232.

NARSS, 1997, Geological and geomorphological studies of Al-Mokattam plateau, Geological Survey of Egypt, Internal Report in Arabic.

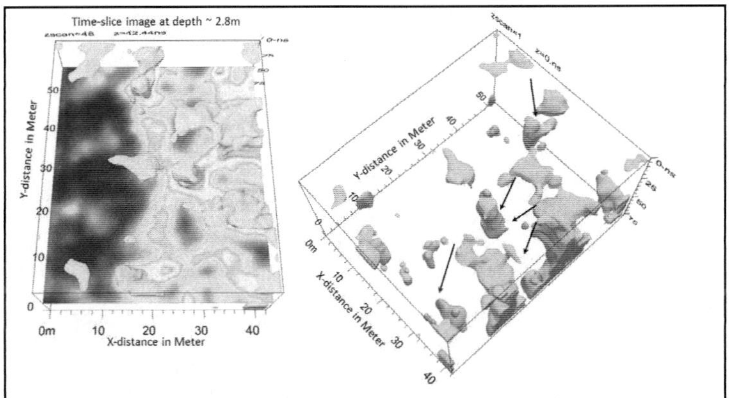

Figure 4. An image showing an isosurface renders generated from 70% strongest reflector overlaid with time slice image for Al-Nadura temple site

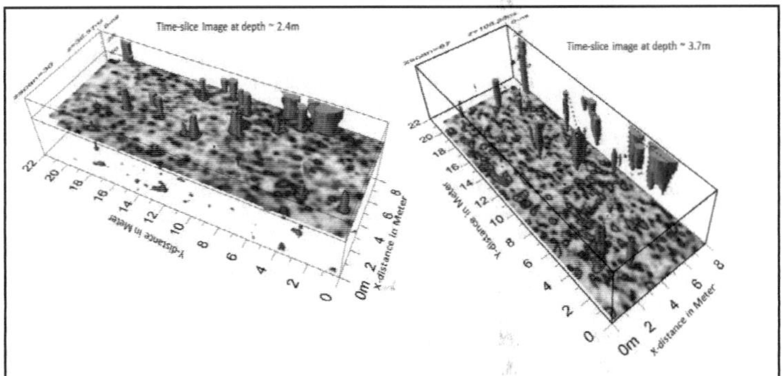

Figure 4. (a)An image showing an isosurface renders generated from 70% strongest reflector overlaid with time slice images for Al-Mokattam geologic site. (b) Interpretation overlay shows how 3D isosurface can effectively characterized subsurface fracture planes.

Sultan, A., 2010, Geophysical investigation for shallow subsurface geotechnical problems of Mokattam area, Cairo, Egypt. Environ Earth Sci (2010) 59:1195–1207.

Moustafa, A.,R., El-Nahhas, F., Abdel Twaab, S., 1991 Engineering geology of Mokattam city and vicinity, eastern Greater Cairo, Egypt. Eng Geol 31:327–344.

REFLEX Software. Sandmeier Scientific Software, Karlsruhe, Germany; 2012.

Exploring the Egyptian Alabaster at the Western Eocene Plateau of Assiut City, Egypt, using ground penetrating radar

Reem H. Sayed [1*], *Gamal Z. Abdel Aal* [1], *Mohamed O. Ibrahim* [1], *Hassan A. Soliman* [1], *Anna A. Klochko* [2] *and Dmitry Shishkov* [2]

[1] *Geology Department, Faculty of Science, Assiut University, Assiut, 71516, Egypt*

[2] *Geology Department, Lomonosov Moscow State University, Moscow, Russia*

Summary

Ground penetrating radar (GPR) is one of the principal geophysical imaging tools for near-surface investigations of physical properties and structure. Georadar survey was conducted in the vicinity of excavated Egyptian alabaster quarry at the western Eocene plateau of Assiut city, Egypt. The main objective of this study was to explore the extend and depth of the alabaster veins and the associated subsurface features (e.g., fractures and cavities). The surface of the study site is absolutely smooth and covered by sandy-gravel layer of Aeolian alluvium origin overlain the Eocene limestone. Several profiles were surveyed using the SIR-3000 GSSI GPR unit and the 200 MHz antenna mounted behind a vehicle. Radan software has been used to process the GPR data and several anomalies were identified. The 2D Georadar sections showed volumetric anomalies with an extremely intensive high amplitude reflection resulting from the alabaster bodies that are associated with different set of fractures and cavities occurred at depth ranges from 2 to 4 meter below the ground surface. A two-layer structure is reconstructed in 3D site simulation. The upper layer of the cross-section represents the fractured limestone up to depth about 3.5 m and characterized by clear and slightly sloping axes of synchronism phase, broken in some spots by area of finely disordered reflections. At depth greater than 4 m, a large branching out body (or group of bodies), creating intensive reflections is determined. It is clear that the results of this study are very promising and a detailed geophysical survey is required that will increase the potentiality of carrying out quarrying and extraction activities of the alabaster bodies at the study area.

Introduction

Ground penetrating radar is a nondestructive geophysical method that produces a continuous cross-sectional profile or record of subsurface features, without drilling, probing, or digging. Ground penetrating radar (GPR) profiles are used for evaluating the location and depth of buried objects and to investigate the presence and continuity of natural subsurface conditions and features. Early utilization of the method for engineering and soils applications as well as mining are given by Morey (1974), Cook (1973), Annan and Davis (1976), Coon *et al.* (1981), and Ulriksen (1982). An extensive overview of the method is given by Davis and Annan (1989).GPR technique was used to investigate the extent of the alabaster veins or bodies and the associated subsurface features (e.g., fractures and cavities) in the vicinity of excavated alabaster quarry at the western Eocene plateau of Assiut city, Egypt (Figure 1a). The alabaster bodies at the study site (Figure 1b) were formed due to hydrothermal mineralization of the Eocene carbonate sequence along fractures and surface faults. The high-quality marble onyx found here has been known under the local name "the Egyptian alabaster" from times of the Ancient Egypt till now and is developed as raw material. The study area is characterized by a flat surface with surficial sediments of sand and gravels (Figure 1c) overlain the Eocene limestone that is exposed at the surface of the western border of the study area (Mansour and Philobbos, 1983).

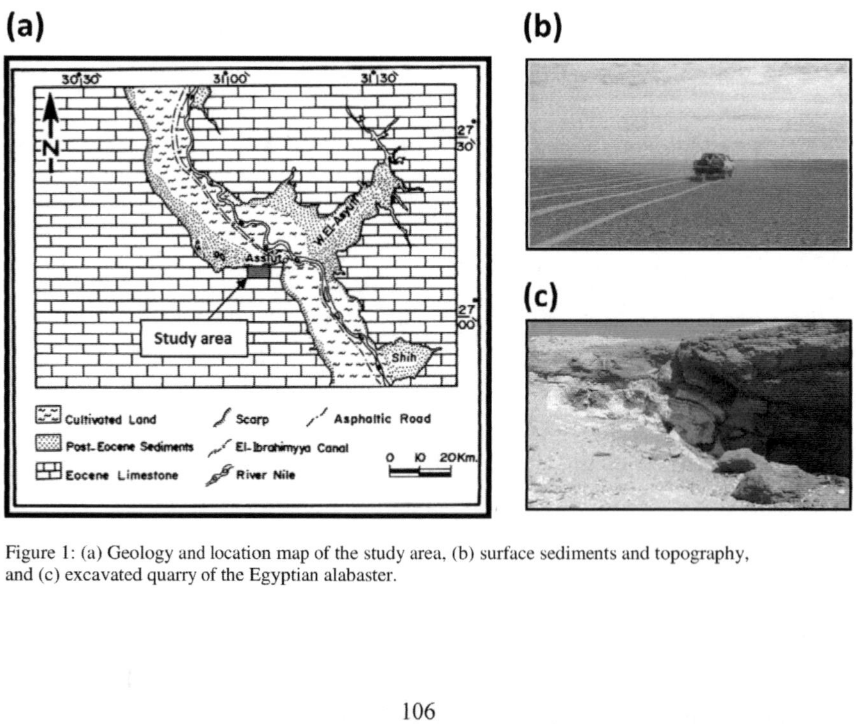

Figure 1: (a) Geology and location map of the study area, (b) surface sediments and topography, and (c) excavated quarry of the Egyptian alabaster.

GPR Principals

Ground penetrating radar operates by transmitting pulses of ultra high frequency radio waves (microwave electromagnetic energy) down into the ground through a transducer (also called an antenna). GPR transmits and receives up to 50 pulses per second. Transmitted wave is reflected at each layer interface, the amount of the energy reflected is primarily a function of moisture content and density. The captured data is an X-Y plot of amplitude against arrival time. To conveniently display many of these traces a color coding scheme is used to convert this into a color display (Figure 3). This is analogous to looking at an X-Ray of the subsurface.

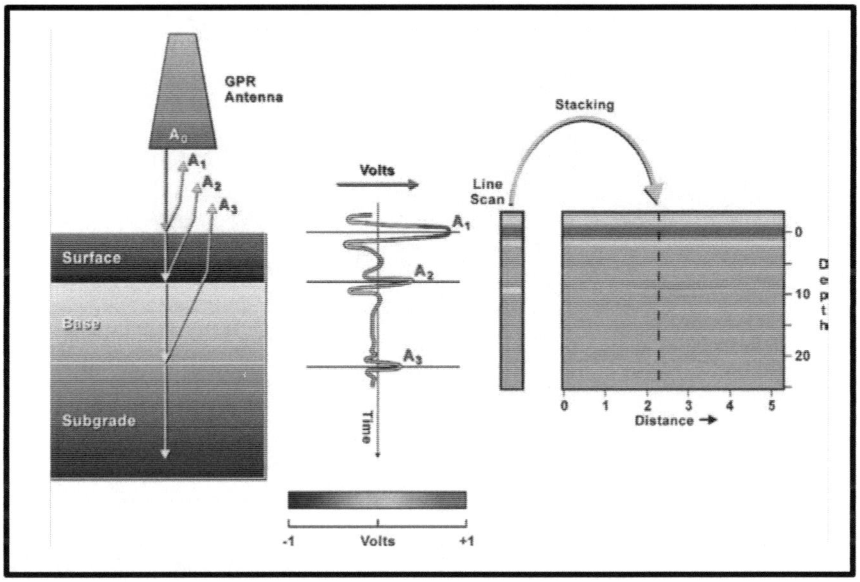

Figure 2: Principals of ground penetrating radar (GPR/Georadar)

Data acquisition and processing

GPR survey in the surrounding area of the Egyptian alabaster quarry was carried out by a method of autoprofiling using a 200 MHz antenna mounted behind a vehicle and the Georadar SIR 3000 GPR unit of the corporation Geophysical Survey Systems, Inc. (GSSI, USA). The GPR profiles (up to 500 m each) were crossing the tectonic mineralized zone of the alabaster from both sides of the quarry. A global positioning system (GPS) was used for accurate and precise location of the GPR profiles. GPR data were processed using the software of Radan 6. The upper part of the geological section was further analyzed by construction of digital three-dimensional (3D) model on the basis of the georadar data. The 3D model was created using the three-dimensional modules Quick Draw to enhance the data visualization and interpretation.

Results and discussion

GPR profile (Figure 3) shows the cross continuation of tectonic mineralized zone of the Egyptian Alabaster from both sides of one of the quarry. The cross-section is saturated with volumetric anomalies with an extremely intensive high amplitude reflection of electromagnetic signal. The largest of them represent areas of chaotic reflection by a diameter up to 20 m, finer ones form parabolic reflection. Probably, the anomaly corresponds to the large clastolites. Environment of clastolite is a mixtite matrix characterized by weakly contrasted reflection. There is no layering determined in the mixtite (Figure 3).

The two-layer structure is reconstructed in 3D site simulation (Figure 4). In the upper part of the cross-section, up to depth about 3.5 m the wave picture is determined on radargrams by quite clear slightly sloping axes of synchronism phase, broken in some spots by area of finely chaotic reflection which represent the upper georadar facies of layered fractured limestones. In this interval of the cross-section the areas of high amplitude reflection of submeridional strike, shifted to the east direction. Fractureness of the rocky massive is shown at the model from depth of 1 m. It corresponds closely to orthogonal system of weakly contrasted reflections that a little more intensive, than those on the background. Analysis of arrangement of fractureness system or even its diagnostics are impossible just by radargrams. On depth more than 4 m the large branching out body (or group of bodies), creating intensive reflections is determined (Figure 4).

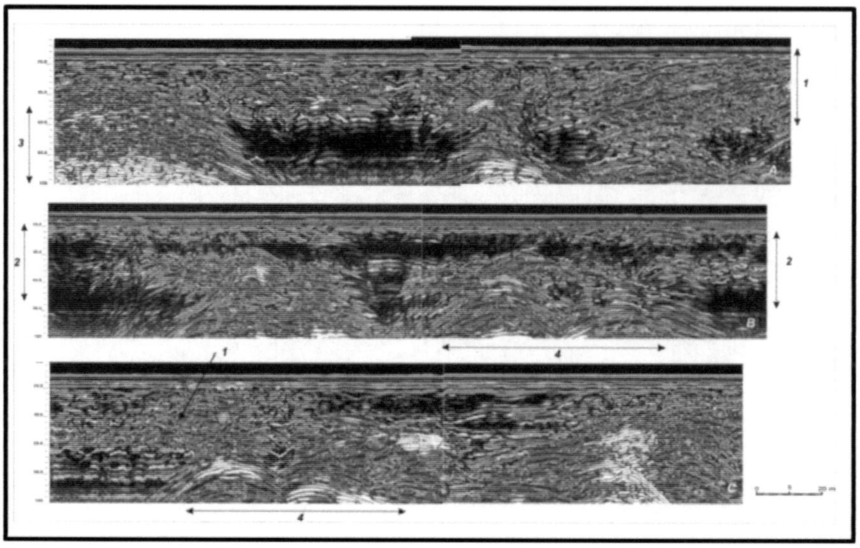

Figure 3: Example of GPR profiles crossing the buried mineralized body of alabaster. 1- chaotic reflection (fractured limestone); 2- Reflections of poor contrast (dense limestone); 3- zones with chaotic intensive reflections (areas of maximum tectonic rock disintegrations) and 4- areas with groups of contrast reflections (large blocks of alabaster

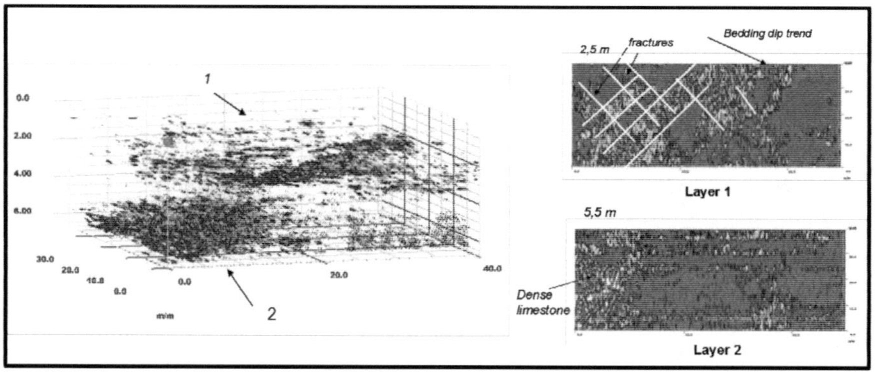

Figure 4: Example of the 3D GPR model at the study area showing two layers with strong reflections; 1- the highly fractured surface layer at a depth of 2.5 m and 2- the dense limestone layer at a depth of 5.5 m.

Conclusions

Ground penetrating radar (GPR) is one of the principal geophysical imaging tools for near-surface investigations of physical properties and structure. A successful application of ground penetrating radar (GPR) was achieved at the vicinity of excavated Egyptian alabaster quarry at the western Eocene plateau of Assiut city, Egypt. It was observed from the GPR data that the alabaster mineralization is associated with a set of subsurface fractures at a depth of 2.5 m from the ground surface. Despite the fact that the GPR results were very promising, detailed geological and geophysical survey is recommended for accurate delineation of the extend and depth of the alabaster mineralization at the study area. Such detailed study will increase the potentiality of carrying out quarrying and extraction activities of the alabaster bodies at the study area

Acknowledgements

This material is based upon work supported and conducted through a scientific and technical collaborative research between Assiut University and Lomonosov Moscow State University. The authors would like to thank Prof. Dr. Moustafa Youssef; Prof. Dr. Ahmed Reda; Prof. Dr. Hamza Ibrahim, demonstrator Happy Salah Eldin at the geology department, Assiut University and Yuriy Kocebinskiy at Moscow State University for their kindly support and help during the field work.

References

Annan, A.P. and Davis, J.L., 1976, Impulse Radar Soundings in Permafrost: RADIO SCIENCE, **11**, 383-394.

Cook, J.C., 1973, Radar Exploration Through Rock in Advance of Mining: TRANSACTION OF THE SOCIETY OF MINING ENGINEERS, AIME, **254**, 140-146.

Coon, J.B., Fowler, J.C. and Schafers, C.J., 1981, Experimental Uses of Short Pulse Radar in Coal Seams: GEOPHYSICS, **46** (8), 1163-1168.

Davis, J.L., and Annan, A.P., 1989, Ground Penetrating Radar for High-Resolution Mapping of Soil and Rock Stratigraphy: GEOPHYSICAL PROSPECTING, **37**, 531-551.

Ulriksen, C.P.F., 1982, Application of Impulse Radar to Civil Engineering: Unpublished Ph.D. Thesis, Dept. of Engr. Geolo., U. of Technology, Lund, Sweden.

Mansour, H.H. and Philobbos, E.R.(1983), Lithosratigrophic classification of the surface Eocene carbonates of the Nile Vally: EGYPTIAN BULLETIN FACULTY OF SCIENCE, ASSUIT UNIVERSITY, **12** (2), 129-153.

Applications of Engineering Seismology in Urban Areas

Oz Yilmaz*, Anatolian Geophysical

Summary

I present two case studies which describe applications of engineering seismology in urban areas (Yilmaz, 2013):

(1) High-resolution shallow offshore 3-D seismic survey for the Bosphorus Subsea Road Tunnel Project, and

(2) Seismic, geotechnical, and earthquake engineering site characterization for a residential project in Istanbul.

In Case Study 1, we conducted a 3-D high-resolution shallow seismic survey in 2011 to confirm the geometry of the future Bosphorus Subsea Road Tunnel traverse. We processed the data to obtain 3-D PSTM and PSDM volumes, and interpreted these image volumes and correlated the interpretation results with the offshore borehole data. With the recoverable signal band of 30-960 Hz, we attained a vertical resolution of less than 1 m. The image volume from 3-D PSTM exhibits the presence of an erosional channel associated with a Palaeo-Bosphorus waterway. The eastern (Asian side) slope of the channel can be delineated in many of the inline sections. Nevertheless, the western slope (European side) can only be faintly inferred from some of the inline sections. The reason is that the western slope is occupied by a zone of highly weathered Trakya bedrock debris with weak impedance contrast with the soil column above and within the channel. Prior to the flooding of the erosional channel by the Palaeo-Bosphorus, the river bed consisted of gravel. There are no faults that can be traced within the seismic image volume with potential capability to generate an earthquake of a magnitude (>6) that would be of concern in the tunnel design. The planned tunnel traverse is within the Trakya Formation except for a middle segment, 806 m in length, within the soil column composed of interbedded sand-clay-silt deposits with varying percentages of fines content.

In Case Study 2, we determined the seismic model of the soil column within a residential project site in Istanbul. To accomplish this, we conducted shallow seismic surveys at 20 locations and estimated the P- and S-wave velocity-depth profiles down to a depth of 30 m. We then combined the seismic velocities with the geotechnical borehole information regarding the lithology of the soil column and determined the site-specific geotechnical earthquake engineering parameters. Specifically, we computed the maximum soil amplification ratio, maximum surface-bedrock acceleration ratio, depth interval of significant acceleration, maximum soil-rock response ratio, and design spectrum periods *TA-TB*. Additionally, we conducted shallow reflection seismic and resistivity surveys to delineate the geometry of landslide surfaces with potential susceptibility to failure.

Case Study 1: High-resolution shallow offshore 3-D seismic survey for the Bosphorus Subsea Road Tunnel Project

A 1200J sparker source and a 24-channel mini-streamer were used for data acquisition. The recording parameters are listed in Table 1.

Table 1. The recording parameters for the Bosphorus high-resolution shallow 3-D seismic survey.

Number of channels per shot record	24
Number of hydrophones per group	6
Group interval	3.125 m
Shot intervval	3.125 m
Total number of shots in the survey	119,485
Total number of traces in the survey data	2,867,640
Sampling rate	0.25 ms
Maximum recording time	500 ms
Bin size	1.5 X 5 m
Extremities of the survey area in EW X NS directions	3.6 X 1.2 km

Figure 1 shows a field record before and after the application of the signal processing sequence. The central frequency of the recoverable signal band is about 400 Hz after signal processing; thus, we attained a vertical resolution of less than 1 m.

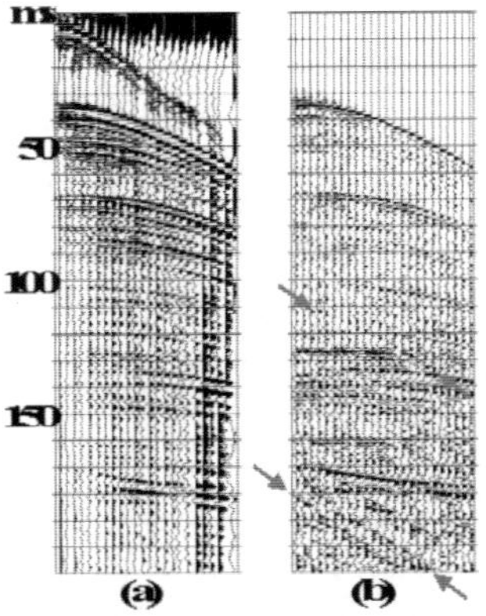

Figure 1. A 24-trace shot gather from the 3-D survey before (a) and after (b) signal processing. The noise associated with point scatterers at the water bottom or marine vessels can have linear or curved moveout as indicated by the red arrows depending on their locations with respect to the streamer. Such noise is difficult to remove and can also manifest itself in the migrated data.

The image volume exhibits the presence of an erosional channel associated with a Palaeo-Bosphorus waterway (Figure 2). Maximum depth of the channel is about 150 m from the sea level. Prior to the flooding of the erosional channel by the Palaeo-Bosphorus, the river bed consisted of gravel.

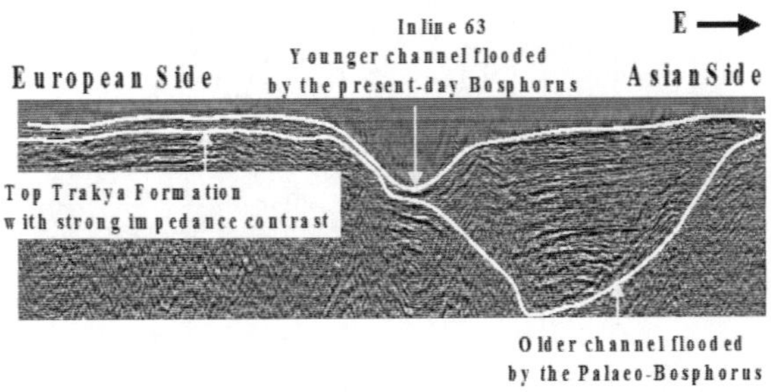

Figure 2. An inline section from the 3-D Stolt migration volume that shows the present-day Bosphorus Channel and the Palaeo-Bosphorus Channel. The lateral extent of the section is about 2,000 m and the time interval shown is 0-200 ms which corresponds to depth interval 0-150 m.

By taking into consideration the geometry of the zone of the highly weathered Trakya bedrock material in the western flank of the channel and the absence of a fault, the geological cross-section constructed from the geotechnical borehole data was modified, accordingly.

The planned tunnel traverse (Figure 3) is within the Trakya Formation except for a middle segment, 806 m in length, within the soil column composed of interbeddings of sand-clay-silt deposits with varying percentages of fines content. The axis of the planned tunnel traverse exits the Trakya Formation and enters the soil column at KM 7+890 measured from the point of origin on the European side, and at a depth of 97 m from the sea level; then re-enters the Trakya Formation at KM 8+696 and at a depth of 58 m from the sea level.

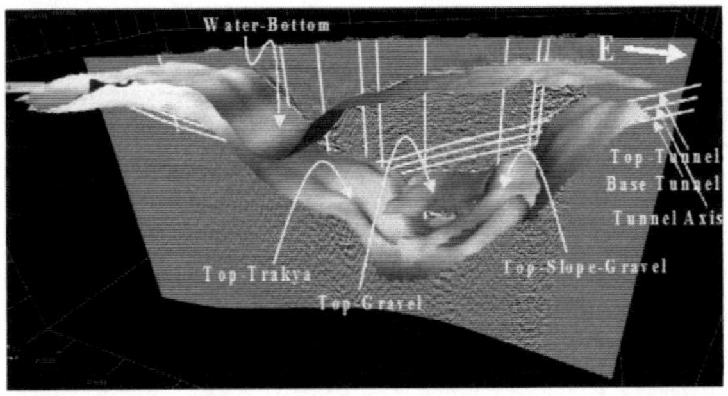

Figure 3. A 3-D view of the depth horizons interpreted from the seismic image volume of 3-D poststack time migration followed by time-to-depth conversion using the well-tops from Borehole BH1-7.

Case Study 2: Seismic, geotechnical, and earthquake engineering site characterization

At each of the 20 locations (KS01-KS20) within the project site, we deployed a receiver spread with 48 4.5-Hz vertical geophones at 2-m intervals. We used a buffalo gun at the bottom of a charge hole with a diameter of 10 cm and a depth of 30 cm, and acquired three shot records with source locations at each end of the spread and at the center of the spread (Figure 4). The sampling rate for recording is 0.5 ms and the record length is 2 s. The orientation of the receiver spread at each location was chosen such that the elevation change along the receiver spread is minimal.

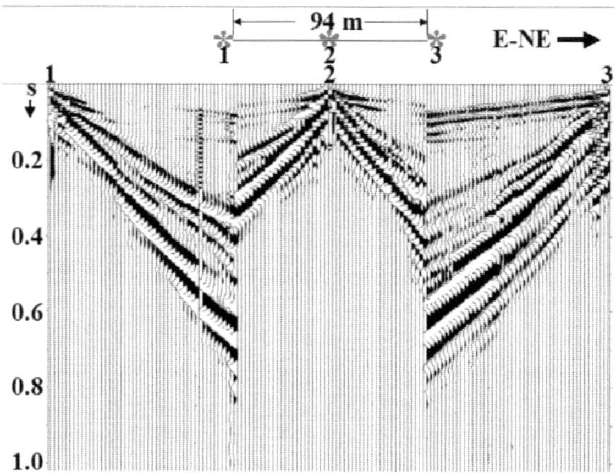

Figure 4. Example common-shot gathers (CSG) acquired by a 94-m receiver spread (the green bar on top). The red asterisks denote the shot locations --- two shots with 2-m offset from the ends of the receiver spread and a shot at the center of the spread.

We applied nonlinear traveltime tomography (Zhang and Toksoz, 1998) to the first-arrival times picked from the three shot records and estimated a near-surface P-wave velocity-depth model along the receiver spread at each of the 20 locations. By lateral averaging after the inversion, we then obtained a P-wave velocity-depth profile down to a depth of 32 m representative of each location. Next, we identified the off-end shot record with the most pronounced dispersive surface-wave pattern and performed Rayleigh-wave inversion to estimate the S-wave velocity as a function of depth (Park et al., 1999; Xia et al., 1999).

Following the analysis of refracted waves to estimate the P-wave velocities and the analysis of surface waves to estimate the S-wave velocities at each of the 20 locations, we generated velocity contour maps for depth levels 0-32 m at 4-m intervals (Figure 5). Additionally, we computed the ratio of the P-wave velocities to the S-wave velocities as a function of depth and generated the *Vp-Vs*-ratio contour maps. Note from the maps shown in Figure 5 the lateral velocity variations, which can be correlated with the surface expression of the landslide masses which we identified in the project site based on the reflection seismic survey.

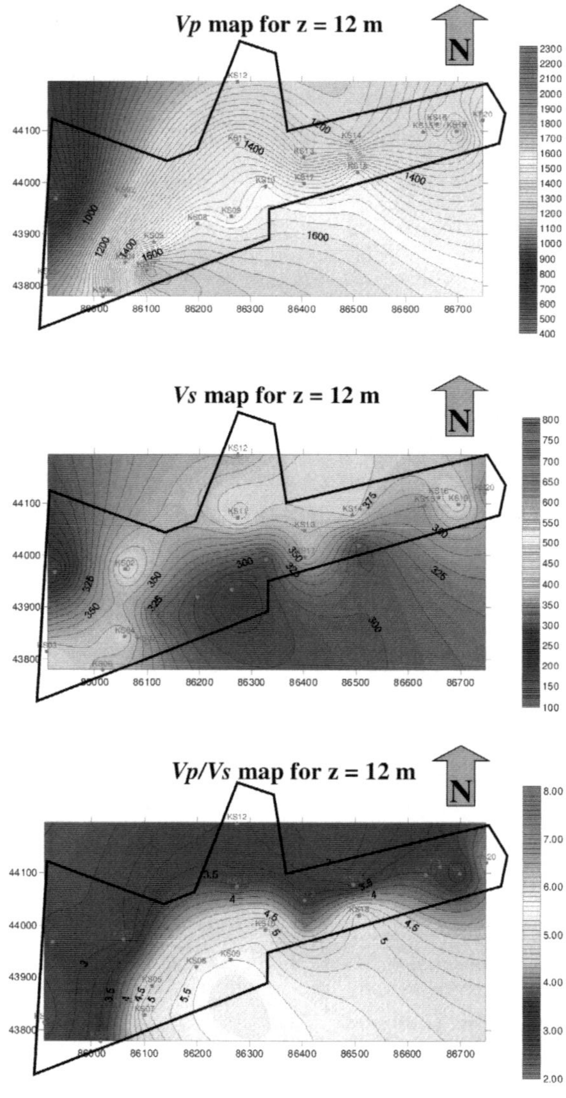

Figure 5. Examples of the velocity maps. The red dots represent the spread centers at the 20 locations where the seismic survey was conducted.

Based on the velocity variations, the site was divided into three sections --- the western section with relatively low S-wave velocities, the eastern section with relatively high S-wave velocities and the transitional central section. We then drilled geotechnical boreholes within each section down to depths of 50, 40, and 30 m at locations coincident with the center of the spreads KS05, KS10, and

KS14, respectively. For each section within the project site, the S-wave velocities within the soil column combined with the borehole lithology were used to determine the geotechnical earthquake engineering parameters (Kramer, 1996; Bardet et al., 2000; Yilmaz et al., 2006) --- *maximum surface-bedrock acceleration ratio, natural period* of the soil column, *maximum soil amplification ratio, as the spectral acceleration ratio*, and *design spectrum periods TA-TB.*

Conclusions

Site investigations require multidisciplinary participation by the geologist, seismologist, geotechnical and earthquake engineers (Figure 6). The seismologist defines the geometry and the sesimic velocities of the soil column, the geomorphologist determines the soil pedology, and the geotechnical engineer determines the dynamic properties of the soil column. The soil geometry, the soil pedology, and the soil dynamics constitute the geotechnical model of the soil column, which is then used by the geotechnical engineer for geotechnical design for soil remediation.

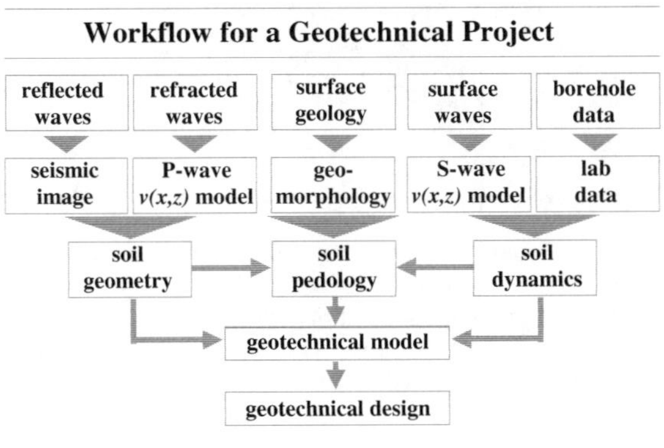

Figure 6. Workflow for a geotechnical project for site characterization.

The two case studies presented in this paper demonstrate successful applications of engineering seismology in urban areas. Specifically, in major urban projects, geophysical methods help the geotechnical engineer design a geotechnical model of the soil column that can be used for soil remediation, and the civil engineer perform structural design of buildings by taking into consideration the seismic properties and, if required, the earthquake engineering parameters of the soil column.

Acknowledgements

We are grateful to Avrasya Tüneli İşletme İnşaat ve Yatırım AŞ (ATAŞ) for granting the permission to publish this paper. We thank the Hanyapi Real Estate Development Corp., Istanbul and Prof. Ahmet M. Işıkara for giving us the opportunity to conduct the integrated site investigation project and granting the permission for publishing the results.

References

Bardet J., Ichii, K., and Lin C., 2000, Manual of EERA: a computer program for equivalent-linear earthquake site response analysis of layered soil deposits, University of Southern California.

Kramer, S. L., 1996, Geotechnical Earthquake Engineering, p. 273, Prentice-Hall, New Jersey.

Park, C. B., Miller, R. D., and Xia, J., 1999, Multichannel analysis of surface waves: Geophysics, 64, 800-808.

Xia, J., Miller, R. D., and Park, C. B., 1999, Estimation of near-surface shear-wave velocity by inversion of Rayleigh waves: Geophysics, 64, 691-700.

Yilmaz, O., Eser, M., and Berilgen, M., 2006, A case study for seismic zonation in municipal areas, The Leading Edge, March issue, 319-330.

Yilmaz, O., 2013, Application of engineering seismology in urban areas, The Leading Edge, March issue, 936-944.

Zhang, J. and Toksoz, M. N., 1998, Nonlinear refraction traveltime tomography: Geophysics, 63, 1726-1737.

Hazard assessment and site characterization at an oil and gas well site using surface wave methods

Steven D. Sloan, J. Tyler Schwenk, Robert H. Stevens, and Benjamin W. Butler, XRI Geophysics*

Summary

Surface wave methods were used to interrogate the subsurface along a proposed natural gas pipeline route to identify potential subsurface hazards prior to construction and around existing sinkholes to determine their lateral extent beyond the surface expression at an oil and gas production facility. Shear-wave velocity cross-sections derived using the multichannel analysis of surface waves method along the pipeline route were used to identify low-velocity areas that might be indicative of voids or structurally weaker zones and to select geotechnical boring locations for a subsequent geotechnical engineering assessment. Small sinkholes are known to exist and develop in the study area and their formation beneath or around pipelines is an immediate threat to public and environmental safety. Potential sinkholes in the immediate area of the well site also pose a threat to personnel, equipment, and infrastructure if not identified before collapsing at the surface. Survey results identified several anomalous zones along the proposed pipeline route to further analyze during the geotechnical assessment. Two sinkholes that have breached the surface were also shown to be larger in extent beneath the ground surface than the surface expression would suggest.

Introduction

The multichannel analysis of surface waves (MASW) method has been used for many different environmental and engineering geophysical applications since its introduction (Park et al., 1999; Xia et al., 1999). Applications have ranged from general site characterization to determine geotechnical properties (Park, 2013), mapping depth to bedrock (Miller et al., 1999), pre-construction site characterization (Suto, 2013; Whiteley and Stewart, 2013), levee interrogation (Ivanov et al., 2004; Ivanov et al., 2005; Ivanov et al., 2006), landslides (Malehmir et al., 2013), seismic hazard classification (Yilmaz et al., 2006), with many others not listed here.

Data were collected at an oil and gas well site in west Texas, USA in an effort to delineate the lateral extent of existing and recently formed sinkholes around the site (Figure 1) and characterize the subsurface along a planned pipeline route to augment geotechnical site assessment and aid in geotechnical borehole placement. The MASW method was a practical selection since it is sensitive to lateral changes in shear-wave velocity and stiffness and could be used to identify low-velocity zones suggestive of less competent material or voids. The roll-along style data acquisition geometry used for the MASW technique also lends itself to utilization of land streamers and accelerated weight drop sources to increase data acquisition efficiency and reduce required manpower in the field.

Figure 1. Photos of the same sinkhole from two different angles. This example is representative of multiple sinkholes that dot the landscape around the site. The black rod in the pictures is 1.83 m (6 ft) long. The depth to the bottom was approximately 2.44 m with near-vertical walls.

Site Description

The site is located in Loving County, Texas, near the border between New Mexico and Texas (Figure 2). Loving County is the least populous county in the United States outside of Alaska, but is also one of the wealthiest counties in the country per capita due to the oil and gas reserves found there. The site lies within the Delaware Basin that stretches across parts of west Texas and southern New Mexico. The Delaware Basin is a part of the larger Permian Basin, which has been a hotbed of drilling and oil and gas production activity in recent years.

The climate is semi-arid with little rainfall and very dry soils. Near-surface material at the site is described as calcareous and gypsiferous loamy alluvium (Rives, 1999). Caliche is common in the area, both at the surface and in the subsurface.

Figure 2. Illustration showing the relative location of the area of investigation

Methods

The MASW method takes advantage of the dispersive properties of Rayleigh-type surface waves to invert for shear-wave velocity (Vs). In the presence of increased seismic velocity with depth, different frequencies of a surface wave travel at different velocities, a property known as dispersion. Lower frequencies sample greater depths due to increasing velocity, while higher frequencies sample relatively shallower depths. This method is sensitive to lateral changes in the Vs of a material, as a portion of the particle motion of a Rayleigh wave is horizontal, producing a two-dimensional Vs plot of the subsurface.

The multi-channel approach uses a common shot gather (Figure 3a) that is transformed into a swept-frequency record (Figure 3b) that allows a fundamental mode dispersion curve to be picked. The dispersion curve is inverted to produce a one-dimensional Vs function with depth (Figure 3c). Each shot gather produces a single dispersion curve and 1D velocity profile. Multiple shots can be combined to produce a 2D Vs profile, as shown in Figure 3d.

Data Acquisition and Processing

Data were collected using a constant-offset, roll-along style acquisition geometry. To improve efficiency in the field, a 48-station land streamer was used, towed by an all-terrain utility vehicle with an accelerated weight drop for the source. The land streamer contained forty-eight 4.5-Hz geophones with a spacing of 1.22 m. The source-to-receiver offset was 4.88 m for all lines, with a source location spacing of 2.44 m. Three weight-drop strikes were collected at each source location to improve the signal-to-noise (S/N) ratio. Each individual strike was recorded and saved for quality control of noisy records and time break delays and subsequent vertical stacking during data processing. Data were recorded using a Geometrics Geode distributed seismograph system with 24-bit A/D, 2 s record lengths, and a 1-ms sampling interval. Station locations were recorded using a real-time kinematic (RTK) global positioning system (GPS). A total of 8.4 km of seismic data were collected at the site.

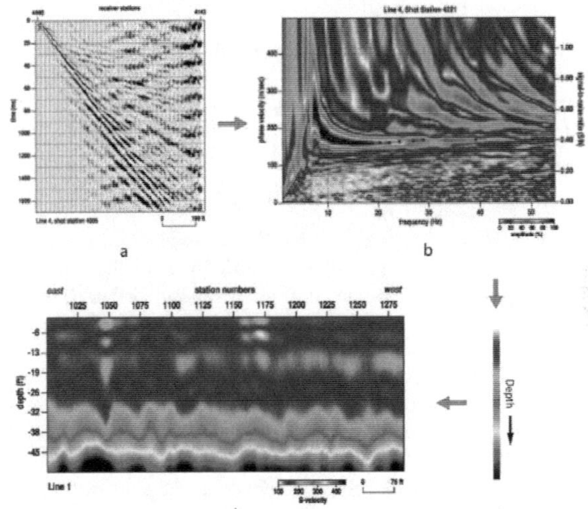

Figure 3. Generic flow typical of MASW processing including a field record (a), overtone image (b), 1D velocity profile (c), and 2D cross section compiled from multiple 1D profiles (d).

Data were processed using the SurfSeis software package from the Kansas Geological Survey. Data processing included geometry definition, vertical stacking, generation of overtone images, dispersion-curve picking, and inversion. The data were especially noisy due to an ongoing hydraulic fracturing, or "fracking", job at the site that included numerous tractor trailers for pumping, tanker trucks coming and going to deliver water, generators for power, and pedestrian traffic. The fracking job ran twenty-four hours a day and time constraints for the project prevented a postponement of the survey until after the well was completed.

Results

Figure 4 shows four parallel lines collected along the proposed pipeline route. The overall length of the lines are slightly variable due to surface obstacles, but are approximately 1.1 kilometers long for scale. The offset between each of the lines is approximately 10 m. The purpose of these lines was to identify areas of subsurface variability and/or anomalous zones that might be of interest from a geotechnical perspective and identify boring locations. An example of such an area would be the low velocity zone between Inline Station Number 10400 and 10600 on each of the four lines (Figure 4).

Additional seismic lines were collected around existing sinkholes to determine the unseen lateral extent in the subsurface beyond the surface expression that may pose risks to personnel, equipment, or vehicles. During the course of shooting one of these lines, abrupt changes were noted in the raw field files during acquisition. Figure 5 shows two such gathers collected with 4.5-Hz geophones. The two shot gathers are separated by approximately 23 m (20 stations). Surface wave backscatter events are prevalent in both shot gathers at station number 16096. In the shot gather on the right, there is a change in the wavelet character of the event sloping from left to right, from approximately 15 ms on the left side of the gather to 160 ms on the right side. Beginning at station number 17021, the frequency of the event appears to decrease and the slope increases (decreased velocity).

The MASW section for the same line is displayed in Figure 6. The shear-wave velocity in the upper 4 m of the section is lower overall between station numbers 16990-17025 compared to that of station numbers 17025-17070. There is also a noticeable decrease in velocity between 4 and 8 m depth from station number 16097 to 17020. Probing at the surface near station number 17000 resulted in a void approximately 30 cm deep that had not yet reached the surface. This line was collected within 5 m of a mobile field office, which had to be relocated as a result. Fortunately, the parking area was on the opposite side of the office, otherwise the void may have been discovered by a vehicle collapsing the roof rather than a geophysical survey.

Conclusions

MASW data were collected along a proposed pipeline route to identify any anomalous areas and potential geotechnical borehole locations. Zones with relatively high or low Vs were identified for further investigation. Geophysical techniques cannot replace engineering assessments, but can be a valuable tool that provides continuous lateral coverage between geotechnical boreholes and to identify anomalous areas that may require further interrogation that could potentially be missed by point measurements. The survey was also successful in identifying a void that was just beneath the surface in close proximity to a mobile field office, allowing the office to be relocated without property damage or loss prior to a surface collapse.

References

Ivanov, J., R. D. Miller, R. F. Ballard, J. B. Dunbar, and J. Stefanov, 2004, Interrogating levees using seismic methods in southern Texas: SEG Annual Meeting, Expanded Abstracts.

Ivanov, J., R. D. Miller, R. F. Ballard, J. B. Dunbar, and S. Smullen, 2005, Time-lapse seismic study of levees in southern Texas: SEG Annual Meeting, Expanded Abstracts, 1121-1125.

Ivanov, J., R. D. Miller, N. Stimac, R. F. Ballard, J. B. Dunbar, and S. Smullen, 2006, Time-lapse seismic study of levees in southern New Mexico: SEG Annual Meeting, Expanded Abstracts, 3255-3259.

Malehmir, A., S. S. Romero, C. Shan, E. Lundberg, C. Juhlin, M. Bastani, L. Persson, C. Krawczyk, U. Polom, A. Adamczyk, M. Malinowski, M. Gurk, and N. Ismail, 2013, A multidisciplinary geophysical and geotechnical investigation of quick clay landslides in Sweden: SEG Annual Meeting, Expanded Abstracts, 1254-1258.

Miller, R. D., J. Xia, C. B. Park, and J. M. Ivanov, 1999, Multichannel analysis of surface waves to map bedrock: The Leading Edge, **18**, 1392–1396.

Park, C., 2013, MASW for geotechnical site investigation: The Leading Edge, 656-662.

Rives, J. L., 1999, Soil Survey of Loving and Winkler Counties, Texas: United States Department of Agriculture, Natural Resources Conservation Service.

Suto, K., 2013, An MASW survey for geotechnical engineering—An application to pre-tunnelling investigation: Near Surface Geophysics Asia Pacific Conference Expanded Abstracts, 185-189.

Whiteley, R. J., and S. B. Stewart, 2013, Applying surface and borehole seismic methods to trenchless construction problems: ASEG Annual Meeting, Expanded Abstracts, 1-4.

Xia, J., Miller, R.D., and Park, C.B., 1999, Estimation of near surface velocity by inversion of Rayleigh waves: Geophysics, v. 64, 691-700.

Yilmaz, O., M. Eser, and M. Berilgen, 2006, Seismic, Geotechnical, and Earthquake Engineering Site Characterization: SEG Annual Meeting, Expanded Abstracts, 1401-1405.

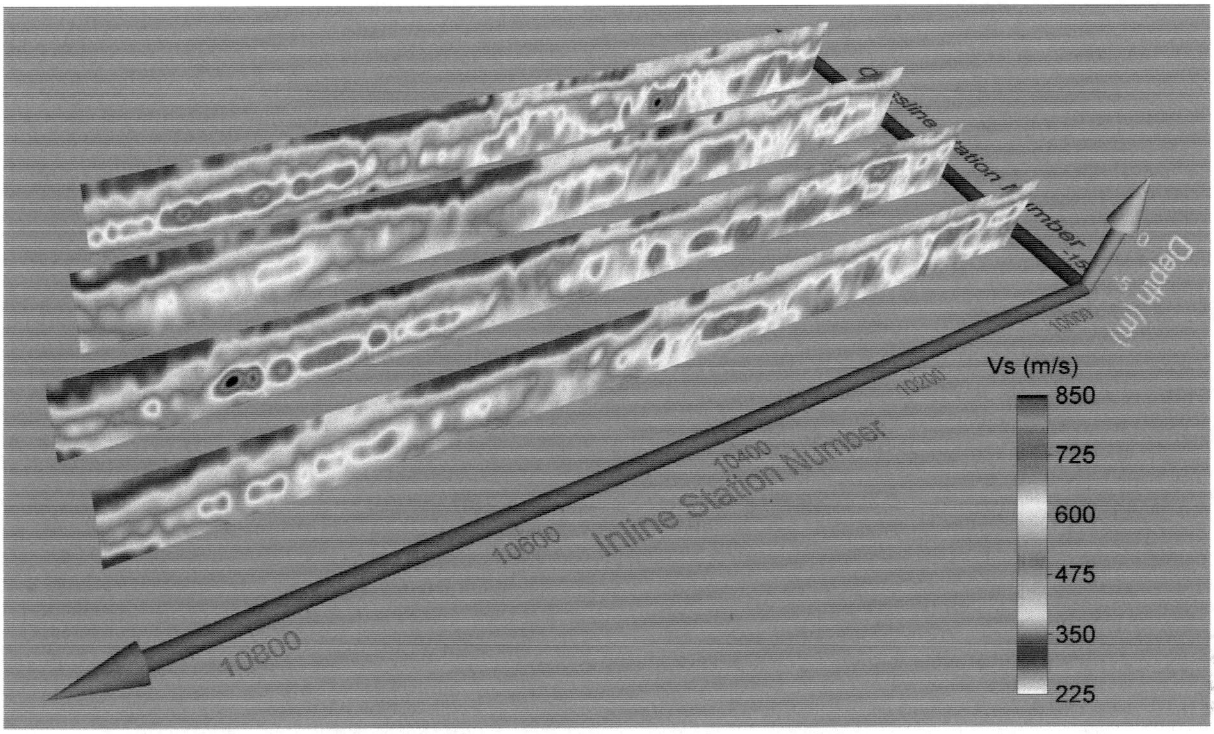

Figure 4. Four parallel MASW lines collected along a proposed pipeline route approximately 1.1 km long. The offset between the lines is approximately 10 m. The crossline and depth scales are exaggerated for display purposes.

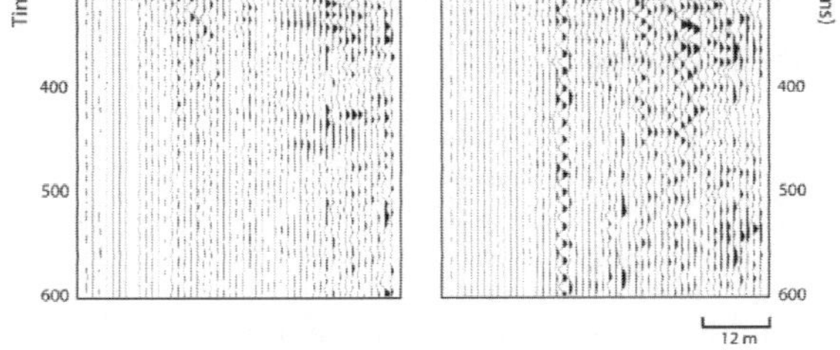

Figure 5. Two shot gathers collected over a potential sinkhole that has not breached the surface. Data are displayed with a low-pass filter to cut everything above 120 Hz due to excessive high-frequency noise at the site.

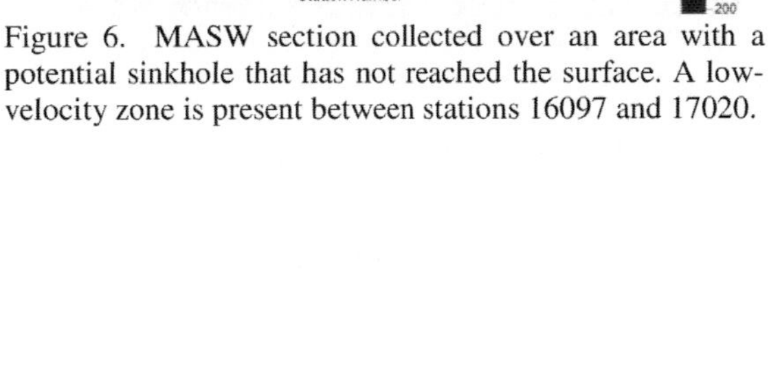

Figure 6. MASW section collected over an area with a potential sinkhole that has not reached the surface. A low-velocity zone is present between stations 16097 and 17020.

Probabilistic Vulnerability Assessment of the Building Inventory in an Extended Seismically Active Area in the UAE

Aman Mwafy, Abdelrahman Ashri and Anas Issa*
Civil and Environmental Engineering Department, United Arab Emirates University, UAE.

Summary

The main outcomes of recent research projects that were devoted to the probabilistic vulnerability assessment of the modern and pre-code building inventory in an earthquake-prone area in the UAE are briefly presented in this paper. The selection and design approaches of a wide range of structures to represent the study area with varying building heights, lateral force resisting systems and construction dates are discussed. The idealization approach of the reference buildings for dynamic response simulations using a detailed modeling approach as well as the selection of forty natural earthquake records to represent the seismo-tectonic setting of the study area are presented. The development of a wide range of three-dimensional simulation-based fragility relationships for the building inventory in the UAE using thousands of inelastic dynamic analyses is introduced. Examples demonstrating the possible impacts of different seismic scenarios on the built environment using the developed fragility functions are finally discussed. While the focus of the presented research projects was on the reliable assessment of physical damage in a seismically active area in the UAE, the outcomes represent the main driving engine for the development of a state of the art loss estimation and hazard mitigation system for the region.

Introduction

The most important components of earthquake loss estimation and hazard mitigation systems are the inventory, seismic hazard and vulnerability. Concerning the inventory of the exposed systems, the contemporary building inventory in the UAE is of great importance when estimating the potential losses from earthquakes. Additionally, pre-seismic code buildings may experience a high risk of damage, and hence their vulnerability should be thoroughly assessed. A vital role in the recovery period following an earthquake is also played by emergency facilities such as hospitals and fire stations. In spite of being constructed according to seismic design provisions, they should receive considerable attention to ensure their readiness and continued operation following earthquakes. Assessment of the vulnerability of the diverse building inventory is significant for the prediction and mitigation of possible earthquake losses.

As for the seismic hazard, although the UAE is generally known to be a region of stable seismic activity, recent earthquakes indicated that the region might be prone to damaging earthquakes (NCMS 2014). Even though no human or notable monetary losses have been reported, recent earthquakes have raised concerns regarding the vulnerability of the existing building inventory in the region and the associated risk. Non-negligible consequences are expected if the seismic risk of the building stock is overlooked, particularly for substandard structures.

Finally, few vulnerability assessment studies have been carried out recently for the UAE (e.g. Mwafy 2012; Mwafy *et al.* 2015). None of the available studies have been carried out based on reliable inventory data; wide range of reference structures with different characteristics representing the building stock; detailed design and modeling approaches; a diverse range of input ground motions representing different seismic scenarios in the study area; reliable limit states or systematic approaches for developing fragility relationships. The above-mentioned discussion underlines the pressing need for comprehensive loss estimation and hazard mitigation studies for the built environment in this region.

The main objective of this comprehensive study is to assess probabilistically the vulnerability of a wide range of reference structures representing the modern, pre-seismic code and emergency structures in a highly populated and seismically active area in the UAE using state of the art methodologies. This enables the direct implementation of a wide range of fragility relationships representing different structures and earthquake scenarios in a loss estimation and hazard mitigation system for the region.

Building Inventory

The modern and pre-code building inventory in the extended area selected in the present study (i.e. Dubai, Sharjah and Ajman, UAE) is collected from twelve zones. The studied area is divided into these zones based on the common building characteristics. Each of the twelve zones is also divided into a number sub-zones to help in collecting the required data from satellite images and site visits. Figure 1 shows the percentage of buildings in different zones. The buildings were classified according to three main criteria due to the significance of the classification criteria of the exposed building stock in risk analysis. The classification criteria are: (i) number of stories, (ii) construction date, and (iii) function.

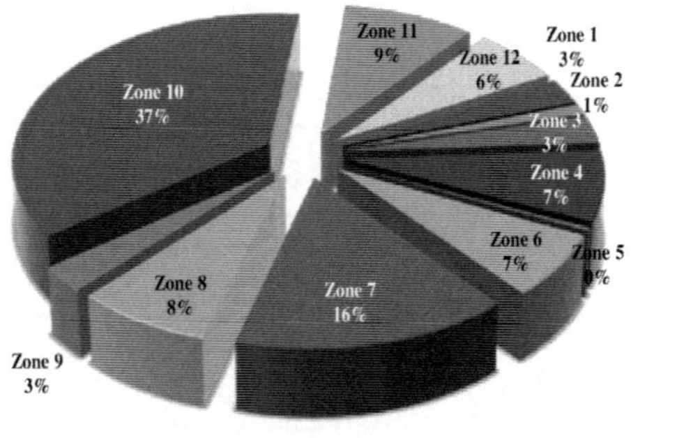

Figure 1: Percentage of buildings in different zones of the study area.

Nineteen structures representing the building inventory in the study area are selected from the collected building database for the vulnerability assessment. The buildings are selected based on different classification criteria, as discussed above. Table 1 summarizes the characteristics of the selected reference structures. Ten of the selected buildings represent modern structures with different heights, ranging from 2 to 100 stories. Moreover, four buildings are selected to represent essential and emergency facilities, while five buildings with different heights are selected to characterize pre-code structures. It is clear that the buildings stock is fairly represented in the set of reference structures shown in Table 1. Figure 2 shows sample of the three-dimensional design models of the reference structures.

Table 1: Reference structures to epresent the building inventory.

No.	Ref.	Classification criteria	description	No. of stories	Structural* System
1	2St	Number of stories	Modern structures	2	MRF
2	8St			8	MRF
3	18St			18	SW
4	26St			26	SW
5	40St			40	SW
6	50St			50	SW
7	56St			56	SW
8	66St			66	SW
9	80St			80	TS
10	100St			100	TS
11	BO-02	Construction date	Pre-seismic code structures	2	MRF
12	BO-08			8	MRF
13	BO-18			18	SW
14	BO-26			26	SW
15	BO-40			40	SW
16	FS	Function	Fire station	2	MRF
17	PS		Police station	2	MRF
18	SC		School	3	MRF
19	HO		Hospital	6	MRF

* MRF: moment resisting frames; SW: structural walls; TS: Tube system

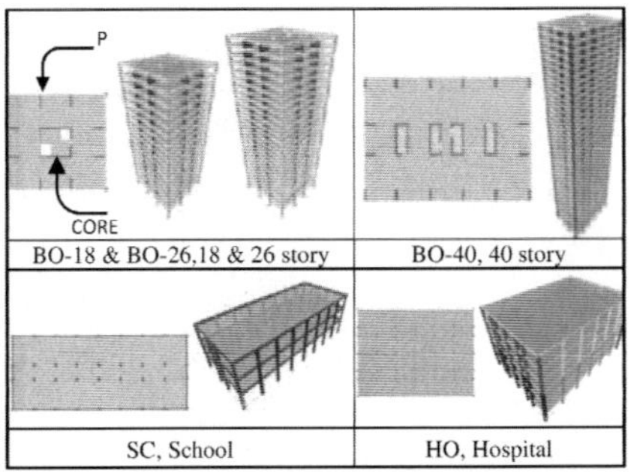

Figure 2: Three-dimensional design models of the reference structures.

Detailed numerical idealizations for inelastic analysis are carried out for the reference structures. The developed models are capable of predicting the large displacement behavior of structures under static and dynamic loading, taking into account both geometric nonlinearities and material inelasticity. In the Inelastic Time History Analysis (ITHA), the reference structures are subjected to transient loads, which consist of accelerations that vary in the real time domain.

Seismic Hazard

For the vulnerability assessment of the building inventory, diverse earthquake ground motions are selected based on previous site-specific seismic hazard studies (e.g. Mwafy *et al.* 2006). The uncertainty in seismic demands due to the variability in input ground motions is accounted for using forty earthquake records representing two earthquake scenarios (i.e. long-distance and near-source events). The natural records are selected from the Pacific Earthquake Engineering Research Center and the European strong-motion databases (Ambraseys *et al.* 2004; PEER 2014). The natural records are selected based on their epicentral distance, magnitude, site class and spectral amplification to match the design spectrum of the study area. Figure 3 shows the elastic response spectra of the far-field earthquake records along with the design spectra for Dubai.

Vulnerability

Due to the diversity of the structural systems investigated in this work and the important role played by limit states in vulnerability assessment, a large number of ITHAs are applied incrementally up to collapse to estimate the performance criteria of the reference structures. Figure 4 shows sample results from this comprehensive analyses. For the sake of brevity, the results of the 19 reference structures are discussed in more detail by Ashri and Mwafy (2014) and Issa and Mwafy (2014).

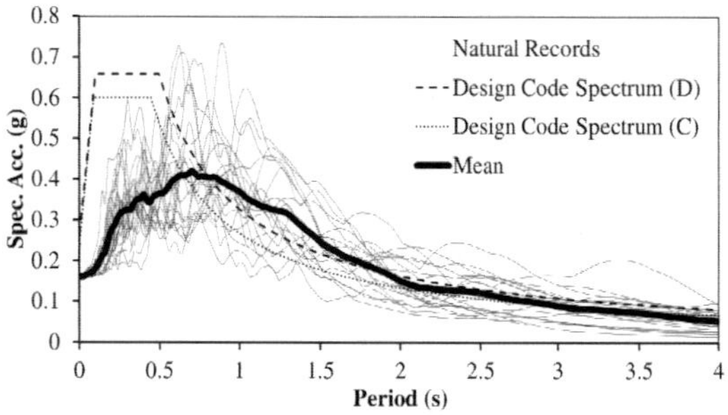

Figure 3: Response spectra of far-field earthquake records along with the design spectra for Dubai for soil classes C and D.

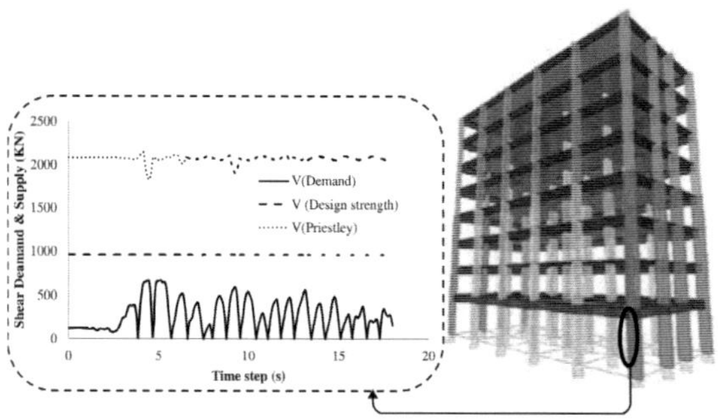

Figure 4: Sample results for the shear response of the 8-story building (BO-08).

The vulnerability relationships of the nineteen reference structures are derived through a large number of ITHAs. Figure 5 depicts sample of the collapse analysis results. Figure 6 shows sample of the derived 3D vulnerability functions of the code-conforming frame structures under the near-source earthquake scenario for different limit states. While the X and Y-axes represent the ground motion intensity and the probability of exceedance, respectively, the third axis represents the number of stories. This versatile 3D format enables the interpolation of results from the derived vulnerability relationships to arrive at the fragilities of a broad range of structures with different heights.

To provide a more informative presentation of this comprehensive vulnerability assessment study, Fig. 7 depicts the limit state exceedance probabilities of the modern code designed buildings and the pre-code structures under the far-field earthquake scenario. These results are presented for the former set of buildings at twice the design intensity, while they are shown for the latter group of structures at both the design and twice design ground motion intensities.

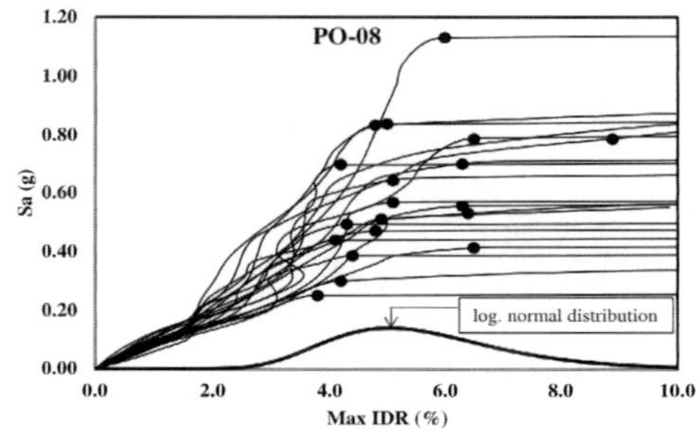

Figure 5: Selected collapse analysis results for the PO-08 building obtained from 20 natural earthquake records.

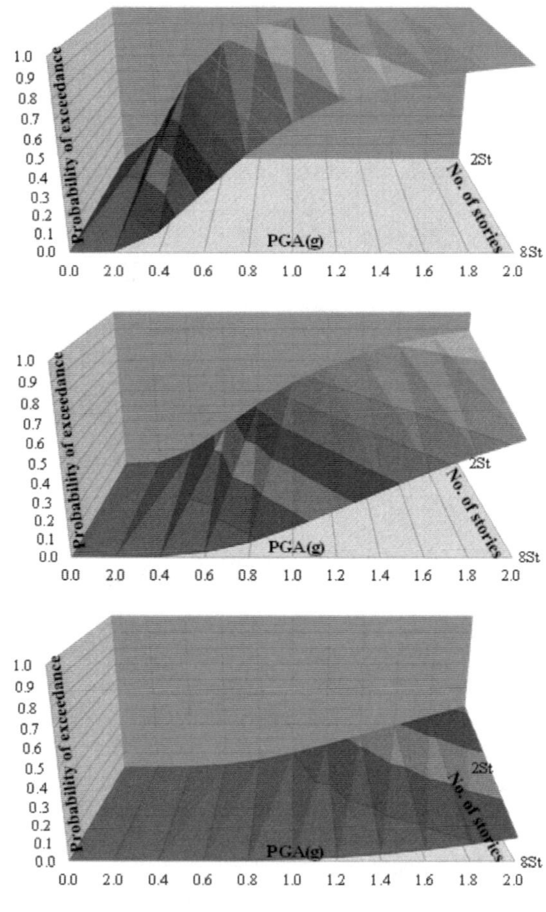

Figure 6: Three dimensional vulnerability functions of modern frame buildings using the near-source earthquake scenario for the immediate occupancy (top), life safety (middle), and collapse prevention (bottom) limit states.

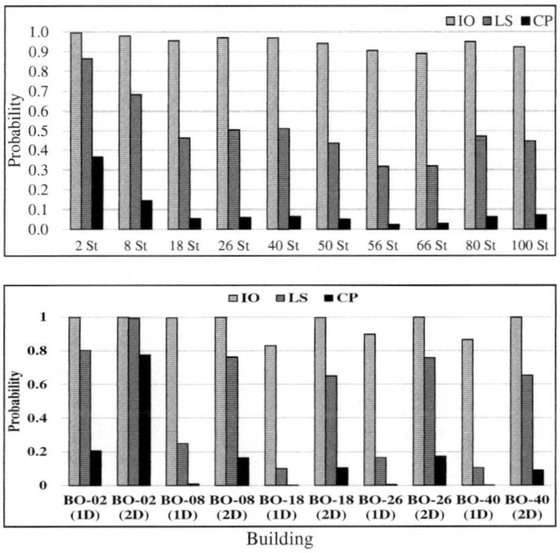

Figure 7: Limit state exceedance probabilities of the modern buildings (top) and the pre-code structures (bottom) at the design and twice the design earthquake intensities.

The sample results of the probabilistic vulnerability assessment shown in Figure 7 confirm the acceptable performance of the contemporary building inventory under twice the design earthquake intensity. This observation is in line with the modern code approach for the seismic design of buildings. On the other hand, the results at the design and at twice the design earthquake levels confirm the vulnerability of the pre-code structures, particularly at twice the design intensities. Figure 7 also confirms the vulnerability of low-rise frame buildings, which is a common system in the study area, to the far-field seismic scenario. The results confirm the pressing need for mitigation strategies to reduce the earthquake losses of the pre-code and low-rise buildings. The developed vulnerability relationships in this comprehensive study are essential elements to arrive at a reliable loss estimation system for the region.

Conclusions

This paper summarized the outcomes of recent research projects that were directed to the probabilistic vulnerability assessment of the building inventory in the UAE. The study involved the selection, design and development of verified fiber-based simulation models for nineteen reference structures representing a wide range of low, medium and high-rise buildings. The reference structures were selected from a collected building database to represent the modern and pre-code buildings and the emergency facilities in a highly populated earthquake prone area in the UAE. The selected structures varied in height from 2 to 100 stories. Forty scenario-based earthquake ground motions were employed in dynamic response simulations for the vulnerability assessment of the building inventory. This comprehensive study enabled deriving a wide range of vulnerability functions that account for different structural systems, building heights and construction date for the direct implementation in a loss estimation and hazard mitigation system for the UAE. While the probabilistic vulnerability assessment study reflected the acceptable performance of the contemporary buildings, it confirmed the vulnerability of the pre-code structures, particularly the low-rise buildings under the effect of the far-field seismic scenario. The developed vulnerability relationships in this comprehensive study are an essential component to arrive at a reliable loss estimation system for this region.

Acknowledgements

This work was supported by the UAE University under research grants No. 31N007 and 31N132.

References

Ambraseys, N. N., Douglas, J., Sigbjornsson, R., Berge-Thierry, C., Suhadolc, P., Costa, G., and Smit, P. M., 2004, Dissemination of European strong-motion data, vol. 2, Engineering and Physical Sciences Research Council, Swindon, UK.

Ashri, A., and Mwafy, A. M., 2014, 3D vulnerability functions for contemporary buildings with varying structural systems and heights: Proceedings of 2nd European Conference on Earthquake Engineering & Seismology, Turkey.

Issa, A., and Mwafy, A. M., 2014, Fragility assessment of pre-seismic code buildings and emergency facilities in the UAE: Proceedings of 2nd European Conference on Earthquake Engineering and Seismology, Turkey.

Mwafy, A. M., 2012, Analytically-derived fragility relationships for the modern high-rise buildings in the UAE: The Structural Design of Tall and Special Buildings, 21, no. 11, 824–843.

Mwafy, A. M., Elnashai, A. S., Sigbjornsson, R., and Salama, A., 2006, Significance of severe distant and moderate close earthquakes on design and behavior of tall buildings: The Structural Design of Tall and Special Buildings, 15, no. 4, 391-416.

Mwafy, A. M., Hussain, N., and El-Sawy, K., 2015, Seismic performance and cost-effectiveness of high-rise buildings with increasing concrete strength: The Structural Design of Tall and Special Buildings, 24, no. 4, 257–279.

NCMS, 2014, http://seismology.ncms.ae/earthquakes, accessed 1 June 2014.

PEER, 2014, PEER NGA database, http://peer.berkeley.edu/nga, accessed 1 June 2014.

GPR study of karst in a carbonate coastal area for evaluating its suitability for construction, Wadi Shab, Eastern Oman.

Hesham El-Kaliouby, Sultan Qaboos University, Earth Science Dept., Oman &*
National Research Center, Geophysics Dept., Egypt.

Summary

Ground Penetrating Radar survey was conducted for the detection of fractures and cavities in the carbonate section at two sites along the coast line in Wadi Shab, Eastern Oman. The survey objective is to determine the location and size of cavities or sinkholes that may cause collapse of buildings of residential area and to assess the suitability of another area for construction and living on it. Two antennas with different central frequency (270MHz and 400MHz) have been used to investigate different depths and resolutions. Results showed a number of hyperbolas that disturbed the continuity of the GPR signals indicating the presence of fracture zones and cavities within the limestone. The GPR waves could not penetrate more than 2-5m depth at the surveyed areas, which could be related to attenuation caused by clays and sea water saturated limestone through fractures filling the limestone secondary pores. Results not only informed of existing subsurface cavities, but also warn of potentially hazardous situation enabling city planners, and civil engineers to take proper precautions before any undesirable situation arise.

Introduction

Sinkholes are geologic features formed by movement of rock or sediments into voids created by dissolution of water-soluble carbonate rocks. Sinkholes are often a major hazard particularly in residential area underlain by limestone containing cavities; such limestone is called karstic limestone. Karst phenomena represent a major hazard features to most of the urban, developed, and infrastructure areas. The cavities form when water flow dissolves and washes away parts of the limestone and can grow until they reach the surface causing sinkholes. Road and highway subsidence, building-foundation collapse, and dam leakage are a few of the problems associated with sinkholes. Structural instability associated with sinkholes can occur as a sudden collapse of the ground surface

Ground Penetrating Radar (GPR) survey was conducted in residential coastal area in Eastern Oman, where rock collapse had recently occurred near from the shore line. The authorities in Oman was concerned that many houses in Wadi Shab town are in danger and wanted to assess the situation about the location and size of cavities or sinkholes that may cause collapse of buildings. In the same time, the authorities requested for a study of another nearby coastal area (about 3 km north from Wadi Shab town) that could be an alternative new town for re-habitation of Wadi Shab residents. The reason for choosing another coastal area for living is to preserve the culture of the residents living there, which is mainly fishing and tourism.

GPR is a quick, easy and non-invasive geophysical method for imaging near surface features. GPR has a great advantage that it can be performed rapidly, relatively inexpensive to run, and has high vertical and horizontal resolution. GPR data is often a useful tool by providing advanced warning of subsurface hazardous situations enabling city planners, and civil engineers to take proper precautions before any catastrophic situation arise. The quality of GPR results obtained in sedimentary rock environments reflects both the lithological nature of sediments and the presence of possible cavities. GPR uses electromagnetic waves to map boundaries between lithologies or objects that have different electrical properties. As the GPR system is towed along a surface, pulses of electromagnetic energy penetrate the subsurface. A fraction of the wave is reflected back to the surface when it encounters a boundary where there is a change in electrical permittivity (dielectric constant) of subsurface materials that is a function of material type, moisture content and pore fluid type. Subsurface structures are mapped by measuring the amplitude and travel time of this reflected energy. Dry Limestone is a favorable media to the penetration of GPR signals and the karstic cavities (filled with air), common in limestone give strong radar reflections due to the contrast in electrical permittivity between the two media (Bristow et al., 2003; Neal, 2004).

The electromagnetic energy transmitted by GPR antenna at a specified central frequency can be recorded with either at the same antenna or a separate antennas. The frequency of the transmitted energy (antenna frequency) is selected with reference to the size and depth of potential targets, the lower the frequency of the transmitted energy, the greater the depth of penetration, and the larger the size of target than can be resolved and vice versa. The low frequency antennas (<200 MHz) are normally used for geological applications while the high-frequency antennas are used for geotechnical and engineering applications (Neal, 2004).

Field site characteristics

The geology of Wadi Shab is mainly dominated by folded and faulted massive bioclastic nodular limestone with subordinate conglomerate, sandstone, marl belonging to the Jufnayn formation and raised Quaternary coral reefs. This formation was deposited in a shallow marine environment during the late Paleocene-Early Eocene time. At the surface, the area is overlain by coastal wadi gravel. The thickness of the Quaternary wadi gravel is highly variable from less than a meter to a few meters. Since the study area is located along a costal cliff, therefore it was subjected to rigorous weathering activities due to transecting wadies and tsunami sea waves. That resulted in intrusive weathering and dissolution of the limestone forming subsurface cavities and sinkholes. Fractures are quite long running for several meters, such fractures played a major role in facilitating water dissolution and subsequent rock fall (Wyns et al, 1992; Hoffmann et al., 2013).

Data acquisition and processing

GPR surveys were conducted at two sites in Wadi Shab denoted as Old and New Wadi Shab (Figure 1) covering an area of about 280m by 130m. The old area is where residents are currently living there, while the new area is vacant where it is proposed to move local residents to live in it. The surface of the study area is of favorable conditions for GPR survey as it is arid dry surface conditions. However, measurements carried out on top of a beach cliff around ~5-10 m above mean sea level, caused the low-frequency noise and the attenuation of GPR waves at depth due to sea water intrusions that disturbed the data. High clay content or saline water content in the subsurface can lead to attenuation of the electromagnetic waves.

At each site, the GPR data were collected with the SIR-3000 system in continuous distance mode using two different antennas (270 MHz, and 400 MHz shielded antenna) along each survey line. Data collected using the higher frequency antenna (400MHz) better defined the stratigraphy and structures than the 270 MHz images. However, any gains in resolution provided by high frequency antenna are constrained by the limitations in depth penetration. Distance calibration was carried out for the two antennas over a fixed distance. Then, the antennas were dragged over the survey area using a four-wheel cart. The length of the lines and distance between the lines varied at each site. A handheld GPS was used for tracking GPR profiles. Since the surface of the study area is almost flat, no topographic correction of the data was needed and the accuracy of the GPS data was sufficient.

Acquisition parameters used are 1024 samples/trace; 40 scan/m representing the lateral resolution with distance. Dielectric constant of 8 was selected since the area is mainly covered by limestone and dolomite. Velocity and depth conversion is calculated based on the dielectric constant value. Total recorded two-way time for the 270MHz antenna is 120ns and 60ns for 400MHz antenna.

Two GPR long profiles were conducted at Old Wadi Shab (figure 1a) due to the limitations of the available surface area to run the GPR antenna. The two profiles were run along the roads with a profile length range from 250-300m. It was not possible to run more profiles between the houses due to the limited space and the noises that affect the measured data due to metallic features. Nine GPR profiles were conducted at the new Wadi Shab (figure 1b) area with a profile length range from 200-260m and profile interval of about 15m. Figure (2) shows the current situation at the two sites, where the houses are constructed at the cliff above the fractured limestone in the old Wadi Shab (Figure 2a) and the new proposed site for re-housing (Figure 2b).

The GPR data was processed using RADAN 7, (v.7, GSSI, Inc.) to process the collected data in the two areas. Usually, radargrams can be interpreted immediately after limited data processing including static correction, background removal, and gain adjustments. The objective was to emphasize only the relevant diffraction-reflections that might indicate the presence of cavities or fractured zones. However, background removal filter was insufficient for removing completely the horizontal banding, due to its horizontal amplitude variation, reducing the horizontal reflection continuity. The GPR sections (figures 3 and 4) are displayed with the horizontal (distance) and vertical (depth) scales in meters using a velocity (~0.106 m/ns) calculated based on the dielectric constant (ε_r) equal to 8.

Interpretation and discussion

After processing, the GPR images revealed several features that yielded information about the boundaries and internal structure of the buried voids/cavities at both sites. The 270MHz provides a greater depth of penetration in the old Wadi Shab area although resolution is less than that which can be obtained with the 400 MHz data. The 270 MHz antenna provides a longer wavelength than the 400MHz antenna where, the signal is less susceptible to scattering but gets attenuated at depth because of the clays and salt water filling the pores and fractures within the limestone.

In general, interpretation is done through marking any interesting subsurface features such as fractures or possible cavities on the radargrams (figures 3 and 4). Voids in the subsurface are visible in the GPR cross sections in the form of diffraction hyperbolae. The procedure followed is to identify and classify reflections and diffractions, recognize changes in reflection patterns, and identify seemingly distorted and truncated reflections. Disruption of reflections from known linear features often indicates anomalous zones (e.g. voids/cavities) in the subsurface. The voids are irregularly shaped with a highly uneven rock-air boundary. This will cause the voids to produce diffraction hyperbolae at the rock-air interface in the GPR cross sections. Analyses of raw and processed profiles show numerous diffraction hyperbolae. GPR signals are scattered by voids, fracturing, inclusions, and other inhomogeneous features in the subsurface and provide a detailed but challenging section to interpret.

Results showed a number of hyperbolas in the radargram indicating the presence of fracture zones that disturbed the continuity of the GPR signals. The radargram also showed ringing (horizontal banding) extending with depth that covers or influences the actual reflectors. The GPR waves could not penetrate more than 2-5m depth at these sites. This could be attributed to clays or to the percolation of the sea water through capillaries or fractures' saturating the limestone pores. The presence of salt water even with small percentage severely attenuates the GPR waves and hence limits the penetration depth.

Figure (3) shows the first 100m of a long profile (300m) from SW towards NE, measured at the old Wadi shab using 270MHz antenna. The small hyperbolas may indicate the presence of small cavities or utilities. The large hyperbolas are due to airwave reflected from surface objects such as buildings or trees. There is a weak signal recorded between 1-2m, which is shown as white (weak) signals along the section indicating no reflectivity or could be due to the presence of moderately conductive layer like infiltrated fresh water. A similar behavior is detected in another parallel profile parallel to the coast line using the same antenna. The maximum effective depth of penetration, where the signal to noise ratio gets weaker, extends up to 6m.

Figure (4) shows the last 120m of a profile of total length of 190m running from SW towards NE and measured at the new Wadi Shab area measured by high frequency antenna (400MHz). The profile shows a boundary between highly fractured zones or Wadi gravel deposits that is characterized by a strong continuous small diffractions. The figure highlights dozens of hyperbolae reflections which reflect the high gravel content. In other parts of the profile, there are typical hyperbolae reflections near the surface that could be related to small cavities within the limestone. The low frequency antenna (270MHz) measurements showed similar results. Unfortunately, the penetration depth of both antennas did not achieve more than 2-3m due to high attenuation caused by clays or sea water partially saturated limestone.

Although the results are promising, there is significant room for improvement through a more detailed characterization of subsidence structures that may be achieved by 3D GPR surveying with denser spacing between profiles to resolve the 3D geometry of subsidence structures, yielding a more detailed 3D subsurface image of the cavities. It is recommended to investigate other near surface geophysical methods like 2D/3D resistivity imaging, microgravity and MASW. Joint or multiple inversions of these techniques might improve the final model. It would be also recommended to conduct a GPR survey after the remediation of cavities/fractures to make sure that all the cavities are well filled with cement after the grouting operation and before any construction at the site to avoid any risks (Booth et al, 2013).

Conclusion

The study provided information regarding the suitability and limitations of the GPR method for investigating cavities and fractures in karsts at two coastal sites in Eastern Oman. In general, GPR was found to be an efficient technique for the identification and characterization of shallow structures in karstic areas that might have an impact on existing or future constructions. The results of the GPR data at the two sites was likely related to the small dimensions of the cavities and vertical fracture planes. The measured data does not show any significant features related to large sinkholes within the depth of investigation that is limited to 2-5m where GPR signals get attenuated after that depth due to clays and saline water within the limestone. Based on the GPR results, the two areas are relatively safe for housing except at the cliff shoreline where the sea water causing rock falls of the limestone rocks. However, grouting is required for the detected cavities at both sites to avoid the formation of sinkholes over time.

Acknowledgements
The author would like to thank Sultan Qaboos University for providing the resources to conduct the field survey. The author also would like to thank the students who helped him during the field work: Musab Alhasni, Yousuf Altobi, Yousuf Alwardi, and Mohamed Alabri. This research is part of a project funded by the Ministry of Interior, Sultanate of Oman, project number 22411002.

References

Booth, A. El-Kaliouby, H., Swidinsky, A., and Styles, P., 2013. Simulation of time-lapse geophysical datasets in cavity remediation applications. EAGE conference on engineering geophysics, Al Ain, UAE, Nov 24-27.

Hoffmann, G., Rupprechter, M. and Mayrhofer, C. 2013. Review of the long-term coastal evolution of North Oman – subsidence versus uplift. German J. Geosci., 164 (2), p. 237–252.

Bristow, C., Jol, H. 2003. An introduction to ground penetrating radar (GPR) in sediments. Geological Society, London, Special Publications 211 (1), p. 1–7.

Neal, A., 2004. Ground-penetrating radar and its use in sedimentology: principles, problems and progress. Earth-Science Reviews 66 (3–4), p. 261–330.

Thierry, P., Debeblia, N., Bitri, A. 2005. Geophysical and geological characterization of karst hazards in urban environment: appliocation to Orleans (Fracnce). Bull. Eng. Geol. Environ., v. 64, p. 139-150.

Rodriguez, V., Gutie´rrez, F. Green, A., Carbonel, D., Horstmeyer, H., and Schmelzbach, C. 2014. Characterizing Sagging and Collapse Sinkholes in a Mantled Karst by Means of Ground Penetrating Radar (GPR). Environmental & Engineering Geoscience, Vol. XX, No. 2, May 2014, p. 109–132

Youssef, A. El-Kaliouby, H. and Zabramawi, Y. 2012. Integration of remote sensing and electrical resistivity in sinkhole investigation in Saudi Arabia, Journal of applied Geophysics v.87, p 28-39

Youssef, A. El-Kaliouby, H. and Zabramawi, Y. 2012. Sinkhole detection using Electrical Resistivity Tomography (ERT) in Saudi Arabia. Journal of Geophysics and Engineering, v. 9, p.655-663

Wyns, S., Bechennec, F., le Metour, j., Roger, j. and Chevrel, 1992. Geological map of Sur. BRGM

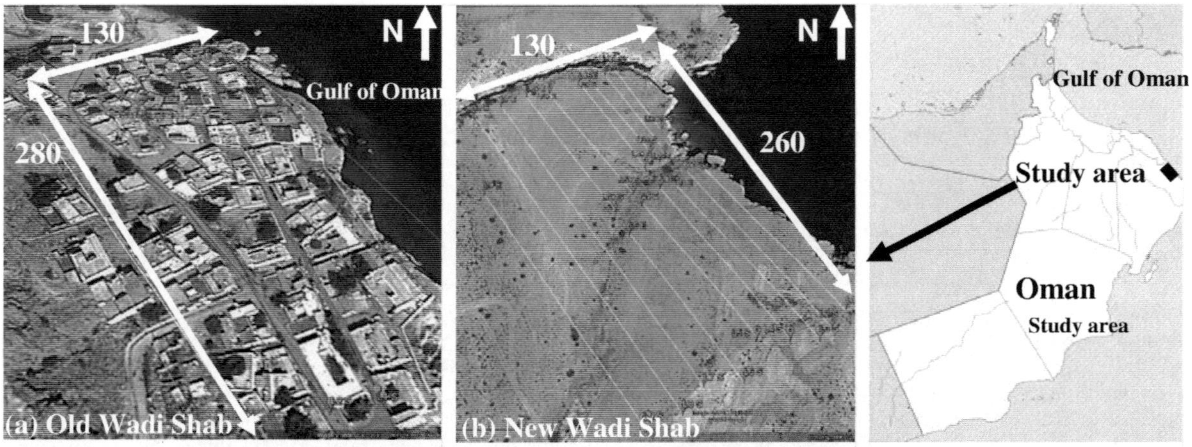

Figure 1. Location of the study area showing the GPR profiles along (a) old wadi Shab abd (b) new wadi Shab.

Figure 2a. Old Wadi Shab

Figure 2b. New Wadi Shab area proposed site

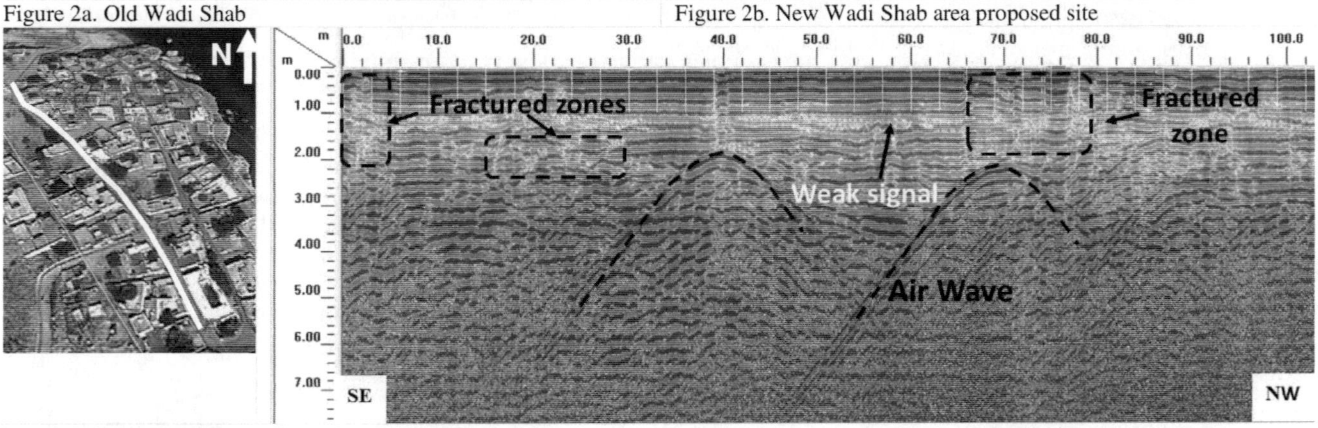

Figure 3. 270 MHz GPR profile at the old Wadi Shab area

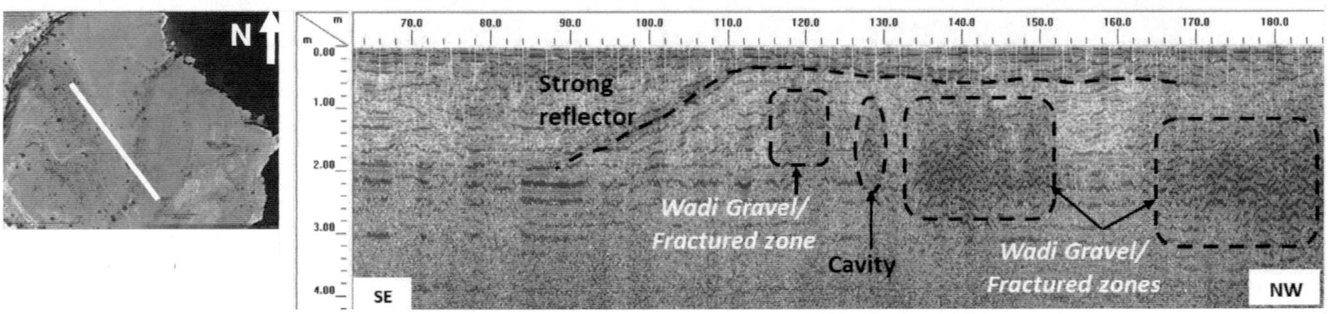

Figure 4. 400MHz GPR Profile at the new Wadi Shab area

Geophysical Study of the mechanism of sinkhole formation in north-east Algeria

F. Khaldaoui[1], M. Djeddi[1,], Y. Djediat[2], A.Ydri[1]*
[1] Laboratoire de Géophysique, [2]Laboratoire de Géodynamique
FSTGAT-Université des Sciences et de la Technologie Houari Boumediene

Summary

The sinkholes are cavities in the ground under whose existence is favored by the dissolution of carbonate rocks. These structures are created by unforeseen events and appear on the surface by collapsing. Carbonate rocks, sensitive karst dissolution, depend on the instability of the geodynamic process which continues to the soil surface.
In the case of presence of the sinkhole, it is important to delineate risk areas in order to define a perimeter of protection for the water resource and geotechnical work.
In this objective a collapse takes place in February 2015 at El Ouldja (North-East of Algeria), a geophysical surveys, including 2D electrical resistivity imaging and electromagnetic induction method (EM31), were employed in the early stages after the initial collapse to determine the risk areas and collapsed features in depth. The work took place on April 2015.

Introduction

At about 50 km northeast of Setif (Algeria), a sinkhole appears in an agricultural area (fig.1). This structure has an ellipsoidal shape, elongated in the north-south direction. The major axis has a length of 17 m. The average depth of this structure is equal to 16 m. Fracturing was observed at the base, oriented NE-SW.
The geophysical methods employed thus make it possible to identify the various geological features and give a better understanding of sinkhole formation mechanisms. Lastly, they can identify zones potentially at the risk for the water resource and geotechnical work.

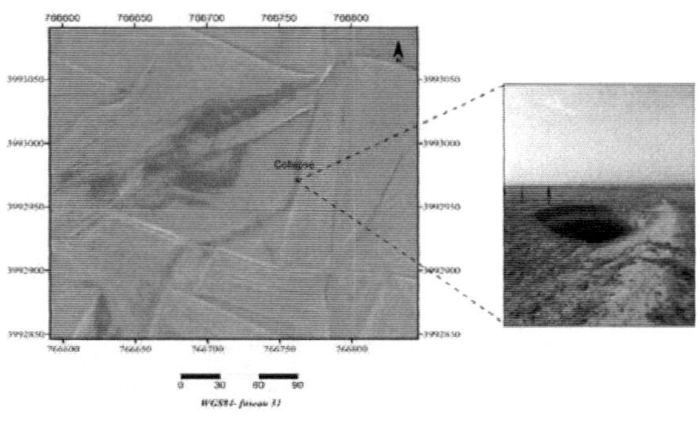

Figure 1: Satellite image of the study area

Geology of the site

Geologically, the sinkhole observed is probably the subsidence of land following the break of karstic vault carved in limestone, appearing in the form of crusts in irregular layers (fig.2).
The sinkhole shows outcrops of carbonate and clastic rocks. We distinguish from the base to the top lacustrine limestones chalky healthy (fig. 2), altered limestone and a succession of fluvial pedogenetic horizons overcome by a horizon marked by white limestone nodules.
The sinkhole is created when chalk dissolves and leads to the development of a karst environment and natural cavities (Klimchouk 2005; Parise 2009; Robert and al 2009; Khaldaoui and al, 2013). Groundwaters charged with carbon dioxide then dissolve the highly porous chalky limestone (Salvati and Sasowsky, 2002, Kaufmann and Quinif, 2002).

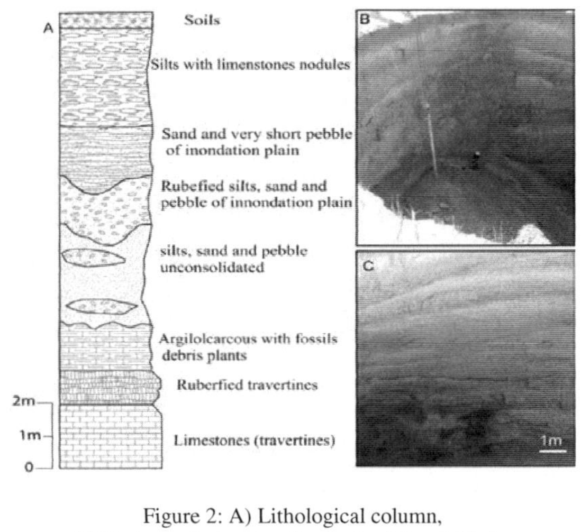

Figure 2: A) Lithological column,
B) Photograph showing the depth of the sinkhole,
C) Dissolution at the base

Materiel and methods

We surveyed the area study using Electrical resistivity imaging (ERI) and controlled source EM system. The resistivity measurements were carried out using multi-electrode system (IRIS instruments) with 48 electrodes. The ERI survey, using dipole-dipole array, was conducted along several profiles (Fig.4) with a inter electrode distance of 5 m. Resistivity measurements were incorporated into the inversion process for create models subsurface resistivity in two-dimensional (Loke and Barker, 1996 a,b). The EM31 system uses a coil transmitter that generates a primary magnetic field at 9.8 kHz. The distance between transmitter and receiver coils is 3.6m. The coils may be used in the HCP mode (with coil axes vertical) which has an effective depth exploration of up to 6m. Measurements were made at 5 m increments along selected traverse and parallel lines.

Results and discussion

Overall, the resistivity model shows heterogeneity in the soil structures. In the middle of ERT1 Profile (fig.3b), at the abscissa x = 120m, a resistive anomaly is observed (100 Ω.m) whose extension is vertical and can reach 20m deep. Specifically, this structure appears vertically a succession of resistive and conductive horizons; probably it is of nested paleochannels. Figure 3b shows the rise of a resistant level at 20 m depth. It is probably the limestone, which is in the form of travertines; a wedging drilling would be desirable at this location. The roof of this resistant level is sub-horizontal except at the point x = 115m, or the observed sinkhole is located near the positions x = 110m and 130m. This result shows the probable extension of the structure of the West. The second layer attributed to deposits alluvial is characterized by resistivity value than 10 Ωm. This geoelectric layer is 8 meters thick.
In this case one could assume that the sinkhole would have formed after the loss of lift of karstic cave vault.

The apparent electrical conductivity map (Fig. 3a) gives an image up to 6m deep. In Figure 3, a resistivity anomaly 30 Ω.m is observed, which is located on the west axis of the sinkhole. The position of both sides of the sinkhole shows a ground conductor (50 mS / m) which correspond to alluvial deposits is also recognized in ERT1. At the abscissa x = 30m (Figure 4a), appears as the circular anomaly which are associated with probable karstic cavities developed within the carbonates rocks

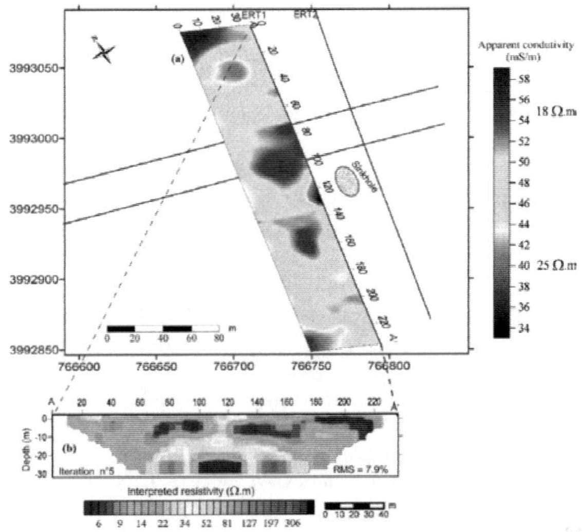

Figure 3: Location of the survey area and geophysical interpretation, a) Electrical resistivity imaging for Profile ERT1, b) Apparent electrical conductivity map (EM31)

Conclusion

The results obtained by the application of the Electrical Resistivity imaging (ERI) and electromagnetic induction confirm the ability of these techniques to delineate a potential risk area. The ERI revealed high resistivity zone is close to the observed sinkhole. This field is also detected by the EM31. All electrical sections have detected the sub-horizontal bedrock at depths of between 15 and 20 m. north of the studied plot, an area at risk is identified by the EM31. The ERI revealed high resistivity zone is close to the observed sinkhole. This field is also detected by the EM31. All electrical sections has detected the sub-horizontal bedrock at depths of between 15 and 20 m. at North of the studied area, an area at risk is identified by the EM31.

References

Kaufmann, O. and Quinif, Y., 2002. Geohazard map of cover-collapse sinkholes in the Tournaisis area, southern Belgium. Engineering geology, **65**, 117-124

Khaldaoui F., Djeddi M., Abtout A. and Baker H.A., 2011. Geophysical investigation to characterize geological risks: Detection zones of dissolution in karst areas, Algeria. 1st Conference on Engineering Geophysics. EAGE, Extended Abstracts, EG37.

Klimchouk, A., 2005. Subsidence hazards in different types of karst: evolutionary and speleogenetic approach. Environmental geology **48**, 287–295

Loke, m.-H. and Barker, r.-d. 1996a. Rapid leastsquare inversion of apparent resistivity pseudosections by a quasi-Newton method. GEOPHYSICAL PROSPECTING, **44**, 131-152.

Loke, m.-H. and Barker, r.-d. 1996b. Pratical techniques for 3D resistivity surveys and data inversion. GEOPHYSICAL PROSPECTING , **44**, 499-523.

Parise, M., De Waele, J., Gutiérrez, F., 2009. Current perspectives on the environmental impacts and hazards in karst. Environmental geology, **58**, 235–237.

Robert, T., Dassargues, A., Brouyère, S., Kaufmann, O., Hallet, V. and Nguyen, F, 2009.Using electrical resistivity tomography and self-potential methods for wells implementations in fractured limestones.15th European Meeting of Environmental and Engineering Geophysics. Near Surface, A32.

Salvati R. And Sasowsky I.D., 2002**.** Development of collapse sinkhole in areas of groundwater discharge: in JOURNAL OF HYDROLOGY **264**, 1-11.

Acknowledgements

The authors thank the local authorities of the municipality of Ouldja W. Setif (Algeria)

Strong motion source model for the 2003 Boumerdes, Algeria, earthquake

Faouzi GHERBOUDJ, [1]* *Hiroe Miyake,*[2] *Toshiaki Yokoi,*[3] *Nasser Laouami,*[1]

[1] *CGS, National Center of Applied Research in Earthquake Engineering, 1 Rue Kaddour* Rahim, Hussein Dey, B.P. 252, Algiers, Algeria.

[2] *Earthquake Research Institute, University of Tokyo, Tokyo, Japan*

[3] *International Institute of Seismology and Earthquake Engineering / Building Research Institute, Tsukuba, Japan.*

Summary

The 2003 M_w 6.8 Boumerdes earthquake caused serious damage to Boumerdes prefecture and the northeast part of Algiers. The strong motions were recorded at several seismic stations by the national accelerometer network. In this study, we have applied empirical Green's function method formulated by Irikura (1986) and Irikura & Kamae (1994) to the 2003 Boumerdes earthquake to determine the strong motion source model that best reproduces near fault records in the 0.2-10 Hz frequency band.

The best source model was obtained by fitting the synthetic accelerations, velocities, and displacement waveforms to the observed ones. Seismic wave propagation and site amplification effects on the strong ground motion are explicitly taken into account considering the aftershocks recorded at same stations as empirical Green's functions.

The source models defined in this study was compared to the scaling laws in terms of rupture area, strong motions generation area and rise time. It shows good agreements with the other source models for crustal earthquakes.

Introduction

Northern part of Algeria has experienced several damaging earthquakes during the last 50 years. The most recent damaging event is the 2003 Boumerdes earthquake (M_w 6.8), located around 50 km northeast of Algiers, the capital of the country. The hypocenter was located in the offshore, causing extensive damage and human casualties (Ayadi et al., 2003; EERI, 2003). An accelerograph network monitored by the Algerian National Research Center of Earthquake Engineering (CGS) recorded a number of waveform data for this event.

In this study, the empirical Green's function method (EGFM) is used in this study for characterizing strong motion source model. This is one of the most important methods for simulating strong ground motions by using ground motion of the small event recorded in the same site to simulate the large event. The synthetic and observed ground motions in terms of acceleration, velocity, and displacement are compared in order to fit the best source model called strong motion generation area (SMGA)

THEORY AND METHODOLOGY: EMPIRICAL GREEN'S FUNCTION METHOD

One of the most effective methods for simulating the broadband strong motion that comes from a large earthquake is to use observation records from small earthquakes that have occurred around the source area of the large earthquake. The use of small events as Green's function attempts to overcome difficulties in consideration of the actual geological structure from the source to the site. In this section, a description is given of source modelling and ground motion simulation applications using the empirical Green's function method for several earthquakes based on Irikura (1986).

In the first scaling relation between small and large events, the parameters studied by Kanamori and Anderson (1975) and by Yokoi and Irikura (1991) are expressed by the following equation:

$$\frac{L}{l}=\frac{W}{w}=\frac{D}{Cd}=\frac{T}{\tau}=\left(\frac{M_0}{m_0}\right)^{\frac{1}{3}}=N \qquad (1)$$

The time series of the target event $U(t)$ can be expressed in terms of the small events as follow:

$$U(t)=C\sum_{i=1}^{N}\sum_{j=1}^{N}\frac{r}{r_{ij}}F\left(t-t_{ij}\right)*u(t), \qquad (2)$$

Where r, r_{ij}, r_0 are the respective distances of sites to the hypocenter of a small event, to the (i, j) subfault and to the starting point of rupture on the fault plane of a large event, $F(t)$ is the filtering *function.*

Results and discussion

The comparison between the observed and the simulated strong motions at stations TZO and ST1 for the M_w 4.7 aftershock and the M_w 4.3 aftershock respectively are shown in Figure 1 and 2, respectively. The comparisons show good agreements.

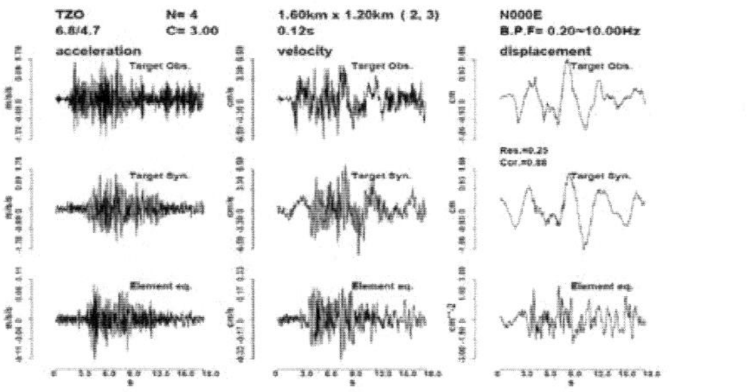

Figure 1. Comparison of the observed and synthetic acceleration, velocity, and displacement waveforms using the Mw 4.7 aftershock for TZO station.

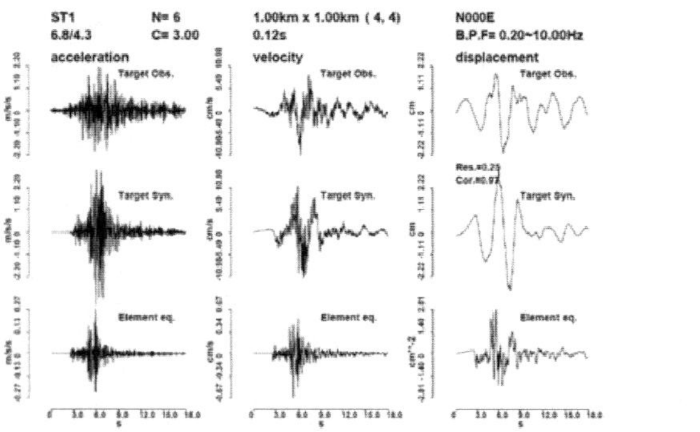

Figure 2. Comparison of the observed and synthetic acceleration, velocity, and displacement waveforms using the Mw 4.3 aftershock for ST1 station.

We focused on estimating rupture starting point, strong motion generation area and rise time by applying trial and modelling that produce a good match between the observed and simulated ground motions. The rupture starting point was selected for all the subfaults and it is found to be near the center of each strong motion generation area.

For the Mw 4.7 aftershock, the SMGA and the rise time are S = 30.72 km^2 and T = 0.48 s respectively. For the Mw 4.3 aftershock, we obtained S = 36 km^2 and T = 0.72 s.

The source models defined in this study was compared to the scaling laws in terms of rupture area, strong motions generation area and rise time. It shows good agreements with the other source models for crustal earthquakes.

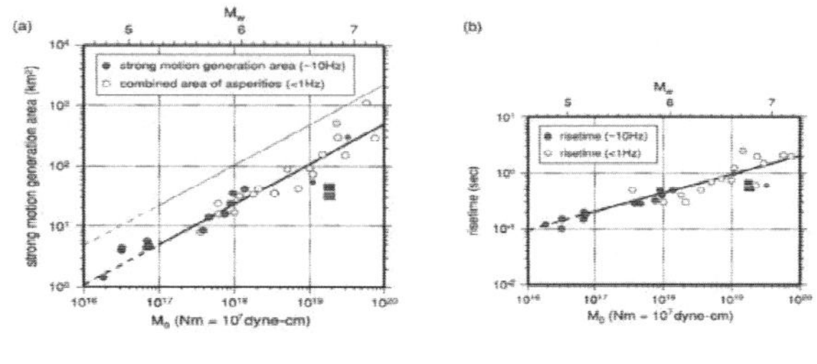

Figure 3. (a) Source scaling of the combined areas of asperities and strong motion generation areas to seismic moment. (b) Source scaling of the rise time with seismic moment. In both panels, the red and blue squares mark the results of this study using the Mw 4.7 and Mw 4.3 aftershocks, respectively. The open and solid circles show the results of Somerville et al. (1999) and Miyake et al. (2003), respectively.

Conclusions

The source model for strong motion was obtained using the two aftershock data. For the M_w 4.7 event, the ground motion generation area was found to be equal to 6.4 km in length and 4.8 km in width, and for the M_w 4.3 event, the source area was found to be equal to 6.0 km in length and 6.0 km in width. The rise time is estimated to 4.8s and 7.2s, respectively. The starting point was found at the northeast bottom for the both models. This model can be used for strong motion simulation in the near field region with sufficient accuracy for engineering purposes.

REFERENCE

Ayadi, A., and 29 co-authors., 2003, Strong Algerian earthquake strikes near capital city, Eos Trans. AGU, 84(50), 561- 568.

Irikura, K., and K. Kamae., 1994, Estimation of strong ground motion in broad-frequency band based on a seismic source scaling model and an empirical Green's function technique, Annali Di Geofisica, 37, 1721-1743.

Kanamori, H., and D. L. Anderson, 1975, Theoretical basis of some empirical relations in seismology, Bull. Seism. Soc. Am, 65, 1073–1095.

Miyake H., T. Iwata, and K. Irikura., 2003, Source characterization for broadband ground-motion simulation: Kinematic heterogeneous source model and strong motion generation area, Bull. Seism. Soc. Am, 93, 2531–2545

Somerville, P., K. Irikura, R. Graves, S. Sawada, D. Wald, N. Abrahamson, Y. wasaki, T. Kagawa, N. Smith & A. Kowada., 1999, Characterizing earthquake heterogeneous slip models for the prediction of strong motion, Seism. Res. Lett, 70, 59-80.

Yokoi, T. and K. Irikura., 1991, Empirical Green's FunctionTechnique Based on the Scaling Law of Source Spectra, Zisin, Ser. 2, 44, 109-122.

Water level variations and earthquakes in the Lake Van Area, Eastern Turkey

*Murat UTKUCU**
Sakarya University, Engineering Faculty, Department of Geophysics, Serdivan, 54187 Sakarya, Turkey

Hasan ARMAN
United Arab Emirates University, College of Science, Department of Geology, Al Ain, United Arab Emirates

Summary

The Lake Van area has been a seismically active region as indicated by historical and instrumental seismicity studies and exemplified by the recent occurrence of the devastating 23 October 2011 Van earthquake (*M*w 7.1). The present study is mainly focus on the seismicity of the area after 1000 AD. The seismicity is compiled from several historical sources, historical seismicity studies and the existing seismicity catalogues. The impact of the large earthquakes and possible source faults are discussed. The seismicity is then compared with the existing water level change reports of the Lake Van for further discussing of the interaction of the earthquakes with the other natural phenomena at work and earthquake hazard in the future.

Introduction

Lake Van is situated on the high plateaus of eastern Turkey, in a closed basin (Figure 1). The origin of Lake Van was principally attributed to the eruption of the Nemrut strato-volcano located NW of the lake (Wong and Degens 1978; Degens et al. 1978). The lake is the biggest lake of Turkey, ranks fourth in volume among the closed lakes of the world and has a surface area of 3713 km^2 (Kadıoğlu et al. 1997). The lake basin has been determined to contain two provinces, Tatvan and Deveboynu basins, from the seismic reflection and bathymetry data (Wong & Degens 1978; Degens *et al.* 1978; Litt et al. 2009; Çukur et al. 2013) (Figure 1). The greatest depth (451 m) was recorded in the Tatvan Basin (TB) (Figure 1).

The present study is concentrated on the water level changes of Lake Van after 1000 AD and their possible relation with the seismicity in the Lake Van Area. The impact of the earthquakes and the relation of the seismicity with the tectonic features lying in the region are discussed.

Tectonic setting

General tectonic setting of the Eastern Turkey and the Lake Van Area as well is mainly controlled by the collision of roughly northerly moving Arabian plate with the Anatolian plate (Figure 1) (McClusky et al. 2000; Reilinger et al. 2006). Anatolian-Arabian plate collision takes place along a deformation zone known as Bitlis Thrust Zone (BTZ). Recent geodynamic studies have clarified that the deformation due to Arabian plate motion is accommodated through distributed strike-slip faults in the Eastern Turkey, which transfers the deformation to the Caucasus thrust zone.

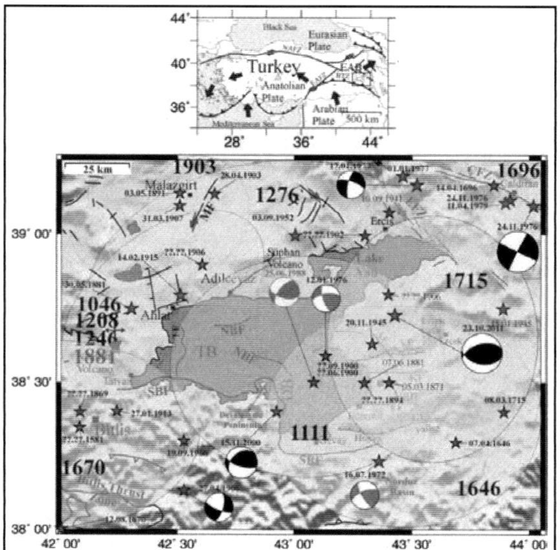

Figure 1: The map showing general tectonic features of Turkey (top) and the tectonic features, seismicity and the earthquake source mechanism solutions of Lake Van area along with the bathymetry of Lake Van (bottom). The area effected by the large earthquakes are indicated by grey shaded ellipses. The $M \geq 5.0$ earthquakes after 1860 are shown with white and black filled stars (macro seismic and instrumental epicenters, respectively). NAFZ: North Anatolian fault Zone, EAFZ: East Anatolian fault Zone, BTZ: Bitlis Thrust Zone, EAB: Eastern Anatolian Block, SBF: Southern Boundary Fault, NFZ: Northern Boundary Fault, ÇFZ: Çaldıran Fault Zone, MF: Malazgirt Fault, TF: Tatvan Fault, EFZ: Erciş Fault Zone, TFZ: Tutak Fault Zone, TB: Tatvan Basin, DB: Deveboynu Basin, AF: Alabayır Fault, KF: Kalecik Fault.

Seismic reflection and bathymetry data (Wong and Degens 1978; Degens et al. 1978; Litt et al. 2009; Çukur et al. 2013) revealed that the lake basin is almost completely bounded by steep boundary faults, namely Northern and southern Boundary Faults, all having downthrown side lake-basin ward (Figure 1).

Seismicity of the lake Van Area

The complex and active tectonic setting of the Lake Van area has caused high seismic activity with damaging earthquakes in the historical (before 1900 AD) and instrumental (after 1900 AD) periods (Figure 1). Historical (Sinclair 1999; Ambraseys 2009; Albini et al. 2012) and instrumental period (Kalafat et al., 2007) seismicity studies and geological data recovered from the offshore and onshore lake sediments (Litt et al., 2009; Üner et al., 2010) indicated many earthquake in the past. The occurrence of the devastating 23 October 2011 Van earthquake (Mw 7.1) was a last and a clear indication of the high seismic activity and hazard in the area (Utkucu 2013).

The existing seismicity catalogues (Kalafat et al., 2007) and historical seismicity studies (e.g. Ambraseys 2009) have been carefully scrutinized for retrieval of the all-damaging earthquakes during the historical period and $M \geq 5.0$ earthquakes in the instrumental period (Figure 1).

Water Level Changes of Lake Van

The fact that the water level of Lake Van has shown alternating rises and decreases in history is another natural phenomenon observed in the area, causing economic, environmental and social problems over the littoral area. Although the water level changes had been obtained to be in the order of tens of meters before 500 A.D. (Figure 2a), it has been relatively stable and fluctuating in the order of a few meters during the past 1500 years (Figure 2b and 2c) (Degens et al. 1978; Sinclair 1999; Utkucu 2006; Reimer et al. 2009; Kuzucuoğlu et al. 2010).

The General Directorate of Electrical Power Resources Survey and Development Administration (GDEPRSDA) has measured the water level of Lake Van instrumentally since 1944 (GDEPR 2002). The monthly change of the mean water level between 1944 and 2010 is shown in Figure 2c along with $M \geq 5.0$ earthquakes after 1944 and $4.0 \leq M < 5.0$ earthquakes after 1970. Notice that episodic water level changes (either increase or decrease) were about 1 m or larger between 1968 and 1970, between 1987 and 1996, between 1998 and 2002 and between 2002 and 2010. The water level change between 1987 and 1996, during which the water level increased episodically about 2 m, brought about economic, environmental and social problems over the littoral area (Utkucu 2006; Deniz and Yıldız 2007).

Discussion

Occurrence of large magnitude earthquake in the Lake Van area is not surprise considering the historical earthquake activity. The notorious 1246 Ahlat, 1275 Erciş, 1646 and 1715 Van and 1696 and 1976 Çaldıran earthquakes can exemplify this fact (Ambraseys 2009). The damage of the 1715 Van earthquake indicates a possible source fault location more northward than the 1646 earthquake in the area with a heavy damage at Van, Erciş and Hoşap and around the Erçek Lake. Regarding their damage distribution pattern, it is much likely that the 2011 Van earthquake is recurrence of the 1715 Van earthquake and the 1646 Van earthquake was produced by a fault lying somewhere south of Van city center, much likely by the Gürpınar fault along the Hoşap valley (Figure 1).

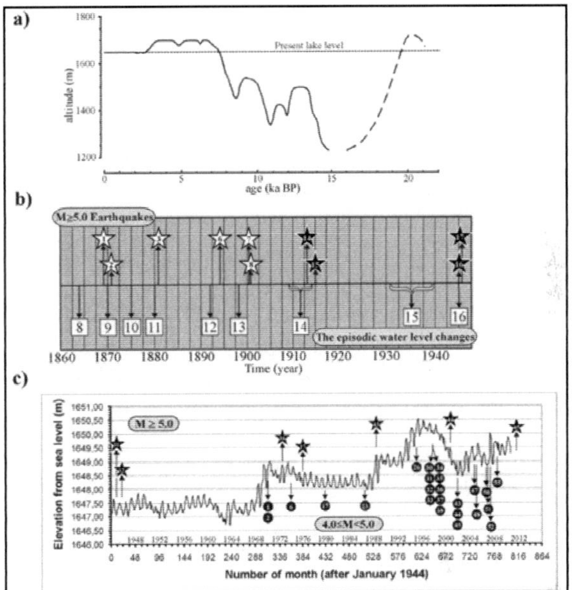

Figure 2: Water level fluctuations of Lake Van in history. (a) Geologically derived early historical lake level changes (Reimer et al. (2009) along with (b) historically reported (after Utkucu 2006) and (c) continuously measured water level changes after 1944 (GDEPR 2002).

Utkucu (2006) claimed that the anomalous change of the annual mean of the water level in the order of 1m had been accompanied or followed by $M \geq 5.0$ earthquakes. As noticed from Figure 2c, episodic water level changes (either increase or decrease) of about 1 m or larger between 1968 and 1970, between 1987 and 1996 and between 1996 and 2002 were all accompanied or followed by the $M \geq 5.0$ earthquakes and increasing numbers of $4.0 \leq M < 5.0$ earthquakes. The 1945 Van earthquake sequence also followed an up rise that was reported to be took place after the late 1930s and the level was anomalously high in 1945 (Figure 2b) (Utkucu 2006). However, analyses of the instrumental catalogue have shown that magnitude of completeness (Mc) is larger than 4.0, between 1970 and 1980 in particular, but not above 4.5. If we take only $M > 4.5$

earthquakes along the basin related faults we find out that 3 $4.5 \leq M < 5.0$ earthquakes occurred in the time period 1970-1988 and 6 $4.5 \leq M < 5.0$ earthquakes occurred in the time period 1989-2011, during which the lake level was generally 1 m above the former period.

Between 2002 and 2010, an episodic up rise of about 1 m in the average lake level took place. This up rise was followed by the occurrence of the 2011 Van earthquake. An ascend of in the order of 1 m in the annual mean again was followed by a $M \geq 5.0$ earthquake along a lake basin related fault, supporting the assertion made by Utkucu (2006). The up rise could have been ended in 2011 but this makes no difference for the lake level change triggering because there is a certain time lag between fluctuations of water loading at surface and resulting pore pressure change at depth.

What was the reason behind the level changes of Lake Van? The recent changes have been pointed out to have a climatic origin rather than a tectonic origin (Kempe et al. 1978; Kadıoğlu et al. 1997) as for many observed in the geologic past (Reimer et al. 2009). It has long been known that natural phenomena on the Sun effect the climate changes in the Earth but the relationship is not well understood. Interestingly, Kempe et al. (1978) introduced sunspot activity as a possible cause of level change regarding good correlation between sunspot number and the lake level changes between 1944 and 1974, in spite of a deviation after 1974. Kadıoğlu et al. (1997) has put forward some other effects, which they considered as the climate change related events, as a cause of the deviation.

It seems that sun spot activity on the Sun induces climate changes in Lake Van area and these climate changes triggers earthquakes in this particular place due to presence of a large natural water reservoir (i.e. Lake Van) with no natural or artificial outlet and high tectonic activity. The sun spot activity is one of the causes behind the climate changes and sometimes dominates as observed during the 1944-1974 period. For example, lasting from about 1790 to 1835, the sunspot activity was notably low in a period known as Dalton Minimum, after which sun spot activity increases to a level comparable to the period 1780-1790. Between 1838 and 1841 the lake level increased 3-3.5 m, which forced the medieval settlement of Erciş town to be displaced further inland (Utkucu 2006). This example shows that the sunspot activity might indirectly effects the lake level. The lake level changes can then trigger the earthquakes along the basin related faults that have near failure threshold. It should finally be emphasized that the sunspot activity is not a unique factor to change the lake level and that earthquakes follow not all the lake level changes.

Conclusions

Occurrence of large magnitude earthquake in the Lake Van area is not surprise considering the historical earthquake activity. The notorious 1246 Ahlat, 1275 Erciş, 1646 and 1715 Van and 1696 and 1976 Çaldıran earthquakes, can exemplify this fact. The 23 October 2011 Van earthquake is much likely a recurrence of the 1715 Van earthquake because of the similarity of their damage patterns. The study suggests that the episodic water level fluctuations of Lake Van in the order of 1 m trigger $M \geq 5.0$ earthquakes along the lake basin related faults and that the sun spot activity is one of the reason behind the fluctuations because it induces climate changes. Nevertheless, the sunspot activity is not a unique factor to change the lake level but rather one of the causes behind that sometimes dominates as observed between 1944 and 1974.

References

Albini, P., M. B. Demircioglu, M. Locati, A. Rovida, K. Sesetyan, M. Stucchi, and D. Vigano, 2012, In search of the predecessors of the 2011 Van (Turkey) Earthquake, Seismological Researc Letters, 83 (5) doi: 10 .1785/0220110146.

Ambraseys, N., 2009, Earthquakes in the Mediterranean and Middle East: a multidisciplinary study of seismicity up to 1900: Cambridge University Press.

Çukur, D., S. Krastel, F. Demirel-Schlüter, E. Demirbağ, C. Imren, F. Niessen, M. Toker and PaleoVan-Working Group, 2013, Sedimentary evolution of Lake Van (Eastern Turkey) reconstructed from high-resolution seismic investigations: International Journal of Earth Science (Geol Rundsch), 102, 571–585, DOI 10.1007/s00531-012-0816-x.

Degens, E. T., H. K. Wong, F. Kurtman, and P. Finkch, 1978, Geological development of Lake Van *in* E.T. Degens, and F. Kurtman, F., eds., The Geology Lake Van: M.T.A. Press, No. 169, 134-146.

Deniz, O. and M.Z. Yıldız, 2007, The ecological consequences of level changes in Lake Van: Water Resources, **34** (6), pp. 707-711.

GDEPR, 2002, Göl seviyeleri, General Directorate of Electrical Power Resources Survey and Development Administration Report, 107 p.

Kadıoğlu, M., Z. Şen, and E. Batur, 1997, The greatest soda-water lake in the world and how it is influenced by climatic change: Ann. Geophysicae, **15**, 273-279

Kalafat, D., Y. Güneş, M. Kara, P. Deniz, K. Kekovalı, S. Kuleli, L. Gülen, M. Yılmazer, and N. Özel, 2007, A revised and extended earthquake catalogue for Turkey since 1900 (M≥4), Boğaziçi University, 553 p.

Kempe, S., F. Khoo, and Y. Gürleyik, 1978, Hydrography of Lake Van and its drainage area, *in* E.T. Degens, and F. Kurtman, F., eds., The Geology Lake Van: M.T.A. Press, No. 169, 30-44.

Kuzucuoglu C., A. Christol, D. Mouralis, A.F. Dogu, E. Akköprü, M. Fort, D. Brunstein, H. Zorer, M. Fontugne, M. Karabiyikoglu, S. Scaillet, J-L. Reyss, H. Guillou, 2010, Formation of the Upper Pleistocene terraces of Lake Van (Turkey): Journal of Quaternary Science, 25, 1124–1137. DOI: 10.1002/jqs.1431.

Litt, T., S. Krastel, M. Sturm, R. Kipfer, S. Örçen, G. Heumann, S.O. Franz, U.B. Ulgen, and F. Niessen, 2009, 'PALEOVAN', International Continental Scientific Drilling Program (ICDP): Site Survey Results and Perspectives: Quaternary Science Review, 28, 1555–1567. doi:10.1016/j.quascirev.2009.03.002

McClusky S, Balassanian S, Barka A, Demir C, Ergintav S, Georgiev I, Gürkan O, Hamburger M, Hurst K, Kahle H, Kastens K, Nadariya M, Ouzouni, A, Paradissis D, Peter Y, Prilepin M, Reilinger R, Sanli I, Seeger H, Tealeb A, Toksöz MN, Veis G , 2000, GPS constraints on plate kinematics and dynamics in the Eastern Mediterranean and Caucasus. J Geophys Res 105:5695-5719.

Reilinger, R., et al., 2006, GPS constraints on continental deformation in the Africa-Arabia-Eurasia continental collision zone and implications for the dynamics of plate interactions: Journal of Geophysical Research, **111**, B05411, doi:10.1029/2005JB004051.

Reimer, A., G. Landmann, and S. Kempe, 2009, Lake Van, Eastern Anatolia: Hydrochemistry and History Aquatic Geochemistry, **15**, 195–222, doi: 10.1007/s10498-008-9049-9.

Sinclair, T., 1999, Two problems concerning the Van Region: Arakel of Tabriz on the earthquake of 1646 and evidence for the rise in the level of the lake. In: Zachariadou E, (ed.), Natural Disasters in the Ottoman Empire, Crete University Press, pp 207-222.

Utkucu, M., 2006, Implications for the water-level-change triggered moderate (M≥4.0) earthquakes in Lake Van basin, Eastern Turkey: Journal of Seismology, **10**, 105–117, doi: 10.1007/s10950-005-9002-y.

Utkucu, M., 2013, 23 October 2011 Van, Eastern Anatolia, earthquake (*M*w 7.1) and seismotectonics of Lake Van Area: Journal of Seismology, **17**, 783-805, doi: 10.1007/s10950-012-9354-z.

Üner, S., Ç. Yeşilova, T. Yakupoğlu, and T. Üner, 2010, Earthquake Induced Soft Sediment Deformation Structures (Seismites):Van Gölü Basin, Eastern Anatolia: Yerbilimleri 31(1), 53–66.

Wong, H. K., and E.T. Degens, 1978, The bathymetry of Lake Van, eastern Turkey, *in* E.T. Degens, and F. Kurtman, F., eds., The Geology Lake Van: M.T.A. Press, No. 169, 6-10.

Detection of subsurface karstic features using electrical resistivity tomography, case study, Al Ain, United Arab Emirates

*Tanzeel Ur Rehman Sabir *, Mahmoud K. Harb, Emad Sharif and Mohd.J.Ahmed, Arab Center for Engineering Studies, Dubai, UAE*
t.rahman@aces-dubai.ae

Summary

Subsurface karstic features are one of the major concerns for the construction sector in United Arab Emirates. Subsurface cavities form an important hazard at Al-Ain city and the suburbs. These cavities have led to ground subsidence eventually leading to damages to infrastructures like buildings, (Shaika Salam Mosque at the center of the Al-Ain city is glaring example) such as sagging and cracking of road surfaces at several locations. This has led to acquiring geophysical information during site investigations which have become a mandatory requirement for the construction projects within AlAin Municipality. These investigations aid in detection and mapping of weak zones, Karst and solution features, further utilized for hazard and risk assessments of the survey area. The interpretation of ERT-2D data shows the existence of several subsurface open and sediment in-filled cavities and zones of weakness in rocks, particularly above the ground water table within the limestone layers. Correlation between ERT results and ground materials are established to prepare risk maps, which identifies high and low risk areas within the site based on specific criteria, which includes resistivity anomaly distribution, results of verification boreholes, structural load and footing design.

Introduction

Near surface geophysical methods have been widely used in geotechnical applications over the last two decades to detect underground cavities, cellars, caves etc. (Anderson et al., 2007). This important information is provided for geotechnical concerns like the detection of boundaries between geological formations, differences in the rock/soil water saturation, the presence of weak features and the depth to bed rock-etc. (Psomiadis et al., 2009). These cavities can be empty or partially filled with water or soil or both. Different geophysical techniques are deployed to detect these karstic features to characterize the variations in the physical parameters of underground formations. To observe the physical parameters, geophysical techniques produce a set of data. Each of these parameters is related to one or more physical properties of the subsurface and to their spatial distribution (Psomiadis et al., 2009). The use of electrical resistivity technique in identification of cavities in buried walls and subsurface rocks at shallow depth has long been a common practice (Sarris and Jones 2000; Drahor et al., 2008; Berge et al., 2011, Leucci and Greco, 2012).

The investigated site is located close to Al-Ain city center, in U.A.E. (Figure 1). The study demonstrates the efficiency of the electrical methods for mapping subsurface conductive zones. In order to calibrate and verify the findings of the geophysical investigation, four verification boreholes were drilled at selected locations. The boreholes were drilled to various depths up to 15m below existing ground surface.

The objectives of the geophysical study was to locate and define the shallow/deep cavities, unconsolidated weak zones and soft soil formations, determine the groundwater depth, understand the nature and composition of the soil as well as the geological features under the site within the depth range of investigation. Such sub-surface geologic hazards and their assessments lead to decide if the site is acceptable for proposed structure and type of footings and determines other measures of risk mitigation such as ground improvement, changing the location and type of footing and load design.

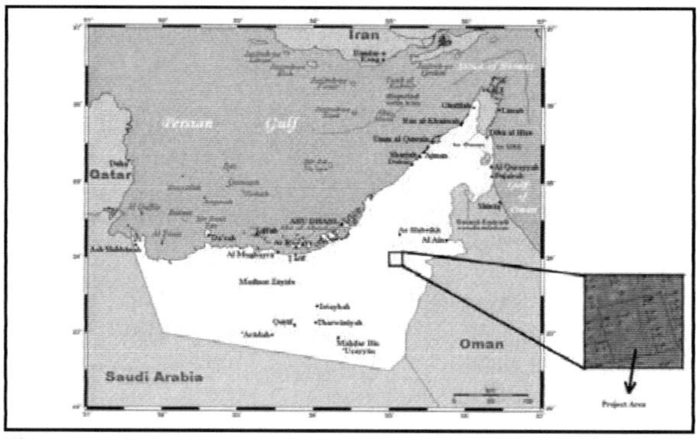

Figure 1 : General location map of the project area

Geology:

The geological evolution of the United Arab Emirates during the Mesozoic and Cenozoic can be directly related to the opening of the southern Neotethys Ocean during the Triassic and its eventual closure in the late Cretaceous to Palaeogene. The majority of the rock units of the UAE were formed within the Neotethys Ocean, but their present-day distribution is largely the result of processes associated with ocean closure

(Alsharhan and Kendall, 1986). This entire carbonate platform succession was deposited on the Arabian continental shelf- the passive margin of the Neotethys Ocean between Permian and early Cretaceous times (Lees, 1928; Hudson et al., 1985b; Glennie et al., 1973; Ricateau and Riche, 1980; Robertson and Searle, 1990). Two main tectonic events have affected the rock units of the UAE as mentioned below:

1) The obduction of the Oman-UAE ophiolite onto the eastern continental margin of the Arabian Platform during the late Cretaceous (Allemann and Peters 1972; Glennie et al, 1974; Lippard et al., 1986); and
2) Later (post-obduction) to thrusting and folding and the formation of a fore land basin, which affected both the Mesozoic and Palaeogene sedimentary sequences (Ricateau and Riche 1980; Searle et al., 1980) in the region.

The general geological succession around the project area is as follows:

Above succession forms a part of the Aruma group deposited in the foreland basin. The late Paleocene to lower Middle Eocene mudstone dominated Muthaymimah Formation was deposited, which developed locally the basal conglomerates, derived from the underlying Simsima Formation. The succession rapidly passes upwards into a sequence of deep marine lime-mudstones (Hamdan and Anan, 1993). The lime-mudstones, wacke stones and pack-stones of the mid to late Eocene- Dammam Formation overlie the underlying Rus Formation. The Dammam Formation was deposited in a more dynamic setting which oscillated between relatively deep water, outer shelf, and a shallow shelf environment (Lippard et al., 1986). The middle Miocene to Pliocene Barzman Formation represents the youngest part of the foreland basin succession and consists of a series of matrix-supported fluvial lithic-sandstones, mudstones and conglomerates. During Quaternary period; there has been a complex interplay between fluvial, aeolian and marine processes, driven principally by climatic changes associated with high latitude glaciations. The summary of Geological succession around the area is given in Table 1.

Table 1: Summary of geological succession around the site

Middle Miocene to Pliocene	Barjman Formation	Sandstone, siltstone, conglomerates.
Mid to Late Eocene-	Dammam Formation	Shale , siltstone
Late Palaeocene to Middle Eocene	Muthaymi mah Formation	Clay mudstones., basal conglomerates
Lower Eocene	Rus Formation	Chalky/ argillaceous limestone, Anhydrites/ evaporites
Upper Cretaceous	Simsima Formation	Limestones/ Lime mudstone

Structurally in Al-Ain area the Cretaceous to Palaeogene sediment cover, postdating the ophiolite obduction, are relatively weakly deformed. Formations, of Cretaceous to Paleocene age, consisting of an interbedded sequence of limestone, marls and conglomerates are predominantly exposed in the periclinal anticline (fold). The structure of the foreland basin sequence is dominated by a series of N-S trending cylindrical folds. The steeply dipping eastern limb is gently folded to open structure that has resulted in local overturning of bedding. The overturning of beds is associated with minor thrusts developed within the marl/ mudstone beds of Dammam Formation. The slickensides developed by flexural movements are orthogonal to the fold axis. The sedimentary rock in between the folds are deformed and manifested by complex NW-SE trending tight anticlines and synclines, cut by small scale thrusts, in Dammam Formation (Dunne et at al., 1990). The foot hills along the hill range have thick deposits of alluvial fan material, under which the cavernous limestone and marls are expected.

The project site located along the western foot hills of the mountain range these thick (>20 m) fluvial/ alluvial fan deposits consisting of silty medium to fine sand with gravels to boulders (up to 1.5 m) of limestone and calc-nodules (these are leached out limestone gravels and boulders) concretions of different sizes. Many of the drainage systems developed on the piedmont fans are, therefore, complex, consisting of braided multi channel networks with widths exceeding several hundred meters (as on the Al Jaww Plain near AlAin) (Noweir and Eloutefi, 1997). Channelized discharge commonly disperses to very broad, shallow sheet floods across these channeled fans. The main flooding hazards, therefore, extend beyond the limits of the main feeder channels and the shifting of drainage lines in the complex channel networks is a major hazard to development on fan surfaces. The extent of these sediments, and the size of the pebbles and boulders found in them, indicate that at the time of their deposition there was much more rainfall over the area than experienced today.

Methodology:

Electrical methods are used to determine the location, thickness of the unconsolidated materials, nature, and extent of the geological materials like clay beds and bedrock and depth of the water table ((Robinson and Coruh, 1988). Electrical resistivity methods have been applied to many groundwater situations where the resistance of the fluid in the ground varies, for example this happens when there is a plume of saline groundwater that may be the result of salt water intrusion or leachate from a landfill (Fetter, 2001). In electrical resistivity methods, one can measure electric current, electric potential and electromagnetic field induced either naturally or artificially. These are classified on the basis of type of energy source, artificially or naturally, these include Self Potential (SP), Electrical Resistivity (ERS), Electromagnetic (EM) and Induced Polarization (IP) method and Electrical Resistivity Tomography (ERT).

A direct investigation utilizing geophysical ERT technique was used to achieve the objectives of the investigation. This method was chosen for this investigation because of its known efficiency for detecting subsurface voids and weak zones in similar site conditions. ERT uses a multi-core cable, that can be connected with a number of electrodes, pre-planted to the ground at fixed distance knowing as electrode spacing (*dx*), one of the ends of the multi-core is plugged into controller box (switch box), which in turn is connected with the main control unit (resistivity meter). To start the measurement, the current (*I*) is injected through transmitting electrodes and the potential difference (voltage *V*) between two receiving electrodes is measured. After the completion of the first reading, the system will switch off some electrodes toward the start of the section and activate other electrodes toward the upper side of the section for the next reading. The system continues this process until using the largest possibility for using the electrodes as potential and current with the first electrode spacing ($dx=x$), which is called Set No.1 ($n=1$). After completing the sequence of measurements with electrode spacing ($dx=x$), the next sequence of measurements with ($dx=2x$) electrode spacing proceeds. Through these steps the depth of the exploration will increase. This process is continued and repeated for measurements until use of the largest possible distance between the electrodes. At the end of the field measurement, there will be hundreds to thousands of resistivity readings which can be used to generate 2D (twodimensional) resistivity section. After completion of the field work and the measurements passed the quality verifications, the acquired data can be converted for 2D - electrical section utilizing appropriate software which is closely correlated with subsurface geology.

Data Acquisition and Processing:

Measurements were obtained using the dipole –dipole array chosen due to its known good lateral resolution which is important for locating isolated targets (voids). The aim of the ERT survey was to detect and map the likely cavities within the site to the specified depth of significance of 30 m. Table 2 shows the summary of ERT survey parameters in general.

Following field data collection, the measured apparent resistivity values were then used to build a pseudo section.

Table 2 : Summary of ERT Survey Design Parameters.

Survey Purposes	ERT (2D)
Spread Configuration	Dipole –Dipole (Max n =8 & Max Dipole =6)
Electrode Spacing	2.5 m
Profile Spacing	5 m
Number of Electrodes	112
Measuring Mode	Resistivity
Cycles	2
Max error	2.0
Max current	1,250 mA / Auto
Measuring time	0.5 sec
Delay Time	0 sec

The pseudo section gives a very approximate picture of the true subsurface resistivity distribution, however it is a distorted picture because the apparent resistivity values depend on the type of array used as well as the true subsurface resistivity. Nevertheless, the apparent resistivity pseudo section is the base to estimate the true resistivity section. This can be achieved using iterative least square inversion methods. The procedure for processing the apparent resistivity values to achieve the true resistivity values can be summarized into 5 basic steps:

- Check the quality of the raw data (measured apparent resistivity) and filter the raw data to be processed.
- Run the least square inversion procedure with the appropriate inversion parameters.
- Remove bad data points (if exist) to decrease the inversion RMS error.
- Run the least inversion procedure again and repeat until the desired RMS error is achieved.
- Display the resulting true resistivity sections in an appropriate color scale to delineate anomalies if exist.

Results and Discussions:

The survey utilized Electrical resistivity Tomography (ERT) method, to identify possible subsurface fractures/voids, less competent rock that may contribute to more complete knowledge of the subsurface conditions of the site. The electrical resistivity tomography survey provided a two dimensional image of the subsurface along the proposed site at specific traverse line locations. The ERT results indicated that the study area of the project site exhibited distinct resistivity conditions and contrasts. Analysis and interpretation of the findings has identified existence of several small isolated subsurface anomalies of geotechnical concern in the investigated area. The detection of features in an ERT survey is dependent on the presence of a significant resistivity contrast between the feature of interest and the surrounding material. A cavity can be identified in the inverted section of ERT (2D) with high reliability, if a small area (high resistivity) is embedded within a zone of low resistivity above or below the water table. Most of these anomalies were encountered within Limestone or Gravel / Cobbles of limestone. None of these anomalies were an index of cavities of a significant size (>2m) but mostly of fractured zones and small open and sediment in-filled cavities, particularly above the (Ground water table) GWT level.

In addition, water loss was also caused by the existing poorly consolidated sand with cemented sandstone fragments materials (e.g. VBH-01). Water loss related to subsurface cavity filled with air (free fall of core barrel) was encountered in VBH-01, VBH-02 and VBH-04, between depths (6-14.8) m as shown in Table 3.

In general, three to four layers were encountered in all ERT profiles. A superficial layer with thickness of 4 to 8 m below ground level of silty and gravelly sand overlay a thin layer (1 to 2 m) of gravel / cobbles and fragment of limestone over a relatively thick layer of limestone most anomalies of high and low resistivity values were encountered in this formation, underneath this layer of mudstone was encountered in this area.

Table 3 : Summary of water loss and free fall of core barrel encountered in verification boreholes

Borehole	Water Loss (100%) Depth in (m)	Free fall Depth in (m)	Geological Formation
BH-01	8.0 – 150.0	12.5 -12.7 14.5 – 14.8	Limestone
BH-02	8.0 – 15.0	10.9 – 11.5	Limestone
BH-04	8.5 – 12.0	6.0 – 6.5 10.3 – 10.8	Limestone/ Calcisiltite

A layer of gravel / cobbles and fragment of limestone with Rock Quality Designation (RQD) values for 0 % was noticed in all boreholes in this subarea, while the (solid core recovery) SCR ranging between 4 – 24 % indicate low rock quality.

A layer of limestone, buff-white/light greenish grey, distinctly weathered, embedded with voids, fractures a very close to close spaced encountered at depth range of (10.00-16.00m)/(5.00-10.00m) with RQD values ranging between 0-59% and SCR 17 -93% as an indication of medium rock quality. A layer of mudstone was encountered at depth >16.00m and extended to the maximum explored depth below natural ground surface. This layer is interlayered with sand and sandstone pieces. The (RQD) values for this layer are 0% - 55% while the SCR is more than 18% indicating low quality of rock. ERT survey results do not immediately identify the composition of anomalies which may be caused by variations in geology and/or moisture content (or other factors). Final data interpretation is greatly enhanced by calibrating or benchmarking the electrical resistivity images against existing site data and/or follow-up confirmation boring data as shown in Figure 2. Furthermore, 3-Dimensional presentation for whole area is shown in Figure 3. Based on the analysis of the geophysical results and the results of the verification boreholes; site geology considerations; risk assessment was established. Figure 4 shows the risk map for the investigated area

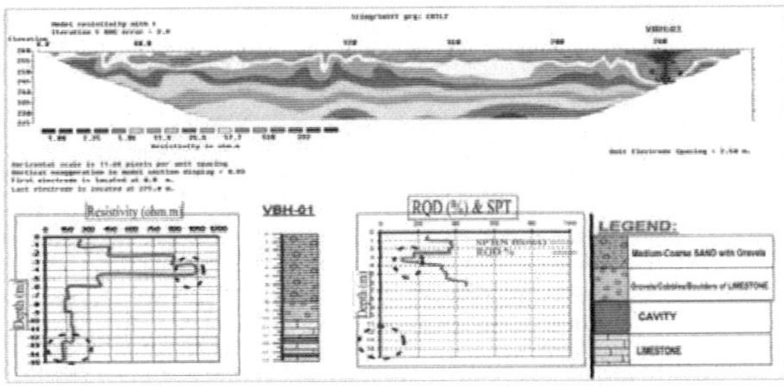

Figure 2 : Borehole log BH-01 imposed on the ERT profile 2.

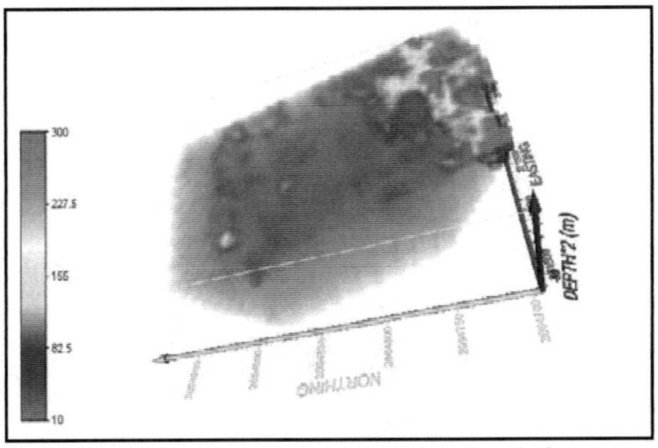

Figure 3 : Three dimensional presentation for whole area.

Conclusion:

Based on the comprehensive analyses of the results, the following conclusions can be drawn from the present survey:

- Only one geophysical technique was used (ERT) and accordingly the encountered high risk anomalies were verified with Four (4) boreholes that provided proven evidence and confirmation of the results of the geophysical findings.
- Correlation between ERT results and geologic materials showed relatively large range of resistivity values for the different material types. This is mostly attributed to differential weathering , fracturing and saturation in the rock mass.
- Subsurface anomalies with different levels of severity have been identified within the site at different depths. A risk assessment map was developed accordingly, which identified areas of high risk zone.
- High risk anomalies exist mostly above or close to the present day GWT and have a channel shape and few isolated locations.

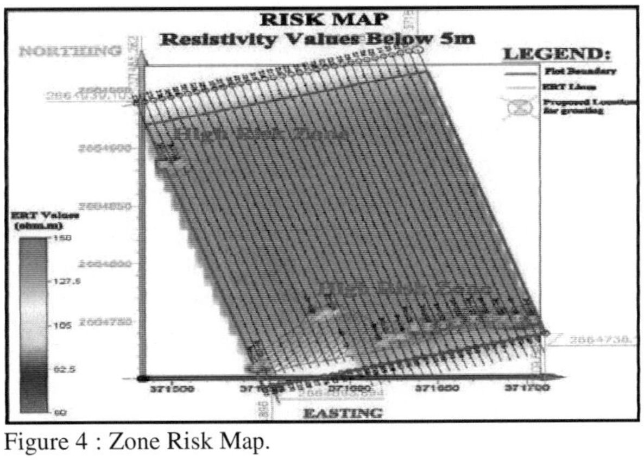

Figure 4 : Zone Risk Map.

Seismic Scenario-Structure-Based Performance Criteria for RC High-rise Wall Buildings

Wael Alwaeli[1], Aman Mwafy[2], Kypros Pilakoutas[1] and Maurizio Guadagnini[1]*
[1]Department of Civil and Structural Engineering, The University of Sheffield, UK.
[2]Civil and Environmental Engineering Department, United Arab Emirates University, UAE.

Summary

The number of RC high-rise buildings is rapidly growing in multiple-scenario earthquake-prone regions around the World as a result of several social and economic considerations. The seismic vulnerability of this category of structures to different earthquake scenarios is poorly understood and proper quantitative limit state criteria to different performance levels are lacking. Motivated by the pressing need to derive more reliable fragility relations to be used in earthquake lose assessments, a methodology is proposed to obtain a new, Seismic Scenario-Structure-Based (SSSB) definition of limit state criteria for RC high-rise wall buildings. A 30-story wall building is utilized to illustrate the methodology. The building is located in Dubai (UAE); a city vulnerable to sever distant earthquakes alongside moderate near-field events therefore selected as a study region in the present study. Multi-Record Incremental Dynamic Analyses (MRIDAs) are conducted to map building local response at different seismic intensities and link it to the global response through a reliable proposed Damage Measure (DM). The study concludes with proposing new, SSSB limit state criteria for the sample building. The proposed methodology and limit state criteria are applicable to a wide range of RC high-rise buildings and seismic regions comparable to those investigated in the current study.

Introduction

The increasing number of high-rise buildings in multiple-scenario earthquake-prone regions over the World necessitates a systematic methodology to conduct seismic vulnerability (fragility) assessment for this population of buildings. The definition of limit state criteria forms the spine to seismic vulnerability assessment and hence to seismic risk analysis programs. Quantitative limit state criteria require mathematical representations of local damage indices (such as deformations, forces, or energy) based on designated structural response levels. Therefore, a suitable damage measure is to be adapted to sufficiently correlate local damage states in the building to their global response. Moreover, wide frequency content in real strong ground motions representative of different earthquake scenarios such as distant and near-field earthquakes can impose different levels of excitation on both the fundamental and higher modes of RC high-rise wall structures. This results in a complex, seismic scenario-based inelastic response. In addition, there are several factors contributing to the failure modes and hence to the definition of limit state criteria in this category of buildings. Among these factors are building total height; axial force level; impact of the supplementary regulations enforced by local authorities; and the local trend in design and construction. Seismic codes and previous studies have proposed a wide range of limit state criteria associated with different performance levels of RC shear walls and wall structures (ASCE 2007; Ghobarah 2004; Ji *et al.* 2009; Mwafy 2012; SEAOC 1999). In these documents, overall Total Interstory Drift (TISD) is used as DM to define limit state criteria with a range of 0.20%-0.83%, 0.52%-1.50% and 1.1%-2.5% for Immediate Occupancy (IO), Life Safety (LS) and Collapse Prevention (CP) performance levels, respectively.

The aim of this study is to propose a methodology to arrive at reliable, SSSB definitions of performance limit state criteria for RC high-rise wall buildings. The proposed methodology is illustrated on a 30-story sample building with RC bearing-wall system located in Dubai (UAE). Figure 1 pictorially summarizes the methodology proposed in the current work.

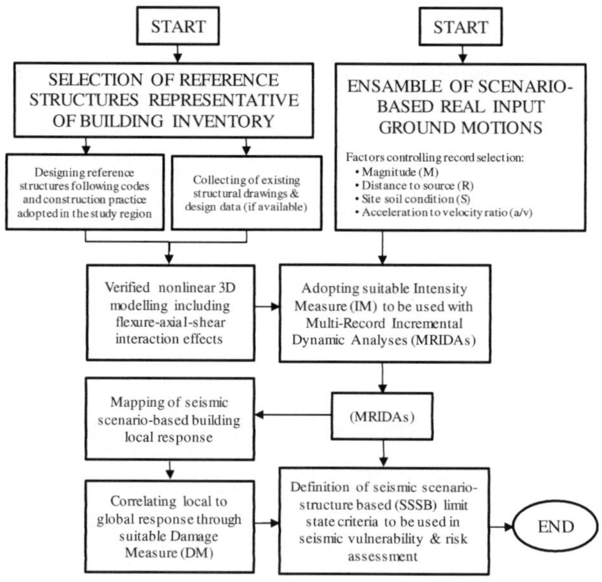

Figure 1: Flowchart of the proposed methodology to define SSSB limit state criteria for RC high-rise wall buildings.

Methodology and Sample Results

The footprint, layout and structural system of the 97.3m sample building are commonly adopted in the study region for this range of height. The building consists of a bearing wall lateral force-resisting system with two basements, a ground story, and twenty-seven typical stories. Figure 2 shows the building typical layout alongside the first 3-transational modes in each orthogonal direction. Based on a number of previous hazard assessment studies for the region, the seismic design loads are calculated using mapped spectral response accelerations of 0.85g and 0.26g at 0.2s and 1.0s, respectively, and soil class S_C (very dense soil and soft rock). The load combinations and design provisions of the ACI building code are followed in proportioning and detailing of all structural members in the sample building (ACI 2011). In the analysis stage, seismic demands are estimated using both the equivalent lateral force procedure (ELFP) and the modal response spectrum analysis (MRSA) and P-Δ effect is accounted for.

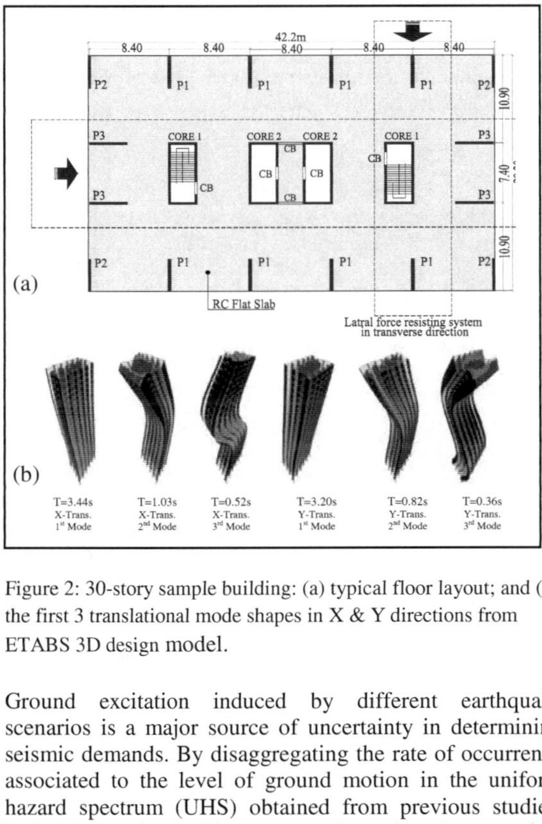

Figure 2: 30-story sample building: (a) typical floor layout; and (b) the first 3 translational mode shapes in X & Y directions from ETABS 3D design model.

Ground excitation induced by different earthquake scenarios is a major source of uncertainty in determining seismic demands. By disaggregating the rate of occurrence associated to the level of ground motion in the uniform hazard spectrum (UHS) obtained from previous studies, Dubai is found vurnable to two main earthquake scenarios: (i) sever distant earthquakes of magnitude 7-8 with 100-200km epicentral distance; and (ii) moderate near-field earthquakes of magnitude 5-6 with site-to-source distance of 10-60km. To account for the uncertainty in ground excitation, the two aforementioned scenarios are represented in the present study with two sets of 20 natural earthquake records (total of 40 records): (i) Record Set#1 representing sever distance earthquake scenario; and (ii) Record Set#2 representing near-field moderate earthquake scenario. Figure 3 depicts the response spectra for the 40 selected records alongside their means, design spectra and the 10% Probability of Exceedance (POE) in 50 years uniform hazard spectra for the study region.

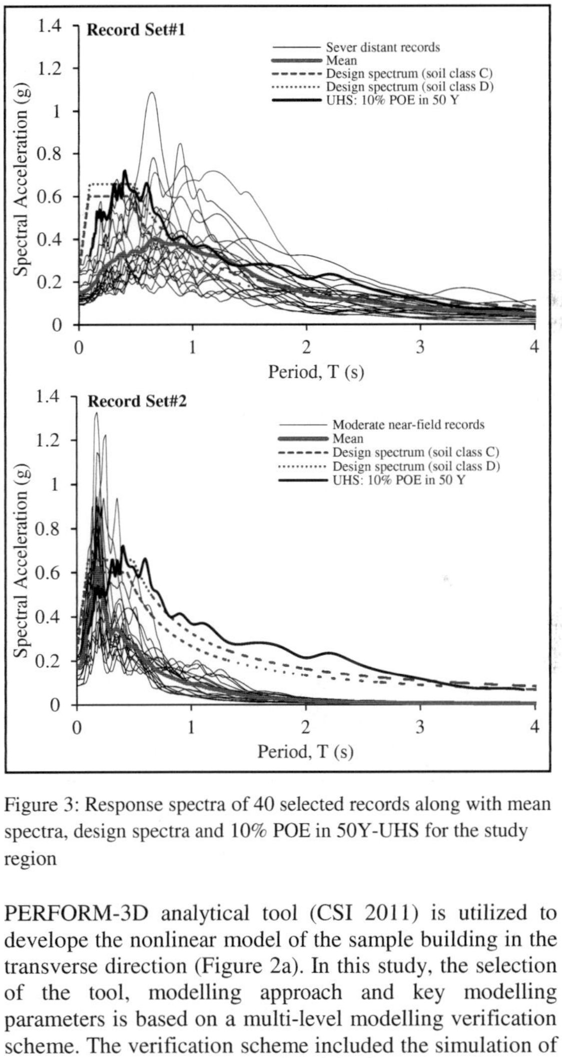

Figure 3: Response spectra of 40 selected records along with mean spectra, design spectra and 10% POE in 50Y-UHS for the study region

PERFORM-3D analytical tool (CSI 2011) is utilized to develope the nonlinear model of the sample building in the transverse direction (Figure 2a). In this study, the selection of the tool, modelling approach and key modelling parameters is based on a multi-level modelling verification scheme. The verification scheme included the simulation of the nonlinear dynamic response of a full-scale seven-story wall building slice tested on the LHPOST at UCSD (Alwaeli 2014; Panagiotou *et al.* 2007).

For the purpose of defining SSSB performance limit states for the sample building, MRIDAs are performed in the present study using the two sets of selected records. To construct IDA curves that can efficiently represent the variable states of response of the structure to different earthquake records with increasing intensities, an improved scalar IM termed spectral acceleration at weighted-average period $S_{a(wa)}$ is proposed. Each of the selected 40 earthquake records associated with its anchored response spectra is then appropriately scaled so the analysis can be carried out up to the satisfaction of all the designated performance levels. R#5 of Record Set#1 and R#3 of Record Set#2 are chosen to map the local response of the sample building. The reason for choosing these records is that their IDA curves are correlated to the 50% fractile of the IDA curves obtained from each of the two sets of records. Figure 4 shows the IDA curves obtained from the two sets of records alongside their 16%, 50%, 50% fractiles with the IDAs for R#5 of Record Set#1 and R#3 of Record Set#2.

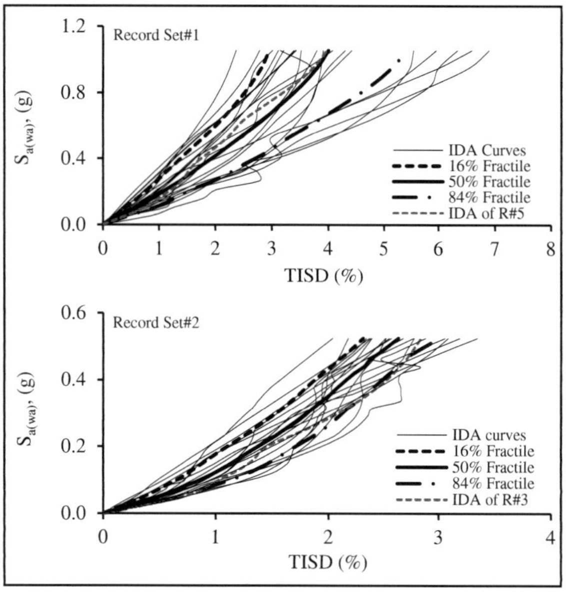

Figure 4: IDA curves for Record Set#1 and Record Set#2 along with their 16%, 50%, 84% fractile curves and IDAs for selected records R#5 and R#3 to represent the 50% fractiles.

To arrive at the SSSB quantitative performance limit state definitions for the sample building, a total of seven local damage indices (rebar yielding/buckling, concrete crushing, RC wall/coupling beam rotation, and wall shear capacity) are mapped throughout the conducted MRIDAs. Among the main observations from the mapping process are: (i) under both seismic scenarios, demands (strain/rotation/shear) at the onset of local damage events reduce with the building height while TISDs increase; (ii) the sequence of local damage propagation differs under the two seismic scenarios. The response of the sample building is flexure-controlled first mode-dominated under sever distant earthquakes while it is shear-controlled higher modes-dominated under moderate near-field earthquakes. Figure 5 shows local damage propagation under R#5 and R#3. Based on the obtained results, the traditional damage measure of TISD is found insufficient to be directly related to structural performance of high-rise buildings and therefore needs to be disaggregated to its main components. Net Interstory Drift (NISD) is proposed as DM in the current study as it directly correlates with local response over the height of the building. The values of the proposed DM which are related to deformation-based local damage events are found to be consistent over buildings with varying heights. Three limit states are defined for the sample building in the current study: (i) IO; (ii) LS; and (iii) CP (ASCE 2007). New SSSB limit state criteria are proposed using the MRIDAs results from the 40 selected records and the mapping/linking of predefined local damages discussed above. The mapped and selected limit state criteria for the building under Record Set#1 and Record Set#2 are listed in Table 1.

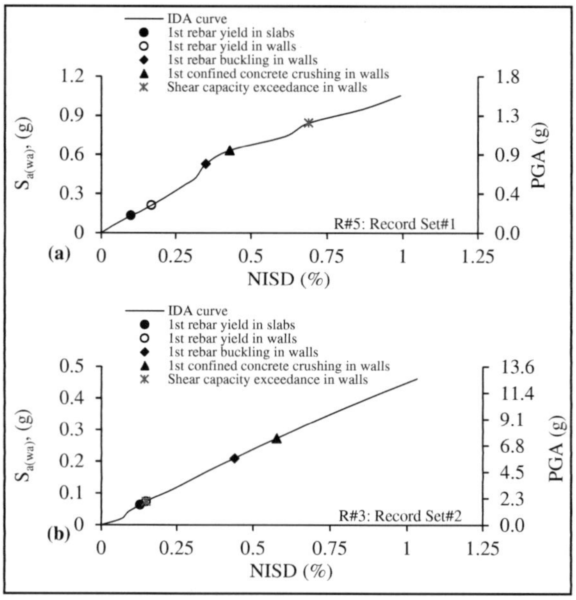

Figure 5: Propagation of local damage events in the sample building: (a) under R#5 of Record Set#1; and (b) under R#3 of Record Set#2.

Conclusions

The results obtained from the present study demonstrate the superiority of Net Interstory Drift (NISD) over Total Interstory Drift (TISD) as a global damage measure to be used in defining limit state criteria for RC high-rise wall buildings. NISD interrelates more directly with local response over the building height and are consistent for buildings with varying heights. The study evidences that the structural system, arrangement and geometry of vertical elements, and axial force level in the lower stories of the sample building have influenced the seismic intensity levels, and thus deformation levels, related to the onset of local damage events. The domination of higher modes on the sample building response under moderate far-field earthquake records renders the shifting of seismic response from being flexure-controlled under Record Set#1 to shear-controlled under Record Set#2. This study thus concludes that the response of RC high-rise wall buildings and consequently the definition of limit state criteria for designated performance levels is Seismic Scenario-Structure-Based (SSSB). The limit state criteria proposed in this study are applicable to a wide range of RC high-rise wall buildings and seismic regions.

Table 1: Mapped and selected limit state criteria for the sample building

Limit State	Damage Index (DI)	NISD (%)			
		Record Set#1		Record Set#2	
		Mapped	Selected	Mapped	Selected
IO	1st yield (slabs)	0.11	**0.11**	0.14	**0.08**
	1st yield (walls)	0.19		0.18	
	Wall rotation limit (ASCE/SEI 41-6)	0.19		0.17	
	All (IO) Dis, combined	0.11		0.14	
	NISD corresponding to value in ASCE/SEI 41-6 for RC walls controlled by shear	N/A		0.08	
LS	Wall rotation (ASCE/SEI 41-6)	0.30	**0.22**	0.32	**0.11**
	50% of NISD from all (CP) deformation-based DIs	0.22		0.25	
	NISD corresponding to value in ASCE/SEI 41-6 for RC walls controlled by shear	N/A		0.11	
CP	1st buckling in walls	0.44	**0.37**	0.49	**0.15**
	1st crushing in walls	0.59		0.67	
	Wall rotation (ASCE/SEI 41-6)	0.51		0.71	
	Shear capacity exceedance	0.34		0.15	
	All (CP) deformation-based DIs	0.44		0.49	
	All (CP) DIs inclusive of shear capacity	0.37		0.15	
	NISD corresponding to value in ASCE/SEI 41-6 for RC walls controlled by shear	N/A		0.14	

References

ACI-318, 2011, Building code requirements for structural concrete (ACI 318M-11) and Commentary, Amarican Concrete Institute, Detroit, Michigan.

Alwaeli, W., 2014, Framework for developing fragility relations of high-rise RC wall buildings based on verified modelling approach, Second European Conference on Earthquake Engineering and Seismology Istanbul, Turkey.

ASCE/SEI-41, 2007, Seismic rehabilitation of existing buildings including Supplement #1, Standard ASCE/SEI 41-06, American Society of Civil Engineers, Reston, VA.

CSI, 2011, Computer and Structures Inc., Perform-3D V5: Nonlinear analysis and performance assessment for 3D structures, 1995 University Avenue, Berkeley, CA.

Ghobarah, A., 2004, On drift limits associated with different damage levels, Proceedings of international workshop on performance-based seismic design, Dept. of Civil Engineering, McMaster University.

Ji, J., Elnashai, A. S., and Kuchma, D. A., 2009, Seismic fragility relationships of reinforced concrete high-rise buildings, The Structural Design of Tall and Special Buildings, Vol. 18, No. 3, 259-277.

Mwafy, A., 2012, Analytically derived fragility relationships for the modern high-rise buildings in the UAE, The Structural Design of Tall and Special Buildings, Vol. 21, No. 11, 824-843.

Panagiotou, M., Restrepo, J., and Conte, J., 2007, Shake table test of a 7 story full scale reinforced concrete structural wall building slice phase I: Rectangular Wall Section, Report No. SSRP-07-07.

SEAOC, 1999, Recommended lateral force requirements and commentary, 7th Edition, Seismology Committee Structural Engineers Association of California, Sacramento, CA.

Recent technologies for near surface imaging in the Arabian Peninsula

A. Vesnaver, Y. Bouzidi and L. Hou, Petroleum Institute, Abu Dhabi, United Arab Emirates*

Summary

Compensating for near surface signal distortions is an evergreen challenge for land seismic imaging in the Arabian Peninsula. Techniques exist that may solve local problems, as high-resolution dedicated surveys, but they are not economically viable at a regional level. In this paper, we review a few recent technologies that contribute to mitigate the near surface challenges with a limited impact to the acquisition budget.

Introduction

The near surface complexities are a major challenge for the hydrocarbon exploration and production in the Middle East, as they may heavily distort the imaging of underlying reservoirs by seismic surveys. Conventional methods based on refraction statics fail when, for example, one or more low-velocity layers are present. This happens quite frequently when the formation is close to the Earth surface. Transition zones pose further challenges as the changing features of the near surface and recording systems. In those cases, methods such as the Multi-Azimuth Surface Wave inversion have been proposed and tested with some success (Socco et al. 2010). Another viable solution is the geostatistical interpolation of upholes and their use to calibrate and constrain reflection/refraction tomography (Al-Ali et al. 2003, Vesnaver et al. 2006 a). The traveltime inversion of first arrivals, often referred to as tomostatics, has been tested in the Arabian Peninsula with mixed success (Vesnaver et al. 2002, 2006 b). Indeed, it can only partially overcome the problem affecting refraction statics, i.e., the so-called "hidden layer(s)", and it works optimally when lateral and vertical velocity variations are mild. When this condition is not met, some recent methods can be used and are reviewed in this paper, such as the integration of remote sensing in the near-surface modelling (Vesnaver et al. 2009), or the use of geostatistics (co-kriging) for extracting information from the Vibroseis response, such as the stiffness and viscosity of the ground (Ley et al. 2006). A case study in Al-Ain (United Arab Emirates) will be presented that integrates seismic refraction, first-arrivals and surface-wave inversion, and compares the results obtained by these methods separately. A major benefit of an accurate near-surface modelling is the removal of propagation effects distorting the images of underlying oil reservoirs and major tectonic structures.

The use of upholes

A large number of upholes were acquired in the past when dynamite sources were current, mainly for very sparse regional surveys composed of long 2D profiles. Reprocessing these data in 3D is a cost-effective way to get a more robust estimate of regional trends. However, the sparsity of available upholes and the possible errors in the velocity estimates do not allow getting high resolution, reliable models for the near surface, except for the lower spatial frequencies.

Figure 1 shows the synthetic example of a velocity field that oscillates between 600 and 2000 m/s with sinusoids in both x and y direction, with a wavelength of 10 km. The black dots in the upper left corner plot correspond to the real location of available upholes in a region of the Arabian Peninsula (Vesnaver et al., 2006a). They are quite irregularly sparse, as they were acquired in a time span of a few decades for regional studies using 2D profiles.

To quantify the accuracy in the actual velocity reconstruction using geostatistical methods as kriging, we build two further models by perturbing the first one, without noise, adding random values in a range of ±100 m/s (central column) and ±300 m/s (right column). In the related variograms (central row) red arrows highlight the value in the first histogram element, i.e., the nugget, which is zero in the noise-less model and increases in the others. The reconstructed velocities (lower row) are obtained by kriging the data sampled at the uphole locations. The estimate of the noise-free data, using a range value of 10 km, is a fair approximation of the actual velocity (top row), while the distortion is not acceptable for the most perturbed upholes (lower row, right). We remark that this higher value is within the experimental error we may expect, mainly because of natural seasonal changes of soil and shallow layers, due to moisture content and temperature.

The use of refraction

Shallow refraction surveys for the near-surface in United Arab Emirates can be very challenging particularly in transition zones (marine/land), sand dunes and areas dominated by alluvium deposits. Refraction methods can be also dependent on natural seasonal changes of soil as mentioned in the previously described co-kriging method. A 500 m long profile of seismic refraction was acquired in Al Ain using a 48 channel recording system and a weight drop source. The study aim was the feasibility in characterizing the complex near-surface, composed mainly of alluvium deposits and outcropping carbonate formations.

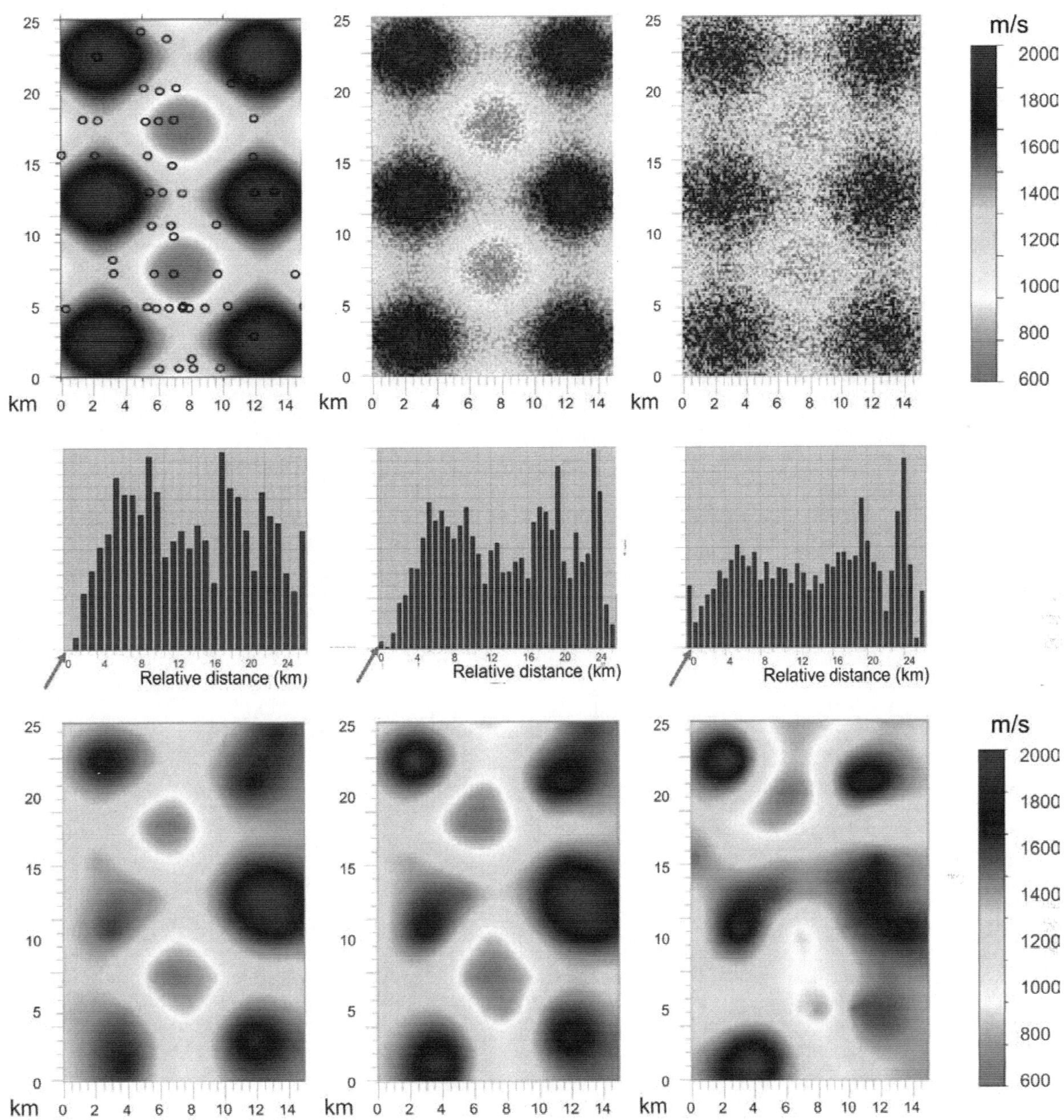

Figure 1: A synthetic example of geostatistical uphole interpolation. Velocity fields with different noise levels: none (left column), 100 m/s (center column) and 300 m/s (right column). The upper row shows the true models, central row shows the related variograms, and the lower row the geostatistical interpolation.

The first arrivals of refracted waves from the recordings are used to invert for the near surface geology. Figure 2 shows the velocity model from selected portions of the data on both sides of the outcrop. The results are quite different, despite the very short distance separating the two locations. The velocities vary significantly along the profile as well as the thickness of the layers, particularly the second layer thickness, despite the very gentle dips of the interfaces between layers. However, the penetration depth is limited and did not reach the water table, which is measured directly at a close well location and found to be about 29 m below surface. These results show that the near-surface can be very heterogeneous and P-wave velocities vary laterally in short distances. These anomalies can cause large fluctuations in arrival times of seismic reflections from deeper targets and should be taken into account adequately to minimize imaging errors.

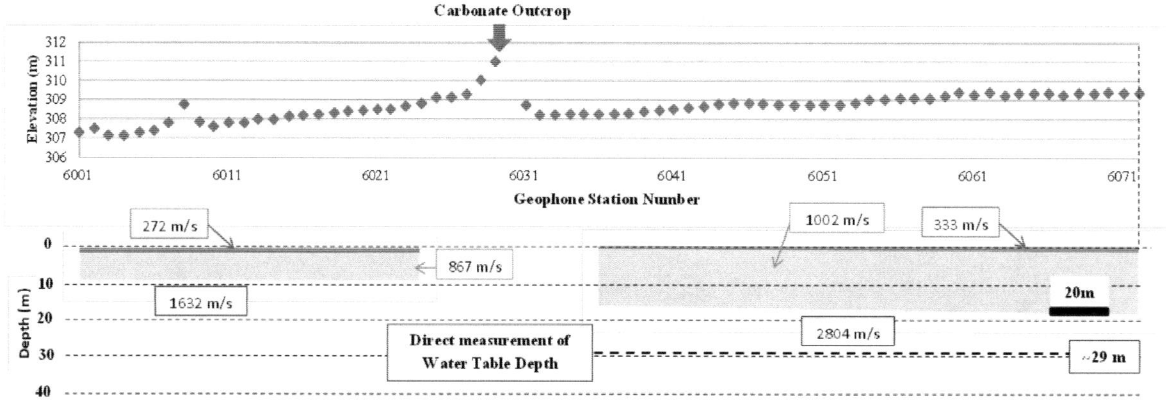

Figure 2: Shallow refraction profile across a carbonate outcrop in the United Arab Emirates. The elevations of each station is given at the top. The inverted seismic velocity model is shown on both sides of the outcrop. The measured water table from a well on the right of the outcrop depth is about 29 m below surface.

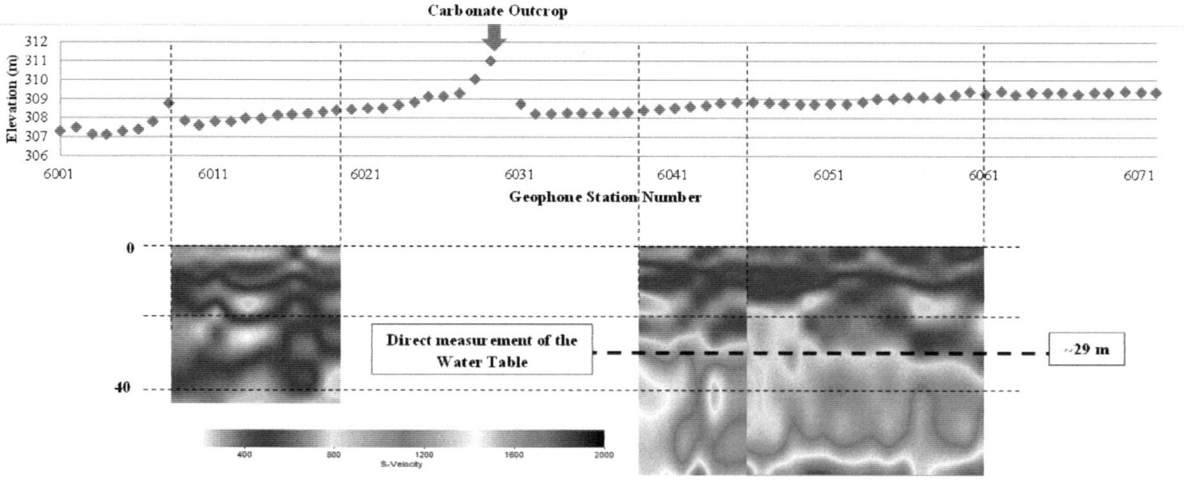

Figure 3: Surface wave profile across a carbonate outcrop in the United Arab Emirates. The elevations of each station is given at the top. The inverted S-wave velocity model is shown on both sides of the outcrop. The measured water table from a well on the right of the outcrop depth is

The use of surface waves

The surface waves extracted from the same records as those used for refraction are used to invert for the near-surface shear wave velocity model. The shear wave model is obtained from the dispersion curves of surface waves. The results are shown in Figure 3 at nearly the same locations as those of the refraction analysis. Again here the two inverted models differ significantly. On both sides of the outcrop the S-wave velocities are nearly the same (~ 500 m/s), but extend deeper on the left than on the right. There is a large S-wave velocity contrast on the right side of the outcrop that coincides with the depth of the water table (~29 m). Obviously the rocks below the water table must differ from those above as S-wave velocity of a porous formation decreases when it is water-saturated. The increase of S-wave velocity in the water-saturated rock indicates a higher rigidity of the porous matrix. These results suggest that the outcrop plays a role in the distribution of the underground water, acting as a barrier for the water to move across to the left.

Conclusions

Near surface complexities remain a major challenge for land seismic data acquisition and imaging in the Arabian Peninsula. Sharp lateral and vertical velocity variations prevent standard methods (as refraction statics and uphole interpolation) to provide generally acceptable Earth models. Additional information as vibrator plate attributes and satellite imagery can delineate shallow features with high precision: when integrated with deeper penetrating methods (as surface wave inversion, gravity and electro-magnetic), significant improvement may be achieved.

Acknowledgments

We thank the Petroleum Institute of Abu Dhabi (United Arab Emirates) for supporting our research work.

References

Al-Ali M., R. Hastings-James, M. Makkawi, and G. Korvin, 2003, Vibrator attributes leading velocity estimation: The Leading Edge, **22**,400-405.

Ley R., W. Adolfs, R. Bridle, M. Al-Homaili, A. Vesnaver, and P. Ras, 2006, Ground viscosity and stiffness measurements for near surface seismic velocity: Geophysical Prospecting, **54**, 751-762, doi: 10.1111/j.1365-2478.2006.00574.x.

Socco V., S. Foti, and D. Boiero, 2010, Surface-wave analysis for building near-surface velocity models - Established approaches and new perspectives: Geophysics, **75**, 75A83-75A102, doi: 10.1190/1.3479491.

Vesnaver A., R. Bridle, J. German, L. Hu, J. J. Kim, R. Ley, and C. Macrides, 2002, Tomostatics for half a million traces: Expanded Abstracts, SEG 72th Annual, 957-960.

Vesnaver A., R. Bridle, B. Henry, R. Ley, R. Rowe, and A. Wylie, 2006 a, Geostatistical integration of near-surface geophysical data: Geophysical Prospecting, **54**, 763-777, doi: 10.1111/j.1365-2478.2006.00573.x

Vesnaver A., R. Bridle, R. Ley, and W. Rowe, 2006 b, Near-surface data clustering for imaging enhancement, SEG Expanded Abstracts, 25, 1451-1455.

Vesnaver A., R. Bridle, R. Ley, and C. Liner, 2009, Painting the near surface using geology, geophysics and satellites, Geophysics: **74**, B61-B69, doi: 10.1190/1.3097886.

Impacts of hydraulic fracturing on the environment and on human health

DJEDDI Mohamed[1]
[1] USTHB – Faculté des sciences de la terre géographie et aménagement du territoire. Laboratoire de Géophysique – Algérie.

Email: dallidjeddi@gmail.com

Abstract

Shale gas and shale oil extraction is currently a hot topic in Europe and recently in North Africa, not only for the potential effects on the environment. Further questions raised relate to the role, if any, these energy carriers can play in the low-carbon economy model towards which the countries intend to shift and their relevance to decrease the dependence on imported energy commodities.

This contribution examines the Impacts of hydraulic fracturing on the environment and on human health. Quantitative and qualitative data impacts are taken from the US experience at the time this method in Europe is in its infancy, while the US has more than 40 years of experience and already drilled over 50 000 wells. Shale gas (GoS) had (especially the US) a boom during the last decade thanks to hydraulic fracturing horizontal drilling that extracts the gas dissolved in the rock at 2 or 3 km deep. This technique requires very high injection pressure of a liquid composed of water (in large quantities), sand and chemicals whose composition varies according to the firms, the rock and the depth of the wells. A portion of this liquid (25 to 50%) occurred and must be stored in surface impoundments before being processed.

Among these impacts we identify which are the main

Geological impact

Impact on waters (groundwater and surface water)

Noxious gas emissions

Impact on Human Health

Geotechnical investigation of Namar dam site using geophysical tools, Riyadh area, Saudi Arabia (King Saud University, Geology and Geophysics Dept.)

Abdullah H. Alhaj and Hussam AL-mudhi*

Summary

Based on the expected hazards that threaten the existing dams in Riyadh area due to Collapse of dams or leakages across them. As a result of defective structure and cavities, the proposed research will employ different geophysical technique on Wadi Namar Dam site which are often the most cost-effective and rapid means. Electric Resistivity Tomography (ERT), Seismic Refraction Tomography (SRT) and GPR techniques will be used for mapping bedrock topography, identifying near-surface karstic sinkholes or cavities and regions of potential weakness (e.g., faults and fracture) in the study site Figure 1.

Introduction

The use of geophysical methods for subsurface characterization study of a dam site is increasingly becoming popular all over the world and has the possibility to give an image of the subsurface to the geologists and geotechnical engineers (Benson et al., 1984; Goldstein, 1994; Benson and Yuhr, 1995, 2002). The subsurface characterization of the dam site including bedrock depth, rock type, layer boundaries, groundwater flow, locating fractures, weak zones, detecting cavities caused by sinkholes will be detected using the electrical resistivity tomography (ERT). Ground Penetration Radar (GPR) and Seismic Refraction Tomography (SRT) Method. By Utilizing ERT and GPR; Seepage pathways through the rock formations under and around the dam foundation can be detected and mapped (e.g. Corwin, 2007; Dahlin and others, 2008) through the use of the low resistivity anomalies associated with active seepage paths. These two methods depends directly on the electrical conductivity that is in direct relation to the ground moisture (McCann and Forde, 2001; Polder, 2001), indicating the areas that are more susceptible to infiltration or leakage. (SRT) method is the most frequently used geophysical means of investigating at dam sites. Multichannel surveys on land produce detailed profiles of depths to bedrock, overburden conditions and the competency of the bedrock

Geological Setting

The Arabian Peninsula is composed of two structures completely different, the first is the Arabian Shield, which covers an area of about 2/5 of the Arabian Peninsula to the west, and the second basin sedimentary rocks to the east of the Arabian Shield. The Mesozoic rocks that most rocks Riyadh region back to it from the thick deposits of successive layers of the continental rocks and sediments shallow water facies dominated by calcareous rocks, where a huge curved form a belt in the center of the Arabian Peninsula. Geological units in Riyadh consist of Jubaila, Arab, Al-Sulay, Al-Kharj and Alluvial Deposits as shown in Figure2. The study Area compose of Alluvial deposits of nested clay with layers of silt and gravel and Arab Formation which has cracked limestone and eclipsed collapsed and became structures and fixtures inside the layers

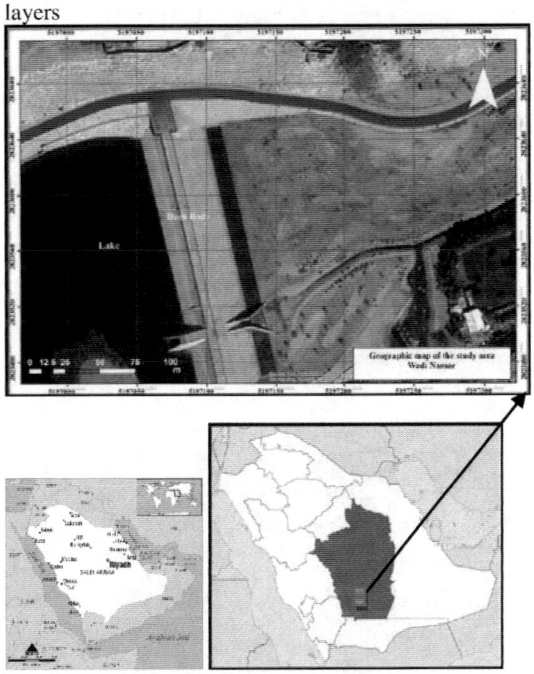

Fig.1: Location map
Of Study Area

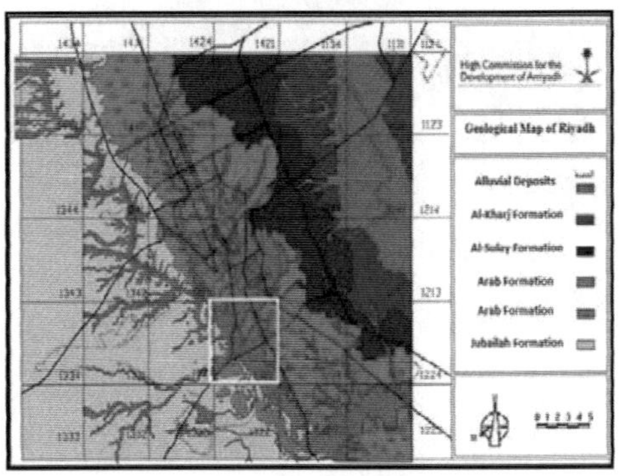

Fig.2: Geological map of Study Area
(High commission Development of Arriyadh)

Theory and/or Method

Geophysical methods are used to detect the subsurface structures and seepages of ground water in the dam site. The Seismic Refraction Tomography method is the most commonly conducted geophysical surveys for engineering investigations, Figure3 and GPR is an electromagnetic (EM) geophysical method for high-resolution detection, imaging and mapping of subsurface soils and rock conditions, Figure 4 as well as electrical resistivity for detection of cavities caused by sinkholes, Figure 5.

Fig.3: Seismic Refraction Survey

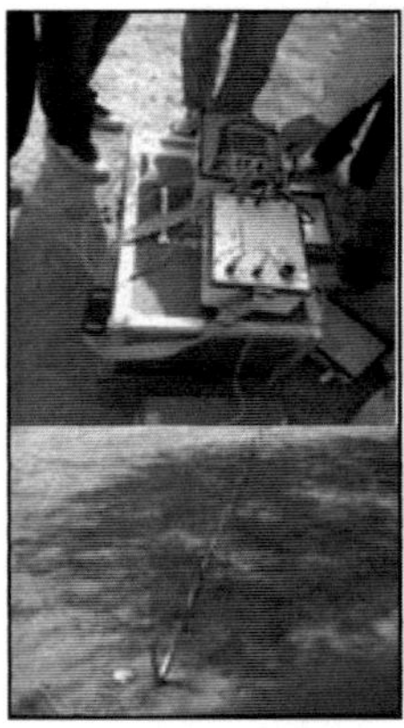

Fig.5: Electrical Resistivity Tomography

Fig.4: GPR Survey

Results

It is obviously demonstrated that the study Area is characterized by the presence of weathered and fractured Limestone layer that underline the alluvial soft sediments. According to the Seismic Refraction Survey technique this layer attain seismic velocity ranged from 1.13 to 3.00 m/ms and thickness ranged from5 to 10 m Figure (6).

The GPR Sections shows some fractures and joints affect this Limestone layer Figure(7).

The electric 2D Resistivity Survey shows the existence of two layers of void and sinkholes structures . The first filled with air that show the resistance circular shape and the other layer attain a relatively low resistance circular shape could be interpreted as sinkholes filled with wet clay Figure(8).

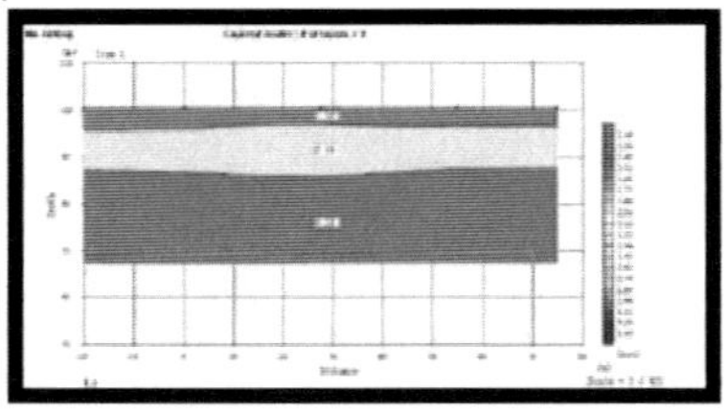

Fig.6: Seismic Section

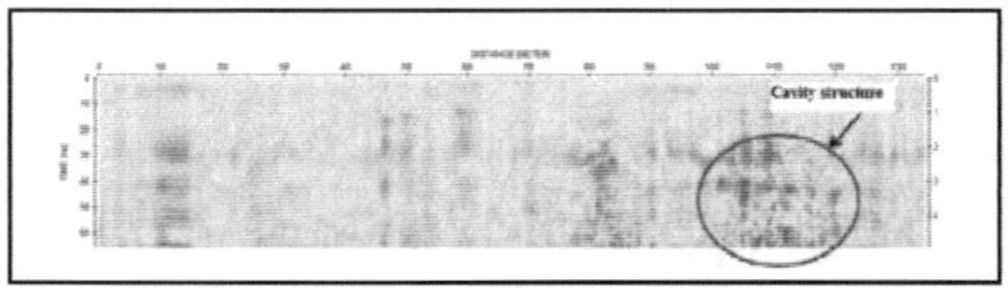

Fig.7: GPR Section

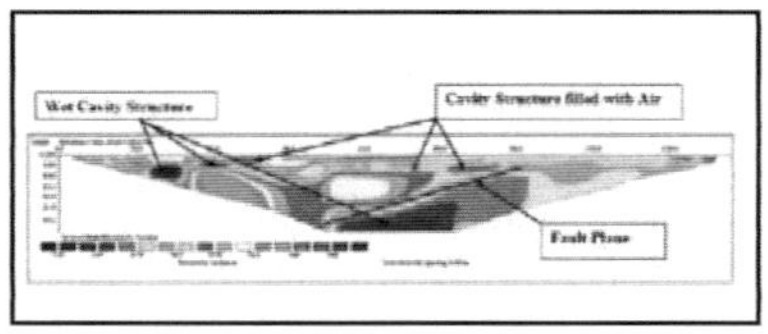

Fig.8: 2D Electrical Resistivity Section

Conclusions

In general, from the previous interpretation of results it could clearly explain the detection of fractures, cracks and voids in the dam structure and delineation the pathways of groundwater seepage in the dam site as well as mapping bedrock topography, faults, weak zones, cavities and near-surface karstic sinkholes in the dam site. It could be concluded that this site is affected by the weathered fractured system and presence of cavities that facilitate the vertical seepages of ground water from the dam site.

REFERENCE CHANGE:

Papers from journals

Abdeltawab, S. (2013). Karst Limestone Geohazards in Egypt and Saudi Arabia. International Journal of Geoengineering Case histories, http://casehistories.geoengineer.org, Vol.2, Issue 4, PP.261, 264. Kosloff, D. D., and E. Baysal, 1982, Forward modeling by a Fourier method: GEOPHYSICS, **47**, 1402–1412.

Tabwassah, C. A. and Obiefuna, G. I. (2012). Geophysical and Geotechnical Investigation of Cham Failed Dam Project, Ne Nigeria. Research Journal of Recent Sciences Vol. 1(2), 2277-2502.

Papers from magazines

Ammar A. Amin And Khalid A. Bankher.1997, Karst Hazard Assessment of Eastern Saudi Arabia.,Kluwer Academic Publishers,Netherlands,PP.22-23

Books

Powers, R. W., Ramires, L. F., Redmond, C. D., and Elberg, E. L.: 1966, Geology of the Arabian Peninsula: Sedimentary Geology of Saudi Arabia, U.S. Geological Survey Professional Paper 560-D, PP.D53-D60.Theses and dissertations

Panthulu, T.V., Krishnaiah, C. and Shirke, J.M. 2001. Detection of seepage paths in earth dams using self-potential and electrical resistivity methods. Engineering Geology 59, p 281–295.

Articles in books

Goldstein NE (1994). Expedited site characterization geophysics: Geophysical methods and tools for site characterization. Prepared for the U.S. Department of Energy by Lawrence Berkeley Laboratory, Univ. of California. 124 p

Web site (or part of Web site)

http://www.arriyadh.com/env-db/edbase2001/NSetting/Geology/Argeol.asp?SectID=100&topicid=120

Lawrence B. Conyers, Basics of GPR Method and Theory, University of Denver. http://mysite.du.edu/~lconyer/

GSSI Web.com

Acknowledgments (Optional)

The Authors wish to thank Pro. Elkheder H. Ibrahim and Prof. Kamal Hassanien for their Supervision and carrying out the field work for this study.

Submit applications form and abstract via email to iceg2015@seg.org.

Workflows for characterization of Shale gas reservoir

Afrah Khalid Ahmed Siddique;NED University of Engineering & Technology.*

Summary

Seismic method plays great role in all aspects of shale-gas-plays starting from identification of promising shales, describing reservoir characteristics (net-to-gross porosity, saturation, fracture, brittleness, etc), optimizing drilling, stimulation & enhancing production. As shale gas resources have emerged as a viable energy source,their characterization has gained significance. However to characterize the shale gas different geophysical workflows are used on 3D surface seismic data and by integrating seismic attributes & logs, source rocks and reservoir properties within shale can be mapped away from the well.We discuss some of these workflows below using well log data and seismic data.

Introduction

As shale gas resources have emerged as a viable energy source,their characterization has gained significance. The organic content in these shales which are measured by their TOC ratings,influence the compressional and shear velocities as well as the density and anisotropy in these formations. Consequently, it should be possible to detect changes in TOC from the surface seismic response. Besides TOC, different shale formations have different properties in terms of maturation, gas-in-place, permeability, and brittleness. We discuss different workflows for characterizing shale formations that involve well log data as well as seismic data.

The development of these shales changed the traditional approach geologists had been following – that of the sequence of gas first being generated in the source rock, followed by its migration into the reservoir rock in which it is trapped. Shale-gas formations are both the source rocks and the reservoir rocks. There is no need for migration and since the permeability is near zero, it forms its own seal The gas may be trapped as free gas in natural fractures and intergranular porosity, as gas sorbed into kerogen and clayparticle surfaces, or as gas dissolved in kerogen and bitumen (Curtis, 2002). The shale gas reservoirs are characterized by low production rates (20-500 Mcf/d) but are usually spread over large areas and are up to 450 m thick. They are organically rich with total organic carbon content (TOC) varying from 1 to 20 wt%, such that the reservoirs contain large gas reserves (2 to 20bcf/km2). Shale gas reservoirs rely on natural fractures for porosity and permeability as the matrix porosity or permeability is low. In the absence of natural fractures, these reservoirs need stimulation by way of hydraulic fracturing. Compared to conventional reservoirs, shale gas reservoirs have low recovery factors (~20%). For the wells drilled into these reservoirs, the initial decline rates are rapid (~70%) but bottom out gradually (~6%) in later years (EIA document, 2011).

For a shale gas reservoir to become a successful shale gas play, the following characteristics need to be considered: (a) organic richness (TOC), (b) maturation (Ro%), (c) thickness, (d) gas-inplace, (e) permeability, (f) mineralogy, (g) brittleness, and (h) pore pressure. Besides all these the depth of the shale gas formation should also be considered as it will have a bearing on the economics of the gas recovery. An optimum combination of these factors leads to favorable productivity.

As different shale gas reservoirs have different properties, it is imperative to study them before any exploitation plan is put in place. In Figure 1 we show a comparison of some of these factors for five different shale gas reservoirs. Notice the differences in each of them. Another important point to consider is that such properties can be determined at the location of the wells where the well log and core data are available. However to characterize the shale gas formations different geophysical workflows need to be used on 3D surface seismic data. We discuss some of these workflows below.

Using well log data:

As resistivity increases significantly in mature rocks due to the generation of nonconducting hydrocarbons, it can be detected easily on the resistivity log curves. Passey et al. (1990) proposed the ΔlogR technique for measuring TOC in shale gas formations which are valid over a wide range of maturities. In this technique the transit time curve and the resistivity curves are scaled in such a way that their relative scaling is -100 μs/ft per two logarithmic resistivity cycles. In this case, the two curves overlie each other over a large depth range, except for organic-rich intervals where they would appear separated by ΔlogR, which in turn is linearly related to TOC, and is a function of maturity. The level of maturity (LOM) is usually obtained from sample analysis where the vitrinite reflectance (Ro%) method is commonly used for the purpose. LOM values derived from R% are seen to vary from 6 or 7 to 12 depending on the type of organic matter, with 7 indicating the onset of maturity for oil-prone kerogen and 12 indicating the onset of overmaturity for oil-prone kerogen. To the left of Figure 2 we show the sonic and resistivity curves indicating the ΔlogR separation.

Attributes generated from the available log curves (sonic, density, resistivity and porosity) need to be crossplotted to gain more insight into the shale reservoir and determine which attributes would be suitable to distinguish the reservoir from the nonreservoir zones.

Løseth et al. (2011) have shown that a relationship between TOC values and acoustic impedance can be obtained by crossplotting measurements made from cores and well logs. Such a relationship can then be used to transform acoustic impedance volumes derived from simultaneous inversion into TOC volumes.

Using seismic data

TOC changes in shale formations influence VP, VS, density and anisotropy and thus should be detected on the seismic response.

1. Carcione (2001) showed that for a given layer thickness and kerogen content, the PP reflection coefficient decreases with increasing angle. This is expected since the acoustic impedance for shale reservoir rocks with TOC > 4% is lower than the same rocks without TOC. Such a simple exercise allows separation of reservoir facies from the non-reservoir ones.

Impedance curves from wells that penetrate the shale strata, when put together on the same plot exhibit reservoir quality variations (Treadgold et al., 2011; Løseth et al., 2011). Zones associated with higher organic content are associated with lower acoustic impedance values and could be picked up from such a display.Such an exercise should form a preliminary workflow.

2. Rickman et al. (2008) showed that brittleness of a rock formation can be estimated from well log curves of Poisson's ratio and Young's modulus. This suggests a workflow for estimating brittleness from 3D seismic data, by way of simultaneous prestack inversion that yields ZP, ZS, VP/VS, Poisson's ratio, and in some cases meaningful estimates ofdensity. Zones with high Young's modulus and low Poisson's ratio are those that would be brittle as well as have better reservoir quality (higher TOC, higher porosity). Such a workflow works well for good quality data.

3. Natural fractures in shale formations can provide permeability pathways and so need to be characterized, if they exist. While many shale formations may be devoid of such natural fractures, others such as the Woodford Shale is hard and faulted. The Eagle Ford Shale as well as the Muskwa-Otter-Park–Evie Shale package in the Horn River Basin also exhibit natural fractures. Such fractures could be characterized by running the discontinuity attributes (coherence and curvature) on post-stack data or by utilizing prestack processes such as AVO/AVAz/VVAz. Natural fractures in the shale formations can be detected by examining azimuthal variations of the velocity/impedance field, e.g. normalized Vfast-Vslow (Treadgold et al., 2011; Zhang et al., 2010).

4. Seismic waveform classification is another quick and easy method that holds promise as an application for shale gas reservoir characterization. It essentially classifies seismic waveforms in the interval of interest based on their shapes.Usually the traces are grouped into classes and displayed in color such that each class has its own color. The resulting map exhibits seismic facies variation and enables demarcation of favorable zones from the others. The waveform classification process could be unconstrained or constrained depending on how the waveforms are subdivided. In Figure 3 we show the result of unconstrained waveform classification and compare it with an equivalent display from the relative acoustic impedance derived from thin-bed reflectivity. The two displays compare favorably.

Conclusions

Geophysical methods can help in characterizing the shale gas resource plays. However, the methodology adopted is in general quite different from methodologies applied to conventional reservoirs. In addition, the characterization of each shale reservoircould require particular types of tools and have to be applied with care. While we have discussed a number of different workflows for characterization of shale gas reservoirs, the choice of such methods would continue to evolve to meet the growing challenges and expectations.

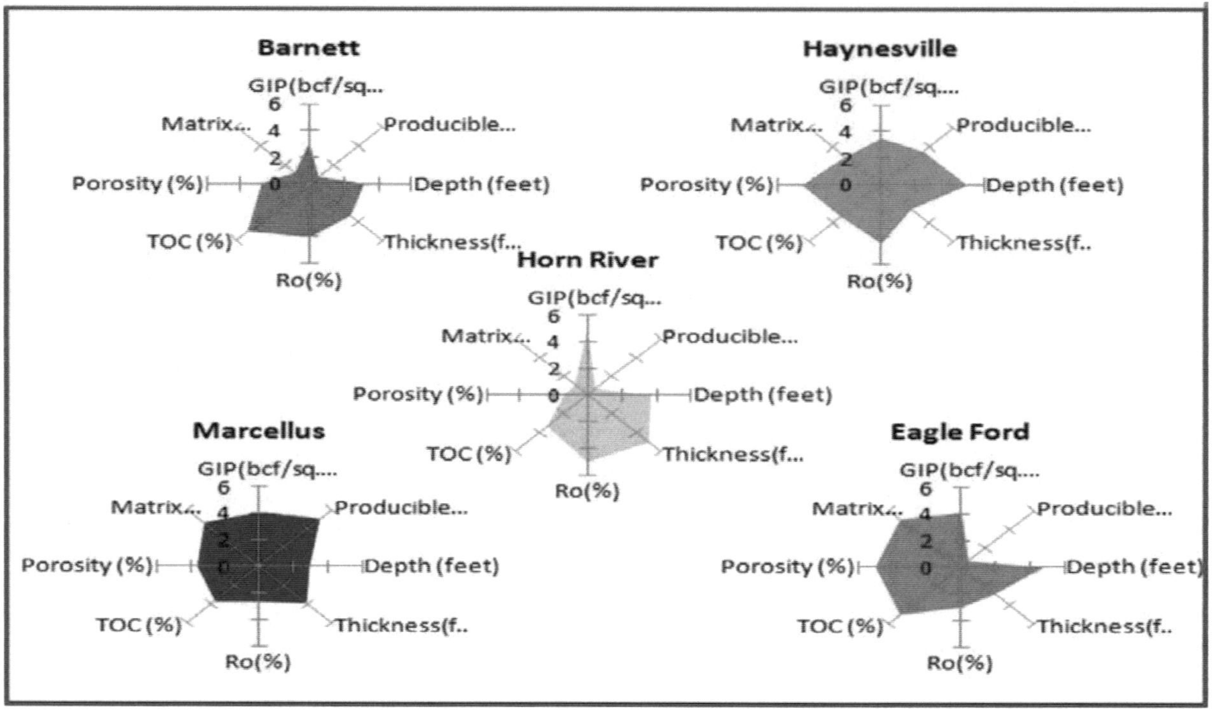

Figure 1: Comparison of shale characteristics for five different types of shale reservoirs. Notice the variation in the eight properties considered. (After Roth, 2010).

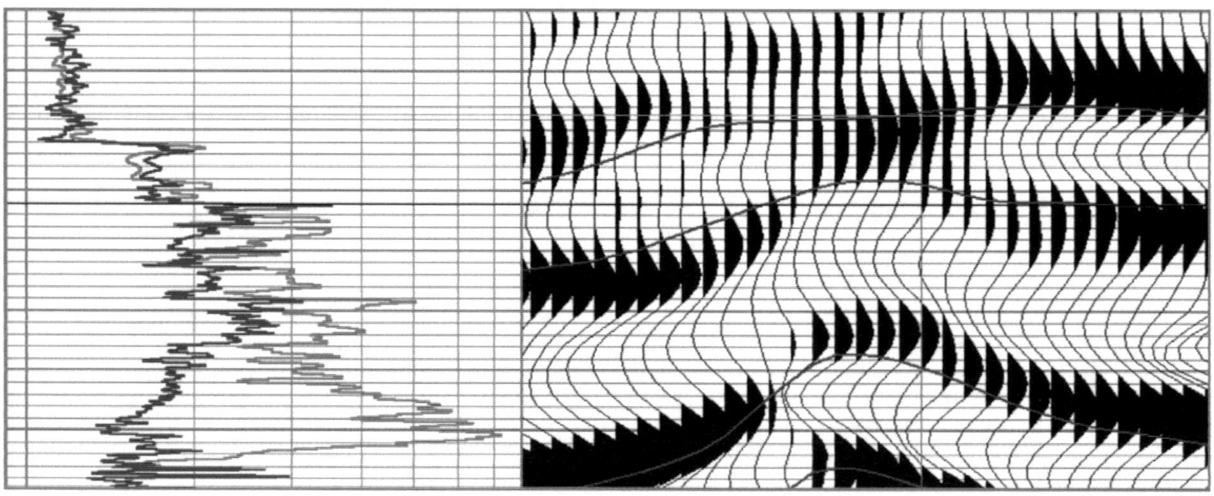

Figure 2: Log data in the form of transit time (blue) and resistivity (red) show a separation for ΔlogR which is indicative of high TOC. The location of the curves is the vertical green line on the seismic section.

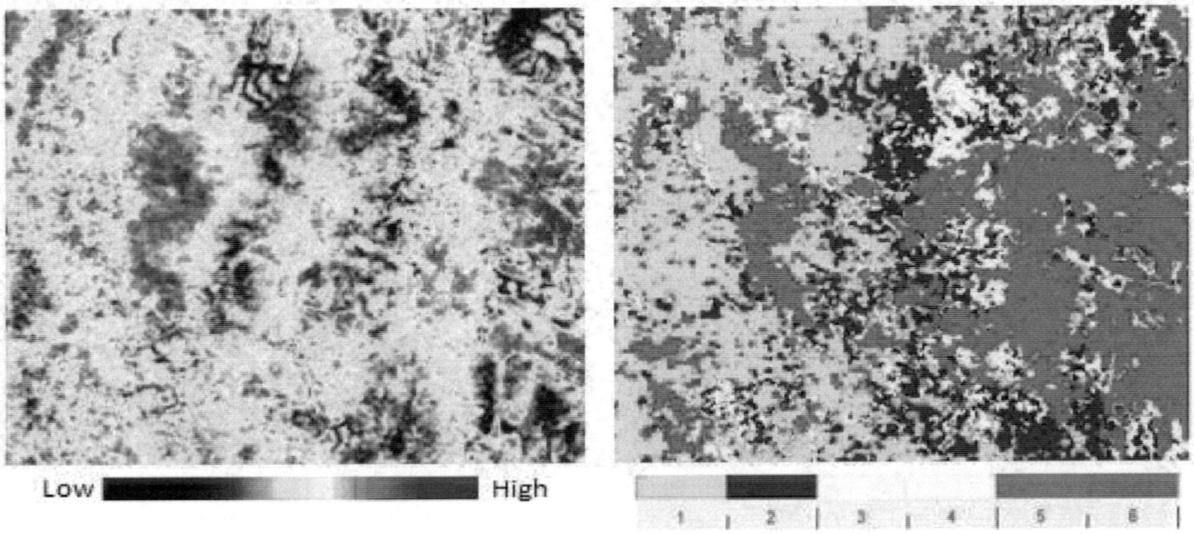

Figure 3: A horizon slice at the Evie Shale level from (left) the relative acoustic impedance volume derived from thin-bed reflectivity inversion of 3D seismic data, and (right) from waveform classification of a window extending 20 ms below the Evie-Shale marker. Notice the overall pattern of the red color seems to follow the pattern seen on the high impedance see on the left.

References

1. Rickman, R., M. Mullen, E. Petre, B. Grieser, and D. Kundert, 2008, A practical use of shale petrophysics for stimulation design optimization: All shale plays are not clones of the Barnett Shale: Annual Technical Conference and Exhibition, Society of Petroleum Engineers, SPE 11528.

2. Treadgold, G., B. Campbell, B. McLain, S. Sinclair, and D. Nicklin, 2011, Eagle Ford shale prospecting with 3D seismic data within a tectonic and depositional system framework: The Leading Edge, **30**, 48–53

3. Thomas Bowman, 2010, Direct Method for Determining Organic Shale Potential from Porosity and Resistivity Logs

4. Harilal and A. K. Tandon, ONGC, 2012, Unconventional Shale-gas plays and their characterization through 3-D seismic attributes and logs

5. Løseth, H., L. Wensaas, M. Gading, K. Duffaut, and M. Springer, 2011, Can hydrocarbon source rocks be identified on seismic data? Geology, 39, 1167–1170.

6. EIA (U.S. Energy Information Administration), 2011, Review of emerging resources: U.S. shale gas and shale oil plays: ftp://ftp.eia.doe.gov/natgas/usshaleplays.pdf, accessed 24 May 2012.

Engineering Geophysics: Human Activities and Humanitarian Applications

Koya Suto, Terra Australis Geophysica Pty Ltd, Brisbane, Australia

Summary

This paper presents an outline of three projects the author is involved for improvement of human living condition: creating better life and recovering from disasters, in particular. Solomon Islands are a Pacific Republic where fossil fuel resource is scarce. A hydroelectric dam project is planned and a seismic survey was conducted across a river as a part of feasibility study. The result led to proceeding construction of the dam. The second project shows a case history of an MASW survey for estimating the extent of damage after a flood and landslide in southern Australia. The third part is a humanitarian aid program currently underway in Serbia and Bosnia and Herzegovina where landslide sites are accessed using geophysical methods, combining promotion and training of local engineers and students.

Introduction

Engineering geophysics may be defined as geophysics in the domain of human activities. It contributes to well-being of human life through construction of safe, comfortable and convenient environment. When such environment is disturbed, geophysics has a role in its recovery.

The first half of this presentation describes an inconvenient life of a developing country and role of geophysics in improving their quality of life by providing stable supply of electricity. Republic of Solomon Islands imports diesel to generate electricity. It is so expensive and most of the residents can afford electricity only part of the day. The nation's first hydroelectric dam is planned and a seismic survey was conducted for its feasibility study.

The second half presents a Geophysicist *without* Borders project recently started in Serbia and Bosnia and Herzegovina. It provides assessment of landslide sites which were damaged by heavy rain in May 2014. An international team has been formed for this project on an volunteer basis. They are specialists in seismic, electric and satellite measurement methods. This project is supported by SEG's Geophysicist *without* Borders, local governments, and local communities. Geoscientists and Environmentalists of Serbia (AGES), one of the SEG's local sections plays a leading role of this project. The project includes training of students of local universities and engineers of local governments, so that the technical knowhow will remain after the project is completed.

DAM PROJECT IN SOLOMON ISLANDS

Background

Republic of Solomon Islands is in the midst of southern Pacific Ocean. It is composed of xxx main islands large and small with the capital Honiara on Guadalcanal Island. Its GDP per capita is 135th among 185 countries, which is one-28th of US (World Bank, 2015). It is said about 15% of the population is engaged in paid employment. The most people grow crops in their garden, catch fish, collect coconuts for themselves and excess is brought to the market for cash.

There is little export industry apart from small mineral deposits and tourism. No fossil fuel energy resources are domestically present. The country imports diesel fuel to generate electricity, hence it is very expensive. In fact its price about US90c/kWh is the most expensive in the world (Wikipedia, 2015).

Figure 1. Tina River, Guadalcanal Island, Republic of Solomon Islands

The nation's first hydro-electricity dam is planned on Tina River on Guadalcanal Island funded by World Bank. The plan is to dam Tina River by a dam 50 meters high and to build a hydro-electric power plant with a capacity of 20 MW. This provides more than 90% of electricity consumed in the capital, Honiara.

Geophysical Survey

A seismic survey for selection of possible dam site took place in December 2013. The same seismic data were analyzed by the refraction and MASW methods.

The site is a narrow gorge with steep slopes both sides: The data acquisition in this environment was difficult. Land seismic line could be laid about 25 meters into the bush on one side (Figure 2, top), but the other side of the river was so steep and a seismic line could not be laid. Marsh geophones were used in the water (Figure 2, bottom) and a sledge hammer for source was carried across the river by swimming.

For this difficult condition with running stream nearby, the data quality was good (Figure 3). It was a pleasant surprise.
Figure 4 shows Vs and Vp profiles of one of the seismic lines. These show the depth of the competent rock is around 5 meters from the surface and its is pretty smooth.

Contract of construction of the dam was recently awarded and work just started.

Figure 2. Data acquisition

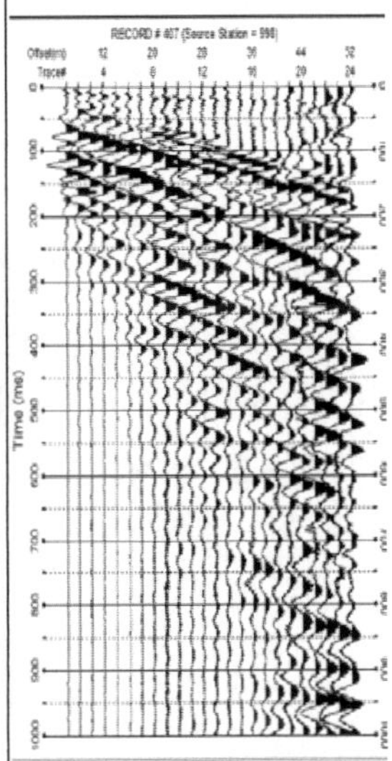

Figure 3. Raw seismic record.

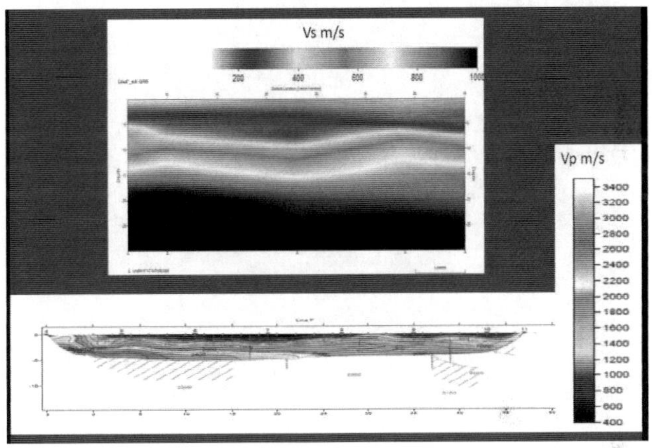

Figure 4. Result of seismic survey: (top) S-wave velocity section by MASW; (bottom) P-wave velocity section by refraction analysis.

ASSESSMENT OF FLOOD DAMAGE

Background

The Grampians National Park is located in the Australian state of Victoria, approximately 300km west of Melbourne (Figure 5). In early 2011, a severe storm and flash-flooding extensively damaged several roads in the park. Land-slips and washouts occurred at ten discreet locations along an important fire-access and tourist road through the park (Figure 6). The road was blocked completely in five separate locations, making much of it inaccessible to vehicles. Repair works involved reconstruction of culverts, fill batter slopes, and construction of gabion-style retaining walls. To minimize construction costs, it was necessary to minimize the height of the walls, and thus in places lower

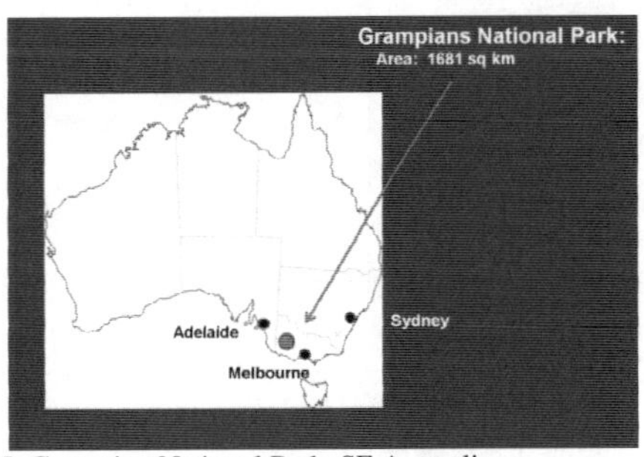

Figure 5. Grampian National Park, SE Australia

the elevation of the road.
Design of the gabion walls, and lowering of the road both required accurate assessments to be made of the depth to bedrock.

MASW Survey

An MASW survey was carried out in January 2012 at six segments on a damaged mountain road totaling 868 meters. Although the road was damaged and closed for public access, the access was possible and the remaining sealed surface made the survey by land streamer possible. However, occasional debris and fallen trees blocked vehicular access, where hand carrying the equipment was necessary.

Figure 6. One of the flood damages

Because of no traffic to cause noise on the closed road, the data quality is quite good. The traces have high amplitude spectra above 20Hz and the data have a broad bandwidth. It suggests that there is adequate signal up to 60Hz even in the far traces (Figure 7).

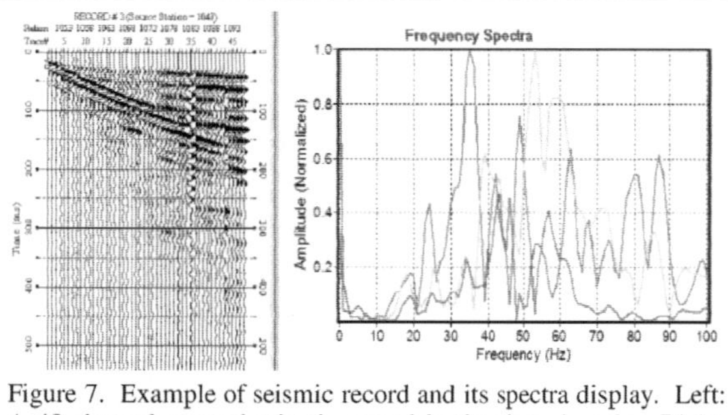

Figure 7. Example of seismic record and its spectra display. Left: A 48-channel merged seismic record in the time domain. Right: Spectra of near (blue), middle (green) and far (red) traces.

An S-wave velocity profile of one of the survey lines is shown in Figure 8. This shows presence of a weak layer at about 5 meters in depth with an S-wave velocity around 300m/s (light blue) above the competent rock below (Vs>700m/s in yellow and faster)

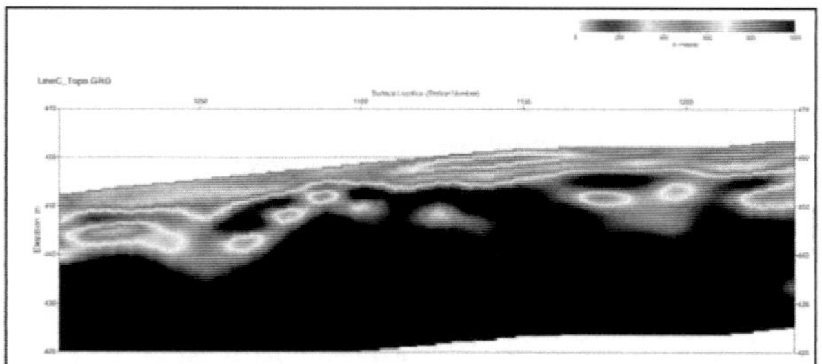

Figure 8. An example of section display adjusted to the surface topography (Line C)

The result was used in designing gabion-style retaining walls of varying height and depth (Figure 9).

Figure 9. One of the gabion-style retaining walls designed based on the MASW survey result.

More details of this application are reported in Suto and Kristinof (2014).

GEOSCIENTISTS *without* BORDERS PROJECT IN BALKANS, 2015-2016

Background

"Geoscientist *without* Borders" is an initiative of Society of Exploration Geophysicists (SEG). It supports humanitarian application of geoscience around the world. Their projects include application of geosciences in the fields of archaeology, earthquake preparedness, landslide preparedness, pollution mitigation, tsunami preparedness, habitat management volcano preparedness water management (SEG, 2015). Naturally many of these projects are related to engineering geophysics.

Balkan Project

One of the projects for 2015 is assessment of flood damage in Bosnia & Herzegovina (BiH) and Serbia. This is led by Association of Geophysicists and Environmentalists of Serbia (AGES), supported by Geoscientists *without* borders of SEG.

The heavy rain in the Balkan region started on 13 May, 2014. It affected countries including Hungary, Croatia, BiH, Serbia and Montenegro (Figure 10). The severe flooding occurred in Serbia and BiH Bosnia and Herzegovina and Croatia, which occurred after a three-months' amount of rain fell onto the region in just three days, devastating impacts in the catchment of Sava River (Komatina S., et al., 2014). The number of landslides amounts to several thousands, damaging roads, bridges, buildings, agricultural lands and other infrastructure. The local community suffered from difficulty in daily life and fear of further landslides (Figure 10). For repair of the roads, the information of the ground condition including distribution of strength is essential.

Figure 10. A landslide caused by the heavy rain.

Aim of the project is to help the local engineering work and mitigation process and to assess/predict the risk of further landslides in the future by combining several geophysical methods. The methods which will be used in this project include seismic reflection, refraction and MASW and electric resistivity methods.

A multinational team of experts are core members of the project. Local students and engineers are invited to participate in the field work and they will be given introduction to geophysical methods and training. The result of the geophysical surveys will be given to the local governments for their use in the recovery process to benefit the local community (Figure 11).

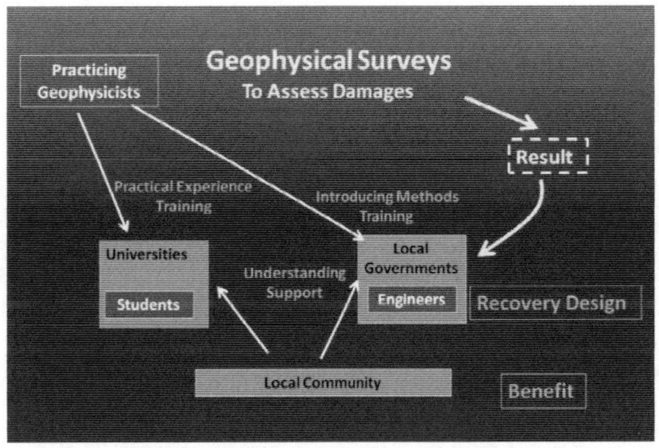

Figure 11. Schematic diagram of organization of the project.

The first phase of fieldwork took place in three landslide-prone areas in Serbia and BiH in June. Nineteen students from four universities, three academics and ten consultants gathered from six countries to carry out the geophysical surveys.

Figure 12. Student at field data acquisition.

Some results from this survey will be presented here.

ACKNOWLEDGEMENTS

The author thanks Entura and Tina Hydro for permission to present the survey in Solomon Islands, and SKM Ltd (now Jacobs) and VicRoads for the Grampion survey. Acknowledgement is due to SEG and a large number of participants and supporters of the G*w*B project led by Prof Snezana Komatina. All the individuals cannot be list here due to the constraint of the space.

A Case Study to Detect the Extent of a Carbonate Outcrop in the Subsurface Using Microgravity

Y. Bouzidi, Petroleum Institute, Abu Dhabi, United Arab Emirates*

Summary

A microgravity experiment is conducted across a carbonate outcrop to detect its extent in the subsurface. An adapted procedure for the case study is used to correct and reduce the gravity measurements. The base station elevation level was used as the datum. The numerical calculation used to approximate and remove the effect of the material above the base station led to a gravity residual anomaly that correlates very well with an east dipping dense carbonate layer. From the interpreted subsurface through gravity modeling it was possible to estimate the dip angle which is found to be approximately 40° towards the east. The carbonate outcrop plays probably a role in the confinement and abundance of the groundwater at the near surface composed by very porous alluvial deposits.

Introduction

Microgravity has long been used as a tool to detect buried features that exhibit a density contrast with their surrounding materials. It is used to detect cavities, mines, and buried bodies of various types (Butler, 1984). Microgravity can be used to detect threatening cavities beneath large structures (Yule *et al.*, 1998). It is also possible to use 4D microgravity for water-flood surveillance (Hare *et al.*, 1999; Ferguson *et al.*, 2008). Repeating a gravity survey can detect the change of the subsurface conditions, such as hydrocarbon depletion during production and water table level change with time. To achieve such results it is necessary to acquire the gravity with the highest precision possible and repeat the survey under the same conditions with time. Microgravity can serve as a tool to map the density redistribution in the subsurface due to fluid recharge as well as depletion associated with aquifer exploitation. One such example is the exploitation of geothermal reservoir (Sugihara & Ishido, 2008). It is also a valuable tool to monitor groundwater reservoir for storage and exploitation such as abandoned mines (Davis *et al.*, 2008).

On the other hand seismic refraction can also be used to map shallow targets (Lankston, 1989). This method can detect buried bodies of various types such as monumental tombs (Tsokas & Papazachos, 1995). It is also used to model groundwater contaminated sites (Zelt *et al.*, 2006). It can be effective in groundwater modelling in certain formations such as those defined by one or more layers of alluvial deposits underlain by denser materials such as limestones and Ophiolites (Haeni, 1986). This is particularly applicable to Al Ain, U.A.E. In this article an experiment was conducted to attempt to detect the extent of a carbonate outcrop using microgravity.

Experimental Setup and Methodology

The Oman Mountains in the east of the border with U.A.E. exhibit many outcrops of ophiolites. Eroded material from these mountains were transported and deposited in the U.A.E. These deposits form funnels that can easily be seen from satellite images. These alluvial deposits serve as a reservoir for fresh ground water in Al Ain. A microgravity

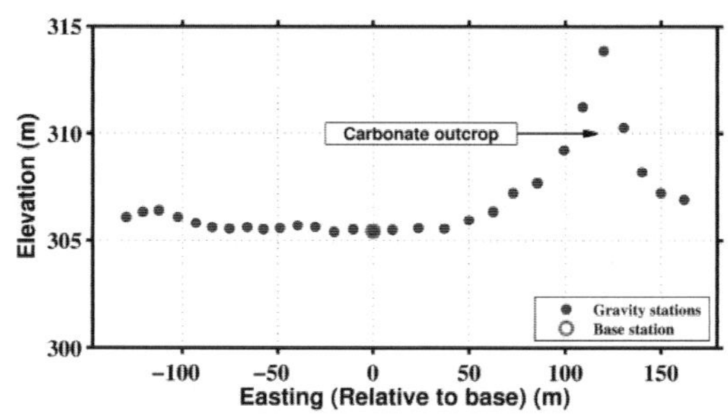

Figure 1: Elevations of the microgravity stations along a short profile (~390 m).

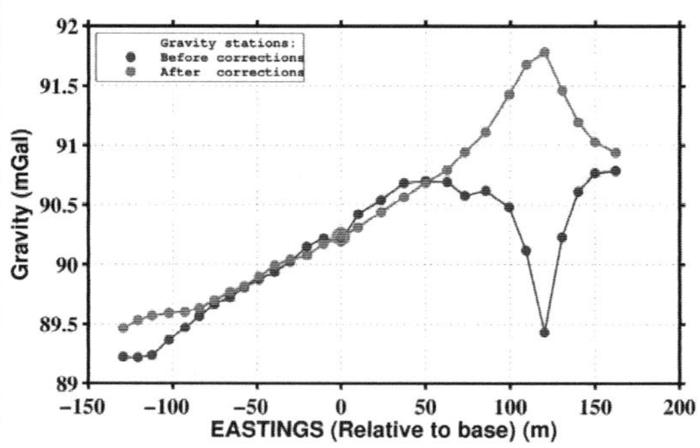

Figure 2: Raw gravity readings are shown in blue and corrected (Free Air Anomaly) readings in red. All standard gravity corrections are applied except Bouguer terrain corrections.

experiment was conducted in Al Ain near the Oman border across a carbonate outcrop (Jebel Auha). The outcrop extends from Oman (south) to U.A.E (North) with varying elevations which are higher within Oman and gradually decrease towards UAE before being completely buried in the subsurface. On each side of the Jebel the near-surface is a mixture of sand and gravel that serves as a ground water reservoir at various depths. Many similar outcrops are present in many areas in Al Ain. These can play a role as barriers to fluid flow and can substantially slow down the recharge of certain parts of the ground water reservoir. The water table level can vary substantially from one area to another and sometimes not far from each other. To map the extent and dip of the carbonate outcrop a gravity profile of about 390 m length is designed to run perpendicular to the outcrop and oriented from west to east with 29 stations at intervals of approximately 10 m. A gravity station in the middle of the profile is chosen to be a base station as shown in figure 1. The stations elevations, relative to the base, are measured with a differential GPS with 1 cm precision and are found nearly small except at the outcrop (Figure 1). The gravity readings were acquired using a SCINTREX CG-5 gravimeter within a time of about 5.5 hours ensuring that the gravity at each station is measured with a standard deviation and an error of 10 and 0 μGal respectively. The raw data is shown in blue in figure 2 with an arbitrary reference value at the base. The raw recorded gravity data includes the tidal effect. The data is then reduced by applying tidal corrections, instrumental drift corrections, latitude corrections and finally the altitude corrections. The Free Air Anomaly result is shown in red in figure 2. In the present case study applying the Bouguer corrections will not be helpful and will not be accurate enough to capture all the small variations of gravity along the profile. This is a consequence of the localized denser material at the outcrop above the base station elevation. Instead an alternative approach is preferred that relies on removing the gravity effect of the material above the base. It is calculated using numerical modeling by assuming the outcrop as a pile of rectangular rods perpendicular to the profile. The gravity of each rectangular rod is calculated based on the analytical solution for the gravity of a cube. An example of the gridded model is shown in figure 3. The z-component of the gravity of each rod is calculated according to the analytical solution given in Chappell et al., (2013). The sand that has accumulated on the flanks of the Jebel was taken into account (Figure 4). The distribution of the excess mass consisting of 3 types of material above the base elevation is shown in figure 4. The result of the corrected gravity is shown in blue in figure 5.

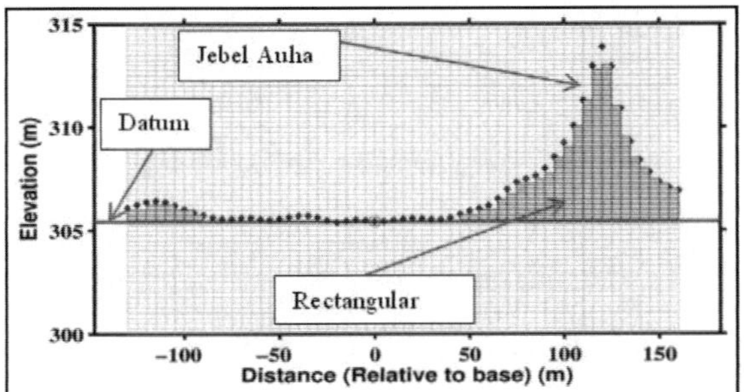

Figure 3: Approximation of the Jebel as a pile of rectangular rods running south to north (along the y-axis that is pointing into the paper. The profile is nearly west-east. The gridding shown here is for illustration only. The calculations were done using a much finer grid.

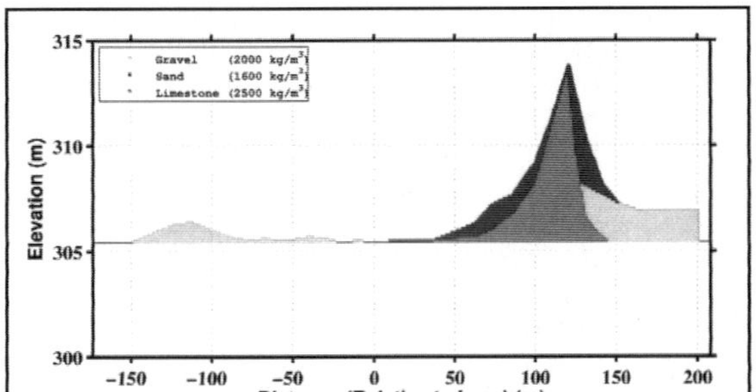

Figure 4: Distribution of the excess mass above the base. Three types of materials were used which are gravel, sand, and limestone.

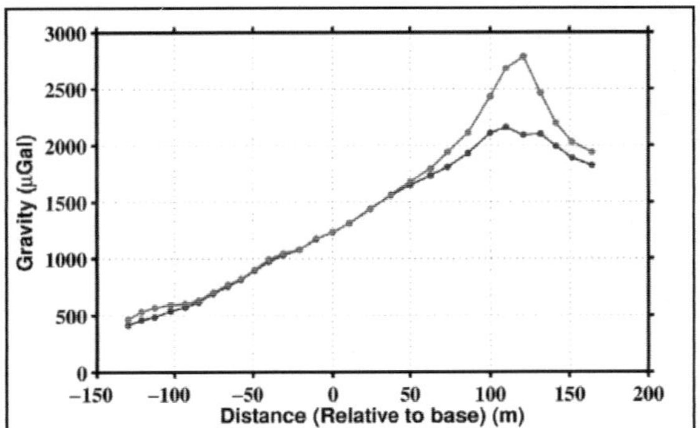

Figure 5: The final gravity anomaly result after removal of the effect of the excess mass above the base elevation level (Blue). The Free Air Anomaly with respect to the base level is shown in red.

Discussion and results

The residual gravity anomaly as shown in blue in figure 5 exhibits a linear decrease of gravity towards the east. Some irregularities are readily noticeable on the anomaly curve around the outcrop. These irregularities are probably due to the irregular shape of the outcrop as well as the sand distribution on the flanks resulting in density variations

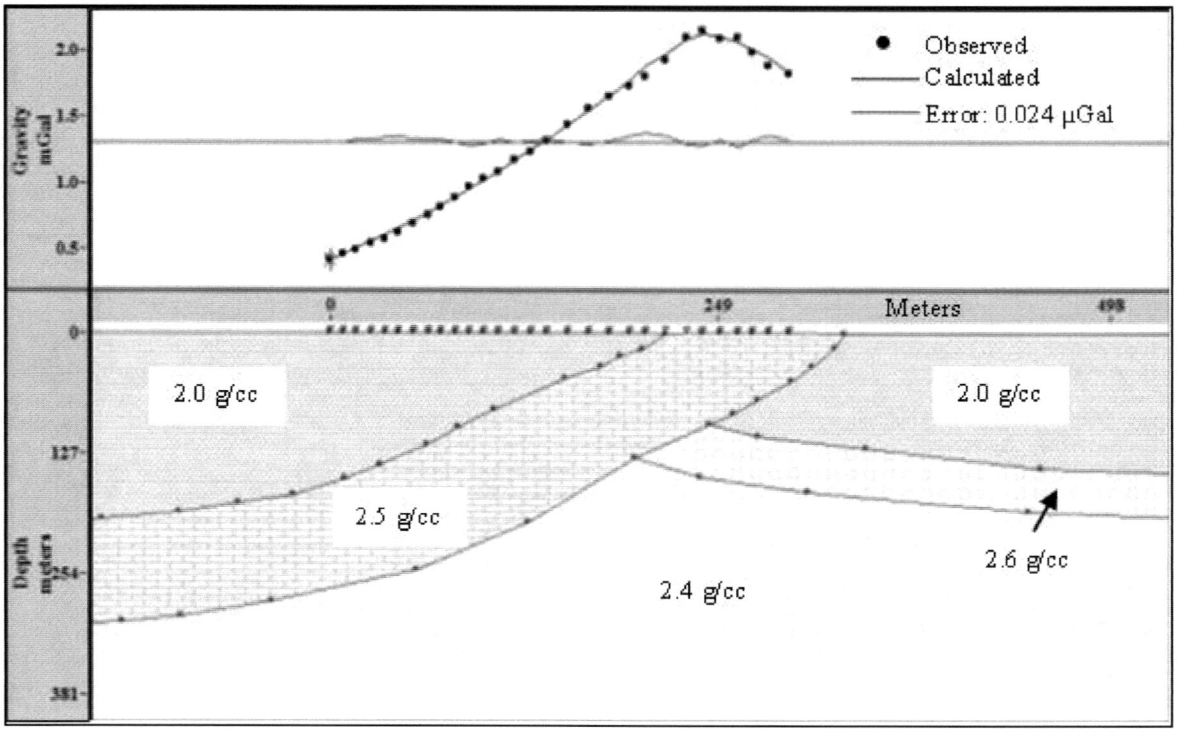

Figure 6: Modeled near-surface geology based on observed residual microgravity anomaly along a short profile across a carbonate outcrop.

across the hill as well as three dimensional effects of the outcrop. However the observed anomaly is consistent with the termination of the outcrop towards the west. It would be interesting to extend the measurements towards the east. Unfortunately it was not possible during the microgravity data acquisition because of the presence of obstacles. Nevertheless the residual gravity is substantial and varies between approximately 400 µGal to 2200 µGal.

The anomaly was modeled using GM-SYS software and the result is shown in figure 6. The model shows a very good match with the observed anomaly with an error of approximately 24 µGal. Four rocks with different densities were used in the model. Near the surface and on both sides of the Jebel one single aggregate material, with an average density of 2000 kg/m^3, is used. It is an overall porous material which is dominated by gravel with high ophiolite and sand content. Carbonates of various thicknesses and densities might also be present. A density of 2500 kg/m^3 is used for the carbonate outcrop which is the main target in the experiment. The other materials are required to fit the calculated gravity anomaly with the observed anomaly. The interpreted model does not focus on the exact nature of the materials that might be present in the subsurface but rather on the different density contrasts. The interpreted results might not be unique but show without ambiguity that the method is capable of detecting the gravity response of a dipping outcrop and can delineate its extent into the subsurface. However this experiment requires high quality microgravity data and special processing steps to achieve good quality results.

Conclusions

A microgravity experiment conducted across a carbonate outcrop in Al Ain, U.A.E was capable to detect its extent in the subsurface as well as estimate the dip angle. This study demonstrates the importance of the processing steps that should be adapted to each situation to better preserve the small residual gravity anomalies. In the present case study a numerical calculation is used to approximate and remove the effect of the material above the base station instead of applying the Bouguer correction. The gravity residual anomaly correlates very well with an east dipping (~40°) carbonate layer which probably plays a role as a confining substrata for the groundwater at shallower depths. This case study shows that gravity can be a valuable tool and can complement other geophysical methods for a better interpretation and characterization of the near-surface geology.

Acknowledgments

This work would not have been possible without the research dedication and support of The Petroleum Institute Abu Dhabi. I would like also to thank the laboratory engineers S. Ullah and M. A. Iqbal for their devoted time during the field acquisition program.
The support of National Drilling Company (NDC) is also appreciated. Special thanks also go to W. Kress (USGS) for his continuous support during all field trips.
Funding for this work was provided by The Petroleum Institute Abu Dhabi through the Research Initiation Funding Program (Code #14335).

References

Butler, D. K., 1984. Microgravimetric and gravity gradient techniques for detection of subsurface cavities, *Geophysics,* 49, 1084-1096.

Chappell, J. M., Chappell, M. J., Iqbal, A. & Abbott, D., 2013. The gravity field of a cube, *Phys. Int.,* 3, 50-57.

Davis, K., Li, Y. & Batzle, M., 2008. Time-lapse gravity monitoring: A systematic 4D approach with application to aquifer storage and recovery, *Geophysics,* 73, 61-69.

Ferguson, J. F., Klopping, F. J., Chen, T., Seibert, J. E., Hare, J. L. & Brady, J. L., 2008. The 4D microgravity method for waterflood surveillance: Part 3--4D absolute microgravity surveys at Prudhoe Bay, Alaska, *Geophysics,* 73, 163-171.

Haeni, F. P., 1986. Application of seismic refraction methods in groundwater modeling studies in New England, *Geophysics,* 51, 236-249.

Hare, J. L., Ferguson, J. F., Aiken, C.L.V. & Brady, J. L., 1999. The 4-D microgravity method for waterflood surveillance: A model study for the Prudhoe Bay reservoir, Alaska, *Geophysics,* 64, 78-87.

Lankston, R. W., 1989. The seismic refraction method: A viable tool for mapping shallow targets into the 1990s, *Geophysics,* 54, 11535-11542.

Sugihara, M. & Ishido, T., 2008. Geothermal reservoir monitoring with a combination of absolute and relative gravimetry,, *Geophysics,* 73 37-47.

Tsokas, G. N. & Papazachos, C. B., 1995. The detection of monumental tombs buried in tumuli by seismic refraction, *Geophysics,* 60 1735-1742.

Yule, D. E., Sharp, M. K. & Butler, D. K., 1998. Microgravity investigations of foundation conditions, *Geophysics,* 63, 95-103.

Zelt, C. A., Azaria, A. & Levander, A., 2006. 3D seismic refraction traveltime tomography at a groundwater contamination site, Geophysics, 71, H67-H78.

Gypsiferous bedrocks and soils of Abu Dhabi and their implications for engineering geozoning

Abdelghany, O., Abu Saima, M., Arman, H., Fowler, A.
Geology Department, UAE University, Al-Ain, Abu Dhabi, U.A.E.

Summary

Much of the bedrock in the Abu Dhabi Emirate consists of Oligocene to Miocene anhydritic to gypsiferous carbonate rocks and mudstones. The Holocene soils above these rocks also have significant sulphate contents. Both bedrocks and overlying soils present a number of engineering challenges related to dissolution cavitation and soil stability.

Introduction

Regionally extensive sedimentary formations of Oligocene to Miocene age exist as shallow bedrock beneath superficial Holocene cover of fluvial sediments and windblown sands in the Abu Dhabi Emirate [1] (Fig. 1). The dominantly mudstone and limestone sediments of these formations are known to host substantial deposits of gypsum and anhydrite in the form of evaporitic layers, nodules and vein systems. Gypsiferous bedrocks and the soils derived from them pose many foundation problems for engineers constructions, e.g. buildings, roads, etc. [2,3]. The principal engineering challenges presented by gypsiferous rocks and soils are: 1) dissolution cavitation and collapse structures in bedrocks, 2) soil heaving, cracking, surface subsidence and drainage barriers. Both potentially lead to settlement irregularities, foundation failures and pipeline breakages. This article first briefly describes the stratigraphy and constitution of the gypsiferous bedrocks of the Abu Dhabi region and their distribution, then documents the main features of engineering concern in these rocks and the soils above them.

Bedrock Stratigraphy

The bedrock is exposed at the surface at only a few locations: east of Jabal Hafit, Al-Ain area, and Jabal Al Dhanna, west of Al Ruwais city. The formations include the Asmari Formation (Early Oligocene age) and overlying Gachsaran Fm (also referred to as Lower Fars Fm, of Early Miocene age). The Asmari Fm consists of up to 140 m of marls with abundant fibrous gypsum veins (Fig. 2a) overlain by coralgal limestones, chalky at the top. The environment of the Asmari was shallow marine inner shelf and reefs to coastal lagoon. The Gachsaran Fm begins with a characteristic celestine ($SrSO_4$) rich layer, followed by at least 300 m of greenish to brownish mudstones with gypsum nodules and thin sandy to gravelly interbeds containing a very sparse shelly fauna **[4]**. The environment of deposition appears to have been initially shallow marine to brackish water, followed by emergence marked by the presence of a calcrete paleo-soil horizons. In our study area east of Al-Ain the Gachsaran Fm contains substantial thicknesses of gypsite beds locally rich in anhydrite nodules (Figs. 2b,3). The Gachsaran Fm is overlain by > 100 m of very coarse conglomerates of the Middle Miocene to Pliocene Barzaman Fm terrestrial conglomerates with numerous calcrete, silcrete and gypcrete paleosoil horizons indicating a continental environment **[1,4]**. The subsurface extent of the Oligo-Miocene units in the UAE is shown in Fig. 1. The Asmari Fm was deposited in the Zagros NW-trending foredeep, while the Gachsaran (Lower Fars) Fm was deposited in the Rub Al-Khali sub-basin, with exceptional thicknesses present in the adjacent contemporary Ras Al-Khaimah sub-basin (Fig. 1) **[5,4]**.

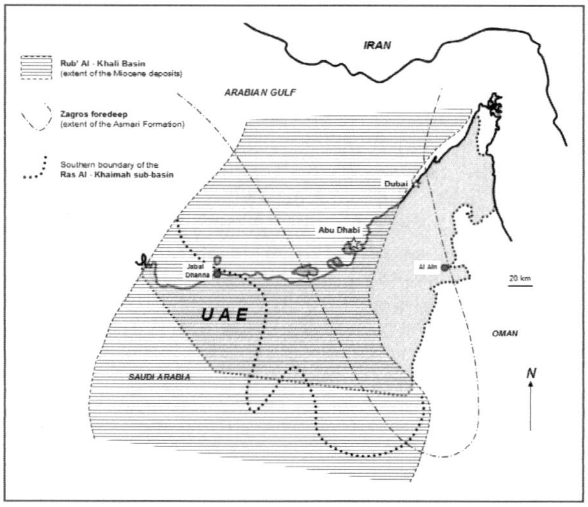

Fig. 1: Map of the UAE and region showing extent of the Oligo-Miocene formations.

Bedrock weathering and regolith formation

Calcite ($CaCO_3$) and the calcium sulphate minerals gypsum ($CaSO_4.2H_2O$) and anhydrite ($CaSO_4$) are highly soluble in the weathering environment especially in climatic conditions favorable for the formation of karst topography **[6]**. Anhydrite dissolves readily while gypsum tends to dissolve in a steady manner **[7]**. Optimum conditions for karstification of carbonate and evaporate rocks existed in the Abu Dhabi region during the warmer interglacial stages of the Pleistocene. The age of caves in the Jabal Hafit carbonate formations (Fig. 4) is estimated at > 1 million years ago **[8]**. Both carbonate and evaporite karst features

include caves, dolines, sinkholes, windows and karren (dissolution grooves and fissures) (Fig. 5).
The gypsiferous bedrocks have contributed to the development of aridosols that are the second commonest soils in Abu Dhabi Emirate **[9]**. These soils commonly have compacted layers of carbonate (calcrete) and gypsum (gypcrete) that strongly affect soil drainage properties, as described below.

Fig. 2a: Fibrous gypsum veins in marls of the Asmari Fm, Al-Ain.

Fig. 2b: White anhydrite nodules and cross-cutting gypsum veins in marls of the Gachsaran Fm, Al-Ain.

Fig. 3: Gachsaran Fm dipping beds of green gypsiferous mudstone and anhydritic gypsite in Al-Ain.

Foundation problems associated with gypsiferous bedrock and soil

Collapse structures and caves are well known in the Asmari Fm limestones in the Al-Ain area **[10,11,12,13]** (Fig. 4). An example of foundation problems caused by these cavities in the Asmari Fm has been reported by Gulf News (2008) **[14]** for the Shaikha Salama Mosque which had to be demolished in April, 2007 due to discovery of foundation damage produced by a large cavity beneath the structure (Fig. 7). Similar collapse structures and dissolved bedding planes, faults and fractures are also found in the overlying Gachsaran Fm (Fig. 5). The dissolution cavities may be completely filled, partially filled or void of later soils.
The existence of gypsum salts as a part of soil solid phase or dissolved within the pore fluid may cause significant effect on hydraulic and mechanical characteristics of unsaturated soils (Fig. 6). Gypsum soils are associated with subsidence (due to the solubility characteristics of gypsum) and heave (due to the hydration of anhydrite). Subsidence may occur progressively or suddenly. The subsidence will also depend on the size of the individual dissolution cavities and the thickness and geotechnical properties of overlying strata **[2]**. Serious structural damage can be associated with heaving of soils containing anhydrite due to periodic wetting. Clay minerals may also expand by exposure to water liberated by dehydration of gypsum. Hydration of anhydrite to gypsum leads to 63% volume increase **[3]**. If gypsum or anhydrite is the soil cementing phase, then dissolution of these sulphates can lead to leaching of the soil fine fraction with consequent

catastrophic weakening of the soil fabric and development of soil cracks.

Fig. 4: Karstic cave in the Asmari Fm, Al-Ain.

Fig. 5: Gachsaran Fm gypsite beds showing soil filled dissolution cavities and dissolved bedding planes.

Conclusions and Recommendations

Karst cavities are important irregularities in limestone basement in Abu Dhabi Emirate, especially the Asmari Fm in Al-Ain. Similar karstic dissolution has affected gypsites and anhydrites of the evaporitic sediments of the Lower Fars Fm that lies above the Oligocene limestones. The hazards produced by these cavities should be carefully assessed in the early stages of building foundation works. Geophysical surveys such as Ground Penetrating Radar (GPR) and 2D Electrical Resistivity imaging have successfully been used in this region to locate subsurface karst cavities **[15]** (Fig. 8). Inclusion of technical assessments for soil plasticity, swelling potential, swelling pressure, soil gypsum solubility, etc., may be included in existing guidelines for engineering geozones in areas of gypsiferous soils **[11,16]**. Improvement of capacity of gypsiferous soil to bear load may be obtained using techniques such as compaction and soaking **[17]**.

Fig. 6: Gypsum-cemented soil above Miocene bedrock, Al-Ain

Fig. 7: Demolition of the Shaikha Salama Mosque in Al-Ain after foundational problems related to karst cavities in the Asmari Fm bedrock.

References

[1] Alsharhan, A.S. and Nairn, A.E.M., 1997, Sedimentary Basins and Petroleum Geology of the Middle East: Elsevier.

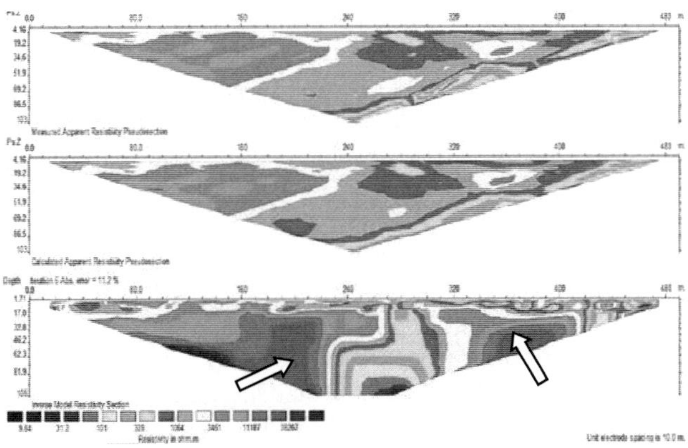

Fig. 8: 2D resistivity pseudosections revealing karst cavities in the subsurface (from Barker and Moore, 1998 **[18]**).

[2] Yilmaz, I., 2001, Gypsum/Anhydrite: Some Engineering Problems: BULLETIN OF ENGINEERING GEOLOGY AND THE ENVIRONMENT, **59**, 227–230.

[3] Azam, S., 2007, Study on the geological and engineering aspects of anhydrite/gypsum transition in the Arabian Gulf coastal deposits: BULLETIN OF ENGINEERING GEOLOGY AND THE ENVIRONMENT, **66**, 177–185.

[4] Styles, M.T., Ellison, R.A., Arkley, S.L.B., Crowley, Q., Farrant, A.R., Goodenough, K.M., McKervey, J.A., Pharoah, T.C., Phillips, E.R., Schofield, D. and Thomas, R.J., 2006, The Geology and Geophysics of the United Arab Emirates. Volume 2: Geology: British Geological Survey.

[5] Zeigler, M.A., 2001, Late Permian to Holocene Paleofacies Evolution of the Arabian Plate and its Hydrocarbon Occurrences: GEOARABIA, **6**, 445-504.

[6] Solis, R. and Zhang, J., 2008,Gypsiferous Soils: An Engineering Problem, *in* L.B. Yuhr, E.C. Alexander, Jr., and B.F. Beck, eds., Sinkholes and the Engineering and Environmental Impacts of Karst, 2008,Geotechnical Special Publications GSP,**183**, 742-749.

[7] Bell, F.G., 2013, Engineering Properties of Soils and Rocks, 3rd Edition: Elsevier.

[8] Fogg, T., Fogg, P, and Waltham, T., 2004, The Magharet Qasir Hafit Cave System, *in*: S. Aspinall and P. Hellyer (Eds.) Jebel Hafit, A Natural History. Emirates Natural History Group, 36-47.

[9] Shabbir, S.A., and Abdelfattah, M.A., 2008, Soils of Abu Dhabi Emirate, *in*: Perry, R.J. (Ed.) Terrestrial Environment of Abu Dhabi Emirate, Environmental Agency, Abu Dhabi, pp. 71-91.

[10] Benhammam K., and Baker H., 2011, Delineating Cavities Underneath Constructed Sites. Case Study in Al-Ain City: Presented at the 1st International Conference of Engineering Geophysics. 11- 14 December, 2011.

[11] Hashem, W., Abdelghany, O., Arman, H., Tokhi, E.M. and Saiy, E.A., 2011, How Do the Geological and Geomechanical Properties of the Carbonate Rocks Affect the Land Development in Al-Ain Vicinity?: Presented at the 1st International Conference on Engineering Geophysics, December 12-14, 2011, 76-79, Al-Ain, United Arab Emirates.

[12] El-Tokhi, M., Arman, H., Abdelghany, O., Hashem, W. and El Saiy, A., 2012, Isotope stratigraphy of Oligocene Limestone in Al-Ain City, United Arab Emirates. ARABIAN JOURNAL OF SCIENCE ENGINEERING,**37**,1439-1449.

[13] Arman, H., Hashem, W., El Tokhi, M., Abdelghany, O., and El Saiy, A., 2014, Petrographical and geomechanical properties of the Lower Oligocene limestones from Al Ain City, United Arab Emirates. ARABIAN JOURNAL FOR SCIENCE AND ENGINEERING, **39**, 261-271.

[14] Gulf News, 2008,"Move to protect buildings from underground cavities" http://gulfnews.com/cmlink/old-gulfnews/news/gulf/uae/housing-property/move-to-protect-buildings-from-underground-cavities-1.91753.

[15] Abdeltawab, S., 2013, Karst limestone foundation geotechnical problems, detection and treatment: Case studies from Egypt and Saudi Arabia. ARABIAN JOURNAL FOR SCIENCE AND ENGINEERING4: 376-387.

[16] Hashem, W., Abdelghany, O., El Saiy, A., Murad, A., Hussein, S., Gabr, A., Baker, H., and Aldahan, A., 2013, Structural and stratigraphic parameters as tools for the Geozoning Project of Al-Ain City, UAE: Presented at the 2ndInternational Conference on Engineering Geophysics, November,24-27, 2013, 242-245, Al-Ain, UAE.

[17] Razouki, S., Kuttah, D.K., Al-Damluji, O., Nashat, I., 2012, Improving fine-grained gypsiferous soil by increased compaction. INTERNATIONAL JOURNAL OF PAVEMENT ENGINEERING, **13**, 32-38.

[18] Barker, R. and Moore, J., 1998, The application of time-lapse electrical tomography in groundwater studies: THE LEADING EDGE, **17**, 1454-1458.

Acknowledgements: The authors would like to express their sincere thanks to the Research Affairs Office, UAEU for financing this project.

The 25 April 2012 Beni Haoua earthquake: instrumental intensity and consistency of ground motion modeling.

Khadidja. Abbes✉[1] , Nacima Benkaci[1], Abdennasser Slimani[1]. Faouzi Gherboudj[1], Mohamed Djeddi[2].

[1] **Centre National de Recherche Appliquée en Génie Parasismique**
Rue Kaddour Rahim Prolongée , Hussein Dey, Alger
[2] **Université des Sciences et de la Technologie Houari Boumediene**
Bp 32 El Alia 16111 Bab Ezzouar Alger.

Abstract

This paper analyzes the instrumental intensity and examines the consistency of ground motion models, used for seismic hazard analysis in Algeria, with the Beni Haoua earthquake. The accelerograms recorded at 11triaxial stations following the April 25, 2012, are used in the evaluation of intensity based on a combined regression of peak ground acceleration (PGA) and velocity (PGV) amplitudes of the horizontal components provided by Tselentis and Danciu (2008). The Arias intensity is computed and compared with the instrumental intensity. To define how the structures react to ground motions, spectral acceleration is presented for a range of periods. The observed PGA of the Beni Haoua earthquake compared to the empirical ground motion prediction (GMPEs) Ambraseys (2005) show an overestimated model.

Introduction

Seismic intensity is considered as a useful and simple parameter describing the damage caused by earthquakes. The building codes of several countries are based on the intensity values assigned to a given seismic region (Paz, 1994). An instrumental intensity serves as a useful link between the ground motion hazard at a particular site and the response of a given structure and facilitates risk-based calculations. The rapid calculation of intensity distribution for an earthquake founded on recorded ground-motion parameters, as well as the hazard assessment for future events, requires a regression relationship between intensity and strong-ground-motion parameters. Many attempts have been made to correlate intensity with recorded ground motions. In this paper we chose to use a combined regression of recorded peak ground acceleration and velocity amplitudes of the mean horizontal components using the Empirical relationship provided by Tselentis and Danciu (2008).

Beni Haoua instrumental intensity distribution

The measure of Beni Haoua earthquake instrumental intensity is based on 11 triaxials recordings of strong ground motion at epicentral distance between ~6 to 80 km applying equation (1) and (2). For the limited range of MM intensities IV $\leq I$MM $\leq$ VIII, Teslentis *et al.* (2008) found that for PGA the intensity is as expressed as:

$$I = 2.355+1.384*LOG10(PGA)+0.297*Mw-0.832*LOG10(Depi)-0.108*S(5) \qquad (1.1)$$

and for PGV within the range V $\leq I$MM $\leq$ IX,

$$Ivel = 5.582+1.397*LOG10(PGV)-0.787*LOG10(Depi)-0.073*S(5) \qquad (1.2)$$

Arias intensity defined as the sum of the energies dissipated during the time duration of strongshaking, which is frequently used as a shaking parameter, being considered the most suitable for characterizing earthquake impact. This measure is given as follow:

$$Ia = (\pi /2g)\int_0^{tf}[\alpha(t)]^2 dt \qquad (2)$$

where g is the acceleration due to gravity, $a(t)$ is the recorded acceleration time-history and tf is the duration of the ground motion.

In this section attention is restricted to present correlations between the Arias with the above proposed Tselentis G-A and Danciu, (2008) instrumental intensity (eq. 1.1 & 1.2), and to show general trends with hypocentral distance.

The figure 1.a and 1.b present the attenuation relations, where Iacc and Ivel are the calculated horizontal instrumental intensities. We obtained for both parameters (PGA and PGV) function of distance :

from PGAas:

$$Iacc=-1.48*ln(D)+9.4914 \qquad (3.1)$$

and from PGV as

$$Ivel=-1.255*ln(D)+8.6895 \qquad (3.2)$$

The correlation coefficient (R) shown in the top of figure 1.a and 1.b inquire that the proposed functions (eq. 3.1 and 3.2) are in perfect agreement with the two instrumental intensities (Iacc and Ivel). Nevertheless the one expressed in terms of velocity gives is the best.

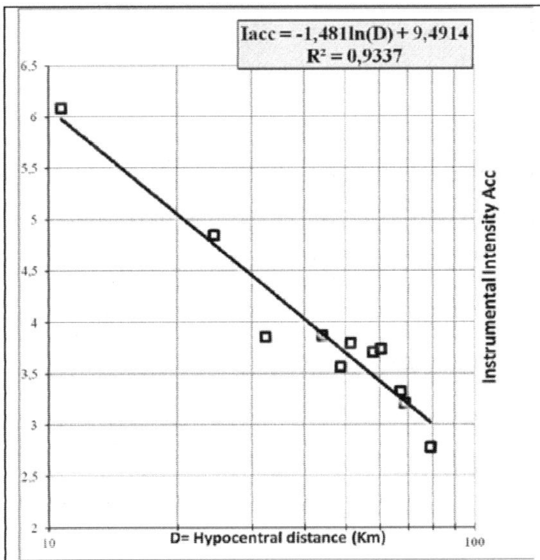

Figure 1.a. Attenuation for (Iacc) Instrumental intensity from PGA as function of D(km).

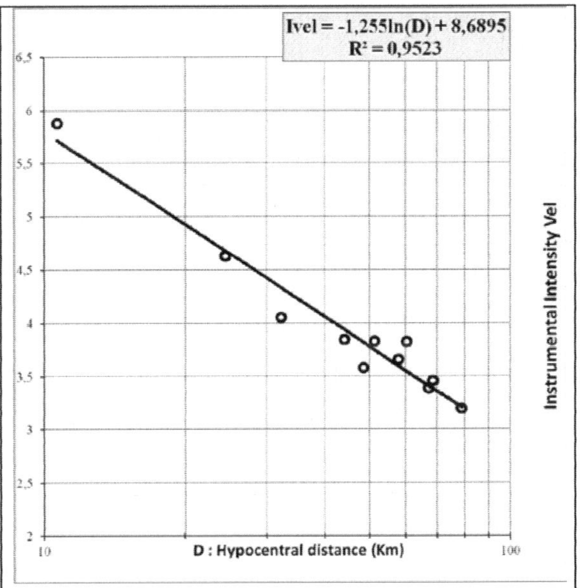

Figure 1.b. Attenuation for (Ivel) Instrumental intensity from PGV as function of D(km)..

The Arias intensity for EW & NS components correlated with hypocentral distance (D) in figure 2.a and 2.b expressed as the follow $\boldsymbol{I_{NS}}$ ***(Arias) = 6748.4D***$^{-2.883}$ and ***I (Arias) = 3341.9D***$^{-2.676}$

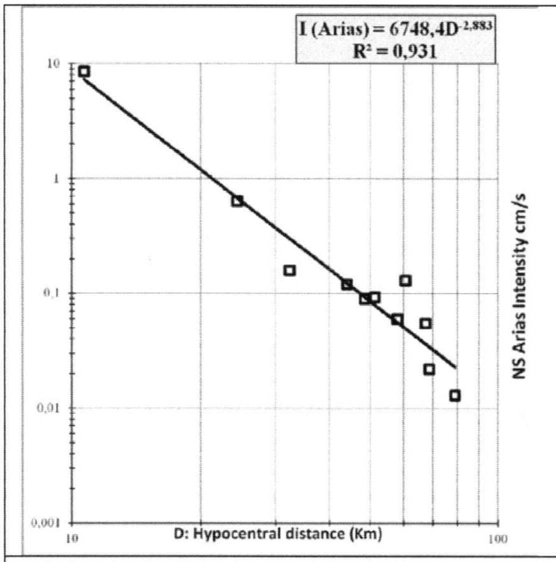

Figure 2.a. Attenuation for (N-S) Arias intensity as a function of hypocentral distance.

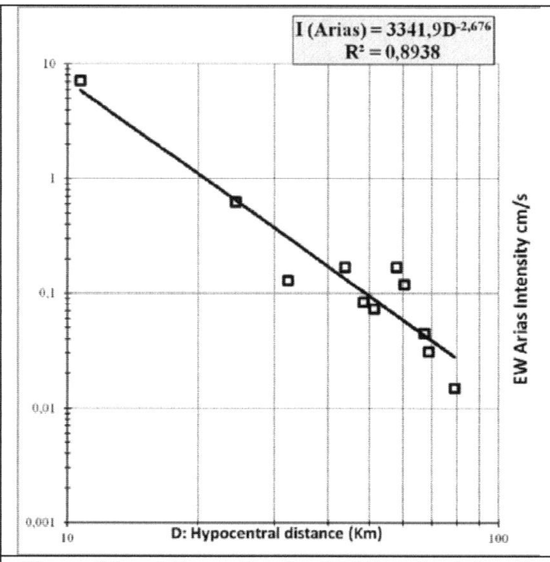

Figure 2.b. Attenuation for (E-W) Arias intensity as a function of hypocentral distance.

Comparaison of the observed PGA with the empirical prediction equation (Ambraseys et al., 2005)

In this part, we proposed to add the contribution of the recorded ground motion (PGA) in the 25April 2012, (Mw 5.0) Beni haoua Eartquake to assess the seismic hazard of Algeria, it is important to know how the seismic the peak ground acceleration (PGA) decays with source-to-site distance (Figure 3.a). Our purpose is restricted to compare the ground motion (PGA) with the empirical attenuation law proposed by Ambraseys (2005). Figure 3.a illustrate that the amplitude of the ground motion of the predicted equation is clearly underestimated the PGA for all distances. Figures (3.b, 3.c and 2.d) show the amplitude of response spectral (SA) calculated from the observed ground motion during the Beni Haoua main shock at different site for 3 period SA(0.5s, 1s and 2s) and then compared to SA of the Ambraseys et al., (2005) ground motion of the predicted equation. The results of Figures (3.b, 3.c and 3.d) illustrate that response spectral of Ambraseys (SA) provides a good prediction of the observed SA for period T=0.5s. The fit is not so good for T = 1s and T = 2s.

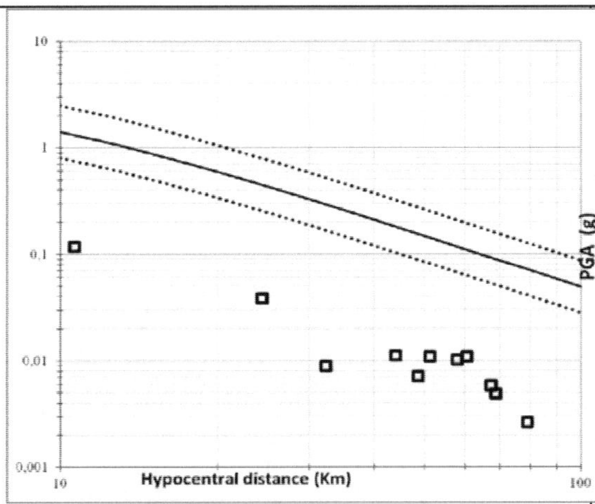

Figure 3.a Black squares is the observed PGA with Ambraseys (2005) GMPE (black line).

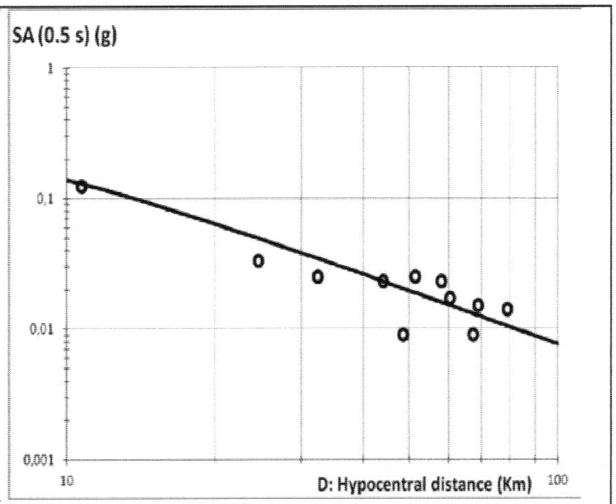

Figure 3.b Black circle is the observed spectral acceleration with (SA) Ambraseys (2005) GMPE (black line) for T=0.5s.

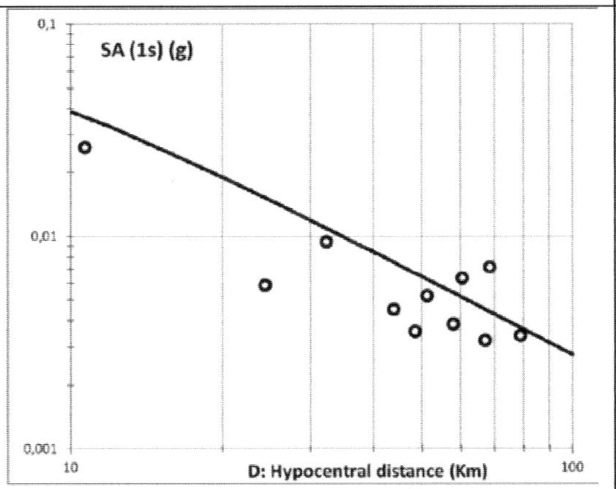

Figure 3.c Black circle is the observed spectral acceleration with (SA) Ambraseys (2005) GMPE (black line) for T=1s.

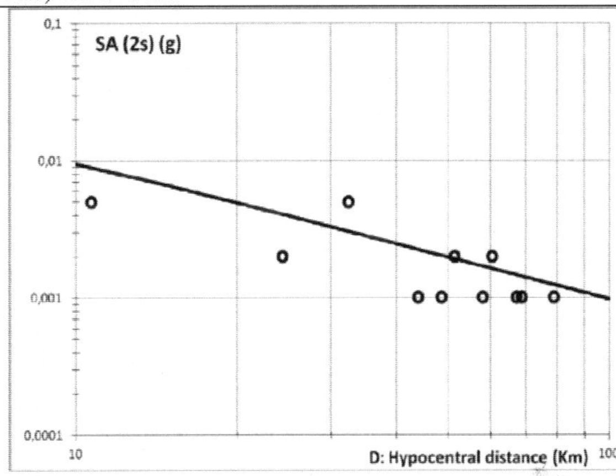

Figure 3.d Black circle is the observed spectral acceleration with (SA) Ambraseys (2005) GMPE (black line) for T=2s.

Conslusion

The Attenuation for (Iacc) Instrumental intensity from PGA and Attenuation for (N-S) Arias intensity as function of D(km) can be used to improve the previous ground motion predicted equation. The Beni Haoua event constitute an contribution for an evaluation of strong ground shaking which may be used in seismic zonation. The results of the comparison illustrate that the predicted Ambraseys (2005) PGME's overestimate the PGA.

References

Ambraseys N.N., John Douglas, Sarma S.K., Smit P. M., 2005. Equations for the estimation of strong ground motions from shallow crustal earthquakes using data from Europe and the Middle East: Horizontal peak ground acceleration and spectral acceleration. Bulletin of Earthquake Engineering, Springer Verlag (Germany), 3 (1), pp.1-53. .

Tselentis, G-A., Danciu, L., 2008. Empirical Relationships between Modified Mercalli Intensity and Engineering Ground-Motion Parameters in Greece. Bull. Seismol. Soc. Am., 98, N°4, 1863–1875.

Geoelectrical Measurements as a Proxy for the Biophysicochemical Properties of Soils Irrigated with Sewage Water at Assiut City, Egypt

Reham S. Abd El-Rhman[1], Gamal Z. AbdelAal[1], Mohamed El-Ameen A. Faragallah[2] and Mohamed H. Abdullah[3]*
[1]Geology Department, Faculty of Science, Assiut University, Assiut, Egypt
[2]Soil and Water department, Faculty of Agriculture, Al-Azhar Univerity, Assiut, Egypt
[3]Botany Department, Faculty of Science, Assiut University, Assiut, Egypt

Summary

At Arab El-Madabegh region, Assiut city, Egypt, many farmers are taking the advantage of the partially treated sewage water discharged from the sewage station for irrigating their agricultural lands as a common practice. Previous studies have shown that the long term use of the partially treated sewage water induced changes in the biophysicochemical properties of irrigated soils. Our objectives were to 1) develop empirical relationships between the geoelectrical measurements using resistivity method and some of the biophysicochemical properties of the soil samples, 2) validate the obtained empirical relationships using random samples. Eleven profiles were selected and twenty seven soil samples were retrieved from the study area at various depths (0-25, 25-60, and 60-90 cm). The soil samples were analyzed for physical (e.g., water content), chemical (e.g., soil ph, EC and some heavy metals), biological (Total Coliform counting) and geoelectrical (electrical resistivity) properties. A cross correlation between the biophysicochemical properties and electrical resistivity of the measured soil samples were performed and the cross correlated data are fitted with exponential trend line with correlation coefficients (R^2) range from 0.53 to 0.79. To validate the obtained empirical relationships, additional three profiles were selected to collect additional samples at similar depths. The resistivity measurements of theses samples are used to estimate the biophysicochemical properties from the previously predefined empirical relationships. A good correlation was obtained between the measured and estimated values of biophysicochemical properties with strong correlation coefficient (R^2) ranges from 0.83 to 0.98. The present study demonstrated the importance of using electrical resistivity measurements as a complementary tool to traditional soil sampling and a proxy to the biophysicochemical properties of soils. Moreover, resistivity measurements can reduce the cost and number of soil samples by providing proper guide for subsurface soil sampling and excavation from the spatial and temporal variations in the geoelectrical properties.

Introduction

Electrical resistivity method is non-destructive and very sensitive to many physicochemical properties of soil and can be used as a proxy for the spatial and temporal variability of these properties. At Arab El-Madabegh region (study area) which is located 3 km southwest of Assiut city (Fig.1), many farmers are taking the advantage of the partially treated sewage water discharged from the sewage station for irrigating their agricultural lands as a common practice (El-Desoky et al., **2010**). The long term use of the partially treated sewage water induced changes in the biophysicochemical properties of irrigated soils (Chen et al. **2004**; Aziz et al., **1999**). The objectives of the study were to 1) develop empirical relationships between the geoelectrical measurements using resistivity method and some of the biophysicochemical properties of the soil samples, 2) validate the obtained empirical relationships using random samples.

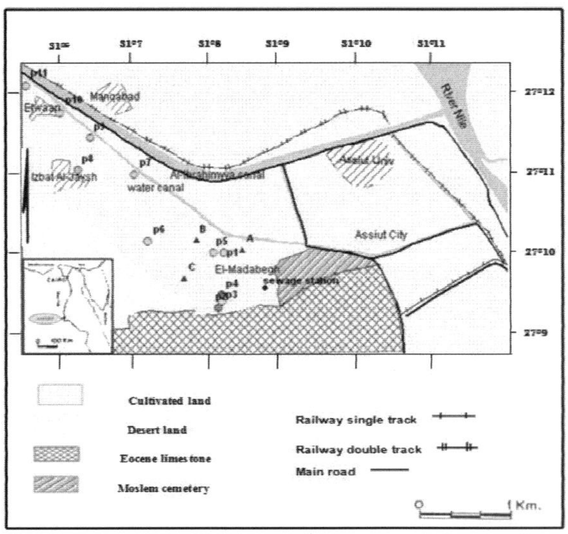

Figure 1. Location Map of the study area and retrieved soil samples

Materials and Methods

Eleven soil profiles were selected at approximately 1 Km interval to retrieve the soil samples at different depths (0-25, 25-60, and 60-90 cm) (Fig. 1). The soil samples were collected in aseptic bags, stored at 4 ^{o}C and carried to the laboratory for analysis. The water content was determined according to the gravimetric method (Gardner, **1986**), whereas the soil pH and EC were carried out by using pH meter and conductivity meter, respectively. Some heavy metals (e.g., Fe, Zn, Mn, Cu, Co and Pb) were extracted from soil samples using the diethylene triamine pentaacetic acid (DTPA) method (Lindsay and Norvell, **1978**) and determined using model 906 AA of GBC atomic absorption spectrophotometer. The bacterial count of the studied soil was carried out according to the pour plate method (Standards Australia **1995**). The electrical resistivity measurements were obtained using the standard four electrode technique and ABEM Terrameter SAS 300. The apparatus used for the electrical resistivity measurement is shown in figure 2. The resistivity (ρ) of the soil sample was then calculated (ρ (ohm.m) = $(\Delta V/I \times A/L)$. The measured electrical resistivity values were cross correlated to the measured biophysicochemical properties of the soil samples and empirical relationships were obtained from the fitted trend lines. To validate the obtained empirical relationships, additional three profiles (A, B and C) were selected to collect additional samples at similar depths. The resistivity measurements of theses samples are used to estimate the biophysicochemical properties from the previously obtained empirical relationships. The laboratory measured biophysicochemical properties of the soil samples were correlated with the estimated values from the empirical relationships.

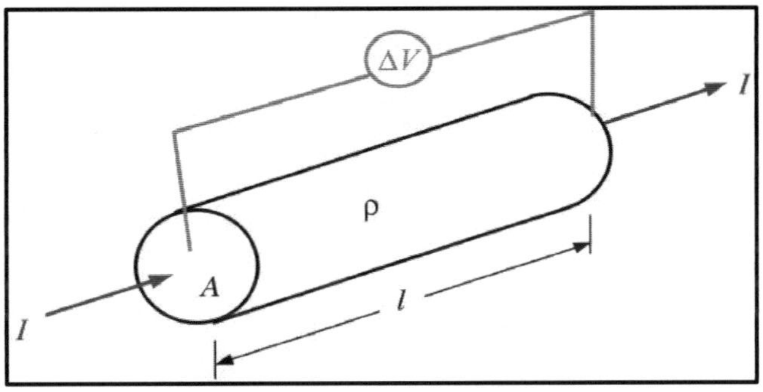

Figure 2. Setup of the laboratory electrical resistivity measurements.

Results and Discussion

The results of the measured biophysicochemical properties and electrical resistivity are shown in figure 3. Figure 3a shows that the water content of the studied soil samples varies 0.227% to 10.799% and increases with depth due to the richness of soil at depth with clay that retain and absorb more water than sandy soils at shallower depths. The EC varies between 0.2 and 1.94 ds/m and are greater at shallower depth (0-25 cm) due to the accumulations of salts and heavy metals (Fig. 3b). The pH values of the soil samples range between 7.44 and 8.92 and are higher at greater depth Fig. 3c). All the studied soils are considered slightly alkaline soils according to the acidity classes of the US Soil Conservation Service (**1962**). Figure 3d shows that most of soil samples extractable metals in the surface layer are higher than the deeper ones. Levels of the DTPA-extractable metals in the soil decreased in the order of Fe>Mn>Zn>Cu>Pb> Co. The total viable counts for total coliform in soil samples range between 27 CFU/g to 265 CFU/g and are higher at the shallower soil samples compared to deeper ones (Fig. 3e). All the measured biophysicochemical properties are reflected on the measured electrical resistivity values as shown in figure 3f. The electrical resistivity values of the soil samples range between 21.24 ohm.m to 391.867 ohm.m and are higher at shallower depth (Fig. 3f).

Figure 4 shows the cross correlation between the biophysicochemical properties and electrical resistivity of the measured soil samples. The cross correlated data are fitted very well with exponential trend line and display correlation coefficients (R^2) range from 0.53 to 0.79. The obtained empirical relationships (exponential equations) can be used to estimate the biophysicochemical properties at any location at the study site from only the measured electrical resistivity.

Figure 5 shows the correlation between the estimated values of biophysicochemical properties of soil samples obtained from laboratory analysis and calculated values obtained from the measured resistivity values and predefined empirical relationships displayed in figure 4. A good correlation was obtained between the measured and estimated values of biophysicochemical properties with strong correlation coefficient (R^2) ranges from 0.83 to 0.98. This indicates that the empirical relationships which were developed are valid and can be used to estimate and monitor the soil properties at the study site.

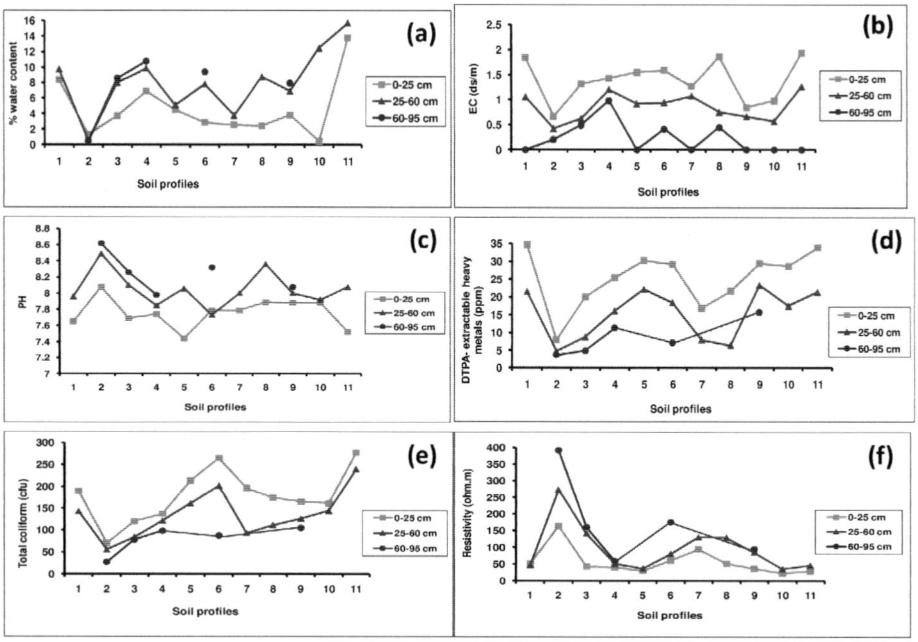

Figure 3. The measured physical (a-water content), chemical (b-EC, c-pH and d-heavy metals), bacteriological (e- total Coliform) and geoelectrical (f-electrical resistivity) of soil profiles at different depths.

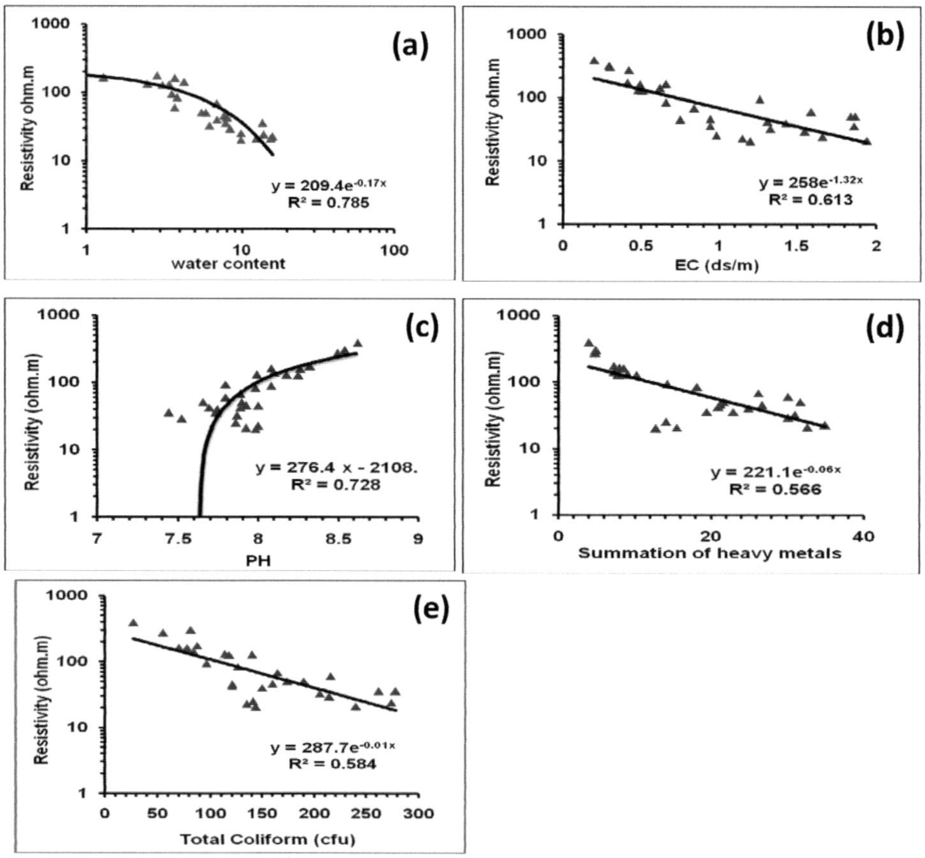

Figure 4. Cross correlation between the resistivity values and a) water content, b) Ec, c) pH, d) summation of heavy metals concentrations and e) total Coliform with exponential fitted equation and correlation coefficient (R2).

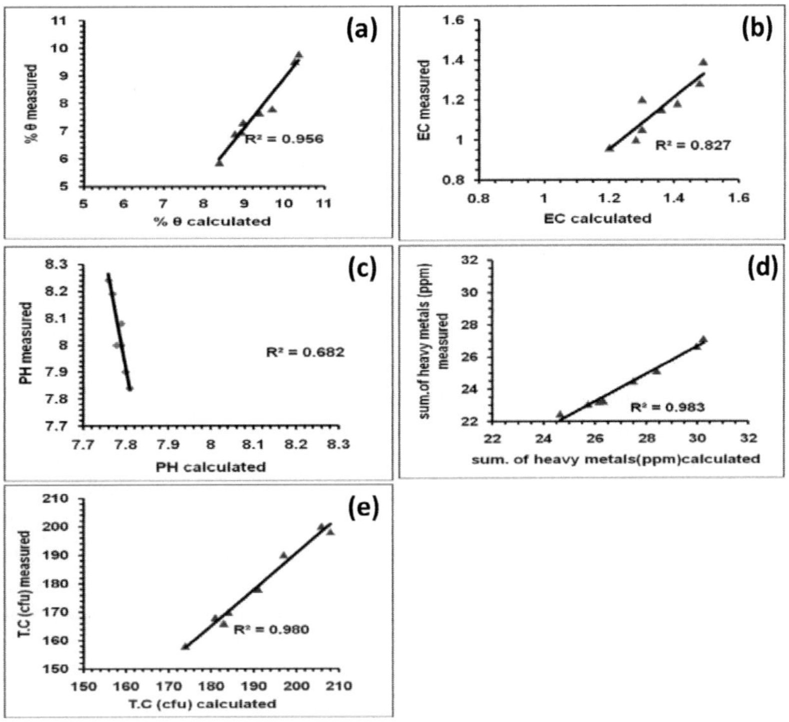

Figure 5. Cross correlation between measured and calculated a) water content, b) Ec, c) pH, d) summation of heavy metals concentrations and c) total Coliform with strong R^2.

Conclusions

The present study demonstrated the importance of using electrical resistivity measurements as a complementary tool to traditional soil sampling and a proxy to the induced changes in the biophysicochemical properties of soils due to irrigation with partially treated sewage water. Moreover, resistivity measurements can reduce the cost and number of soil samples by providing a proper guide for subsurface soil sampling and excavation from the spatial and temporal variations in the geoelectrical properties.

References:

Aziz, O., Inam, A., and Samiullah (1999), Utilization of petrochemical industry waste water for agriculture: WATER, AIR AND SOIL POLLUTION, **115**, 321–335.

Chen, Y., Wang, C., Wang, Z., and Huang, S. (2004), Assessment of the contamination and genotoxicity of soil irrigated with wastewater: PLANT AND SOIL, **261**, 189–196.

El-Desoky, M.A., Faragallah, M. A., Khalifa, E.M. and El-Ansary, M. A., (2010), Soil fabric change induced by prolonge irrigation with sewage effluent at Assiut, Egypt: JOURNAL OF AGRICULTURAL SCIENCE ASSIUT UNIVERSITY, **15** (16-17), 297-321.

Lindsay, W.L. and Norvell, W. A., (1978), Development of DTPA soil test for zinc, iron, manganese and copper: SOIL SCIENCE SOCIETY OF AMERICA JOURNAL, **43**, 966-972.

Standards Australia (1995), Water Microbiology – Heterotrophic Colony Count Methods – Pour Plate Method using Plate Count Agar: AS 4276.3.1–1995. Committee FT/20.

U.S. Soil Conservation Service (1962), Soil Survey Manual, US Government Printing Office, Washington, DC.

Can high-resolution marine geophysical data provide quantitative information useful for engineering design?

Mark E Vardy[1]

1. Marine Geoscience Group, National Oceanography Centre Southampton, European Way, Southampton SO14 3ZH, UK.

The amplitude, phase and frequency content of seismic reflection data contain quantitative information regarding the nature of the seafloor and sub-seafloor sediments. The petroleum industry has made significant developments over the past 20+ years to extract this information from reservoir-scale surveys using seismic inversion, allowing them to build detailed 3D models of hydrocarbon reservoirs. While some techniques (e.g., waveform inversion) are prohibitively computationally expensive to permit widespread application across all targets, other less expensive variants (e.g., impedance and amplitude-versus-angle inversion) have become a standard component of most interpretation workflows. In contrast, there has been very little progress toward the remote classification of near-surface sediments through the inversion of high-resolution marine geophysical data, with both academia and industry relying on extensive coring and cone penetrometer (CPTU) measurements. These provide very-high-resolution records of key subsurface properties (e.g., undrained shear strength, stiffness, grain size variation), but are expensive and time consuming to acquire and, more importantly, are only representative of the subsurface conditions at an isolated location at the time of sampling. High-resolution marine geophysical data can be acquired relatively quickly and cheaply, and at most infrastructure sites are standardly acquired at multiple stages of development across large potions of the site. These data are, however, commonly only used to produce a structural model. The construction of a reliable and detailed engineering ground model is a critical stage in the development of all offshore infrastructure, in deep or shallow water. In this talk, I will explore methods for maximizing the usefulness of high-resolution marine geophysical data, providing better calibration with existing borehole/CPTU data and deriving quantitative information (e.g., bulk density, porosity, Gmax). I will show that we can identify metre-scale stratigraphic changes, as well as subtle decimetre-scale structure, such as the grading within a glaciogenic sequence and a 40 cm thick landslide glide plane. The remote derivation of such high-fidelity soil properties has significant applications, both within academia and the offshore services/exploration industry.

Preliminary Test Investigation to Compare the Vertical Resolution of Different Geophones for Near Surface Shallow Seismic Investigation of an Abandoned Coal Mine Site

Ahmed I. Isiaka[*], *R.J. Durrheim*[**] *and M.S.D. Manzi*[***]
**University of the Witwatersrand, South Africa: Ahmed.Isiaka@students.wits.ac.za*
***University of the Witwatersrand and CSIR, South Africa: Raymond.Durrheim@wits.ac.za*
****University of the Witwatersrand, South Africa: Musa.Manzi@wits.ac.za*

Summary

High resolution shallow reflection seismic data was acquired across a borehole located within an abandoned coal mining site using three different types of geophones. The main objective of the study is to compare the vertical resolution of these geophones in order to choose the appropriate survey design for subsurface delineation of sinkholes or subsidence that might be present in the area. Two vertical geophones with resonant frequencies of 48 Hz and 100 Hz, and a horizontal geophone with resonant frequency of 14 Hz were employed in the investigation. Acquisition geometry was the same for all the geophones, and the energy source was provided by a sledge hammer hitting an aluminium plate. The data processing sequence was similar for all the geophones, and involves pre-stack, stacking, and post-stack stages. Signal attenuation with depth and dominant frequency vary among the geophones. The effect of multiples was noticed to be more severe in the data acquired with the 48 Hz vertical geophones, but less prominent in the 100 Hz vertical and the 14 Hz horizontal geophones. Post-stack, time-migrated sections generated for the three geophone types also show several reflections with depth for the 100 Hz vertical and 14 Hz horizontal geophones, as compared to the 48 Hz geophones.

Introduction

In near surface investigations involving high resolution shallow seismic reflection imaging method, choosing the appropriate survey parameters can be very challenging. This is because; achieving the desirable minimum resolution depends on several acquisition parameters (e.g. instrumentation and acquisition geometry) and sub-surface geological conditions (e.g. heterogeneity of the overburden layer, thickness of the weathered layer, and water table condition). The recognition of a geological event in a seismic data is largely dependent on the quality of the seismic data, which is determined by the signal to noise ratio. It is therefore of importance in seismic exploration, or any geophysical investigation, to acquire high quality data, as there is often limit to what data processing can do to improve signal quality. Several documented report of different acquisition techniques to improve the quality of shallow seismic reflection data include Hunter et al. (1984), Pullan and Hunter (1990), and Steeples and Miller (1990).

Acquiring good and quality seismic data depends on a number of considerable factors, including seismic equipment and instrumentation. The choice of equipment or instrumentation may be site dependent (e.g. Feroci et al., 2000) or, more importantly, on the resolution target (e.g. Kalweit and Wood, 1982; Widess, 1982; Singh, 1984; Palmer, 1987; Maxwell et al., 1994; Burger et al., 2006). While the equipment may be site dependent, it is often necessary that preliminary test investigations are carried-out to determine the most appropriate acquisition parameters to be employed in a particular geological condition.

Geophones are essential part of seismic exploration, because of their primary functions as detectors of ground motion. They are capable of detecting the amplitude of a ground motion as low as 0.0025 mm/s, and can measure the frequency of seismic data up to 500 Hz (Alcudia et al., 2008), depending on the resonant and spurious frequency. The resonant frequency (also refer to as natural frequency) of a geophone determines the limit of the recordable seismic frequency that is obtainable in a seismic data, which also limit the amount of seismic information that can be obtained from a seismic data. This parameter makes the choice of geophone considerably important in seismic exploration, especially in near surface investigation, where detail information of shallow sub-surface (< 20 m) is usually required.

We conducted a test investigation at abandoned coal mine site using a high resolution shallow seismic reflection imaging method. The main objective of the study is to compare the vertical resolution of three geophone types of different resonant frequencies, in order to choose the appropriate survey parameters for future seismic investigation of sinkholes and subsidence in the area.

Geology of the study area

The study area is an abandoned coal mine site that is being proposed for commercial and industrial development. The site is located within Benoni Agricultural Holdings,

Johannesburg, South Africa. History of coal workings and sand mining were reported to have taken place within the vicinity of this area in the early 1930s, with no documented report of the extent and depth of these activities. Water filled subsidence, with some measuring about 2 -10 m in diameter, and collapsed existing structures were noticed to have developed in some part of the area, as a result of the roof of the colliery collapsing into the workings. Preliminary geotechnical investigation, conducted through borehole samplings, also revealed subsurface underlying dolomites, which raises concerns over potential developments of dolomite related sinkholes.

The subsurface geology of the study area is well known to, at least, depth of 40 m, based on the information available from the borehole that were drilled for preliminary geotechnical assessment. The underlying geology consists of sand, clay, coal (which is about 7 m thick), and dolomite of the Transvaal Supergroup. One of the boreholes encountered dolomite at a depth of about 24 m. Intrusive rock suspected to be dolerite sill was also encountered in borehole 004 (figure 1). It is, however, difficult to establish the thickness of this dolerite sill, due to the limited depth penetrated by this borehole. It is believed, if this is a sill, and not a dike, that dolomite probably occurs below it.

Water table is very shallow in the study area, as some of the holes that have developed from subsidence in the eastern part of the study area were seen to be filled with water.

Data acquisition

High resolution reflection seismic data was acquired along a profile crossing one of the boreholes (figure 1) with the aid of a Geometric digital seismograph. Two vertical geophones with resonant frequencies of 48 Hz and 100 Hz, and a horizontal geophone with resonant frequency of 14 Hz were employed in the investigation. The same acquisition geometry was maintained for all the three geophone types, with maximum of 24 geophones per geophone spread. The geophones were spaced at 2 m apart with an offset of 8 m. Shots were fired with the aid of a sledge hammer striking an aluminium plate at 2 m interval within the offsets and the geophone spread. The total recording time of 500 ms was employed in the data acquisition at a sampling rate of 0.25 ms.

Figure 1. Satellite image of the study area showing the boreholes in the area and the seismic profile across borehole 004

Data processing

The quality of the data is generally good with high signal to noise ratio, as several reflections signal could be seen in the raw shot gathers. Comparison of raw shot gathers from the three geophone types shows variations in signal attenuation with depth (figure 2). Similarly, dominant signal frequency also varies for all the geophones, with 50 Hz for the 48 Hz vertical geophone, 100 Hz for the 100 Hz vertical geophone and 110 Hz for the horizontal geophone.

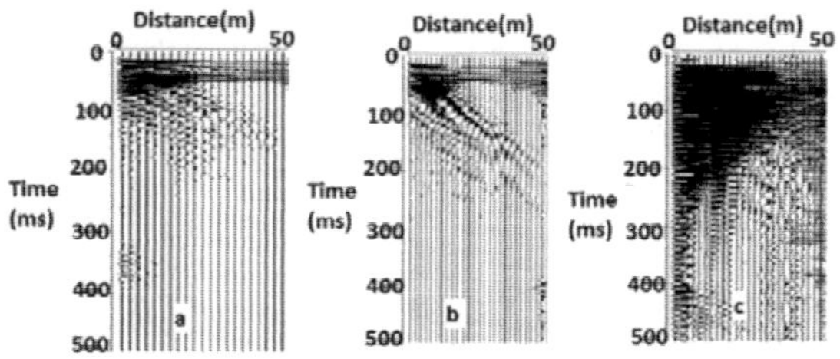

Figure 2. Raw shot gathers showing comparison of signal attenuation for the three geophone types (a) 48 Hz vertical geophone (b) 100 Hz vertical geophone (c) 14 Hz horizontal geophone

The seismic data processing employed in this study is similar to conventional processing steps described by Yilmaz (2001). Similar processing sequence was employed for the data acquired for all the three geophones, and involves three different processing stages: pre-stack, stacking and post-stack stage. The effect of multiples was severely observed in the 48 Hz geophone, therefore, predictive deconvolution, using time-lags of 50 ms, was applied to the data to remove the multiples prior to stacking (figure 3).

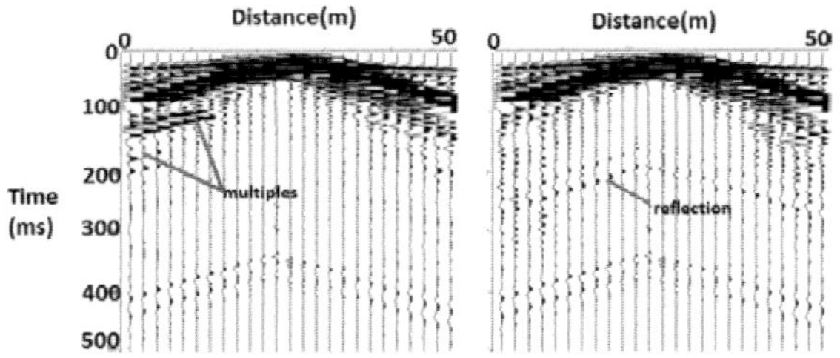

Figure 3. Shot gather before and after applying predictive deconvolution. Note the suppression of the multiples and the enhancement of the reflection event in the figure on the right.

The NMO correction and stacking were carried-out simultaneously for each of the geophone, using optimum velocity that is sufficient to flatten the hyperbolas and stack the traces. The optimum velocity was calculated using un-normalized cross-correlation described in details by Yilmaz (2001), and requires picking the velocity that best corresponds to correlative amplitudes along a hyperbola. This method also requires several trials to determine the optimum velocity for NMO correction and stacking, as inadequate stacking velocity may give rise to velocity artefacts in stack sections (figure 4). The data were best stacked at an optimum velocity range of 2200-2800 m/s.

The stacked data were later migrated using Kirchhoff migration method. An average medium velocity of 2500 m/s was used in the data migration as determined from the refraction analysis. The migrated data was then followed by the application of 2D median filter to smoothen the data.

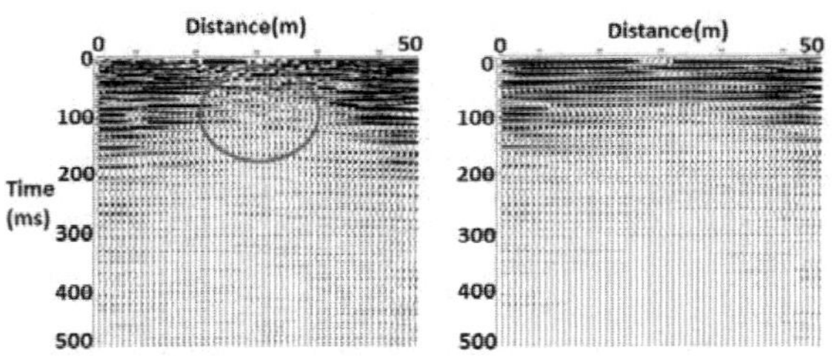

Figure 4. Velocity artefacts due to inadequate stacking velocity (circled region in left figure) and a well stacked section using optimum stacking velocity (right figure). Note the lateral continuity of the beds in the figure on the right.

Data Interpretation

The layer thickness obtained from the hyperbola analysis using Dix's method (Dix, 1955) makes it possible to correlate the migrated sections with the data from the borehole. The correlation shows variation in vertical resolution for all the three geophone types. Reflection events can only be seen to the depth of about 17 m in the 48 Hz vertical geophone, while several reflections occur beyond the depth of 17 m in the 100 Hz vertical and 14 Hz horizontal geophone. Figure 5, 6 and 7 respectively shows the correlation of the time-migrated section with borehole data for the 48 Hz and 100 Hz vertical geophones, and the 14 Hz horizontal geophone.

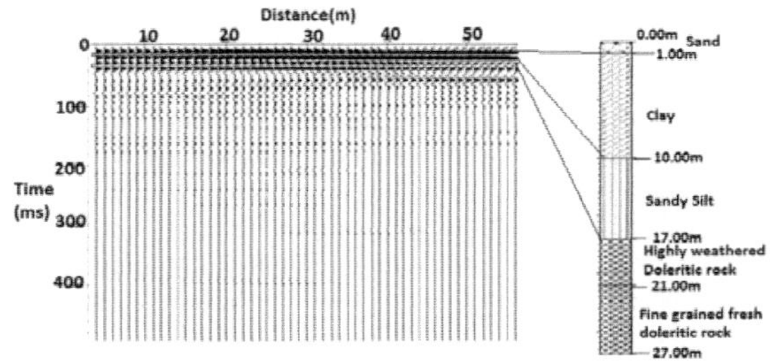

Figure 5. Time-migrated section with borehole correlation for the 48 Hz vertical geophone

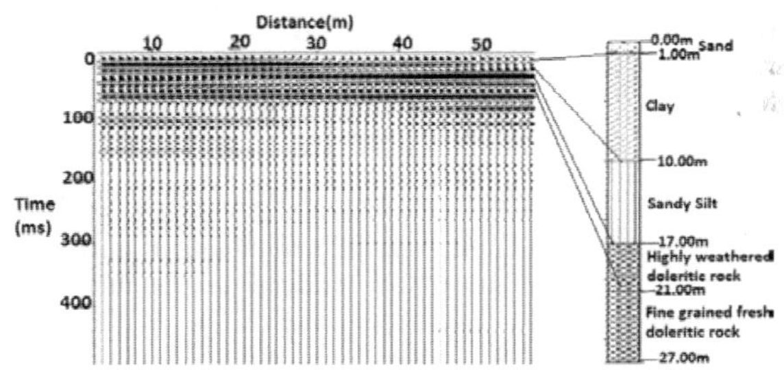

Figure 6. Time-migrated section with borehole correlation for the 100 Hz vertical geophone

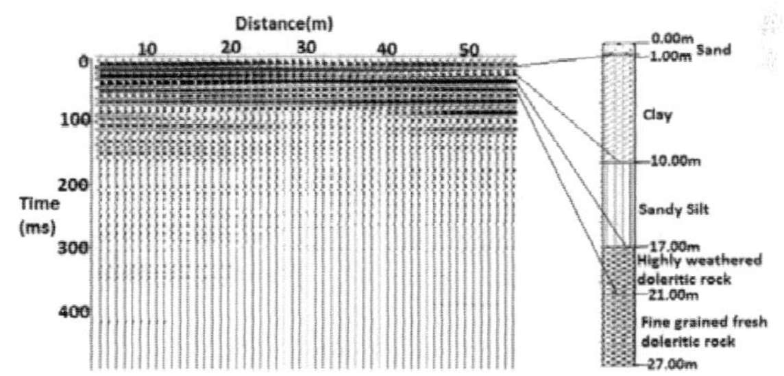

Figure 7. Time migrated section with borehole correlation for the 14 Hz horizontal geophone

Conclusions

Though no significant subsidence or void was observed below this profile, however the preliminary investigation has revealed the three geophone types to vary in their signal attenuation with depth, as well as dominant frequency contents and vertical resolution. The three geophone types will be suitable for shallow seismic investigation targeting depths below 17 m, while the 100 Hz vertical and 14 Hz horizontal geophone will provide better resolution for shallow investigation to depth of 30 m in the area.

References

Alcudia, A. D., R. R. Stewart, K. W. Hall and E. V. Gallant, 2008, Field Comparison of 3-C Geophones and Microphones to high-precision blasting sensors: CREWES Research Report, **20.**

Burger, H. R., A. F. Sheehan and C. H. Jones, 2006, Introduction to Applied Geophysics- Exploring the shallow subsurface: W.W. Norton and Company Inc.

Dix, C. H., 1955, Seismic velocities from surface measurements: GEOPHYSICS, **20**, 68-86.

Feroci, M., L. Orlando, R. Balia, C. Bosman, E. Cardarelli and G. Deidda, 2000, Some considerations on shallow seismic reflection surveys: Journal of Applied Geophysics, **45**, 127-139.

Hunter, J.A., S. E. Pullan, R. A. Burns, R. M. Gagne and R. L. Good, 1984, Shallow seismic reflection mapping of the overburden–bedrock interface with the engineering seismograph — some simple technique: GEOPHYSICS, **49**, no. 8, 1381–1385.

Kalweit, R.S. and L. C. Wood, 1982, The limits of resolution of zero phase wavelets: GEOPHYSICS, **47**, no. 7, 1035–1046.

Maxwell, P. W., K. Faber and H. A. K. Edelman, 1994, Modern geophones: do they meet the demands of shallow seismic measurements? 64th Annual Meeting, SEG, Expanded Abstracts, 559–597.

Palmer, D., 1987, High resolution seismic reflection surveys for coal: Geoexploration, **24**, 397–408.

Pullan, S. E. and J. A. Hunter, 1990, Delineation of buried bedrock valleys using the optimum offset shallow seismic reflection technique: In: Ward, S.H. _Edition, Geotechnical and Environmental Geophysics. Society for Exploration Geophysics, **3**, 75–87.

Singh, S., 1984, High-frequency shallow reflection mapping in tin mining: Geophysical Prospecting, **32**, 1033–1044.

Steeples, D. W. and R. D. Miller, 1990, Seismic reflection methods applied to engineering, environmental and groundwater problems: Society for Exploration Geophysicists Series; Investigation Geophysics, **5,** 1–30.

Widess, M.B., 1982: Quantifying resolving power of seismic system: GEOPHYSICS, **47**, no. 8, 1160–1173.

Yilmaz, O. Z., 2001, Seismic Data Analysis: Investigation in Geophysics; Society for Exploration Geophysicists Series, **1**, 274-299.

Comparison of dispersion curves from horizontal and vertical components of particle motion

Zahid Khan, Assistant Professor, American University of Sharjah
Magdi El-Emam, Associate Professor, American University of Sharjah
Mousa Attom, Professor, American University of Sharjah
Aqeel Ahmed, Instructor, American University of Sharjah

Summary

Multichannel Analysis of Surface Waves (MASW) is increasingly being used to evaluate the shear wave velocity profile of sites. A typical test configuration of MASW involves recording of the vertical component of particle motion using vertically oriented geophones. Shear geophones, although pricier, are also being used in geophysical studies; however, their use in MASW has not been extensively evaluated. This study presents the simultaneous deployment of horizontal (shear) and vertical geophones to measure both the horizontal and vertical component of particle motion during the propagation of surface waves. The results indicate excellent correlation between the dispersion curves obtained from the recordings of horizontal and vertical geophones. The inverted velocity profiles also exhibit agreement up to a certain depth.

Introduction

Dynamic properties of subsurface soils such as shear wave velocity and damping ratio are generally required and measured to solve problems involving dynamic loadings. Most building codes require the characterization of the first 30 m of the site for classification purposes; whereas, in the case of site response analysis and liquefaction analysis, the shear wave velocity and material damping ratios of individual layers are usually required.

Dynamic properties can be measured in the laboratory at both low and large strains (Khan et al. 2008, 2011); whereas, only low strain properties can be measured in the field. Field measurements such as seismic refraction, DHST, and MASW are considered better representative of the character of the soils due to scaling effects and sample disturbances that might occur in laboratory specimens. Laboratory tests are required if dynamic properties as function of strain level are required in addition to the measurement of material damping ratio.

Background

In field methods, seismic waves are generated by using an impact source at the surface such as a hammer operated manually or hydraulically. The amplitude of propagating waves is recorded by sensors (receiver array) along a line embedded at predetermined spacing (Figure 1). The MASW technique which is based on the analysis of Rayleigh waves (Rayleigh 1885) is preferred due to its ability in detecting weaker layers underlying higher velocity layers.

The depth of investigation in MASW depends on the wave lengths of the stress waves. Consequently higher resolution at shallow depths require seismic energy propagating at smaller wave lengths. The reliability of analyzing and resolving different wavelengths depends on configuration of the test setup and the energy of propagating Raleigh waves (Stokoe et al. 1988; Al Hunaidi 1993). The particle motion during the propagation of Raleigh wave approximates an ellipse (Figure 1) and the decreases in the amplitude of particle motion occurs with depth (e.g. Nasseri et al. 2007). The horizontal component of motion; however, decreases at a greater rate compared to the vertical component. The MASW tests typically rely on the vertical component of motion during the propagation of Raleigh waves by using vertical geophones. The vertical geophones have been traditionally prevalent because of cost and because they can be used in other geophysical tests such as DHST and refraction. The horizontal geophones are also used in many seismic applications such as DHST and shear wave refraction; however, the use of shear geophones in MASW testing has not been tested and established.

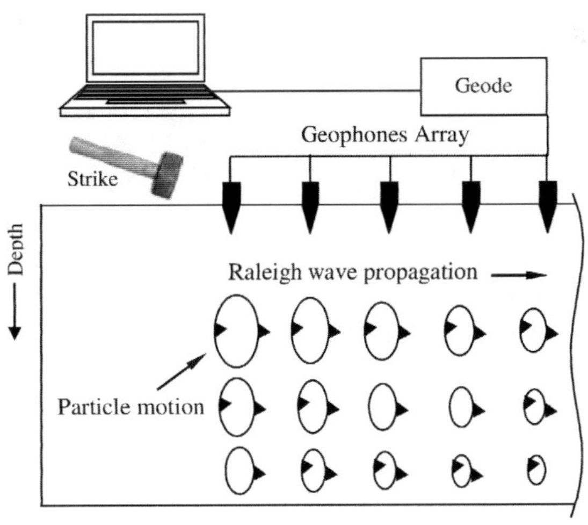

Figure 1: Typical layout of the MASW testing

Typically 12 or more geophones (sensors) are deployed in MASW testing to increase the effectiveness of site characterization in addition to significant saving in time (Park et al. 1999). Different source offset distances and receiver spacing are typically used to better characterize different depths at a site by analyzing different wave lengths (Park et al. 1999, Nazarian and Stokoe 1984). The phase velocity (Raleigh Velocity VR) is calculated by measuring the phase difference ($\Delta\phi(f)$) at different frequencies (f) as function of distance. The dispersion curve (variation of V_R with f or wave length λ) is developed for inversion process. The phase velocity (VR) for receiver spacing of Δx is typically calculated by using phase unwrapping method (Stokoe and Nazarian 1983) from Equation 1.

$$V_R(f) = 2\pi f \cdot \frac{\Delta x}{\Delta\phi(f)} \qquad (1)$$

The change in phase corresponding to different frequencies can be affected by spurious sources such as traffic and environmental noise. The MASW testing typically involves stacking of the signals and other measures to ensure predominant control of Raleigh waves over the phase function at different wave lengths (Heisey et al. 1982; Gucunski 1991, Gucunski et al. 1996). Equation 2 presents a simple criterion proposed by Heisey et al. (1982) to reliably analyze range of wavelengths as function of receiver spacing.

$$\frac{\lambda}{3} \le \Delta x \le 2\lambda \qquad (2)$$

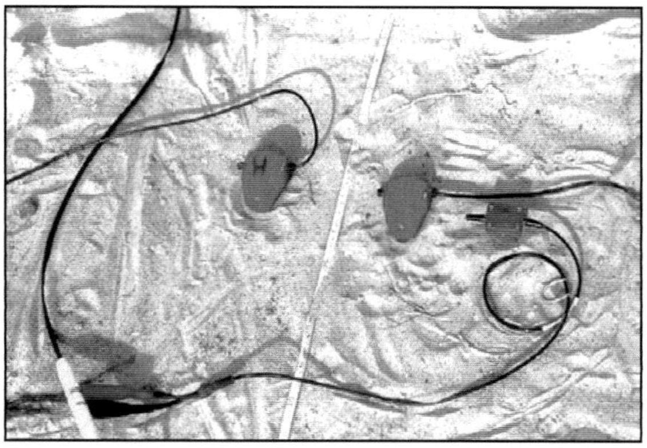

Figure 2: Typical positions of geophones at one location

Commercial software (e.g. SWAN and SurfSeis) typically uses either the frequency wave number (F-K) or frequency phase velocity (F-V) fields (McMechan and Yedlin 1981) to compute the experimental dispersion curves in the frequency domain. The dispersion curve is selecting by identifying spectral magnitudes corresponding to Raleigh modes of vibration. The velocity profile of a tested site is then computed by fitting a theoretical dispersion curve to the experimental dispersion curve using iterative procedure (Thomson 1950, Haskell 1953, Kausel and Roësset 1981, Kennet 1974). The theoretical dispersion curve is computed using Knopoff's method (Schwab and Knopoff 1972). The bisection method proposed by Press et al. (1992) is used to achieve the fit between the theoretical and experimental dispersion curves using the initial guess of ground profile.

Experimental Program

The field testing program consists of conducting MASW tests on a site to evaluate and compare the shear wave velocity profiles computed from the horizontal and vertical geophones. A 12 channels geode supplied by geometrics, twelve horizontal (shear) geophones (4.5 Hz), twelve vertical geophones (4.5 Hz), a spread cable of 12 connectors, a 10 kilo hammer, and a triggering cable with sensor is used in the tests.

Three test configurations are used to conduct the MASW tests on the site. In the first round of testing, only vertical geophones are connected to the spread cable (Figure 2). The shot offset distance and receiver spacing are changed accordingly to analyze different wavelengths. Horizontal geophones are connected in the second round to the spread cable to measure the horizontal component of motion. In the final (third) test configuration both the horizontal and vertical geophones are connected to the spread cable to measure the vertical and horizontal components of motion simultaneously. In this configuration only six recording points could be utilized due to limitations in the 12 available channels of the geode.

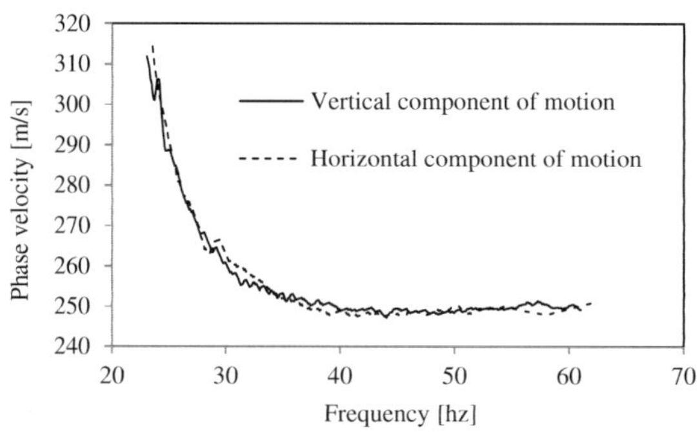

Figure 3: Comparison of dispersion curves from vertical and horizontal geophones

Each recording is verified by GEOPSY on site before performing another test. MASW tests are performed on weekends during the early hours of the morning to reduce contamination of higher frequency phases from traffic vibrations. The analysis of the field data is performed by using the commercially available software of SWAN licensed from Geoastir using frequency-wave number (F-K) analyses. Few data files are also analyzed randomly in MathCAD using phase unwrapping method to verify the results of SWAN. During a typical analysis F-K plots, dispersion curves, and inverted subsurface velocity models are produced.

Results and Discussions

Figure 3 presents the comparison of dispersion curves obtained from the particle motion recorded in horizontal and vertical direction during the propagation of Raleigh waves. The curves agree very well at all frequencies. The agreement indicates that the dispersion curves can be obtained reliably independent of the recording axis of geophones. The founding principle of converting the time domain data into the frequency domain data relies on the frequency content of the signals. The Rayleigh waves create a particle motion that resembles a retrograde ellipse. The increase in disagreement is expected due to greater decrease in the amplitude of vibration with depth for horizontal component of particle motion (Nasseri et al. 2007).

Conclusions

The important conclusions from the present study reveals that horizontal geophones are capable of not only detecting the horizontal component but can be used to measure the dispersion curves. Likewise the velocity profiles from both the horizontal and vertical geophones agreed very well at most depths.

At larger depths the deviations are due to issues with the generated wave lengths and diminishing amplitude of horizontal motion compared to the vertical motion; however, high sensitivity geophones can be used to compensate for this effect. Similarly, the transmitted energy from the shot impact can be increased to substantiate the particle motion.

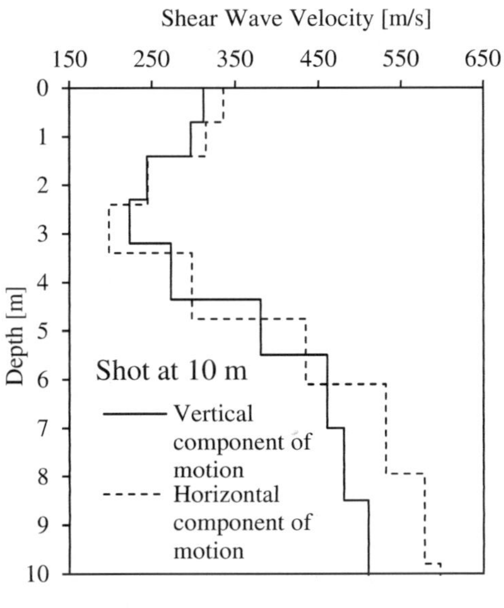

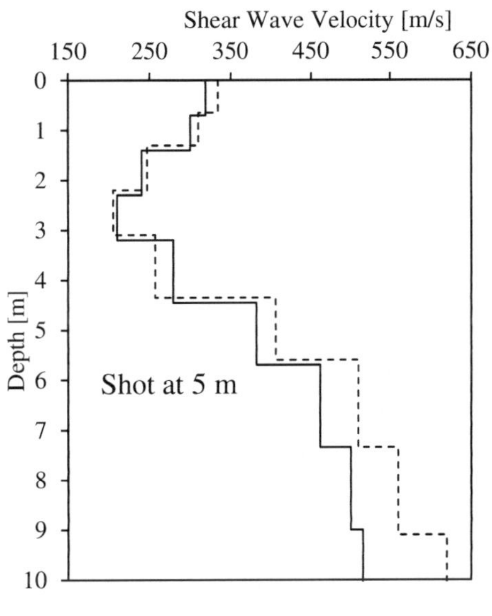

Figure 4: Typical positions of geophones at one location

Figure 4 presents the velocity profiles obtained from dispersion curves of horizontal and vertical component of motion at shot distance of 10 m. Similarly the velocity profiles at shot distance of 5 m is also presented. The velocity profiles at respective shot distances agree well for the horizontal and vertical component; however, the deviation becomes significant as the depth increase.

Acknowledgments

This research has been conducted with equipment funded by the Office of Research and Graduate Studies at the American University of Sharjah.

An Improved Iteration of Density Interface Inversion

Chong Zhang, Danian Huang, College of Geo-Exploration Science and Technology, Jilin University, Changchun, China*

Summary

This article first introduces the background of the research, and then puts forward the theory of the improved iteration inversion, to verify the new method, a numerical example is implemented, and a geophysical example is tested too, both examples show good viability of the given method.

Introduction

Inversion is an important technology in potential field exploration. The shape of an underground mass distribution, the geometry of an interface density contrast and other geological structure can be inverted by many kinds of methods. Oldenburg (1974) rearranged Parker's (1973) forward calculation of gravity anomalies and gave the determination of the underground interface. His method is a classical algorithm in inverting the density interface, and called P-O method in this article. But there is a divergent term in P-O algorithm. To get convergent results Oldenburg filtered while the iteration, this would remove the high frequency components and might obtain inaccurate results which deviate from the real value. This article is based on Zhang's (2015) improved method. No filter is needed along with the iteration, after the inversion filter the results and get a better estimation than the original P-O method.

Method of inversion

Parker's equation of calculating potential anomalies caused by an uneven layer is written as

$$\mathsf{F}[\Delta g] = -2\pi G \rho e^{-|\vec{k}|z_0} \sum_{n=1}^{\infty} \frac{|\vec{k}|^{n-1}}{n!} \mathsf{F}\left[h^n(\vec{r})\right]$$

(1)

where $\mathsf{F}[\ ]$ is Fourier transformation, Δg is the gravity anomaly, G is the gravitational constant, ρ is the contrast of density across the interface, k is the wave number, λ is wavelength, z_0 is the mean depth of the interface and $h(\vec{r})$ is the interface variety to the mean depth.

Oldenburg rearranged formula (1) and obtained an expression to calculate the depth of the interface as following

$$\mathsf{F}\left[h(\vec{r})\right] = -\frac{\mathsf{F}[\Delta g]}{2\pi G \rho} e^{-|\vec{k}|z_0} - \sum_{n=2}^{\infty} \frac{|\vec{k}|^{n-1}}{n!} \mathsf{F}\left[h^n(\vec{r})\right]$$

(2)

Equation (2) can provide the depth of interface, but there is a divergent term, which is tantamount to downward continuation factor in frequency domain of potential field. Xu (2006) presented a method for potential field downward continuation by using iterative calculation. Use the known upper field as the initial value of the unknown lower field, and then do upward continuation using the initial value of the lower field, compared with the known upper field and correct the lower field until a small difference was obtained. Potential field continuation is similar to potential field inversion in form in Fourier domain. Xiao (2007) did an excellent work, which combined P-O method with Xu's iteration to avoid calculating the divergent term, but he ignored the second term of the right hand side of formula (2) and did not give the iterative equation.

This article would put forward the equations of the combined method including the second term of formula (2) and filter the results only in the end of the inversion iteration. Fourier transformation of upward continuation is:

$$\mathsf{F}[u(x,y,z_0)] = e^{-|\vec{k}|z_0} \mathsf{F}[u(x,y,0)]$$

(3)

Rewrite formula (1) to following form,

$$\mathsf{F}[\Delta g] = e^{-|\vec{k}|z_0} \mathsf{F}\left[-2\pi G \rho \sum_{n=1}^{\infty} \frac{|\vec{k}|^{n-1}}{n!} \left(h^n(\vec{r})\right)\right]$$

(4)

Formula (3) and (4) are in similar form, and according to the iteration method of formula (3), we could iterate formula (4) in the way of Xu's approach to (3). And we (in Zhang's paper (2015)) obtained,

$$-2\pi G \rho \sum_{n=1}^{\infty} \frac{|\vec{k}|^{n-1}}{n!} \left(h^n(\vec{r})^{(n+1)}\right) = -2\pi G \rho \sum_{n=1}^{\infty} \frac{|\vec{k}|^{n-1}}{n!} \left(h^n(\vec{r})^{(n)}\right)$$
$$+\mathrm{s}\left(\Delta g - \mathsf{F}^{-1}\left[e^{-|\vec{k}|z_0}\left(\mathsf{F}\left[-2\pi G \rho \sum_{n=1}^{\infty} \frac{|\vec{k}|^{n-1}}{n!} \left(h^n(\vec{r})^{(n)}\right)\right]\right)\right]\right)$$

(5)

Rearrange (5) and get the following iterative equation,

$$h\left(\vec{r}\right)^{(n+1)}=h\left(\vec{r}\right)^{(n)}-\frac{1}{2\pi G\rho}\mathrm{s}(\Delta g-\mathsf{F}^{-1}\left[e^{-\left|\vec{k}\right|z_0}\left(\mathsf{F}\left[-2\pi G\rho\sum_{n=1}^{\infty}\frac{\left|\vec{k}\right|^{n-1}}{n!}\left(h^n\left(\vec{r}\right)^{(n)}\right)\right]\right)\right]) \tag{6}$$

The depth of the interface is defined by,

$$\mathrm{D}(\vec{r})=z_0+\sum_{n=1}^{\infty}h(\vec{r})^{(n)} \tag{7}$$

Numerical Examples

Figure 1 A is the contour map of the underground interface, at the depth of 60 km there is a subsurface consisted of two opposite bumps. One bump is 40 km wide and the center is (40, 40) at the base with the height of -10 km compared with mean depth, another is 30 km wide and the center is (90, 90) at the base with the height of 7 km compared with mean depth. The density contrast across the interface is 0.6 g/cm^3. The columns and rows of this area is 128×128, and the interval of the columns and rows is 1 km. Figure 1 B is the gravity anomaly map of the density subsurface with 5% white Gaussian noise. Figure 1 C is the inversion geometry map of density subsurface by original Parker-Oldenburg's method with 5% white Gaussian noise and filtered during calculation. The result is gained from the application of the improved 3DINVER.M function provided by David etc. (2005). At the 3rd iteration, convergence of the iterative procedure is met when the convergence criterion is established in 0.02 km. The maximum depth of the inverted subsurface is about 63.77 km located at (49, 49) and the minimum depth is about 58.07 km located at (106, 106). Figure 1 D is the inversion geometry map of density subsurface by this article's method with 5% white Gaussian noise and filtered after mapping. The result is gained from this article's inversion function but the same numerical data in Figure 1 C. In a similar way, at the 37th iteration, convergence of the iterative procedure is met when the convergence criterion is established in 0.02 km. The maximum depth of the inverted subsurface is about 66.38 km located at (49, 49) and the minimum depth is about 55.32 km located at (89, 89). The detail comparison can be seen in the Table 1. With the times of iteration, the RMS errors between the synthetic data (without noises) and the final calculated data of the two methods are shown in Figure 1 E. As we can see, compared with the original inversion method, the maximum and minimum center locations obtained from this article's method are remarkably similar to the geometry of the synthetic density subsurface. And the maximum and minimum depth calculated from this article is better than the original one. Though the iterative times of this article's method is bigger than the original one, with the iteration of the two methods, the RMS errors become small and this article's method provide a smaller value than Parker-Oldenburg's. It is very uplifting because the outcomes are much near the synthetic example.

Geophysical Example

The gravity anomalies used are the data caused by the crust-mantle boundary in Brittany, France (David, 2005). This area is 300×300 km^2. The grid interval is 6 km. The density contrast is 0.4 g/cm^3 and the mean depth of the Moho is 30 km. The map below was extended to 666×666 km^2 which the number of columns and rows are 112×112 to get rid of edge effects. And other parameters are according to David's application example. Figure 2 A is the Bouguer anomaly in Brittany. Figure 2 B is the Moho relief inverted from the Bouguer anomaly in Brittany by this article's new method. Figure 2 C is the gravity anomaly obtained by the Moho relief of Figure 2 B. As can be seen, the gravity anomaly obtained by the calculated Moho depth is very similar to the observed anomaly not only by the size of the geologic body but also by magnitude of the anomalies. It indicates the depth gained by this method is receivable in the real case and this method can be promoted to the other gravity inversion of actual geophysical prospecting.

Conclusions

This article's inversion is valid by using on both synthetic data and real geophysical data. The algorithm does not ignore the second term in Xiao's article and get a better result than P-O method. In this method filter is applied after the output file of the inversion but not used during inversion iteration. This helps to reduce cutting off the high frequency components of the data and to obtain better results. The numerical example demonstrates the new method is more accuracy in inverting the geometry of the subsurface than the P-O method and is convergent in calculation. A convincing interface of Moho is obtained by applied this method to an actual geophysical problem.

Acknowledgments

The authors thank the reviewers for their constructive suggestions and comments on the manuscript to improve it. The real data this article used was from David's code of 3DINVER.M. The authors thank David Gomez-Ortiz for his dat. This work was supported by a grant from the National High Technology Research and Development Program of China (863 Program, No. 2014AA06A613) and by Graduate Innovation Fund of Jilin University (No. 2015025).

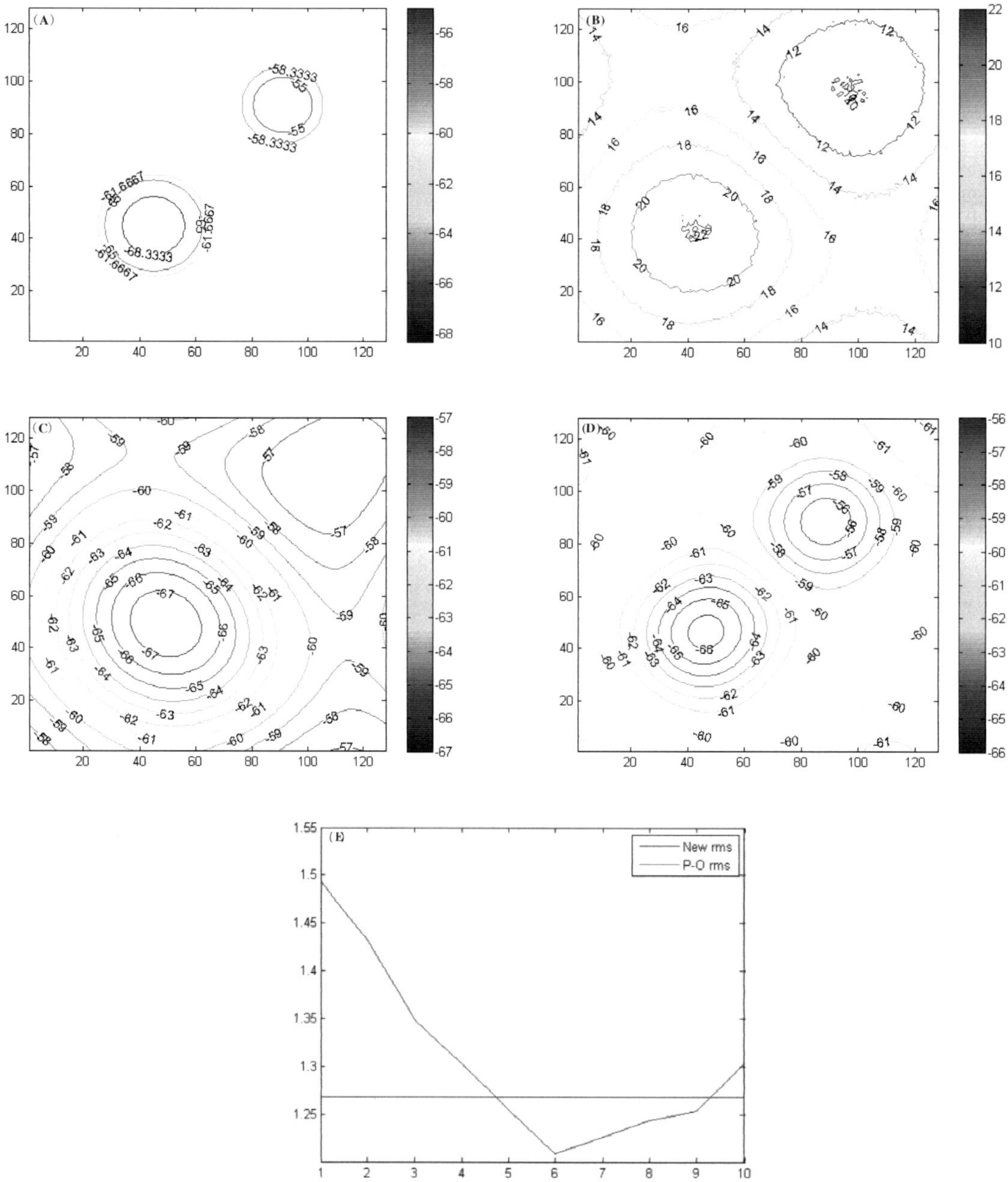

Figure 1 A is the contour map of the density subsurface. B is the gravity anomaly map of the density subsurface with 5% white Gaussian noise. C is the inversion geometry map of density subsurface by original Parker-Oldenburg's method with 5% white Gaussian noise and filtered during calculation. D is the inversion geometry map of density subsurface by this article's method with 5% white Gaussian noise and filtered only after mapping. E is the comparison of the RMS between these two methods.

	Max depth	Max location	Min depth	Min location	criterion	Iterative times
Synthetic model	70.00	(44,44)	53.00	(90,90)	null	null
P-O inversion	63.77	(49,49)	58.07	(106,106)	0.02	3
New inversion(s=0.9)	66.38	(46,46)	55.32	(89,89)	0.02	37

Table 1 *The results and parameters for synthetic model, Parker-Oldenburg's method and this article's method*

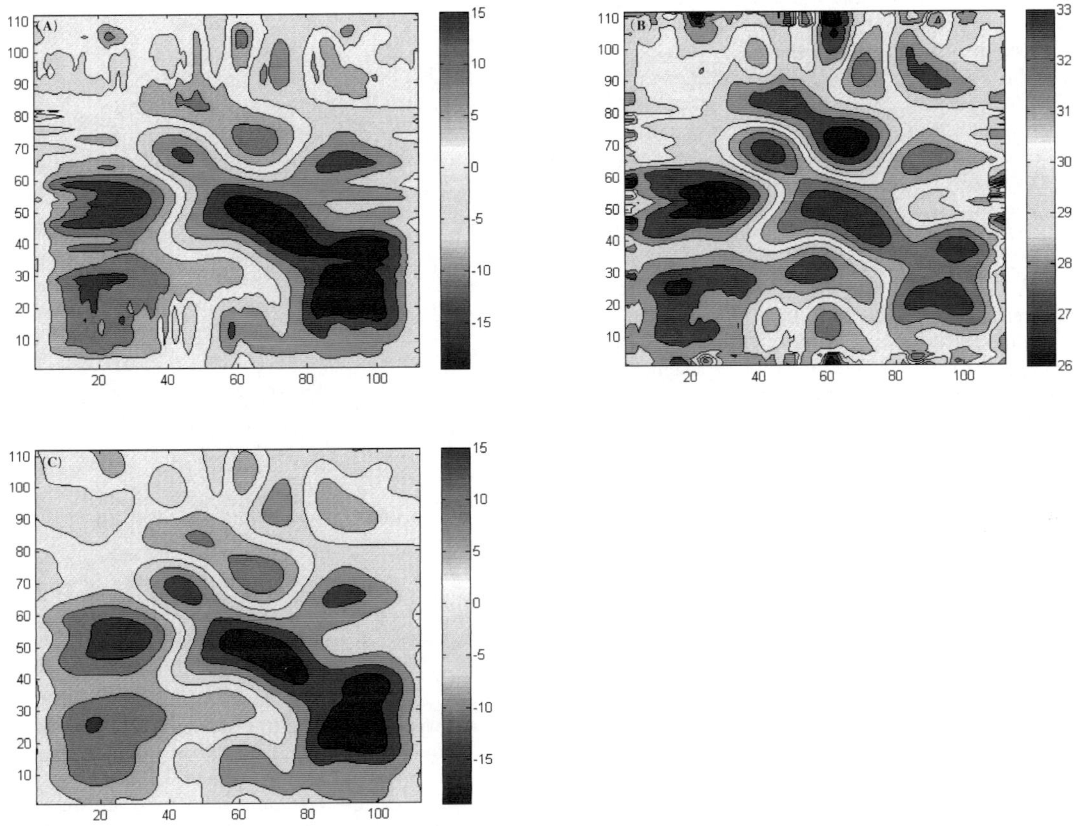

Figure 2 A is the gravity anomaly in Brittany. B is the Moho relief in Brittany by this article's forward calculation. C is the gravity anomaly obtained by the Moho relief of Figure 2 B by this article's method.

Regional – residual gravity anomaly separation – A new processing technique

S. Barala (Indian School of Mines, Dhanbad); N. L. Mohan (Osmania University, Hyderabad); R. R. Mathur (Osmania University, Hyderabad)*

Summary

The most difficult problems in gravity data interpretation is the separation of regional and residual gravity anomalies from gravity data. There are many techniques to separate the regional – residual anomalies. It is quite interesting that speech signals are similar to geophysical signals. In our paper a method is used to separate the regional residual anomalies which is generally used to separate the vowels in a given speech signal. The method is adapted to simulated anomalies and field data sets and proved that the method works for significant enhancement of the S/N ratio. This method would act as separation of regional and residuals of gravity data.

Introduction

The gravity method is a geophysical technique that measures differences in the earth's gravitational field at specific locations. It is very useful in locating the subsurface geological objects and different geological structures. The success rate of the gravity and magnetic method depends on the different earth materials having different parameters like-bulk densities (mass) or magnetic significance that produce variations in the measured field. The important step in gravity data processing is to separate the regional anomaly from the data. There are many methods which are generally used for separation of regional - residual anomaly.

In our paper we extend and implemented a method which is generally used to separate the vowels in a given speech signal. We illustrate the methodology to the response of different geological objects and structures after adding sufficient amount of noise and then we show how our technique is more effective in delineate the subsurface geological objects and in separating the regional-residual anomaly.

Method

The gravity field data is transformed into frequency domain, using Fourier transform. The gravity data is segmented into several units and a particular unit preferably the 1st unit can be considered as reference unit. It may be noted that it would act as regional field entity. The Fourier amplitude of each unit is computed and the Fourier amplitude of the 1st unit is subtracted from each unit.

After taking the Fourier transform of gravity data it is segmented into several units say - $S(w)$.

If $S'(w)$ is the reference window then -

The difference between two windows or units on a log magnitude versus frequency scale is defined by

$$V(w) = \log S(w) - \log S'(w)$$

The difference between two windows or units on frequency scale is defined by

$$V'(w) = S(w) - S'(w)$$

Where S (w) and S' (w) are the two window.

Now the log spectral distance (d values) is defined by

$$d = \frac{sum(V(w)}{2\pi}$$

Where n is the window length

Now we compute the d values using this method. This method is applied on different data sets both synthetic as well as real and obtained results are shown below.

Application on Synthetic data

(a) Sphere:

Gravity effect of a sphere buried at a depth z is given by-

$$G(x) = \mathbf{m}\frac{z}{(z^2 + x^2)^{\wedge 3/2}}$$

Where: $m = \frac{4}{3}\pi p y R^3$ p is the density, y is the universal gravitational constant and R is the radius.

The gravity response due to a sphere is shown in figure 1a. After adding a sufficient amount of noise in data, the d values are computed, and the results are shown in figure 1b. The results which are obtained from our method are compared with the gravity response due to a sphere and shown in figure 1c which is quite nicely matching with the original anomaly. Before and after applying the method, the noise level is shown in figure 1d and we can see that the noise level is reduced.

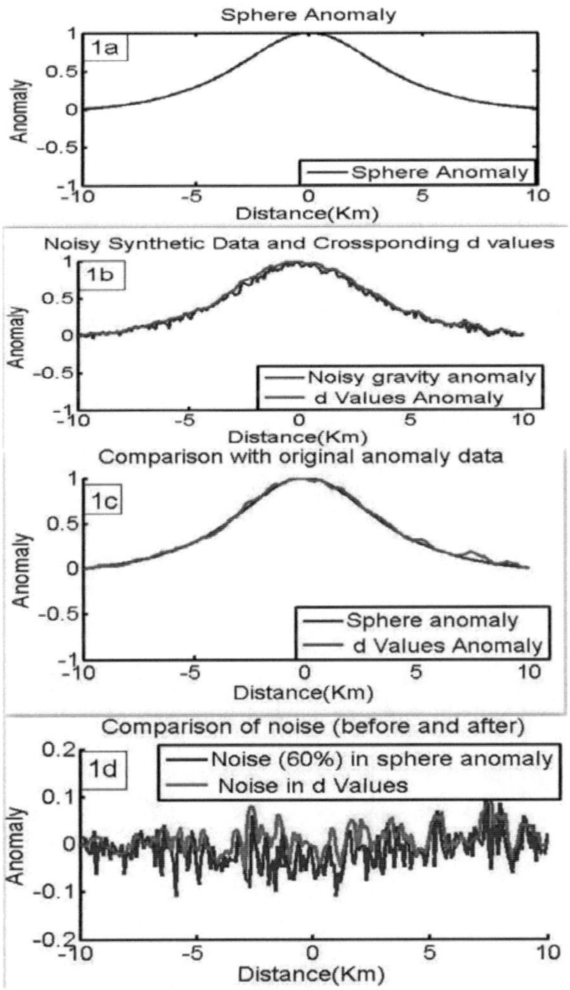

Figure 1 (1a) Gravity anomaly over a sphere. (1b) Noisy Gravity anomaly and computed d values. (1c) d values compared with original anomaly. (1d) Noise in gravity data and d values.

(b) Pull apart basin (synthetic data):

The synthetic gravity anomaly data over pull apart basin is shown in **Fig.2a,** after adding the 30% noise in given anomaly data, the distance function (d values) is computed and shown is **Fig.2b**. The obtained d values are compared with the real anomaly (**Fig.2c**) and the results are nicely illustrate the accuracy of our method and the noise level is highly reduced which is shown in **Fig.2d**.

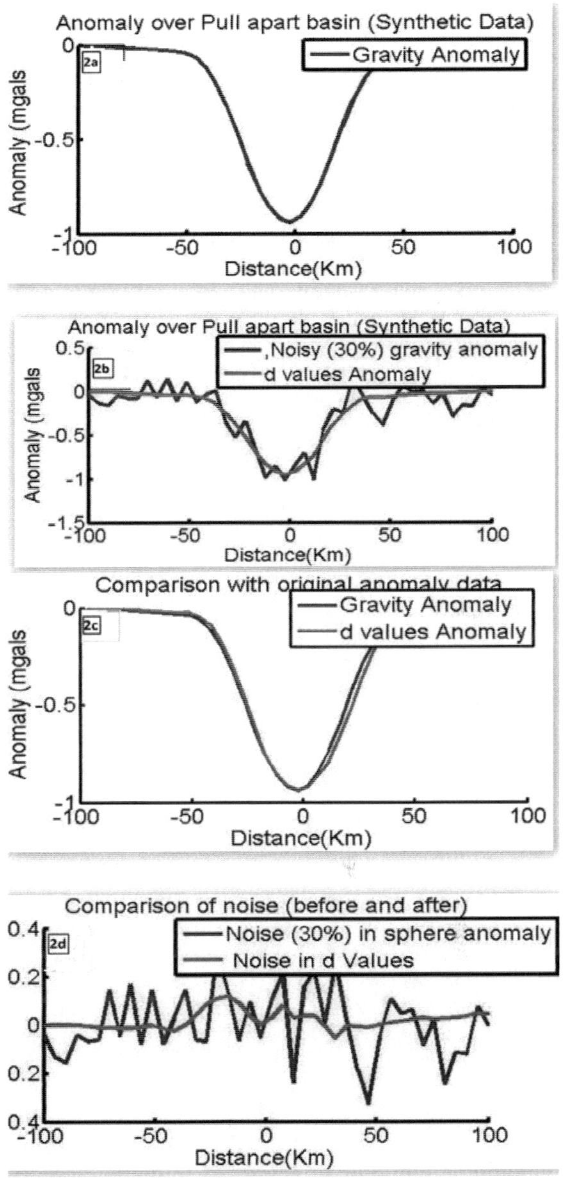

Figure 2 (2a) Gravity anomaly over pull apart basin. (2b) Noisy Gravity anomaly and computed d values. (2c) d values compared with original anomaly. (2d) Noise in gravity data and d values.

Application on field data

In order to examine the technique a field example is considered and results are also computed using **Analytic signal technique** in order to determine the accuracy of our method.

The gravity field data (after adding necessary corrections) that was collected in some part of Madhya Pradesh state (India), is used for analysis. These field data are collected from three profiles (PL1, PL2, and PL3) which are nearer to each other and almost parallel, shown in **Fig.3**. The station spacing is almost 50 m (a few observations are slightly varied in spacing interval and hence is 50m).

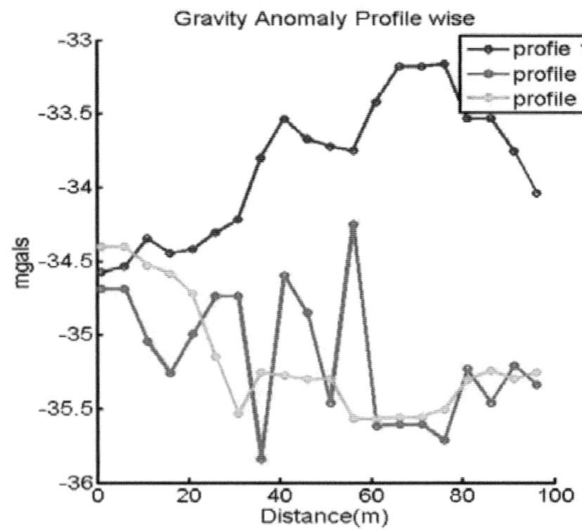

Figure 3 Gravity anomaly profile wise

Through the analysis of the analytical signal concept it would be possible to identify the signal and location of the subsurface geological object. In fact, almost all the potential field studies the concept of analytic signal is being used. It is evolved when any physical phenomenon, like heat, gravity field, magnetic field etc., is continuous and differentiable in a given domain and occurs at least either a maximum or minimum in that part of domain then it is called analytic.

If $G(x,z)$ is the gravity anomaly then its analytic signal is given by

$$A(x,z) = \left(\frac{\partial G}{\partial x}\right) - j\left(\frac{\partial G}{\partial z}\right)$$

$\left(\frac{\partial G}{\partial x}\right)$ =First horizontal derivative of the gravity field, $\left(\frac{\partial G}{\partial z}\right)$= First vertical derivative of the gravity field.

Then the field data are computed using our method. The Analytic signal of the field data are shown in **Fig.4a** and the d values for each profile is computed and shown in **Fig.4b.**

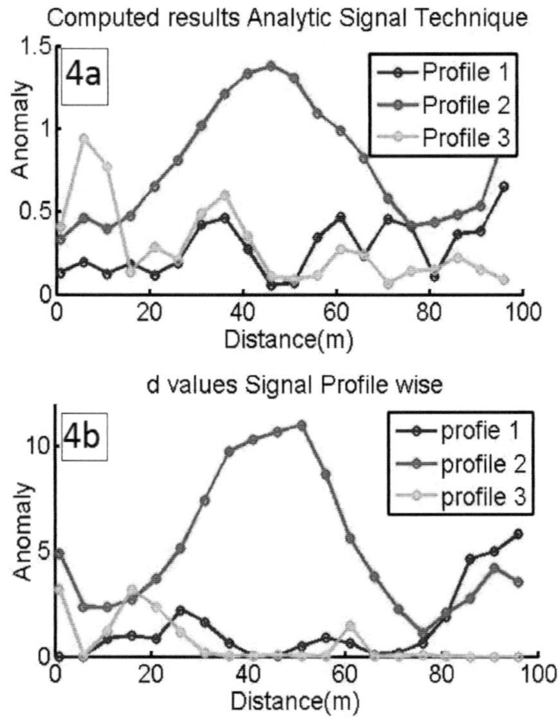

Figure 4 (4a) Analytic Signal Amplitude. (4b) d values signal amplitude

It has been clearly observed that gravity anomaly caused by the subsurface geological object is significantly indicated in d values signal compare to analytic signal. If we compare both the results obtained using Analytic signal technique and d values then it is clearly indicated that the signal to noise ratio is highly improved and regional gravity anomaly is separated from the gravity data.

Since the gravity data are collected from three profiles PL1, PL2, and PL3, which are nearer and almost parallel and therefore a contour map is constructed and shown. The amplitude of the analytic signal is shown in the form of contour map in **Fig. 5a**, the contour map of d values is shown in **Fig.5b**.

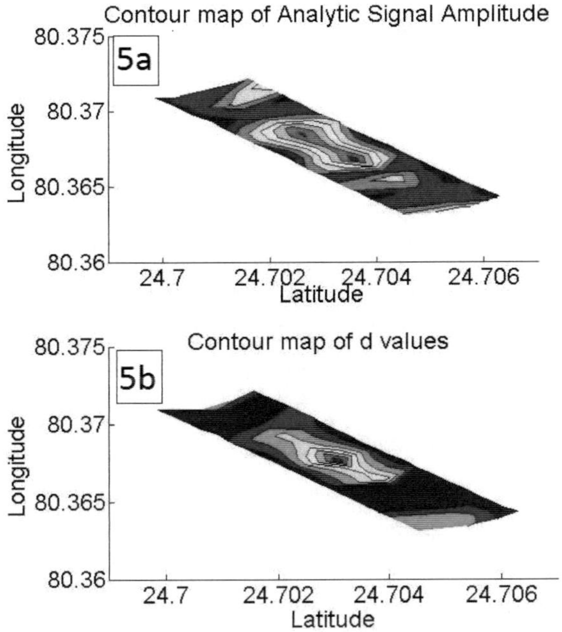

Figure 5 (5a) Contour map of Analytic Signal amplitude (5b) Contour map of d values

It is clearly indicated in contour map (**Fig.5a**) that the subsurface geological object is only in a fixed part of entire area. And the object is at shallow depth in that location with compared to the other locations of the area. The above distance function (d values), computed from all the 3 profiles and shown in contour image (**Fig 5b**), significantly demarcates the regional and the residual field of the gravity field. The gravity anomaly caused due to subsurface object is clearly identified and sharply indicating in very specific portion of entire area then the Analytic Signal technique.

Conclusions

In our paper we purposed a method which is very useful in the separation of regional- residual anomaly and it is also a more accurate approach to delineate the subsurface geological object from noisy data. We apply this method on synthetic data (after adding sufficient amount of noise which may referred as regional anomaly) as well as a real field data and show that how this method can be more effective in locating the subsurface geological objects. In fact it can be applied on almost all the potential field data and helps more effectively in reducing the noise. We also apply Analytic Signal technique to the field data to confirm the results which are obtained using our method. We also show that this is technique gives more accurate results then the Analytic signal technique and it is very effective in locating the subsurface geological objects.

Acknowledgments

I would like to thank Prof. N. L. Mohan from Osmania University, Hyderabad for giving me a wonderful opportunity to work and learn under his guidance. I really enjoyed working in Osmania University. I would also like to thank Dr. R. R. Mathur for granting permission for presenting the paper.

Understanding salt tectonics in Northern Tunisia - Insight from gravity data analysis

Imen Hamdi Nasr[1,2], Kawtahr Sebei[1], [1], Haifa Boussigua, Benen Sarsar[1], Amiri Adnen[1], Oussama Abidi[1],., Mouna Brahmi[1], Said Tlig[1], Inoubli Mohamed. Hedi[1].*
[1]Unité de recherché de géophysique appliquée aux minerais et aux matériaux, département géologie, Faculté des sciences de Tunis, Tunisia
[2]Faculty of Science of Bizerte, Tunisia

Introduction

Introduction
Salt tectonics is one of the key fundamental problems of structural analysis in petroliferous basins, and is of broad practice prospecting in petroleum exploration and development stages. Many authors have dealt with salt dome folding and typology, especially after the discovery of 'diapir plays' within gas and oil-fields from the Gulf of Mexico and Iran (Talbot and Jaravis 1984; Talbot 1993; Jackson, Vendeville and Schultz- Ela 1994; Jackson 1995; Vendeville 2002; Krzywiec 2004; Talbot and Aftabi 2004). The study on salt tectonics in extensional basins (such as the Gulf of Mexico) has made great achievements (McBride et al., 1998), and salt tectonics research on compressive regime of foreland fold belt has also made progress, such as in the Persian Gulf Basin (Letouzey et al., 1995). The salt beds can change the occurrence of overlying horizons, forming various structural traps (Rowan, 1995). The salt bed itself is the most effective caprocks (Liang-Jie et al 2004). Focuses were performed on salt rising mechanism and geologic parameters effect such as: tectonic regime, basement faulting, initial salt layers thicknesses, overburden thickness, rate of sediment aggradations and salt dissolution.
The beginning of salt movement is related essentially to extension in the subsalt basement (thickskin) or to extension in post-salt sedimentation (thinskin). Density-driven diapirism is probably not the initiating mechanism (Perthuisot, 1978), as these diapirs are not developed in stable regions even with the presence of a thick initial salt layer. Regional tectonics play more important role in the initiation of salt movements than stratigraphic density inversion.
Northern tunisian Atlas is characterized by numerous NE–SW trending Triassic salt outcrops (Fig.1) in contact with younger rocks.
Opinions are divided concerning the mechanism of emplacement of Triassic edifices and their dynamics.
For some authors, Perthuisot (1978), Hammami (1999), Chikhaoui (2002), Jallouli (2005), Triassic salts series have pierced their cover as mushroom-like structures whereas some others Ghanmi (2001), Vila et al. 2002) have mapped "salt glaciers" bodies on a regional scale.
Gravity data analysis based on recently collected gravity data, observed surface geology and assisted by analogous salt rising models help constraining the Triassic salt structure of outcrops in northern tunisian Atlas and allow deriving the mechanism of their presumably polyphasic motion. It permits the identification of a Triassic dome covered by Quaternary series.

Geological Setting

The study area is located in northern Tunisian Atlas which is characterized by numerous NE-trending outcrops of Triassic rocks (fig.1).
Triassic outcrops herein are bounded by Jurassic-devoid, Meso-cenozic marls and limy sequences all of which are reduced in total thicknesses. Triassic crop outs bear gypsum, clays, plus/mines metasomatized basalts (sills, dikes), sands as well as lime- and dolostones.
Furthermore, among the most important Triassic alignments in study area are the Fedj el Adoum-Ain Jemmala, Jebel Cheid-Knana and the North-Eastern Thibar complexes. The Jebel Cheid edifices strike to the Northeast as the depressed anticlinorium with flanking, rolled-over Cretaceous to Miocene marly and limy series, Jebel Touila and J. Kebbouch to the southwest. To the Northeast, the elongated Djebba-Souk es Sebt Triassic band (Fig.2) is bounded on its northern flank by Neogene series; its southern flank is limited by Cretaceous sediments. On the western part of the map,
Cretaceous sequences vary from compact limestone to marls, whereas the Cainozoic sequence consists
of clays, marls and sands. These sequences are affected by numerous faults trending NW and NS.
During Late Cretaceous and Cainozoic, Northern Tunisia was affected by a series of compressional events caused by the northward movement of the African plate with the most prominent event occurring during the Late Miocene. Such tectonic events played a role in the rising of the Triassic evaporitic rocks and controlled the overall structure of the region.

Gravity data analysis

High-resolution gravity data in Northern Tunisian Atlas collected by the "Office National des Mines" (ONM)

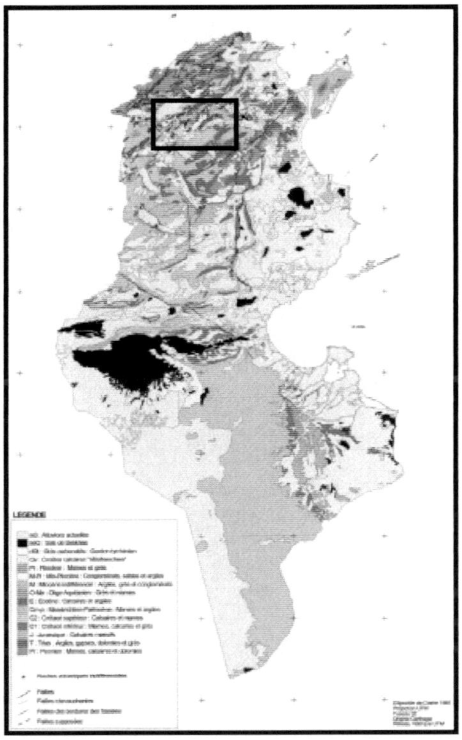

Figure. 1: Geological map of Tunisia

were used. Acquisition was carried out in 1997 by ValDor Sagax Company.

Gravity data cover the 1/50,000 scale maps of Jendouba, Nebeur, Teboursouk and Gaafour. The studied area is covered by gravity stations on an almost 1x1 km2 grid. Station coordinates were determined using global positioning system GPS (Leica) in a differential mode. Adopted projection is Lambert North using the Clark ellipsoid 1880 and Carthage datum. Free-air and Bouguer gravity corrections were performed using sea level as a datum and a reduction density of 2.4 g.cm3. The choice of this value results from the comparison of the results from several methods (ONM, 2000).

All gravity values were gridded using minimum curvature technique (Briggs, 1974). The complete Bouguer gravity anomaly values range from -21mGal to the South in the Oued Tessa Bled Elgorfa, to +9mGal in Argoub Ennaoua, and Tabet ech Chrif at the Nord-Est. (Fig.2). The outcrop pattern of the Triassic rocks has been digitized from the 1/50,000 scale geological maps and superimposed onto the complete Bouguer gravity map. Positive anomalies are situated in the central-eastern part of the map; they coincide with Triassic and Cretaceous outcrops at Jebel El Zitoun, Sidi el Mahdi, Jebel Argoub Naoua and Jbel Sfa Boubakrer, J Aksab, J Cheid, J, Touila zones. The maximum amplitude occurs over the lower cretaceous outcrops.

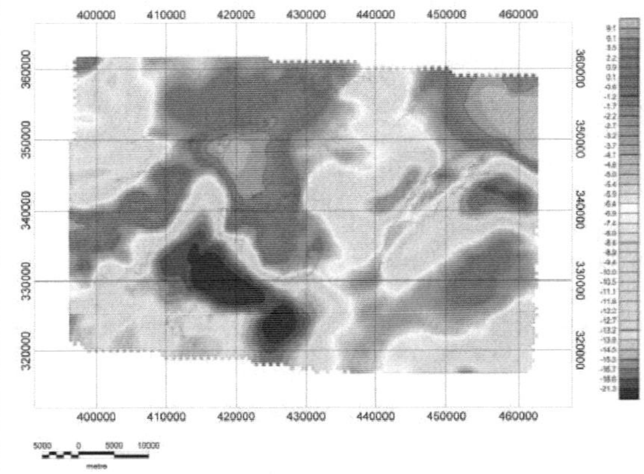

Fig. 3Figure.2 Bouguer anomaly map. Black polygons represent Triassic outcrops

The anomaly sited over El Merdja zone expresses less amplitude and extension than the previous one.

Residual gravity anomaly shows that Triassic outcrops are associated with positive gravity anomalies, which are better individualized (J. Cheid, J. Zitoun Sidi Mahdi, Tibar, Fedj Elhdoum, J, Touila and J. Boukhil). To the North a positive circular anomaly corresponds to El Mardja zone. This large scale gravity response occurs mostly over Quaternary deposits and is located at the western edge of the Majerda River. This anomaly expands laterally by thin elongated EW to NE directed positive gravity response. As the surface quaternary series cannot explain such anomaly, this suggests the presence of dense and deeper units

In order to delineate a lateral boundary originated the main

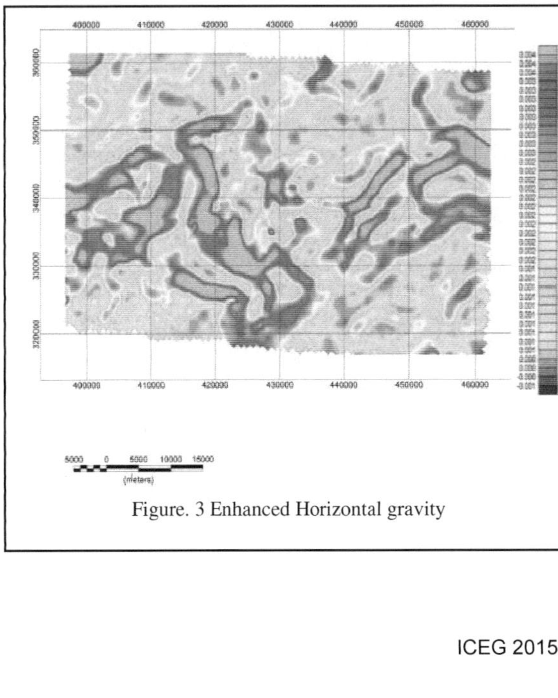

Figure. 3 Enhanced Horizontal gravity

sources of gravity responses, edge enhancement techniques based on gravity signal derivatives was performed. These techniques are utilized to locate the lateral boundaries of density contrasts, to provide information on the geological units location (Blakely and Simpson, 1986). The amplitude of the enhanced horizontal gradients (Fedi and Florio 2001) was used to analyze the gravity anomalies; the previously defined geological details are recognizable in the EHD map. The map shows important well defined NE alignments covering the elongated Triassic band Tibar - J. Sidi Mahdi - J. Zitoun - J. Rhedir El kalba - Kt. Bou Rokba. The southern area mostly expresses regions with gravity maxima. The maximum value of the horizontal gravity gradient tends to be located on the horizontal edges of the gravity sources marked by rapid changes in density values (Blakely, 1996). The horizontal gravity gradient shows a stretched out alignments bordering J. Chedid; thus, there is tentative evidence that these might reflect a lateral contrast in density between the Triassic predominately evaporitic bodies and their Cretaceous marly and limy envelope. The gradients are stronger (i.e., higher contrast in density between Triassic and Mesozoic enveloping rocks) at the northwestern border of the local Triassic forms than at their southeastern boundary. Field observations of the Triassic assemblages in contact with the Cretaceous rock complicate the result; however, its general attitude is sub-vertical while showing overturned Cretaceous layers in some localities, especially at the northwestern edge of the diapiric alignments.

The Euler deconvolution was applied to gravity field of the area of study. In this work we have solved the equation with changing at each time a parameter while keeping the other constants. The correct parameter is then selected by inspecting the maps deduced from each one and compared to the others, giving the tighter solutions clustering. Euler solutions presented in Figure 4, were obtained for a structural index of zero; this value provides usually a reliable indication of large-scale faulting, a window size of 11x11 cells and a threshold tolerance of 8. The colored dots indicate the source edges, and each color is related to one of the estimated depths which are spread out in eight ranges. In the central eastern and southern regions, results correlate with the known geological contacts and faults. The positive anomaly of J. Sidi Mahdi is bordered to the North by aligned solutions corresponding approximately to the Northern limit. To the South there is a NS aligned fault system solutions that limits the Triassic outcrop of Fej Elhdoum to the East. J. Cheid is bordered by two alignments marked by the aligned and stretched solutions which reflect the two fault system delineating the diapir structure. The south-central part of the area, is contoured by a succession of solutions which indicate the contact between this negative anomaly and both bordering positive ones. These solutions are indication of fault systems which orientation is well known. In the southwestern part at J. Dardouria Triassic outcrops a NE deep fault is outlined. In the northern region, the solutions indicate the presence of deep geological structures beneath Quaternary deposits. Therefore, tightest clustering solutions outline new faults affecting geological series under quaternary deposits. Thus, El Mardja anomaly is delineated by a series of faults behaving as an arched system.

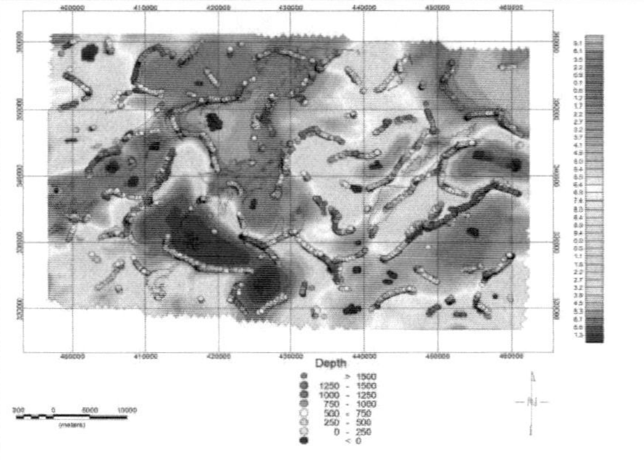

Fig. 4: Results obtained with Euler deconvolution superimposed to residual gravity anomaly

Conclusion

Our geological observations aided by gravity compilations fit the model of a diapir-like extrusive mass in J. Chehid triassic outcrops. The overall geometry of Triassic outcrops displays a somehow flat crest with steepened flanks delineated by subvertical to turn over strata, mainly Cretaceous marls and limy rocks. The latter are frequently cut by radial faults in conjunction with some piercing Triassic rocks across the envelope. The fold/diaper-like antiform supported by this study might be emplaced by means of master shear zones, trending majorly NE–SW, clearly retraced by our isogal curves in the horizontal gravity gradient map. Oblique slip movements of wrenching faults organized as the stepping tectonic system, pulled up Triassic evaporite-bearing masses that tore to pieces the Jurassic and Cretaceous overburden. The results testify the presence of abnormal positive gravity anomaly generated by a higher density of the diapir-like Triassic mass contrasting with a lighter Meso-Cainozoic rock envelope. Bouguer anomaly and basically residual gravity anomaly show narrow and parallel isogal-curves bordering the Triassic outcrops. Also, the gravity field derivatives allow distinction of alignments extending NE-SW, the contact between the Triassic and Cretaceous systems appears to correspond to a part of steep gravity gradient. These faults are thought to have given issue to ascending evaporite-bearing masses.

El Mardja anomaly with its circular form and the fact that it is bordered by a "curved" fault system presents the response of a Triassic dome, well expressed by the positive density contrast of the Triassic material. Triassic material average density approximates 2.51g.cm-3 (ONM, 2000).

References

Blakely R. and Simpson R. 1986. Approximating edges of sources bodies from magnetic or gravity anomalies. Geophysics 51, 1494–1498.

Briggs I. 1974. Machine contouring using minimum curvature. Geophysics 39, 39–48.

Chikhaoui M. 2002. La zone des diapirs en Tunisie: cadre structural, évolution géodynamique de la sédimentation méso cénozoique et géométrie des corps triasiques. PhD thesis, Es Sciences Université de Tunis el Manar.

Fedi M. and Florio G. 2001. Detection of potential fields source boundaries by enhanced horizontal derivative method. Geophysical Prospecting 49, 40–58.

Ghanmi M., Ben youssef M., Jouirou M., Zargouni F. & Vila J. M. (2001)- Halocinèse crétacée au Jebel Kebbouch (Nord-Ouest tunisien): mise en place à fleur d'eau et évolution d'un « glacier de sel » albien, comparaisons. Eclogae geol. Helv, pp 153-160.

Hammami M. 1999. Tectonique, halocinèse et mise en place de la minéralisation dans la zone des diapirs (Tunisie septentrionale). PhD thesis, Université de Tunis el Manar.

Hamdi Nasr. I., Ben Salem A., Inoubli M. H., Dhifi J., Alouani R., Chaqui A. & Perthuisot V. 2008. Apports de la gravimétrie dans la caractérisation des structures effondrées dans la région de Nebeur (Nord Ouest de la Tunisie). Swiss J. Geosci. 101 (2008) 17–27.

Hamdi-Nasr, I., Inoubli, M. H., Ben Salem, A., Tlig, S., and Mansouri, A., 2009, Gravity contributions to the understanding of salt tectonics from the Jebel Cheid area (dome zone, Northern Tunisia). Geophysical Prospecting, 57, 719–728.

Hamdi Nasr I., Amiri A., Inoubli M. H., Ben Salem, A., Chaqui A., Tlig S., 2010, Structural Setting of Northern Tunisia Insights from Gravity Data Analysis, Jendouba Case Study. Pure and Applied Geophysics,

Jackson M.P.A. 1995. Retrospective salt tectonics. In: Salt Tectonics: A Global Perspective (eds M.P.A. Jackson, D.G. Roberts and S. Senelson), pp. 1–28. AAPG Memoir 65.

Jackson M.P.A., Vendeville B.C. and Schultz-Ela D.D. 1994. Structural dynamics of salt systems. Annual Review of Earth and Planetary Sciences 22, 93–117.

Jallouli C., Chikhaoui M., Brahem A., Turki M.M., Mickus K. and Benassi R. 2005. Evidence for Triassic salt domes in the Tunisian Atlas from gravity and geological data. Tectonophysics 396, 209–225.

Krzywiec P. 2004. Triassic evolution of the Klodawa salt structure: basement controlled salt tectonics within the Mid-Polish trough (central Poland). Geological Quarterly 48, 123–134.

Letouzey, J., Colletta, B., Vially, R., Chermette, J.C., 1995. Evolution of salt – related structures in compressional settings. In: Jackson, M.P.A., Roberts, D.G., Snelson, S. (Eds.), Salt Tectonics, A Global Perspective. AAPG Memoir, vol. 65, pp. 41–60.

Liang-Jie et al 2004. Salt tectonic evolution and hydrocarbon accumulation of Kuqa foreland fold belt, Tarim Basin, NW China. Journal of Petroleum Science and Engineering 41, 97– 108.

Magdalena S., Ulf B. and Bjorn L. 2003. Salt movements in the northeast German Basin and its relation to major post-Permian tectonic phases-results from 3D structural modelling, backstripping and reflection seismic data. Tectonophysics 316, 277–299.

McBride, B.C., Rowan, M.G., Weimer, P., 1998. The evolution of allochthonous salt systems, Western Green Canyon and Ewing Bank (offshore Louisiana), Northern Gulf of Mexico. AAPG Bulletin 82 (5), 1013– 1036.

Office National des Mines. 2000. Campagne gravimétrique CG3: Coupures 1/50000 de Nebeur, Jendouba, Saline et Teboursouk.

Perthuisot V. 1978. Dynamique et pétrogène se des extrusions triasiques en Tunisie septentrionale. PhD thesis, Ecole Normale Supérieure, Paris. Rowan, M.G., 1995. Structural styles and evolution of allochthonous salt, Central Louisiana Outer Shelf and Upper Slope. In: Jackson, M.P.A., Roberts, D.G., Snelson, S. (Eds.), Salt Tectonics, A Global Perspective. AAPG Memoir, vol. 65, pp. 199– 228.

Talbot C.J. 1993. Spreading of salt structures in the Gulf of Mexico. Tectonophysics 228, 151–166.

Talbot C.J. and Aftabi P. 2004. Geology and models of salt extrusion at Qum Kuh, central Iran. Journal of the Geological Society London 161, 321–334.

Talbot C.J. and Jaravis R.J. 1984. Age, budget and dynamics of an active salt extrusion in Iran. Journal of Structural Geology 6, 521– 533.

Vendeville B.C. 2002. A new interpretation of Trusheim's classic model of saltdiapir growth. Gulf Coast Association of Geological Societies Transactions 52, 943–952.

Vendeville B.C. and Jackson M.P.A. 1992. The rise of diapirs during thin-skinned extension. Marine and Petroleum Geology 9, 331– 353.

Vila J.M., Ghanmi M., Ben Youssef M. and Jouirou M. 2002. Les "glaciers de sels" sous-marins des marges continentales passives du nord-est du Maghreb (Algérie-Tunisie) et de la Gulf Coast (USA): comparaison, nouveau regard sur les "glaciers de sels" composites, illustré par celui de Fedj el Adoum (Nord-Ouest tunisien) et revue globale. Eclogae Geologicae Helvetiae 95, 347– 380

Gravity for hydrocarbon exploration promotion

M.H. INOUBLI, I. HAMDI, O. ABIDI, A. AMIRI*

Mohamed Hedi Inoubli: mh.inoubli@planet.tn
University of Tunis-El Manar – Geoplus Sarl (Service Company)

Summary

Petroleum systems are often described as play perception within a region and used as a specific model and approach to recognize new petroleum accumulation intervals.
The identification of each play-component engages the use of seismic surveying, assessment techniques and geological synthesis to outline and situate prospective structures. Each model is labelled through its specific mark pattern; which develop into a measure, often used to look for comparable response model across the studied area. This develops into chiefly significant need as exploration moves to identify minor, more intricate, and less accessible goals. The richness of seismic data makes this method as the devoted tool for hydrocarbon exploration. So, from the total expenditure in geophysics huge amounts are paid for reflection seismic and almost nothing for the other methods as well as the other sectors. However seismic has its own limitations.
Fortunately, the development of new exploration techniques has improved exlorationists' understanding and increased the efficiency of exploration. So, although targets are getting smaller, exploration and appraisal wells can now be sited more accurately and with greater chance of success through the use of multi-source information.
One of the successful tools of choice to assist with this endeavour has been the usage of Gravity data. Further more, the increasing complexity of the exploration targets is the driving force behind the integration of seismic and non-seismic methods into the subsurface geological modelling process. The availability of high quality digital data, and truly integrated workstation software, would enable interpreters to simultaneously model seismic and non-seismic data sets.

This paper reports the obtained results using gravity data for oil and gas exploration in various areas. Although gravity is less accurate than the universally used acoustic methods, measurements are inexpensive and are not time-consuming for data acquisition; gravity can be used in irregular and unreachable areas. Measurements are employed to delineate structural highs and lows which represent sedimentary thickness and facies change which constitute important aspects in the recognition and delineation of hydrocarbon exploration systems. This remains significant both during reconnaissance stage as well as mature exploration of a permit.

Key Words: Hydrocarbon Exploration, Gravity anomaly, Enhanced gradient, Euler Deconvolution, fault termination, seismic limitations, Tunisia.

Introduction

Hydrocarbon exploration is qualified as a high risk investment. The exploration risk is assessed and may be reduced through the recognition of the key elements of the petroleum system as well as the hydrocarbon play of the prospected area. Geophysical Methods are used to determine and locate geological structures suitable for the accumulation of HC. Reflection seismic data richness makes the seismic method the dedicated tool for exploration. Nevertheless, seismic has its limitations. Gravity is presented here as one of the successful tools of choice to assist with this endeavor.

Geology

The experimented area pertains to the larger Atlassic domain of Tunisia. Its geological history obeys to the main geodynamic events which have long structured and conditioned the Tethyan geological history at the North-Eastern margin of the African plate. The area includes a variety of megastructures with distinct orientations developed by successive orogenic phases. The overall building of the main edifices ended in collision between the Atlassic and the Northern Tellian domains with the European plate from Senonian to Miocene.
Helpful 1/50,000 scale topographic and geological sheets together with properly computed complete Bouguer anomaly resulted in an elaborated consistent information system. Field checking of local portions of maps proved to be reliable, but somehow limited due to chronostatigraphical mapping of units rather than applied lithostratigraphical compositions; this sometimes hampers subsurface correlations of sedimentary series and corresponding lithology. The central Tunisia inherited irregular topography with major erected NE-SW folds and hosting Triassic crop-out cores with thick evaporite-salt series, some of which exist and were visited in study area.
Geological crop-outs with alignments directed NE-SW, NW-SE, NS and E-W. Detailed study indicates tectonic instability and syndepositional fault plays at the transition from a Southern platform to a Northern basin

environmental settings. Transtensive, mostly reactivated shear zones prevailed in the Mesozoic interval, cut across the Mesozoic cover, and presumably express the effects of buried in-depth discontinuities of the local basement.

Gravity

A detailed high-resolution land gravity survey was carried out over the northern and central Tunisia, which entailed a total of 35000 stations of data acquired over an area of approximately 36000 square-kilometres, at an average of almost one station per square kilometre. The area covered by the gravity data generally extended beyond that of the seismic data sets, and thus offered the opportunity to extend the structures interpreted solely from the seismic data.

Data analysis and interpretation

It is sensitive to point out that the range of densities generally observed in the region, were found to be able of accounting for the observed gravity field over the interpreted structures from surface geology as well as from seismic data. The measurements delineate structural highs and lows which represent sedimentary thickness, an important aspect in hydrocarbon exploration.

However, there is an indication that interesting structures may exist outside the existing seismic data coverage. It would suggest that there is sufficient density contrast across the various stratigraphic sections and that more detailed gravity data should be able to add useful structural information where the seismic method is at a disadvantage, particularly when applied to mapping of the footwall structures.

Despite the number of drilled wells, the available quantity of recorded 2D seismic data and the quality of the available surface geological maps, Gravity has been used, without much conviction in the beginning, in order to look for discriminating subsurface structural response.

Interpretation, including modelling of representative seismic lines, reveals the particular usefulness of these data for mapping shallow faults. Longer wavelength features, typically the major tectonic elements such as basin bounding faults, and relative sediment thickness are also well expressed. Comparisons of various gravity Bouguer anomaly and its bounded components showed to be powerful to delineate subsurface geometry and distribution and locate promising areas.

Focused gravity data processing leads to the recognition of more than thirty structures of potential hydrocarbon exploration interest. Gravity helped to appraise areas without seismic data. It helped evaluating and ranking areas with well expressed structures.

Gravity data are used also to constrain the 2D seismic program recorded in 2008 and 2009. A number of defined structures were tested and showed to be water bearing. Gravity showed that some of these holes were not located in the crests.

The relative size of the observed anomalies compared the spatial sampling rate of gravity data may require additional measurements with smaller inter-measurements distance. This should lead to a better discrimination and would end up to an enhanced delineation and definition of the key and economic subsurface structures. Available seismic data integrated and subsequent subsurface geology interpretations aided by borehole data help classifying the computed anomalies based on size as well as definition and quality of petroleum system(s) throughout the region.

Non-seismic geophysical exploration techniques are important strategic components of the exploration toolkit when properly calibrated and applied. This low-cost oil and gas prospecting technology helps identifying potential structures from the ground.

The Bouguer anomaly map is made of two components: a deep one which is roughly in relation with deep bodies (i.e. the geometry of the mohorovicic for extended areas) and a residual component that images the shallower intervals that can be the sedimentary layers representing the objectives for hydrocarbon explorationists.

The Values of the residual gravity field range from hot colors which are related to relatively dense structures or masses to cold colors expressing the response of relatively light material.

Superposition of these structural elements on gravity map lead to the recognition of anticlines, synclines, depressions and troughs and faults. So, it is comfortable to state that gravity images very well the structural elements here.

The enhanced horizontal gradient was very powerful to express lineaments. Their maximum amplitudes are superimposed to the identified fault patterns on surface geology. Quite some faults are expressed by EHG map and not surface geology.

Euler deconvolution was used for linear solution. Optimal well organized solutions were superimposed to the EHG map. So, both methods converge and produce comparable solutions.

Conclusions

Gravity imaging represents effective means adding new insightful information and helping investigation of the potential play models of a region of interest both during a reconnaissance and mature exploration period. Non-Seismic methods are to be considered as important strategic components of the Oil and Gas exploration toolkit when properly calibrated and applied.

Gravity is the unique geophysical method which, on its essence, expresses the 3D distribution of the all geological bodies. Meanwhile each of the available exploration methods would provide valuable information in all steps of the exploration stage.

It is recommended to use non-seismic techniques in addition to seismic technology, in order to get a comprehensive view of the subsurface geology of a region. It is believed that it is crucial to get as much information as possible from exploration so that the impact of needless drilling can be reduced. It's worth spending more money in Gravity. Integrating Gravity would enable explorationists to focus on frontier exploration, define new hydrocarbon plays and to identify new geological targets.

Acknowledgments

The authors express their appreciation to 'Office National des Mines" of Tunisia for permission to use the information on discussed areas, and also for their support and access to the various gravity data sets.

REFERENCE

Bain, J.E., Weyand, J., Horscroft, T.R., Saad, A.H., and Bulling, D.N., 1993. Complex Salt Features Resolved by Integrating Seismic, Gravity, and Magnetics. EAEG/EAPG 1993, Annual Conference and Exhibition, expanded abstracts.

Gardner, G.H.F., Gardner, L.W., and Gregory, A.R., 1974. Formation velocity and density – the diagnostic basis for stratigraphic traps; Geophysics v.39, p.770-780.

Inoubli M.H., Amiri A., Ben Salem A., Hamdi I., Tlig S. Chandoul S., Mansouri A., 2012. Non -seismic methods as an insight to promote hydrocarbon exploration. 69 Conference: Istanbul 2012 - International Geophysical Conference and Oil & Gas Exhibition. DOI: 10.1190/IST092012-001.

Mavko, G., Mukerji, T., and Dvorkin, J., 1998. The rock physics handbook – Tools for seismic analysis in porous media, Cambridge University Press.

Mukerji, T., Avseth, P., Mavko, G. Takahashi, I. And Gonzalez, E.F., 2001. Statistical Rock Physics: Combining rock physics, information theory and geostatistics to reduce uncertainty in seismic reservoir characterization, The Leading Edge, March, p.313-319.

Nafe, J. E., and Drake C. L., 1957. Variation with depth in shallow and deep water marine sediments of porosity, density, and the velocity of compressional and shear waves, Geophysics, 22, 523–552, 1957

Pawlowski, R., 1994. Emerging Workstation-based Potential Field Methodologies. The Leading Edge (SEG), June, 1994, 687-689, 1994.

Saad, A.H., 1993. Interactive Integrated Interpretation of Gravity, Magnetic, and Seismic Data: Tools and Examples. Offshore Technology Conference Proceedings, Paper 7079, 35-44, 1993.

FITZGERALD D., MCINERNEY P., 2003. Achieving better geology through better geophysics: Non-seismic methods, First break, vol. 21, NOV, pp. 69-75.

Gravity data interpretation for structure studies at a coastal area, south west saudi arabia

Eslam Elawadi[(1)]*, *Aseem Sulaiman*[(2)], *and Saad Mogren*[(1)]
(1) King Saud University, Riyadh, KSA.
(2) King Abdulaziz City for Science and Technology, Riyadh, KSA.

Summary

This study aims to map the basement depth and structure of the coastal area in the south western part of Saudi Arabia, as an aid to the groundwater potentially assessment. The gravity survey data was filtered and analyzed using different techniques finalized by 2-D modeling conducted along representative profiles to provide the topography and depth variations of the basement surface along the area. The depth to the basement rocks range from 0 m at the east and increase up to 2200 m at the west through local emulations related to the tilted-faulted blocks structure dominant in the Red Sea rift zone. Two faulting systems were recognized in the area; the first is normal faulting system trending in NNW-SSE direction and is related to the Red Sea rift, and the second representing the cross-cut oblique faulting system trending in NE-SW direction. The interaction between these faulting systems resulted in formation of a set of closed basins elongated at the NNW - SSE detection and terminated by the NE-SE faulting system. The geomorphology and sedimentary section of these basins qualify them for potentially groundwater accumulation.

Introduction

The study area is located in the Eastern coast of the Red Sea, southwest Saudi Arabia, including many of the relatively big cities along the Red sea Coast and major wadis. They depend on the desalination of Sea water and treatment of rain water accumulated by dams to provide the drinking water.

Topographically, the eastern part of the area is dominated by Red Sea escarpment which trends northwest with maximum altitude of about 1000 m and decline steeply to the west. The area below and to the south of the escarpment is marked by small mountains and ridges which ranges in altitude from 300 to 100 m. The relatively flat coastal plain 25 to 40 km wide, extends from the mountainous area southwest to the Red Sea.

Geologically, most of the area is covered by the coastal plain sediments while the eastern part of the area is covered by the Arabian Shield rocks represented by metamorphosed volcanic and sedimentary rocks and intrusion rocks.

The main objectives of this work are:

- Estimating the depth to the basement surface under the coastal sediments.
- Mapping basined areas formed due to structurally variation of the basement depths.
- Relating these basins to the groundwater potentialities that might exist at the study area at the study area

General geology

The Study area is a part of the south western zone of the Arabian Shield at the eastern cost of the Red Sea. The geology of the area was compiled within Al Qunfudhah quadrangle by Prinz, 1984 from the older reconnaissance 1:100,000 geologic maps. The eastern strip of the area is covered by the Arabian Shield rocks represented by metamorphosed volcanic and sedimentary rocks of late Proterozoic age (Baish and Baha groups, 1000 to 800 Ma ago). These older rocks were folded and intruded by Proterozoic plutonic and hypabyssal intrusive Tonalite and Diorite batholiths (800 to 890 Ma).

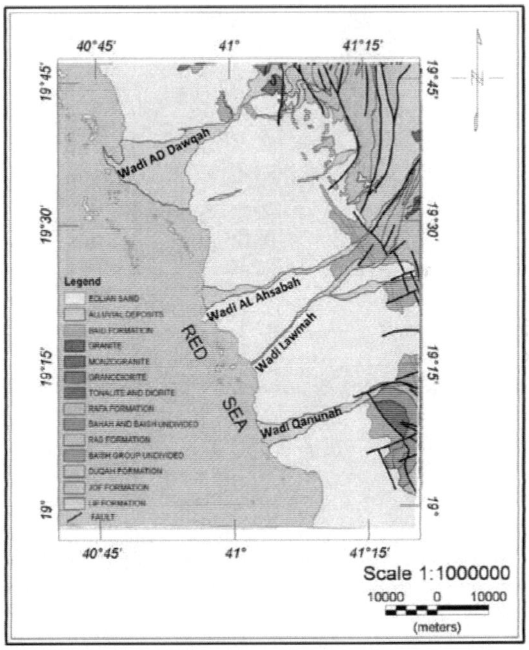

Figure 1: Geologic map of the study area.

These events were followed by deposition of Ablah group ended 763 Ma ago with the Diapiric intrusion of Monzogranite to Granodiorite gneiss. This event was accompanied by regional metamorphism to Greenschist amphibolite faces, folding and faulting.

In Miocene time, northwest trending faults related to rifting along the axis of the Red Sea was formed and some of them was intruded by gabbro. Basaltic rocks were flowed from fissures near to the Red Sea escarpment, eastward into low areas and westward down several wadis below the escarpment (Prinz, 1984). Overlain the aforementioned rocks, the central and western parts of the area is covered by Quaternary coastal sediments, wadi deposits and Eolian sediments composed of sand, gravel and silt.

Gravity data and analysis

ARAMCO has conducted a detailed seismic and gravity survey covering most of the southern coastal plan including the study area. The gravity survey was conducted along lines parallel and perpendicular to the Red Sea Coast with 10 km line spacing and 200m station separation.

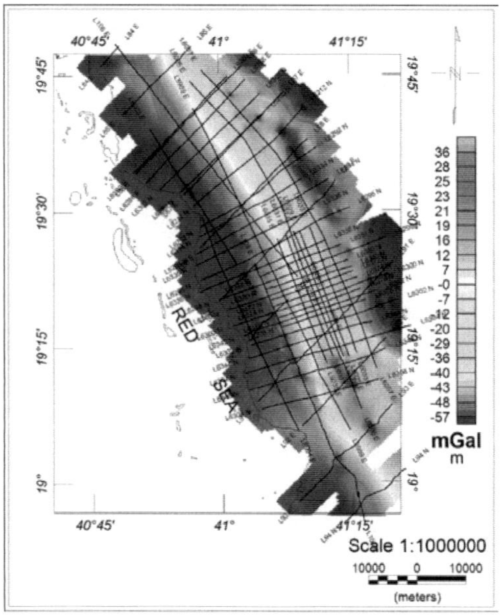

Figure 2: Bougure gravity map showing the gravity survey lines.

The gravity data was corrected to eliminate the inherent effect of the crustal thickness variation in the costal zones (isostatic correction). Afterword, the data was enhanced using band-pass and the high-pass filter in frequency domain to emphasize the shallow structure affecting the top of the basement rocks (Bird, 1997).

Furthermore, source edge location and depth estimation techniques such as source edge detection, power spectral analysis, and Euler deconvolution have been applied to the filtered gravity data to locate and outline different causative sources as well as to estimate their depths and trends.

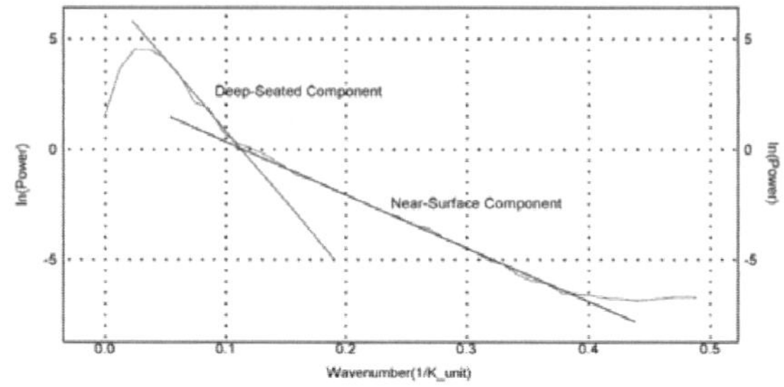

Figure 3: Radially averaged power spectrum of gravity data.

The source edge detection technique (Oasis montaj™) is an automatic technique apply the horizontal gradient to the gravity data to image the edge of the causative sources. The (SED) technique was applied to delineate the deep-seated and near-surface structures. The estimated strike direction for the mapped structures enabled direct analysis of structural trends. Deep-seated (older) structures trend in NNW-SSE (Red Sea trend) and the shallow structures has (NNW-SSE) as well as minor trends at NE-SW trend.

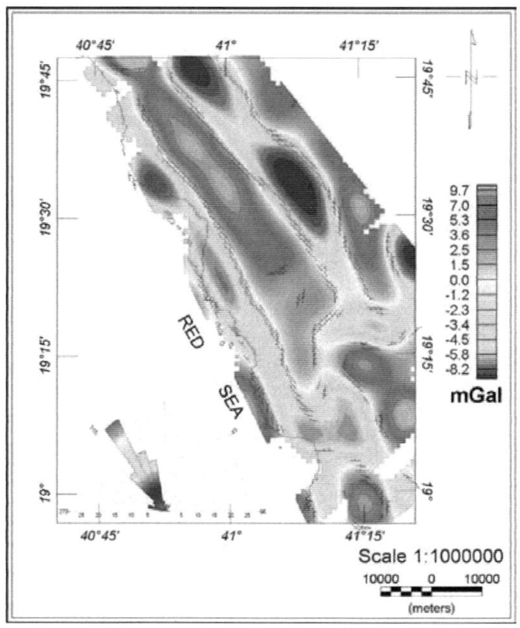

Figure 4: Source edge detection and the trend analysis for the regional Bouguer gravity data.

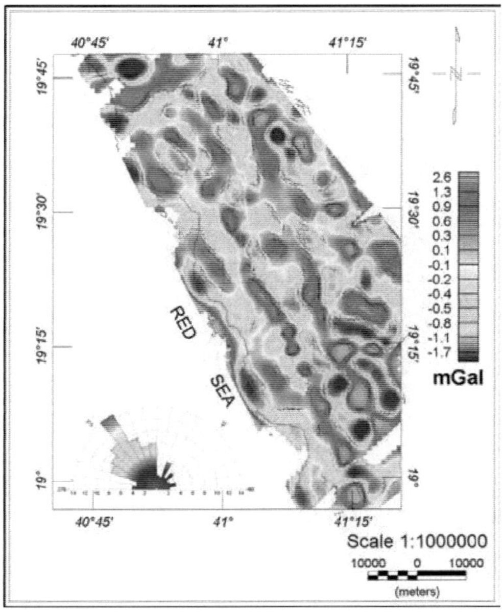

Figure 5: Source edge detection and the trend analysis for the residual Bouguer gravity data.

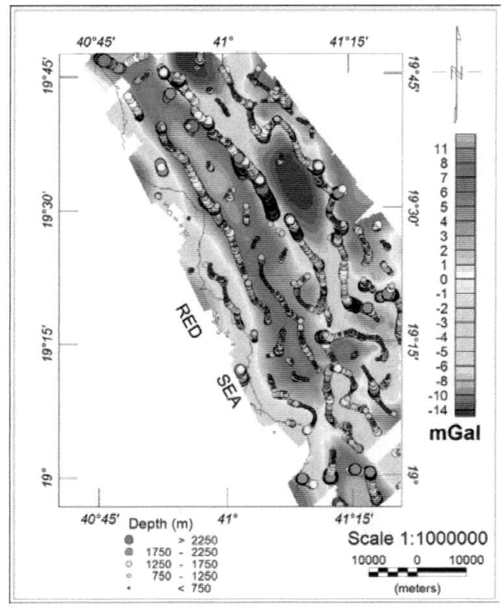

Figure 6: Euler deconvolution solutions for the regional Bouguer gravity data using contact structure index.

Euler deconvolution Tompson (1982) and Reid et al. (1990) has been applied to the first vertical derivative of the filtered gravity data. Suitable window size (3000 m) and structure index (SI=0) to map and estimate the depth to the contact (fault) structures. The method provides worthy location of the fault locations and reasonable depths to these structures ranging from 750 to 2250 meters. The Euler depth solutions (Fig. 6) also support the idea of tilted faulted blocks, where it provided depth variation matches well with the blocks boundaries and vertical movements. The maximum depths are obtained in the northeastern part of the map coincides with basin zone of down fault area.

The obtained depths and structures were used to initiate the primary models for thirteen represented gravity profiles to construct a set of 2-D models using the GM-SYS package. Pursuing simplicity and due to the lack of drilling information concerning the basement depth and type as well as the sedimentary sequence, the basement rocks was considered as uniform rock type with density around of 2.67 g/cm3 and the so for the sedimentary section with constant density around 2.3 g/cm3. The initial models were refined and optimized through iterations of forward modeling and the inversion was applied to the final fine optimization till attaining the lowest R.M.S error.

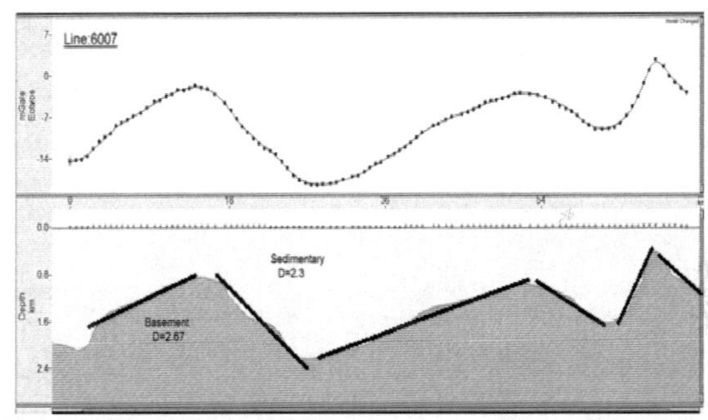

Figure 7: 2D gravity model for an example profile parallel to the Red Sea coast.

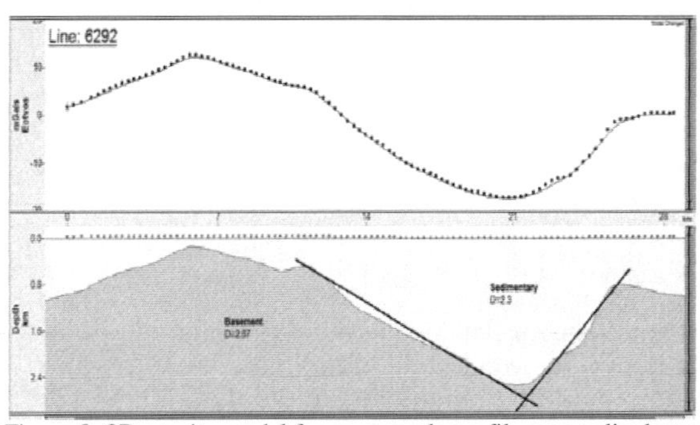

Figure 8: 2D gravity model for an example profile perpendicular to the Red Sea coast.

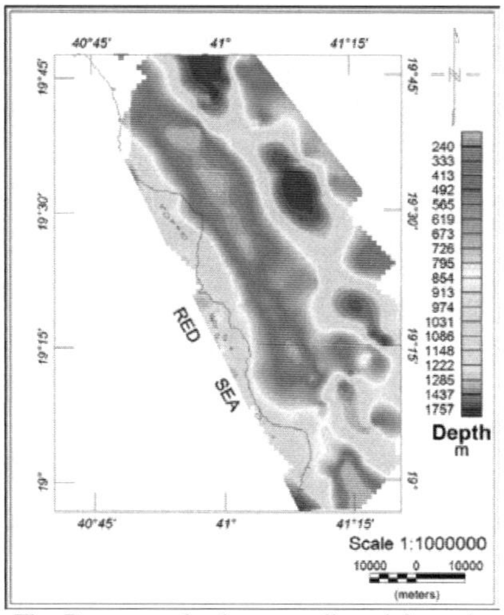

Figure 9: The Basement depth map collected from the modeling profiles.

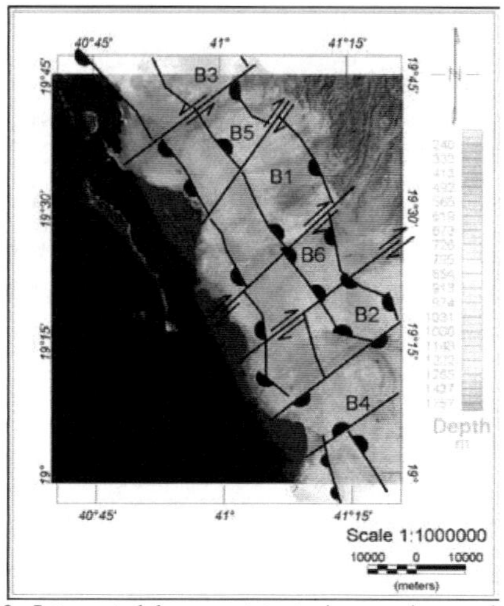

Figure 10: Interpreted basement tectonic superimposed over the satellite image.

The constructed models depict the variation of the basement depth along the modeled profiles from 0 to 2200 m under the ground surface. In general, the depth to the basement increase toward the west through local emulations related to the tilted-faulted blocks structure dominant in the Red Sea rift zone. The depth to the basement obtained along the modeling profiles is gridded and displayed in the basement depth map (Fig. 9). This map show excellent confirmation between the modeled lines in terms of depth to the basement and density contrast. On the other hand, the depth map reveals very good agreement with the gravity variation of the filtered gravity map.

Integration of the analysis and interpretation techniques was presented in the interpreted basement tectonic map showing the summary of the structures affecting the area (Fig. 10). Two faulting systems were recognized in the area, the first is normal fault system trending in NNW-SSE direction and related to the Red Sea rift, and the second representing the cross-cut oblique faulting system trending in NE-SW direction. The interaction between the aforementioned fault systems resulted in formation of a set of closed basins elongated at the NNW-SSE detection and terminated by the NE-SE faulting system. The geomorphology and sedimentary section of these basins qualify them for potentially groundwater accumulation.

Conclusions

Integrated interpretation of the gravity data reveals depth to the basement (sediment thickness) range, from few meters up to 2200m. This basement depth variation occure through local emulations controlled by the tilted-faulted blocks structure dominating the Red Sea rift zone.

The main depressions of the basement surface elongated at the eastern part of the investigated area parallel to to the Red Sea coast. These depressions are controlled by two faulting systems affecting the area; the first is normal faulting system trending in NNW-SSE direction and is related to the Red Sea rift. The second faulting system representing the cross-cut oblique faulting system trending in NE-SW direction. These NE trending faults cut and displace the NNW faults prove younger age or movements. The interaction between these fault systems resulted in formation of a set of closed basins elongated at the NNW - SSE direction and terminated by the NE-SE faulting system. These basins located to the west of the Red Sea mountain escarpment with thick The sedimentary section and, hence, could be suitable for potentially groundwater accumulation.

Refferences

Bird, D., 1997, Interpreting magnetic data: GEOPHYSICAL CORNER, EXPLORER, AAPG and SEG, May, 1997.

Prinz, C, 1984, Geology map of the Wadi Haliy Quadrangle, sheet 18E, Kingdom of Saudi Arabia, Ministry of Petroleum and Mineral Resources Deputy Ministry for Mineral resources. (U.S. Geol. Surv., Saudi Arabia Mission).

Reid, A.B., Allsop, J. M., Granser, H., Millet, A. J., and Somerton, I. W., 1990, Magnetic interpretation in three dimensions using Euler deconvolution: GEOPHYSICS, 55, 80-91.

Thompson, D. T., 1982, EULDPH a new technique for making computer assisted depth estimates from magnetic data: GEOPHYSICS, 47, 31-37.

Abstract

Near-Surface Geophysics – Defining What , Why and How

A.P.Annan

Sensors & Software Inc.

Applied geophysics supports of many human endeavours. Historically, the most common use has been in oil and gas exploration closely followed by mineral exploration. Both resource-based communities constantly advance geophysical technology and mapping techniques on depths scales of hundreds to thousands of meters.

Near surface geophysics has been a late comer to the applied geophysics world although there is nothing fundamentality different. Engineering and infrastructure needs are driving rapid advancement in high-resolution geophysics. These applications are diverse and range from assessing engineered structures, ground water search and evaluation, environmental site assessment, archeology, forensics, unexploded ordnance, and many others. Many of the applications fall into the classification of "near-surface geophysics".

Answering the question "What is near surface geophysics?" is always challenging because of the diversity of applications and technologies. There are 3 key defining features that characterize near surface geophysics.

- applications demand much higher spatial resolution, typically on the sub meter to 10's of meters scale and often on them scale
- traditional instrumentation and data acquisition must be transformed to make geophysical surveys practical
- simpler faster and often real-time data analysis and interpretation are demanded..

Achieving this level of detail requires higher spatial data sampling density, more rapid data acquisition and more scale sensitive analysis and processing. Ancillary control information require the same level of detail to aid with data evaluation and interpretation. Accomplishing all this at an affordable cost is extremely challenging!

Examples drawn from real-world infrastructure investigations will be used to illustrate these concepts and demonstrate the challenges. Over the last five to ten years, steady advancement of sensor technology, field deployment computering capacity and communications are overcoming the challenges. Applications of geophysics, felt to be unattainable because of cost, are now becoming common place

Seismic Reflection for Near-surface Characterization of Earthen Dams and Dikes: Upstream, Downstream

Richard D. Miller, Kansas Geological Survey, University of Kansas, Lawrence, KS; Richard D. Markiewicz, U.S. Bureau of Reclamation, Denver, CO; Julian Ivanov and Shelby Peterie, Kansas Geological Survey, University of Kansas, Lawrence, KS; Jianghai Xia, China University of Geosciences, Wuhan, Hubei, China

Summary

High-resolution seismic reflection has been used successfully to characterize material and investigate a variety of engineering problems and hazardous settings associated with earthen dams, dikes, and abutments. Limitations and challenges of seismic reflection when interrogating these structures and lithologies are nontrivial and require critical thinking and unique approaches to acquisition and processing to allow meaningful and necessary interpretations and analysis. Seismic reflection has proved an effective tool for a range of applications and in this paper we examine mapping confining units for integration into engineering and design of impermeable structures; mapping lithologies and bedrock structures during safety analysis for earthquake retrofitting; delineating karst in bedrock beneath the dam core responsible for subsidence on the upstream side of structure; Vp/Vs in near-surface lacustrine sediments within 1 km of a major active fault scarp; and detecting high permeability zones within a glacial outwash embankment. Extreme geometries and material variability associated with any man-made structure are the most formidable challenge to seismic imaging. Inconsistent source wavelets, out-of-the-plane energy, extreme statics (topography and velocity based), and source noise (disproportionately high percentage of surface waves) are all problems that are not unique to earthen dams, dikes, and levees, but they are certainly more prevalent with those types of structures. Success of the technique in each of these settings relied on confident and defensible interpretation of reflection energy at each step in the processing flow.

Introduction

Water retention structures built of earthen materials are susceptible to failures related to overtopping, seepage, and structural degradation. Scouring and erosion are generally responsible for overtopping failures. When water is allowed to move through or beneath the structure, generally the result of boiling and/or piping failure is gradual and classified as a seepage failure. Structural failures generally result from instability of the foundation, but they can also be the result of seepage. Cracks and settlements are generally indicators of increased failure potential. Physical characteristics are key to a structure's breach potential and there are a variety of very effective tools to assess a structure's breach potential once those parameters are known (Wahl, 1997). Defining the key characteristics for these three-dimensional structures in a fashion that does not adversely affect the capabilities of the structure is the difficult part. Regardless of the failure type, the key to dam/levee safety lies in the integrity and characteristics of the core.

Dams and levees have undergone extensive evaluations using electrical and EM methods for decades (Dobecki et al., 1988). In cases where seismic has been used, refraction or first arrival analysis has been the preferred method (Dutta, 1984). Seismic reflection has proven to be an effective and efficient method of interrogating the internal geometries and associated anomalies in seismic material characteristics in earthen structures when the conditions and materials are ideal (Miller and Steeples, 1991; Dakoulas, 1993; Llopis et al., 1995; Nie et al., 2001; Woolery et al., 2001; Miller and Markiewicz, 2000, 2001; Miller et al., 2004, Bailey et al., 2012; Miller et al., 2012). Reduced strength related to anomalous zones within or below the earthen structures could represent unsafe conditions.

Understanding and quantifying the properties of shallow subsurface materials are critical for characterizations and associated models used to guide design and retrofitting of structures to withstand expected ground motions. Dam safety is always a high priority for any dam owner. Concerns about performance during potential earthquakes and flood events have prompted many owners to embark on extensive drilling and geophysical site characterization programs to provide improved properties for site response models. Prioritizing the risk of individual structures has generally been based on potential of liquefiable foundation materials, increased flow in weirs, sinkholes, or changes in local seismicity making characterization a reaction rather than a preemptive activity.

New and/or adaptations of existing technologies are being developed and evaluated at sites with both physical characteristics conducive to those technologies and failure potential. Proven correlation between seismic properties (especially shear wave velocity) and stiffness/rigidity is the basis for developing and implementing field-efficient, laterally continuous, non-invasive methods to accurately measure the seismic wavefield. As well, routine non-invasive appraisal of dam/dike core integrity (time lapse study) could prove quite valuable in some settings. Ultimately, the goal of these kinds of research efforts is to define technologies and parameters for implementing those technologies that allow identification of localized anomalous material zones—indicative of either dissolution activity, piping, or non-uniform compaction/settling—prior to surface expressions. Seismic techniques hold vast potential for imaging and measuring materials in a fashion applicable to evaluations of dam integrity.

Study Sites

Clearwater Dam, Clearwater, Missouri

Clearwater Lake Dam, located in southeastern Missouri, was designed and constructed in the early 1940s as a control structure across the Black River. During the design and building of the cutoff trench, two distinct and unique sets of enlarged joints were discovered in the limestone bedrock. Additional drilling suggested the joints were extremely localized and inferred to only extend several tens of feet beyond the cutoff trench excavation. Based on the trench and drill findings, the joints were filled with concrete consistent with the dam building practices of that time.

On January 14, 2003, project personnel working at Clearwater Lake found a 3 m diameter by 3 m deep sinkhole. A sinkhole of this nature and location represented a threat to dam integrity and prompted a response by the dam owner. On January 17, 2003, a backhoe excavated the sinkhole down to 7 m below ground surface where the diameter of the subsidence feature narrowed to 1 m. The 7-m-deep excavation was backfilled with a clay plug, returning the dam surface to original grade. On February 15, 2003, with the first sizeable rainfall (2 cm) since the installation of the clay plug came subsidence of the clay plug (3 cm around the perimeter and 7 cm at the center). The backfilled sinkhole has been under constant surveillance since it formed with no evidence of downstream seepage, upstream slumps, cracks, or whirlpools. Soil engineers describe the 2-m-thick clay blanket covering the upstream slope as being very compacted and probably hiding a void that likely formed during the summer of 2002 and then finally collapsed in January, forming the sinkhole.

A.V. Watkins Dam, Utah

A.V. Watkins is located in north central Utah, on the shore of the Great Salt Lake and immediately adjacent to Interstate 15. The dam is about 20 km miles long, roughly rectangular shaped, and is constructed of about 17 million cubic meters of material. The structure encloses a reservoir (called the Willard Reservoir) with 215,120 acre-foot high pool capacity. This reservoir is an off-channel storage facility that is fed by the Willard Canal, which diverts water from the Slaterville Diversion Dam on the Weber River located 10 km south of the reservoir.

Willard Reservoir was designed to sustain minimum flow rates on the Weber River during the dry/irrigation season. This 10,000-surface-acre reservoir is confined and controlled by the 12-m-high earthen dam and two pumping stations. Since this is an off-channel reservoir, water is not released into a natural drainage system, but rather the reservoir is both filled and drained through the Willard Canal. The basically U-shaped zoned earthfill dam was completed in 1964 and repaired in 2007 to remediate and terminate foundation soil erosion. The effects of piping and internal erosion, evident in surface seepage, reached a point of near dam failure. Piping of the foundation soils beneath the dam core harvested significant quantities of fine-grained, silty and sandy soils, dramatically weakening the structure.

A complete understanding and functional model of site response to possible seismic events with the maximum expected ground movement originating from anywhere along the Wasatch Fault system are key to insuring previous repairs to the dam embankment and toe drain system along the southeast side of the dam are long term and commensurate with the site characteristics. These models are critical tools for future construction planning at the site. Failure of A.V. Watkins Dam could result in loss of life. The loss of life and level of property damage would depend on the location of the dam breach. Site response to seismic events is key to dam safety at this facility.

Mud Mountain Dam, Enumclaw, Washington

Mud Mountain Dam was originally designed in the 1930s for flood control on the White River which heads at Carbon Glacier on the flanks of Mt. Rainier. Springs, seeps, and increased material saturations were observed during periods of elevated pool levels. Considered changes to pool storage and use would require elevated pool levels more frequently and sustained for longer periods of time. The potential of a seepage-induced failure of the reservoir rim similar to the 1918 landslide at Masonry Pool on the Cedar River prompted concern for the integrity of the Mud Mountain Dam.

A key imaging need was a long path channel (Big Springs) suspected to possess at least 45 m relative elevation difference in the top of bedrock representing an interglacial course of the ancestral White River (U.S. Army Corps of Engineers, 1986). Of equal interest is any subsurface expression of a localized outcrop of boulders in the sidehill approximately 6 m below and 45 m northwest of the main access road. A chimney sink exposed on the top of a nearby ridge could be suggestive of deeper sediment erosion.

Data Acquisition and Processing

Each survey employed data acquisition equipment and parameters unique to the requirements of the study and consistent with the state-of-the-art at that time. Determining the spread lengths, receiver and shot spacings, and roll-along strategies were all based on well documented procedures (Hunter et al., 1984; Steeples and Miller, 1990).

- For ***Clearwater Dam***, a pair of 120-channel lines with two 40-Hz L28E geophones per receiver station recorded three 10-second, 25 to 250 Hz IVI minivib sweeps at each shot station. All ground stations were separated by 4 ft with data recorded on a 24-bit, 240-channel Geometrics Strataview seismograph with a StrataVisor NZC controller.

- At ***A.V. Watkins Dam***, both compressional and shear (SH) methods were used along two profiles, each approximately 1.5 km long and separated by about 3 km. The source was an IVI minivib occupying source stations every 6 m and triple 28-Hz Sensor geophones for the P-wave and single 14-Hz perpendicular and horizontal oriented GeoSpace geophones both on 3 m centers. For the source energy, P-wave energy revolved around a 5-second upsweep from 30 to 350 Hz and S-wave energy was confined to a 10-second upsweep between 20 and 200 Hz.
- The ***Mud Mountain*** data were acquired on a 48-channel EG&G Geometrics 2401x seismograph. The sources were an 8-gauge auger gun (Healey et al., 1991) and an IVI minivib1, while the receivers for all the data were three Mark Products L28E 40-Hz receivers offset from 20 ft to 204 ft from the source with 8 ft station spacing.

High-resolution seismic reflection data, by its very nature, lends itself to over-processing, inappropriate processing, and minimal involvement processing. Interpretations of high-resolution shallow-reflection data must take into consideration not only the geologic information available, but also each step of the processing flow and the presence of reflection events on raw unprocessed data. Processing for the reflection portion of this study included only operations or processes that enhanced signal-to-noise-ratio and/or resolution as determined by evaluation of high confidence reflections interpreted directly on shot gathers. For the most part, processing of high-resolution shallow-reflection data is a matter of scaling down conventional processing techniques and methods; however, without extreme attention to details, conventional processing approaches will produce undesirable artifacts. In-field processing of the reflection data resulted in correlated shot gathers that were subject to a variety of scaling and filtering operations. In-field processing was coincident with data acquisition and did not impact the full-day field schedules.

Results

Unequivocal identification and verification of reflections on shot gathers is not only necessary, it is mandatory for meaningful interpretations of near-surface seismic data. The most conclusive means to both verify and analyze reflections is to match modeled NMO curves (based on uphole velocity information when possible) with reflection hyperbola interpreted on shot gathers. This combination incorporates ground truth (borehole-determined velocity), geometric curve fitting (forward and inverse modeling), and event identification directly from single-fold shot gather data. Too many times the power of seismic processing software and lack of careful attention to detail results in stacked seismic-reflection data that inaccurately represent the subsurface, and is justified by happenstance, consistency with existing drill data, and geologists' models.

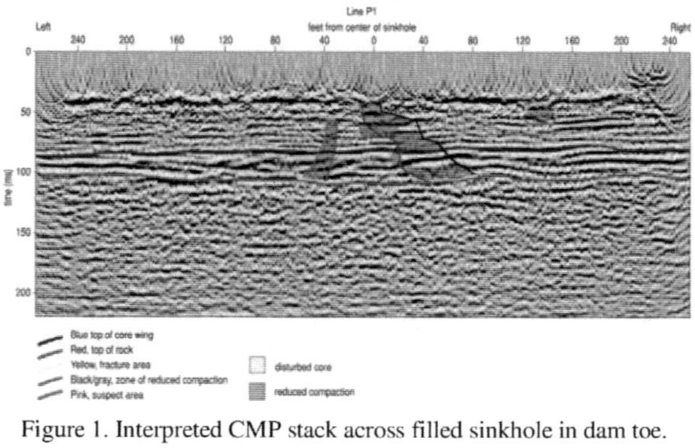

Figure 1. Interpreted CMP stack across filled sinkhole in dam toe.

Clearwater
Starting at the top of the section beneath the sinkhole, a downdropped reflection can be seen at about 50 ms, which equates to about 14 m (Figure 1). This zone is about 4 m wide and extends from about station 2.5 m right to 2.5 m left of the sinkhole; from here it drifts strongly to the right side. At about 22 m it appears to widen from 3 m left of the sinkhole to over 12 m right of the sinkhole. Looking deeper into the section the "chimney" or subsidence-altered zone appears to meander around stronger or more resistant layers, with these stronger layers forming bridges or, in some cases, subsiding intact. At the top of the clay core, interpreted at about 80 ms (~30 m), the subsiding materials have left either the top or a reflective layer near the top offset a bit. The void or less competent area in the subsurface continues to move right, where at the bedrock surface it correlates with a set of what appears to be fractures located between 25 m and 6 m right of the sinkhole. These inferred fractures correlated quite closely to the projection of the enlarged joints observed during dam construction.

A.V. Watkins
Correlating reflections from the same reflectors between compressional (Figure 2) and shear -wave (Figure 3) shot gathers

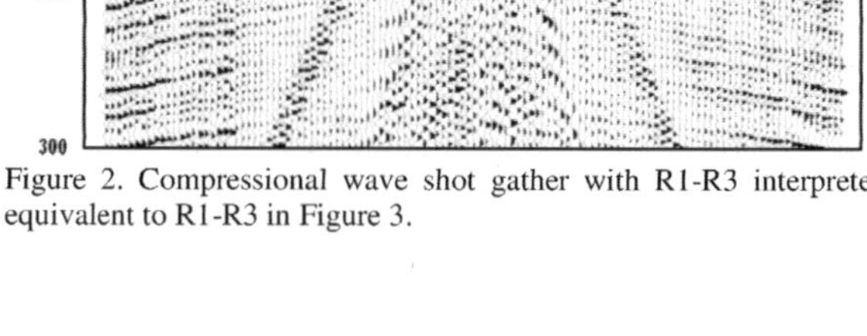

Figure 2. Compressional wave shot gather with R1-R3 interpreted equivalent to R1-R3 in Figure 3.

is challenging but can be done with good confidence on these data since data quality (signal-to-noise and resolution) is very good. With the range in Vp/Vs ratios extending from just over 2 to more than 9 through the depth interval of interest (1 km) the potential for liquefaction at this site is very high. Clearly

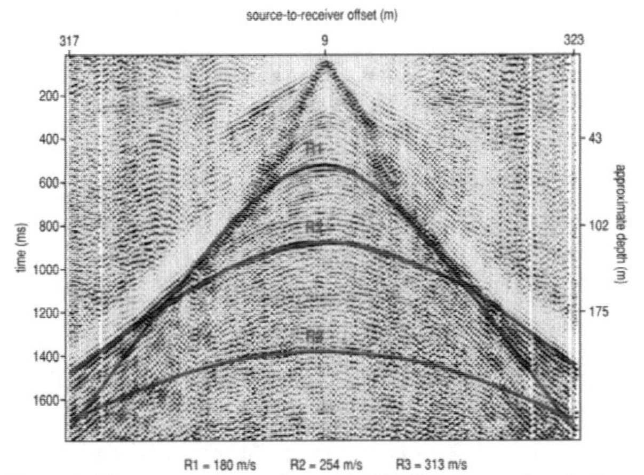

Figure 3. Shear wave shot gather with R1-R3 interpreted equivalent to R1-R3 in Figure 2.

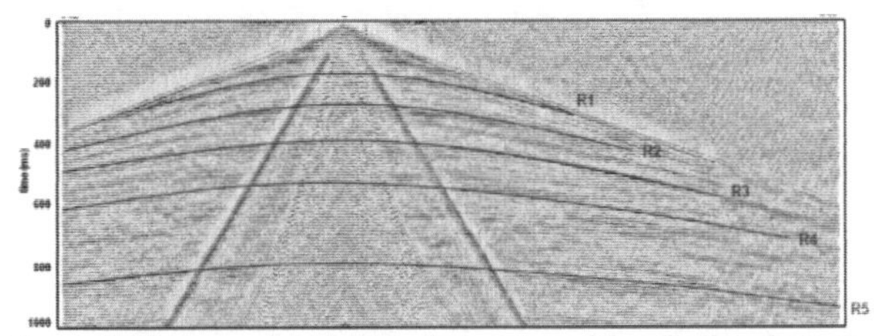

Figure 4. Compresssional wave shot gather with reflection down to over 1 km interpreted.

the greatest risk is in the upper 10 m with the Vp/Vs ratio at depth increasing to values in the 3 to 4 range throughout the interval from 30 m to over 1 km (Figure 4).

Mud Mountain

Cut-and-fill features in the Vashon Lake Bed (VLB) sequences appear to extend from stations 1040 and 1110 and from stations 1210 and 1280 on the CMP stacked section with the potential for very unsorted piles of material (possibly similar to boulder outcrops on both flanks of the embayment) deposited in channels between stations 1075 and 1105 (Figure 5). The depths of these channels are around 15 m to 22 m. Cuts into the VLB across the expanse of the embayment is likely more common than the presence of complete VLB and deltaic sequences, as observed in borehole 34, and identified on the CMP stacked section.

A basin or closed low in the bedrock surface is evident beneath well 38 with a low extending westward off the end of the seismic profile. The closed low or basin beneath well 38 could act as an isolated conduit for the movement of fluid.

Conclusions

High-resolution seismic reflection has proven an effective tool in these traditionally challenging near-surface settings. Dominant frequencies were observed clearly exceeding 200 Hz to depths over 300 m on some of the data presented here. Dramatic bedding geometries, completely unexpected based on drill data alone, were imaged at these sites. Incorporation of these seismic-reflection data with other information provided an excellent compliment to existing borehole data, providing a solid scientific basis for interpretations and projections of structural response to various pool scenarios.

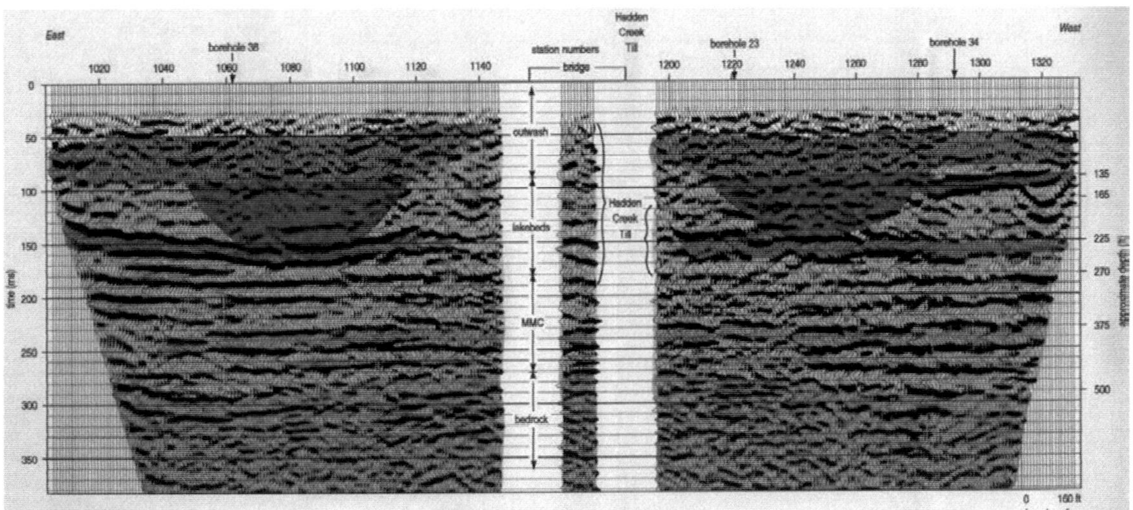

Figure 5. Interpreted CMP stack Mud Mountain Dam.

References

Bailey, B.L., D.W. Steeples, R.D. Miller, S.L. Peterie, J. Ivanov, and R.D. Markiewicz, 2012, Evaluation of extended correlation on shallow Vibroseis data [Exp. Abs.]: Soc. of Expl. Geophys., Las Vegas, NV, November 4-9, 3 p. (e-Poster.)

Dakoulas, P., 1993, Earth dam-canyon interaction effects for obliquely incident SH waves: *Journal of Geotechnical Engineering*, v. 119, n. 11, p. 1696-1716.

Dobecki, T.L., R.F. Corwin, and J.L. Llopis, 1988, Integrated geophysical investigations of seepage at an existing dam [Exp. Abs.]: Society of Exploration Geophysicists, v. 1, p. 338-339.

Dutta, N.P., 1984, Seismic refraction method to study the foundation rock of a dam: *Geophysical Prospecting*, v. 32, n. 6, p. 1103.

Healey, J., J. Anderson, R.D. Miller, D. Keiswetter, D.W. Steeples, and B. Bennett, 1991, Improved shallow seismic-reflection source: Building a better Buffalo [Exp. Abs.]: Soc. Explor. Geophys. v. 1, p. 588-591.

Hunter, J.A., S.E. Pullan, R.A. Burns, R.M. Gagne, and R.S. Good, 1984, Shallow seismic-reflection mapping of the overburden-bedrock interface with the engineering seismograph—Some simple techniques: *Geophysics*, v. 40, p. 1381-1385.

Llopis, J.L., M.K. Sharp, D.K. Butler, D.E. Yule, 1995, Applied geophysical techniques to evaluate earth dams and foundations: Proceedings of SPIE–The International Society for Optical Engineering Conference, Society of Photo-Optical Instrumentation Engineers, v. 2457, p. 195-204.

Miller, R.D., 1992, Normal moveout stretch mute on shallow-reflection data: *Geophysics*, v. 57, p. 1502-1507.

Miller, R.D., and R.D. Markiewicz, 2000, Shallow seismic reflection survey at Wickiup Dam in Central Oregon: Proceedings of the Symposium on the Application of Geophysics to Engineering and Environmental Problems (SAGEEP), Arlington, Virginia, February 20-24, p. 1245-1254.

Miller, R.D., and R.D. Markiewicz, 2001, High resolution seismic reflection survey at Keechelus Dam: Symposium on the Application of Geophysics to Engineering and Environmental Problems (SAGEEP), Denver, Colorado, March 4-7. Proceedings published on CD.

Miller, R.D., and D.W. Steeples, 1991, Detecting voids in a 0.6-m coal seam, 7 m deep, using seismic reflection: *Geoexploration,* Elsevier Science Publishers B.V., Amsterdam, The Netherlands, v. 28, p. 109-119.

Miller, R.D., B.L. Bailey, S.L. Peterie, J. Ivanov, and R. Markiewicz, 2012, Shallow P- and S-wave reflections to characterize the near-surface [Exp. Abs.]: Soc. of Expl. Geophys., Las Vegas, NV, November 4-9, 4 p.

Miller, R.D., J. Ivanov, S. Hartung, and L. Block, 2004, Seismic investigation of a sinkhole on Clearwater Dam: Symposium on the Application of Geophysics to Engineering and Environmental Problems (SAGEEP), Colorado Springs, Colorado, February 22-26, Paper KAR01, p. 1082-1098.

Nie, Y., H. Chen, and J. Wang, 2001, Study on a reflection technique for detecting the thickness of a rock-filled dike in Shenzhen, China: *Chinese Journal of Rock Mechanics and Engineering*, v. 20 (suppl.), p. 1893-1896.

Steeples, D.W., and R.D. Miller, 1990, Seismic reflection methods applied to engineering, environmental, and groundwater problems: Soc. Explor. Geophys. Investigations in Geophysics no. 5, Stan H. Ward, ed., *Volume 1: Review and Tutorial,* p. 1-30.

U.S. Army Corps of Engineers, 1986, *Mud Mountain Dam, Dam Safety Assurance Program, General Design Memorandum No. 26, Supplement No. 1, Core of Dam, Seepage Control Measures*: U.S. Army Corps of Engineers, Seattle District, Seattle, WA, 269 p. 16 plates.

Yilmaz, O., 1987, *Seismic Data Processing*: Society of Exploration Geophysicists, Investigations in Geophysics no. 2, S.M. Doherty and E.B. Neitzel, series eds.

Wahl, T.L., 1997, Predicting embankment dam breach parameters–A needs assessment: Twenty-seventh IAHR Congress, San Francisco, August 10-15, 7 p.

Woolery, E.W., R.L. Street, and S.T. Hornbeck, 2001, SH-wave seismic reflection imaging of anomalous foundation conditions–Mississinewa Dam, Indiana: Symposium on the Application of Geophysics to Engineering and Environmental Problems (SAGEEP), Denver, Colorado, March 4-7.

Hydrogeological and Geophysical investigations of a Pilot Area between Jabal Hafit and Oman Border, Al Ain, UAE

A.Murad, S. Hussein, A. Gabr, H. Baker, K. Al Bloushi , A. Aldahan*
Geology Department, United Arab Emirates University, Al Ain, UAE

Summary

A pilot area in Al Ain City, United Arab Emirates (UAE), located between Jabal Hafit to the west and Oman border to the east has been selected for the study of variability in groundwater conditions. Many wadies run through the area and are originated from the Oman Mountains ranges, which form the main catchment for both surface and groundwater. Subsurface and subsurface investigations of these wadies are vital for better understanding of the aquifer system condition and groundwater sustainability. It is believed that the recent wadies alluviums follow the configuration of ancient paleochannels thus provides interesting extended aquifers at depth. In these Quaternary alluviums, which are composed of a mixture of gravel, sand and clay beds, the groundwater level depth varies from 5 m to 80 m having maximum depth nearby Hafit Mountain and the minimum in the eastern side closed to Oman Mountains. Transmissivity of the alluvium aquifer varies between about 1837 m^2/day in the unconfined and 2449 m^2/day in the confined with corresponding permeability values at 30 m/day and 40 m/day respectively. The two dimensional (2D) Electrical Resistivity Tomography (ERT) profiles measured across selected wadies confirm the spatial extension of the Quaternary deposits. Qualitative and quantitative interpretation of the processed data indicated different hydrostratigraphic units of the alluvium section which may represent ancient paleochannels.

Introduction

Groundwater is the main natural water resource in the UAE where the Quaternary aquifer represents a major shallow groundwater system in Al Ain City (Figure 1). Rainfall in Oman Mountains and Jabal Hafit contributes to the recharge of these Quaternary deposits through mainly the wadies. From the hydrogeological point of view, subsurface investigations of these wadies are necessary in order to characterize the aquifer system in terms of water flow, water quality and management of water abstraction. Paleochannels may at depth occur along the recent configuration of the wadies (Menges and Woodward, 1993). These paleochannels are promising targets for fresh groundwater,
as they may contain appreciable thickness of water-bearing beds that are recharged from the surrounding mountain region.

Hydrogeological and Geophysical measurements:

Depth to groundwater level from the surface and the total dissolved solids were measured in field using electric sounder and multi-parameter meter. The wells left to pump water for at least one hour before sampling. Moreover, the lithological composition and specific capacity of the deepest groundwater well (W80) in the study area are recorded (Figure 1).

Two (2-D) Electrical Resistivity Tomography (ERT) profiles have been acquired, using Wenner electrode configuration along some selected profiles crossing Wadi Muraykhat and Wadi Sa'a (Figure 1). The Syscal Pro-Resistivity Meter with 96 electrodes and 10 m spacing is used at profiles extension of 600 m (Figure 1). The apparent resistivity data inverted to create a model of the resistivity of the subsurface using the software Res2dinv, ver. 3.54. Res2dinv is iterative smoothness-constrained least-squares method (Loke and Barker 1996).

Figure (1): Location map of the study area, used wells

Results and Discussions

Depth to groundwater level and the total dissolved solids variations are shown in figures 2 and 3. The groundwater level varies from 5 m to 80 m depth from the ground surface. The depth of groundwater increases toward the southwest nearby Jabal Hafit and shallows moving toward the north and east approaching the Oman Mountains. This variability is influenced by factors such as changes of the ground elevation especially along the western and eastern borders due the existence of Hafit and Oman Mountains and the complex nature of the subsurface lithological constituents.

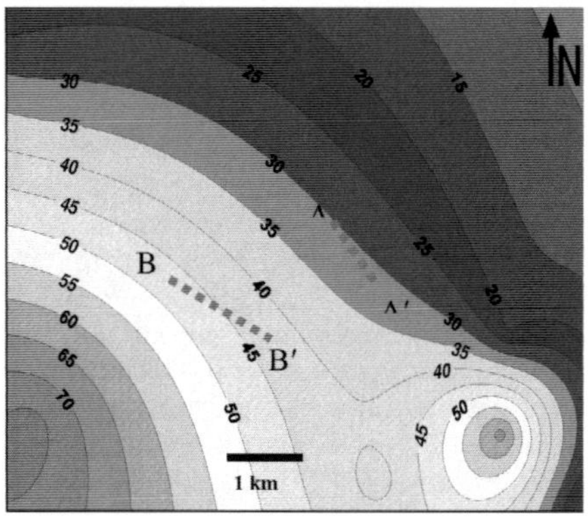

Figure (2): Groundwater level in the study area

Variation in the groundwater salinity (TDS) reveals low values (down to 200 mg/L, fresh water) in the central part of the study area and high values (up to 6200 mg/L, brackish water) in the northeastern part. This variation trend in the total dissolved solids likely relates to the mainly different types of the Quaternary clastics in the Jawwa plain and the carbonate rocks in the proximity to Oman Mountains (north and northeast) and Jabal Hafit in the west. Other factors that may affect the TDS of groundwater on local scale can be mineralogical nature of the aquifer body (carbonate-dominated clasts), amounts of easily dissolved minerals such as salts, and mixing between different groundwater types due to extensive abstraction. The available lithological data from the drilled wells, though not completely accurate, indicate Quaternary deposits as the major aquifers in the central part of the area down to a depth of about 80-100 m. In such aquifers, calculated transmissivity values using Driscoll (1986) method are 1837 m2/day for unconfined and 2449 m2/day for confined aquifers respectively. The corresponding permeability is 30 m/day and 40 m/day respectively, assuming aquifer thickness of about 60 m. In both cases, the two parameters reveal reasonable values consistent with the aquifer lithological cohesive nature that is composed of gravel, sand and mud, intercalations.

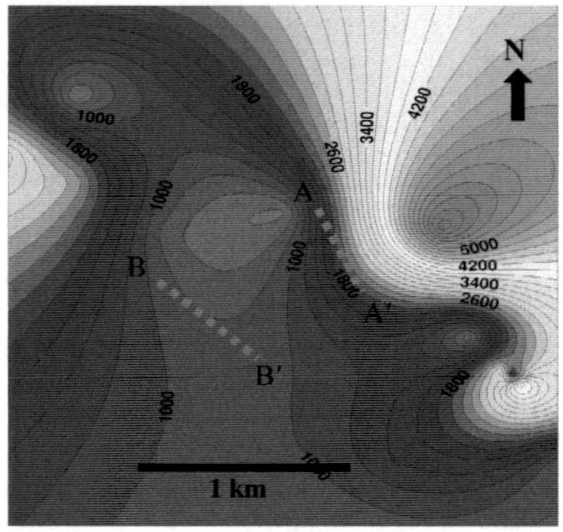

Figure (3): Total Dissolved Solids (TDS) in the study area.

The lithogical variability and indications of water table level and TDS were further confirmed spatially by the resistivity tomograms guided with the available borehole information. Interpretation of the resistivity data (Fig. 4) of the two ERT profiles A-A′ and B-B′ suggest existence of three major geoelectric- lithological units which are:

Unit-1 is a surficial zone corresponds to the upper part of the very resistive (> 100 ohm-m) layer. The resistive nature of this layer is indicative of dry conditions in the upper part of the alluvium of loose to weakly consolidated sand and gravel. Furthermore, occurrence of localized resistivity in this layer is comparable to the presences of sand-gravel bars within the braided river wadi.

Unit-2 is a moderate resistivity thick zone suggesting partial saturation and/or the presence of a clay-rich matrix within the gravel and sand. At certain locations in this unit there is a zone of coarse gravel of a varying thickness at the bottom of the alluvial section. Groundwater level is indicated at about 30 m depth, which is confirmable with the measurements obtained from the available wells.

Unit-3 has a resistivity range of less than 10 Ohm-m suggesting existence of fine-grained beds such as mudstones. In some places, this zone would have resistivity of less than 5 Ohm-m probably due to the increase of groundwater salinity with depth.

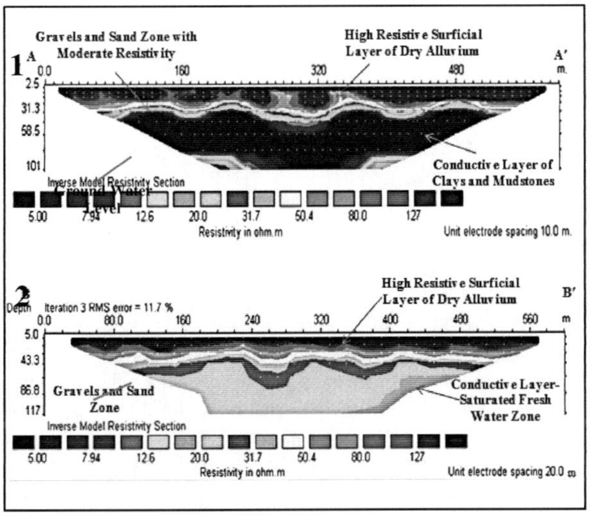

Figure (4): Electrical Resistivity Tomograms (ERT) of 1) A-A′ and 2) B-B′ profiles.

Conclusion

Data on groundwater level, TDS and lithological composition of the Quaternary aquifer system in the area between Jabal Hafit and the Oman Mountains (Al Ain, UAE) together with Electrical Resistivity Tomograms (ERT) suggest existence of different hydrostratigraphic units. These units have characteristic restrictive values that are changing with depth. Variation trend in the total dissolved solids (TDS) likely relate to differences in the Quaternary clastics such as mineralogical nature of the aquifer body (carbonate-dominated), amounts of easily dissolved minerals such as salts, and mixing between different groundwater types due to extensive abstraction.

References

Driscoll, F.G., 1986, Groundwater and Wells (2nd ed.): Johnson Division, St. Paul, Minnesota, 1021 p.

Loke, M. H. and Barker, R.D., 1996. Rapid least-squares inversion of apparent resistivity pseudosections by a quasi-Newton method. Geophysical Prospecting, 44, 131-152.

Menges and Woodward, US Geological Survey and United Arab Emirates National Drilling Company, 1993. Ground Water Resources of Al Ain area, Abu Dhabi Emirate. Unpublished administrative report 93-001, National Drilling Company, Abu Dhabi, UAE, 315 p.

Estimating aquifer hydraulic parameters from electrical resistivity measurements; a case study at Khuff Formation Aquifer, Al Quwy'yia Area, Central of Saudi Arabia

Mohamed Metwaly[*1], Mohamed El Alfy[3] ,Eslam Eawad[4], Ahmed Ismail[2] ,Gad El-Qady[2]*
1) Archaeology Dept., College of Tourism and Archaeology, King Saud University, S.A.
2) National Research Institute of Astronomy and Geophysics (NRIAG), Cairo, Egypt
3) CIPW, Prince Sultan Institute for Environmental, Water and Desert Research, King Saud University, S.A.
4) Deanship of Scientific Research, King Saud University, S.A.

Summary

Measuring the aquifers hydraulic parameters is an essential process for groundwater management and sustainable developments. These parameters are typically measured from the expensive and longtime pumping test experiments that require a number of water wells. Surface electrical resistivity measurements are recently used as less expensive alternative to estimate aquifers hydraulic parameters. A number of 20 vertical electrical resistivity sounding (VES) were acquired at Khuff aquifer in Al Quwy'yia area, to estimate the hydraulic parameters of Khuff aquifer. Based on the electrical resistivity measurements, the formation factor, porosity, hydraulic conductivity and transmissivity were estimated from the electrical resistivity measurements. These estimated parameters were compared to the parameters obtained from the previous pumping test experiments. The transmissivity values showed a reasonable comparison with the values measured from a previous pumping test experiments. According to the results of this study, surface electrical resistivity proved to be a successful tool in characterizing the Khuff aquifer and estimating most of the aquifer's hydraulic parameters.

Introduction

Determination of groundwater hydrological parameters that controls the movement and storage of the water in sediment pore spaces are greatly important to facilitate the management of water resources in arid regions. Evaluation of the groundwater resources for the purposes of sustainable development requires a clear understating of some aquifer parameters such as ; porosity, hydraulic conductivity, transmissivity, aquifer thickness and type of sediments forming the aquifer (Niwas et al., 2011). The classical methods used in defining such aquifer parameters rely on geotechnical measurements from borehole, surface samples and pumping test experiments. However, such methods are costly and only provide subsurface information at the measured location. Therefore, there was a strong motivation to find fast, cost effective and reliable alternative techniques to estimate the hydrological parameters. Surface geoelectrical resistivity is one of the most used geophysical techniques in the hydrological applications (Utom et al., 2013). During the last decade, many attempts had been made to estimate aquifer transmissivity and hydraulic conductivity from surface resistivity data (Khalil et al., 2010), based on the empirical relationships between the electrical resistivity (ρ) and hydraulic conductivity (k). Such relationships between these two parameters are controlled by Darcy's law, which specifies the discharge is proportional to the hydraulic gradient between two points. In this work, we are attempting to use the electrical resistivity measurements for estimating the hydrological parameters of Khuff aquifer at Al Quwy'yia area and compare the results with those obtained from a previous pumping test experiments conducted for some few wells in the area. After validating the hydrological parameters estimated from the electrical resistivity, the parameters can be used to enhance the hydrological information that is required for constructing and validating the hydrological model of the multi-layered aquifer system in that area.

Study area

Al Quwy'yia area is part of a large plain extending northwest-southeast over several hundred kilometers, east to the basement complex rocks exposure of the Arabian Shield (Fig. 1). Urban and industry developments are growing rapidly in this area during the past few years, which are significantly increased the stress on the limited groundwater resources. Most of Al Quwy'yia area is covered with alluvium and sand dunes Quaternary deposits. The Quaternary deposits are underlined by the calcareous units of the Khuff Formation (Permian), which is exposed, as a hilly belt 20 km wide, at the eastern side of the basement outcrop. The Khuff Formation overlies the basement complex as recorded in wells Qap2-1 and Qap1-1 at the western and southern parts of Al Quway'yia area (Metwaly et al., 2012). Because of these hyper-arid conditions, the groundwater reserve in the Al Quway'yia area is thought to be very limited in general, and even smaller than that of the regional catchment basin of Riyadh. At Al Quway'yia, the Khuff aquifer has less groundwater potentialities because it exists at the upper boundary of the catchment basin, where the only recharge source is locally limited to the infiltrated rainfall. The aquifer has poor-moderate hydraulic parameters; which slows down the groundwater flows within the limestone sediments (BRGM, 1982).

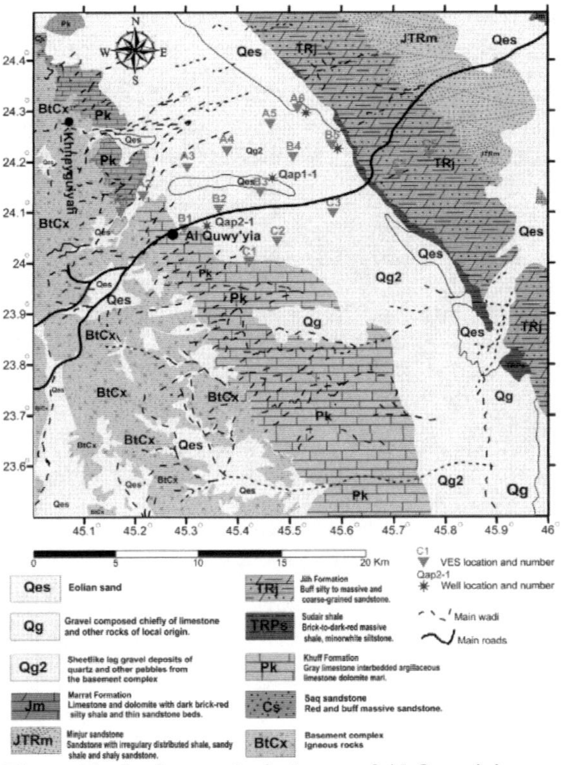

Figure (1): Surface geological map of Al Quwy'yia area and its surroundings.

The aquifer Khuff Formation has low hydraulic parameters, however, the fracture system in the aquifer assists the water flow downstream to the east and southeast. Based on the depth to water table and ground elevations measured from the available wells (22 wells) at the study area, a water table contour map for the Khuff aquifer was prepared (Fig. 2). The general trend of groundwater flow is from the west and the southwest towards the east and the northeast directions. These flow directions are coincident with regional topographic trends and correspond to the low dip of the underlying sedimentary layers and relief of the underlain basement rocks.

Geoelectrical Resistivity measurements and results

Sixteen Vertical Electrical Sounding (VES) data sets were recorded along three profiles passing through Al Quwy'yia

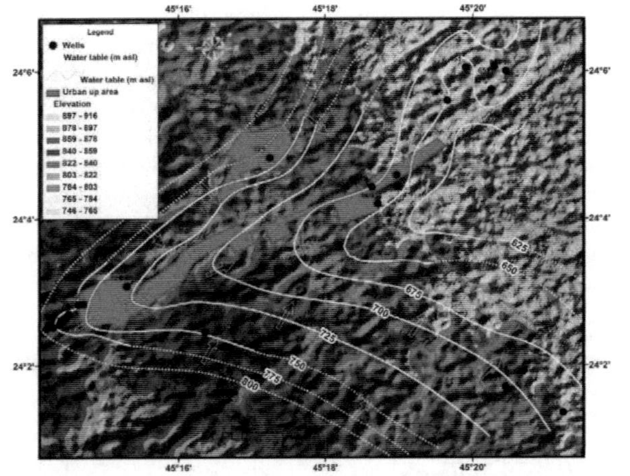

Figure (2): Water table contour map of Khuff aquifer in Al-Quwy'yia area.

area (Fig. 1). Two VES (B2 and B3) have been carried out close to two boreholes (Qap2-1 and Qap 1-1), respectively. The geological data obtained from the wells were used in calibrating the geoelectrical models, minimize the uncertainties of the 1D inverted models and constraining the interpretation (Fig. 3).

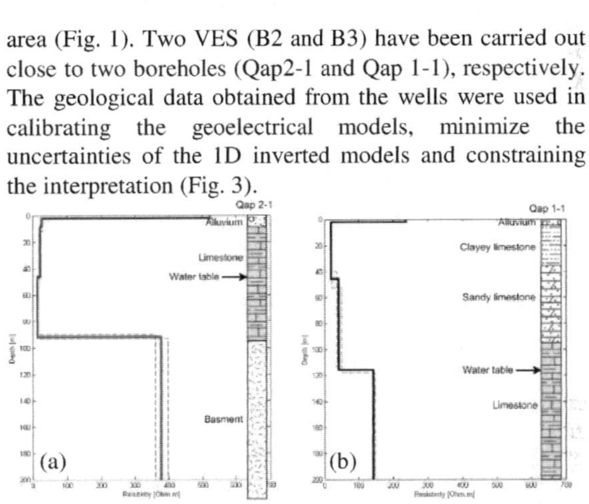

Figure (3): The layered resistivity model compared to the geological controls from two wells along profile B; (a) VES (B2) and well (Qap2-1); (b) VES (B3) and well (Qap1-1).

Applying the Uchida's (1991) algorithm, 2D resistivity models were derived for the available VES profiles. The subsurface medium was represented by numerous rectangular blocks shown by the 2D mesh. Resistivity values were assigned to each block of the mesh individually. Based on the resultant 2D inversion of VES data, the electrical resistivity data are set of multi-layered models; each of them fits the observed field curve and describes the electrical properties of the subsurface medium. Underneath the most topographically elevated area in the south western side, the resistivity values are relatively high (> 500 Ohm.m) and mainly due to the occurrence of the basement rocks at shallow depths in this area (Fig. 4). The sedimentary cover overlying the basement rocks, shows resistivity values much lower than the basement (<70 Ohm.m). The geoelectrical layer corresponding to this sedimentary cover shows variable thickness that correlates to the topography of the basement complex rocks. The eastern side of the area which is topographically flat shows relatively low resistivity values. This low resistivity body represents the Khuff limestone aquifer (Fig. 4).

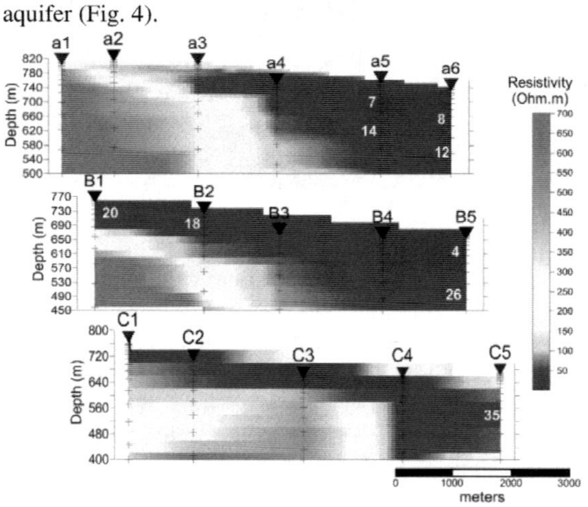

Figure (4): 2D Geoelectrical cross sections generated from a 2D inversion of a number of 1D VESs.

The static water table is coincident with the gradient of the limestone strata, giving an indication that the aquifer depth and thickness are increasing eastward.

Estimating aquifer hydrolauic parameters:

Archie (1942) introduced a formula that established a relationship between the formation factor, resistivity of sediments and its brine filled sediments in clay free sediments

$$F = \rho_a / \rho_w \qquad (1)$$

Where (ρ_a) is the bulk resistivity and (ρ_w) is the fluid (pore water) resistivity.

However,

$$F = a / \emptyset^m \qquad (2)$$

Where (φ) is the porosity of the medium, (a) and (m) are the cementation factors. The quantities (a) and (m) have been reported to vary widely for different formations. However for Khuff Formation, it has a limited cementation range (with an average of a=1 and m=2.2). The second Archie's law expresses the relation between the formation factor, porosity, bulk resistivity and fluid resistivity as follow:

$$F = \frac{R_o}{R_w} = a\varphi^{-m} S_w^{-n} \qquad (3)$$

where R_o is the formation resistivity, R_w is the pore water resistivity, φ is the porosity, S_w is the water saturation. From equations (2 and 3) and measured bulk resistivity from the surface electrical resistivity measurements and using the standard values for fluid resistivity for the water samples from the wells in Al Quwy'yia area (water conductivity at 25° is 8.25 mmhos/cm), it can estimate the formation factor at the measured stations.

The porosity values can be calculated as well at each values for bulk resistivity. Average hydraulic conductivity can be calculated using the Kozeny-Carman-Bear equation (4) (Domenico and Schwartz, 1990):

$$K = \left(\frac{\delta_w g}{\mu}\right) \cdot \left(\frac{d^2}{180}\right) \cdot \left[\frac{\varphi^3}{(1-\varphi)^2}\right] \qquad (4)$$

Where:

δ_w: is the fluid density (supposed to be 1000 kg/m3).

μ: is the dynamic viscosity (0.0014 kg/ms, Fetter, 1994).

d: is the grain size (0.001mm).

The transmissivity (T in m2/s or m2/day) is estimated for the aquifer saturated layer of thickness (h) for the cross sections interpreted from electrical resistivity measurements from the equation ; T = K.h,

The upper part has lower resistivity values (7, 8 and 4 Ohm.m) due to increase of the water contents in the fractures. The lower part is massive and unweathered bedrock, possessing relatively higher resistivity values (14, 12, 26 and 35) due to the lower fracture density and, hence, the less water saturation. The transverse resistivity values, which are the product of formation resistivity and thickness (R*T) have been calculated for the measured stations and plotted against the transmissivity values (Fig. 5). The regression line fitted to these data (R^2 = 0.98) indicate that the higher transmissivity values are related to the high resistivity values and/or high layer thickness. The locations with high transmissivity are considered the locations of favorable target for groundwater exploration.

After calculating the hydraulic parameters for the Khuff limestone aquifer at Al Quwy'yia area, it was essential to compare the results particularly the tranmissivity with that measured by a previous pumping test in the area conducted by BRGM (1982). It was observed that the calculated values from the pumping test and those estimated using the surface resistivity data are in good match. The transmissivity values are estimated from the resistivity measurements at each VES location and compared to transmissivity values measured from a previous pumping test at the corresponding locations. The transmissivity values measured from the two methods seem to be well

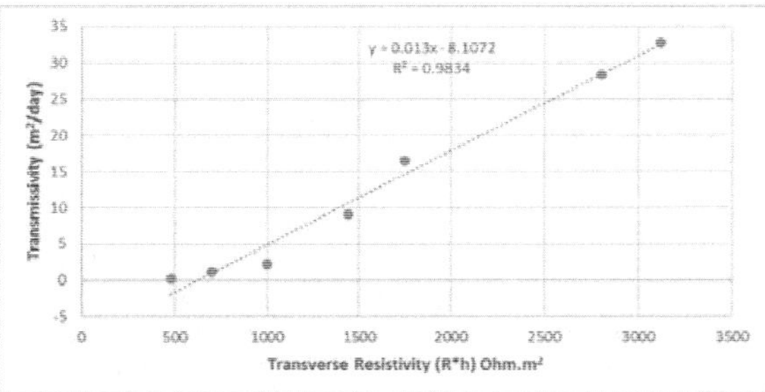

Figure (5): Relation between estimated Transmissivity and transverse resistivity for Khuff formation.

correlating with regression fitting ($R^2 = 0.85$), proofing that the electrical resistivity method can be, efficiently, used for estimating the aquifer hydraulic parameters, whereas, there is no available boreholes or even there is no pumping test measurements along the whole site.

Conclusion

In Al Quwy'yia area, the sustainable development projects require a better understanding of the hydrological aspects of the Khuff limestone aquifer. This requires accusation of many hydraulic parameters, which need different kinds of subsurface measurements to be applied. However those measurements are very costly, time consuming and need a lot of processing steps. In this work, we have attempted to estimate some of the essential hydraulic parameters from the surface electrical resistivity measurements. Therefore, vertical electrical sounding (VES) have been acquired along longitudinal profiles running East-West in the study area. The data have been processed in terms of 1D and 2D modeling schemes for getting the true formation resistivity values and portraying the aquifer geometry.

Applying Archie Formula using the resistivity values for the Khuff limestone rocks and pore water sediments, the formation factor was estimated. Then the porosity also was calculated at each values for bulk resistivity along the 2D resistivity sections. The Average hydraulic conductivity was calculated using the Kozeny-Carman-Bear equation. Then the transmissivity was estimated along the saturated thickness of the cross sections interpreted from electrical resistivity measurements. Transverse resistivity values were plotted against transmissivity and the regression line fitted to these data indicates acceptable accuracy of the estimated values of the transmissivity. The estimated values of transmissivity have been compared with those calculated from the pumping tests and calibatred hydrologic model, and the results are promissing for getting the transmissivity values for the areas whereas no pumping test exist, or even no borehole exists. Locations with high transmissivity are considered the favorable target of groundwater exploitation. The results of this study indicate that the surface resistivity measurements can be a reasonable alternative to estimate hydraulic parameters of groundwater aquifers at Al Quwy'yia area.

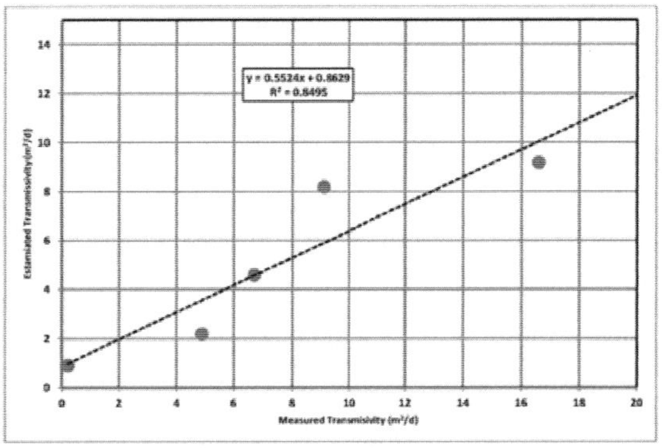

Figure (6): Correlation between the Calculated and measured Transmissivity.

References

Archie G.E., 1942, The electrical resistivity log as an aid in determining some reservoir characteristics. Trans. Am. Inst. Min. Metall. Eng. **146**, 56 – 62.

Bureau de Recherches Geologiques et Minieres Agence D'interventions (BRGM) 1982, Hydrogeological investigations in Al Quwayyah area (Pumping test) internal report submitted to Ministry of Agriculture and Water.

Domenico, P.A. and Schwartz, F.W., 1990, Physical and Chemical Hydrogeology (New York; Wiley).

Fetter, C.W., 1994, Applied Hydrogeology, 3rd ed., 600 pp., Prentice-Hall, Inc., New Jersey

Khalil M., A.,, Hafez M., A., Santos F., M., Ramalho, E., C., Mesbah, H., S., and El-Qady G., M., 2010, An approach to estimate porosity and groundwater salinity by combined application of GPR and VES: a case study in the Nubian sandstone aquifer. Near Surface Geophysics, **8**, 223-233.

Metwaly M, Elawadi E, Moustafal SSR, Al Fouzan F, Al Mogren S, Al Arifi N., 2012, Groundwater exploration using geoelectrical resistivity technique at Al-Quwy'yia area central Saudi Arabia. Int J Phys Sci **7**:317–326

Niwas S, Tezkan B, Israil M., 2011, Aquifer hydraulic conductivity estimation from surface geoelectrical measurements for Krauthausen test site, Germany. Hydrogeology Journal **19**, 307–315.

Uchida T., 1991, Two resistivity inversion for Schlumberger sounding, Geophs. Explor. of Japan (ButsuriTansa), **44**: 1-17.

Utom, A.U., Odoh, B.I., Egboka, B.CE., Rgboka, N.E., Okeke, H.C., 2013, Estimation of subsurface hydrological paramters around Akwuke, Enugu, Nigeria using surface resistivity measurements, J. Geophys. Eng. **10**, 1-8.

Groundwater Exploration Using Airborne Electromagnetic Data

Craig Annison, Simon Wetherley*
CGG Multiphysics

Summary

Airborne electromagnetic surveys are a rapid, effective and economical way of determining the geometry, structure, and quality of ground water aquifers. Here we present three examples of airborne EM applied to ground water exploration applications.

The first example shows how a DIGHEM helicopter frequency-domain EM (HFEM) survey in northeast Brazil was used to identify ground water associated with fractures and faults in a gneiss terrane. DIGHEM's extremely high spatial resolution and ability to detect subtle variations in subsurface conductivity allowed it to identify the fractures filled with ground water.

In the United Arab Emirates, TEMPEST data was used to distinguish fresh water from saline water in the main conglomerate aquifer. Good correlation of the interpreted distribution of fresh ground water from the EM data to the distribution derived from boreholes in the area shows the EM data is able to map the fresh ground water. Borehole conductivity logs compare favourably to the derived conductivity from the EM survey.

The final example illustrates how airborne EM data can be used to map palaeochannels, basement/regolith contacts, aquifer geometry and salt water intrusion in the coastal district of the Pilbara in northern Western Australia.

Introduction

In recent years there has been a growing urgency in identifying and understanding groundwater resources. Airborne electromagnetic surveys can provide useful, rapid, cost effective information about the geometry of aquifers and their host rocks. Both fixed wing and helicopter systems have been used extensively for ground water applications. Frequency domain systems (RESOLVE® and DIGHEM®) and time domain systems (TEMPEST® and HELITEM®) are operated by CGG.

Here we show three examples airborne electromagnetic (EM) data applied to a range of ground water applications around the world.

Water-Bearing Fractures in Brazil

Northeastern Brazil is home to 25 million people and is plagued by problems of uncertain water supply and regular drought. Ground water in the area is generally slightly brackish, and often hosted in near-vertical fractures making it an ideal EM target to a system that detects subtle conductivity variations. DIGHEM is a frequency-domain EM system with 5 frequencies over a range from 900 Hz to 33 kHz, including two coaxial (horizontal axis) transmitter-receiver pairs that provide the highest lateral resolution of any airborne EM – especially to steeply-dipping conductive zones. As depth of exploration is a function of EM frequency, the range of frequencies provide measurements of both shallow and deep geology, and the high number of frequencies provides for detailed sampling within that range of depths. A DIGHEM survey was flown to assist in the understanding of the subsurface water resources.

The area comprises orthogneiss and migmatite in the southeast (light colour on Figure 1) and paragneiss and metasediment throughout the remainder. Note the northeast trending quartzite unit which bisects the area signified by the dashed line on Figure 1. The area (including the quartzite) is cross cut by north and northwest trending faults and fractures.

Previous drilling in the area produced productive wells to the north of the quartzite and non-productive wells to the south. It was suspected the quartzite unit acts as a 'dam' wall preventing ground water flowing to the south causing pooling of water to the north.

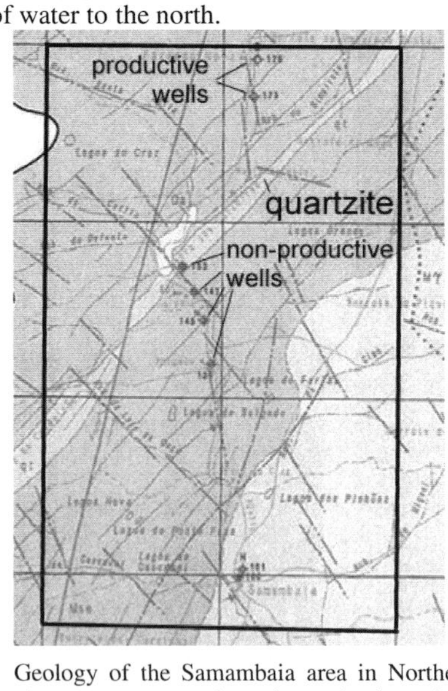

Figure 1: Geology of the Samambaia area in Northeast Brazil. Pink in southeast represent orthogneiss and migmatite and grey represents paragneiss and metasediment. Red circles indicate location of water bores drilled in the area.

Data collected during the DIGHEM survey clearly shows high conductivity to the north of the quartzite unit which is interpreted to be pooled subsurface ground water (Figure 2). To the south of the quartzite unit, conductivities are generally low except in areas associated with mapped fractures and faults. These areas are often higher in conductivity indicating that water may be associated with faults and fractures within the basement. In particular, a large north-northeast oriented fault cross cutting the quartzite has significant high conductivity zones indicating ground water may be present within this fault system. This study shows that DIGHEM is able to identify ground water resources associated with fractured basement.

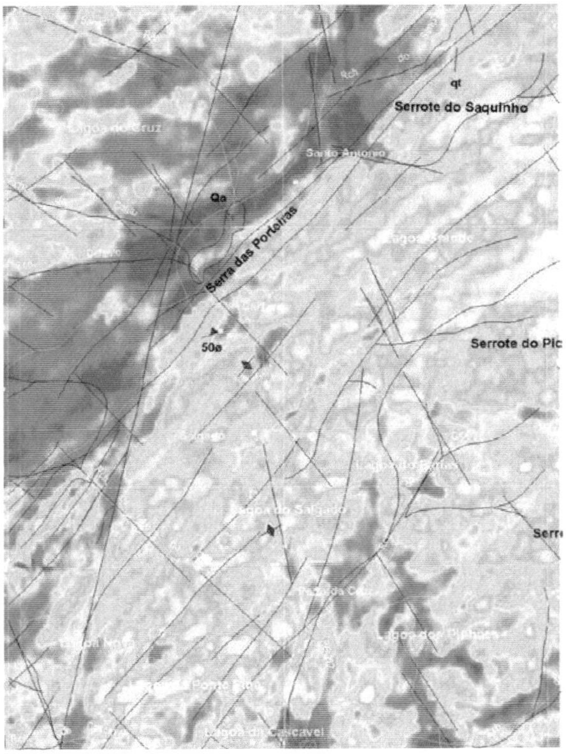

Figure 2: Apparent resistivity map (4,443 Hz Coaxial) overlain by geology. Red/purples are high conductivity areas and grey/blue are low conductivity areas.

Aquifer Mapping in the United Arab Emirates

TEMPEST is a broad band, fixed-wing TEM system primarily developed for soil salinity studies and mineral and groundwater mapping. Its broad bandwidth allows subtle conductivity contrasts to be detected making it ideal for use for ground water mapping.

A TEMPEST survey was flown in the United Arab Emirates to compare available water bore information to acquired EM data. Boreholes in the area indicate a sequence of dry, dune sand cover (0-30 metre) in places resting on a 20 metre thick sandstone layer that overlies a sedimentary sequence comprising interbedded layers of conglomerate, caliche, and siltstone layers that overlie claystones at a depth of approximately 150m.

Where sand dunes are not present, the water table occurs at depths of a few metres. Fresh water is weakly conductive and saline water is highly conductive. Generally the conglomerates form the main aquifer whose resistivity is interpreted to be dependent on water quality and salinity.

The recorded variations in conductivity are interpreted to be directly related to the salinity of the groundwater rather than to differences in lithology.

Figure 3 shows conductivity depth images that display highly resistive (dark blue) sand dunes as topographic highs at the surface. A thin conductive layer (red) occurs near the surface. Below this conductor, in some parts of the area, is "a shallow resistor" (light green), as identified by the vertical arrows. These are the target zones for fresh water with less than 500 ppm total dissolved solids, occurring at depths of 30-70 metres and with typical resistivity of 9 ohm-m.

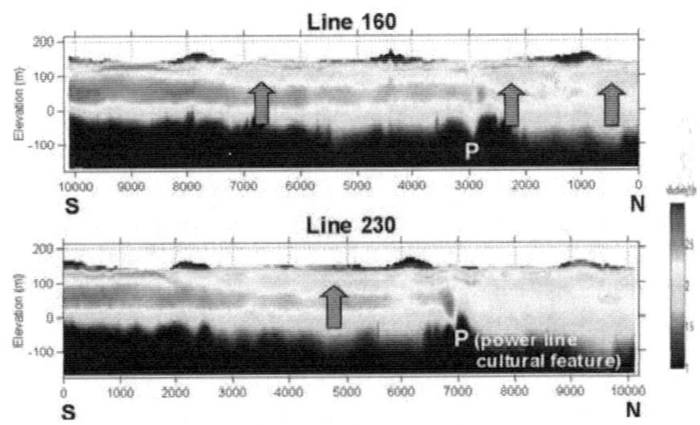

Figure 3: Conductivity depth images for two lines through the survey area. Arrows highlight blue-green shallow resistive zones where fresh water occurs.

Figure 4 shows a plan view of the near surface 9 ohm-m resistive zone that is interpreted to contain the freshest water in the area. The shape of this resistive zone is generally in very good agreement with the 3,000 ppm salinity contour in the central and southern part of the area, as shown in Figure 5.

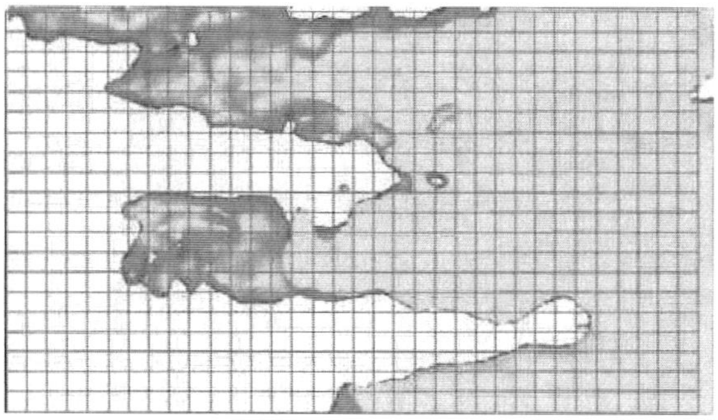

Figure 4: Top view of the 9 ohm-m isosurface through the conductivity volume interpreted to represent the extent of the fresh water.

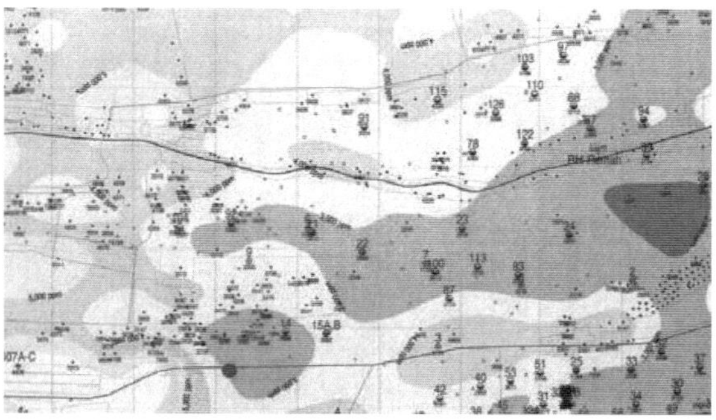

Figure 5: Water quality contour map showing the total dissolved salts obtained by analysing groundwater samples from boreholes (indicated by circles).

One of the more obvious differences between Figure 4 and Figure 5 is the extensive resistive zone in the north of Figure 4, extending from east to west across the area. Here there are noticeably fewer boreholes (Figure 5) and the contour values on the salinity map have largely been interpolated. This suggests that greater confidence should be placed on the resistive zone as mapped by TEMPEST, rather than the contours depicting salinity values derived from sparse borehole data.

Resistivity logs are available for many of the boreholes in the area (Figure 6). A colour bar has been added to the resistivity log using the same colour scale as the conductivity section (Figure 6). Overall the match between the resistivity log and the TDEM conductivity data is remarkably good considering the difference in the scale of sampling, the time difference between measurements, and that the locations are not exactly coincident.

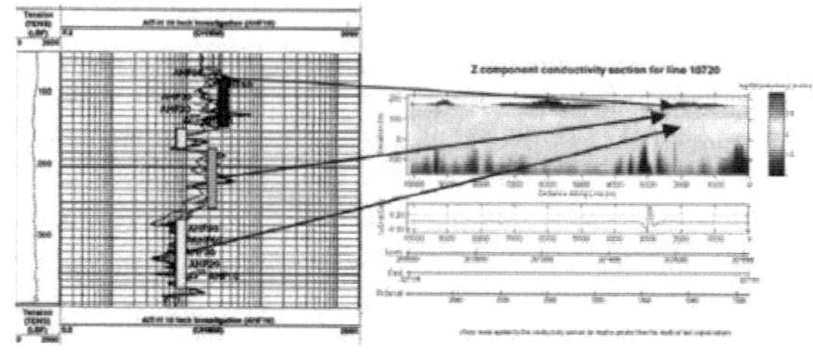

Figure 6: Resistivity log (left) from a borehole near the conductivity depth image (right). The same colour bar has been applied to the resistivity log and the conductivity image. Note the good agreement between the resistivity log and the EM derived conductivity image.

This proof of concept survey shows that data derived from the TEMPEST survey has good agreement with the available borehole data and is an effective tool to assist in ground water exploration.

Palaeochannel Mapping in Northern Western Australia

Additional water resources are required for the expanding towns in the Pilbara region of northern Western Australia and to supply the many mines in the area. A TEMPEST survey was flown over four areas to assist with understanding the geometry of the aquifers including palaeochannels and the near surface stratigraphy.

The principal aquifers within each survey area are the shallow alluvial deposits, though within the De Grey and Yule Rivers these deposits can locally reach a thickness of 75m within the basement palaeochannel of the De Grey River and 70m adjacent to the Yule River. In the eastern half of the De Grey River survey area the Broome and Wallal Sandstone aquifers associated with the West Canning Basin, are also present.

Conductivity depth images, in conjunction with available borehole data, were interpreted to identify the alluvial aquifers, palaeochannels, near surface stratigraphy, weathered basement contact and seawater wedge (Figure 7). The modern day river channels are generally low in conductivity presumably due to higher sand content and flushing with fresh water whereas the alluvial deposits are more conductive due to the presence of clay. On the northeastern end of the line, the Broome Sandstone aquifer is underlain by the Jarlemai Siltstone which forms an upper constraint on the deeper Wallal Sandstone aquifer (Figure 7).

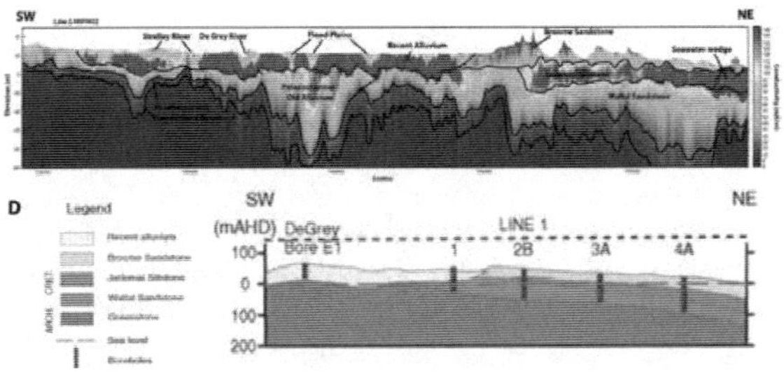

Figure 7: Interpreted conductivity depth image (top) and stratigraphy interpreted from boreholes (bottom). Note the good agreement between the conductivity depth image and the borehole interpretation.

Several palaeochannels were identified in the survey areas (Figure 8). These channels were identified by interpreting each of the conductivity depth images and correlating the channels between them (Figure 7; Figure 8).

Figure 8: Perspective view showing an interpreted palaeochannel in red.

Due to the large electrical conductivity contrast between seawater and freshwater, airborne EM data can be effectively used to identify and map the regional extents of seawater intrusion within coastal aquifers.

In order to map the seawater intrusion, a conductivity value was assigned that defined the leading edge of the seawater intrusion. Both the 3-D voxel model and conductivity depth images were examined to identify zones of high conductivity gradient that could represent the boundaries of the saline zone. High conductivity gradients were identified within the data, focussed around a number of conductivity values. These values were compared with published conductivities for seawater saturated lithologies to determine a conductivity value representing the boundary of the seawater intrusion. Using a conductivity value of 300 mS/m, isosurfaces were generated for each of the survey areas to represent the extent of the seawater intrusion (Figure 9).

Figure 9: Perspective view of the seawater wedge interpreted from conductivity depth images shown in red.

Conclusions

We have shown that both TEMPEST fixed wing time domain and DIGHEM helicopter frequency domain electromagnetic surveys can be used to help constrain the geometry of ground water aquifers and their host rocks.

Ground water associated with fractures and faults in northeast Brazil was identified using a DIGHEM survey. The near surface stratigraphy and associated ground water aquifers were mapped in the United Arab Emirates allowing areas of fresh and saline water to be distinguished. Good agreement between the geometries derived from borehole and EM data derived interpretation was found.

In northern Western Australia, near-surface stratigraphy, palaeochannels, alluvial deposits and the seawater wedge were mapped to provide a better understanding of the ground water systems. These examples show that airborne EM is a valuable, cost effective tool that should be used for ground water exploration.

Acknowledgments

The authors thank CGG for permission to publish this paper.

Mapping coastal aquifer and sea water intrusion using time-domain electromagnetic method, North Oman.

Hesham El-Kaliouby, Sultan Qaboos University, Earth Science Dept., Oman & National Research Center, Geophysics Dept., Egypt.*

Summary

Time Domain Electromagnetic (TDEM) has been widely used in mapping and assessment of groundwater resources and sea water intrusion along coastal alluvial plains. TDEM soundings were used to define aquifer zones in the alluvial aquifer in Al Khoud coastal area in North Oman. The main aquifer zones, composed of clean gravels, clayey gravels, and cemented gravels, could be identified, along with the basal contact of the aquifer. The TDEM data also used to map the depth to the interface between freshwater and saline-water along the coast. The inland extents and thickness of the saline water were mapped along two NW-SE TDEM profiles. The depths to the intrusion zone was investigated using different transmitter loop areas. The resistivity distribution derived from TDEM data agrees closely with the electrical conductivity and resistivity logs from available well data in the study area. TDEM proved to be a successful tool for mapping saline zones and fresh water in arid environment with high depth of penetration and speed of operation.

Introduction

The coastal plain in northern Oman, that extend between Oman mountains and the Gulf of Oman, represent the main source of fresh water. The alluvial coastal aquifer provides an important freshwater resource for both domestic and agricultural use. Almost 50% of the agriculture production is coming from the northern coastal plain, which depends on groundwater as the primary source of agriculture water. However, over pumping from the coastal aquifer resulted in large water-level declines and associated environmental impacts, including groundwater quality deterioration and sea water intrusion, which represents a growing potential threat to water supplies. The salinity increased to levels that affected the crops and human consumption. The yield of the crops, including the date palms which usually tolerate higher salinities, has been affected. Farms have been abandoned as soil salinity has increased. Many wells have been abandoned, all within 6km of the coastline. This highlights the need for urgent conservation measures to sustain this precious economic wealth (Young, et al. 1998, 2004).

The study area (figure 1) is located at Al Khoud fan, northern coastal area of Oman, represent the lower reaches of Samail Catchment, which extends from the mountains region into the coastal plain. Groundwater recharge in the Al Khoud fan occurs from flood runoff originating in the upper Samail catchment. As the volume of abstracted water has increased, and the water level has declined, sea water has moved landwards. This hazard is highlighted by the shallow depth of the water table on the coastal plain, which is 10-20m from the ground surface. This hazard is highlighted by the shallow depth of the water table, which is less than 1m above mean sea level and often below sea level. For these reasons, there is a great concern and a pressing requirement to assess the available groundwater reserve in the northern coastal plain aquifer (Young, et al. 1998).

It is important to have a fast and reliable tool for regular monitoring of salinity changes along the coastal aquifer in order to get better future prediction and to improve management plans to limit or prevent saltwater intrusion. This is conventionally achieved through drilling a network of monitoring wells along the coastal plane to monitor changes in water salinity over time. However, installation and operation of these wells is expensive and provides limited spatial and vertical coverage. An alternative approach to determine the existence or potential of groundwater resources and saline water intrusion is to use noninvasive geophysical methods (Nenna et.al., 2013).

TDEM is a relatively low cost, rapid, and noninvasive reconnaissance tool to locate potential freshwater resources. TDEM method is well suited to near-surface hydrological studies as it can be used to differentiate between fresh, brackish, and saline water. TDEM method investigates the electrical conductivity of the subsurface through measurements of electromagnetic field over time. The characteristics of the TDEM responses of the earth due to a pulsed primary signal from a large wire square-loop laid on the ground surface yield information about the variations of the electrical conductivity with depth. The conductivity of soils and rocks are controlled by mineralogy, clay content, water content, salinity, and porosity. Changes in the conductivity of soils and rocks, produce variations in the electromagnetic signature. The exact position of the saline water/ fresh-water interface in most coastal areas is poorly constrained on a local level. A combination of monitoring well water quality measurements borehole data, and surface TDEM can be used to accurately map the position of the interface (e.g. Al-Garni and El-Kaliouby, 2011; El-Kaliouby and Abdallah, 2015).

Field site characteristics

Major groundwater flows in Northern Oman are from the mountains towards the sea. The coastal alluvial deposits in wadi beds form excellent aquifers and contain renewable resources of good quality groundwater by regular recharge from rainfall, surface water flow and locally the inflow of groundwater from adjacent hard rock aquifers. Recharge from local rainfall appears to constitute an important component in the areas, where coarse gravel deposits provide favorable infiltration conditions. Many streams originating in the mountains may flow for an extended period in downstream to recharge coastal aquifer (Stanger, 1986; Young, et al. 2004).

The study area is covered by Quaternary unconsolidated to semi-consolidated sands and gravels with interbedded clay lenses or layers, which rest on clayey conglomerates or directly upon fissured Ophiolites. The Quaternary cover has generally a wedge shape, thickening from a few meters close to the mountains to more than 100 m along the coast. A thick sequence of marls and clays underlies the coarser Quaternary sediments in some areas near the coast, the thickness of this sequence may reaches up to 600 m (Stanger, 1986). Three major units dominate the lithological succession of the alluvium: upper, middle and lower gravel units. The upper gravel unit is composed predominantly of large-size gravels and sand including boulders (figure 2). This unit is loose and very poorly sorted of permeable alluvium with good water quality that makes it the main aquifer zone. The middle unit is discontinuous clayey gravel, which is found in the form of lenses and inter-bedded claystone bands between the upper and lower gravel units. The lower unit is cemented gravely marl and clays, which is more compacted and conglomeratic, due to the cementation caused by diagenesis associated with a fluctuating water table. (Stanger, 1986; Young, et al. 1998).

TDEM Method

TDEM made by transmitting an alternating current into square loops of insulated wire deployed on the land surface. The transmitter loop current consists of equal periods of time-on and time-off base frequencies, which produce an electromagnetic field near the loop of wire. Termination of the current flow is not instantaneous but occurs over a few microseconds, known as the ramp time (or turn-off time), during which the magnetic field is time-variant. The time-variant nature of the induced electromagnetic field creates a primary electromagnetic field in the ground underneath the loop in accordance with Faraday's Law. This resultant field instantly begins to decay, generating additional eddy currents that spread downward and outward into the subsurface like a succession of smoke rings. These eddy currents then produce a secondary magnetic field that is recorded as voltage measurements throughout the time-off period by a receiver coil positioned in the center or at an off-set from the transmitter loop. The voltage magnitude received from the eddy currents at specific times and depths is determined by the overall conductivity of subsurface rock units and fluids. From these voltage measurements, apparent resistivity values can be calculated. Depth of exploration depends on the transmitter loop dimensions and the injected current intensity (McNeill, 1994).

Data acquisition and processing

A Geonics Protem-47 system (figure 3) using a square transmitter (Tx) loop was used to collect the TDEM soundings. A 1-meter-diameter multi-turn coil receiver (Rx) with a 31.4-square-meter area measured the secondary magnetic fields with central loop and an off-set loop of 25m from the center of the transmitter loop. The Protem-47 transmitter used an injection current of 3.5 amps for the various soundings with preset base frequencies of 25Hz; and 2.5Hz. The Protem TDEM unit samples signal amplitudes in 20 subintervals (time gates) during each measurement and to collect three voltage data sets. Data collected with TDEM method, provided significant contributions to image the extent of saltwater intrusion at many sites in relatively short time. The depths and inland extend of the sea water intrusion was mapped through two profiles perpendicular to the coastline using different loop sizes (20m, and 300m-loop side). The two profiles extend from Al Khoud dam northward towards the sea with a 3 km and 4 km in length (figure 1). The depths of investigations range from tens to hundreds of meters below the ground surface, depending on Tx loop dimensions.

The apparent resistivity data were plotted as a function of time on a log scale. Outliers that deviated severely more than 10 percent above or below other data points were removed or masked before inverse modeling. Inversion is the process of creating an estimate of the true distribution of subsurface resistivity from the measured apparent resistivity obtained from TDEM soundings. Smooth and layered-earth models were generated for each sounding. A smooth model is a vertical delineation of calculated apparent resistivity that represents a non-unique estimate of the true resistivity (Constable, et al., 1987). Ridge regression and Occam's inversion were used by the inversion software (IX1D, Interpex) in a series of iterations to create a smooth model consisting of 20–30 layers for each sounding site. Layered-earth models were then generated based on the results of smooth modeling. The layered-earth models are simplified to fewer layers to represent geologic units with depth. Throughout the area, the layered-earth models range from 3 to 4 layers, depending on observed changes in the raw data and smooth model inversions.

Interpretation and discussion

The results shown in figures (4 and 5) are for two main NE-SW profiles starting behind Al Khoud dam northward towards the sea. These two profiles (figure 1) were measured by different TDEM loop sizes to be able to penetrate at different depths of investigation. TDEM models show an upper dry and fresh-water saturated gravels followed by an intermediate resistivity that may account to clayey gravel and a bottom low resistivity layer that is attributed to brackish and saline water intrusion. The thickness of the intruded zone increases towards the sea.

Small TDEM loop (20m by20m) survey covered a depth of 100m showing upper high resistivity layer (100-200 Ohm.m) of dry sand and gravel, that extends to about 15-20m followed by and fresh-water saturated gravels of resistivity (40-90 Ohm.m) and extends to 30m depth . A deeper layer of resistivity (10-40 Ohm.m) that corresponds to clayey gravels extending to a 60-80m depth. A very low resistivity bottom layer (<2 Ohm.m), towards the sea at a depth of more than 80m, is detected and could be attributed to saline interface at the northern stations, since their distance from the sea is about 2km. The small-loop transect showed only the top of the saline zone with a high resolution of the upper layers including the water table but the base of the saline zone is not well defined by this loop dimensions (figure 4).

Larger loop dimensions (300m by 300m) provided a good definition of the base of the saline water at a depth of 250m below ground surface. However, the near surface layers are poorly resolved due to the averaging of resistivities over the area of the loop. This prevent accurate estimation of the resistivity of the overlying aquifer or the other resistivity boundaries of the upper layers that only provide qualitative estimate of the variation of water quality. Below the intrusion zone, an intermediate resistivity layer (20-50 Ohm.m) that could be attributed to cemented gravel composed of low porosity limestone and marl with clay or due to water of lower salinity confined by impermeable clay layer (figure 5).

The results from the two profiles correlates well with water Electrical Conductivity (EC) logs (figure 4), which reflects the total dissolved salt (TDS) within the wells in the study area.

Conclusion

TDEM has proven to be a cost-effective method tool in mapping the fresh/saline water interface and for locating the depth of fresh water aquifer within depths from a few tens of meters to several hundred meters. TDEM offers important advantages over the conventional resistivity depth sounding for mapping the electrically conductive saline zones in the arid environment with improved depth of penetration, depth resolution and speed of acquisition. The TDEM data collected were considered as good results on the basis of root mean square errors calculated after inversion, comparisons with borehole logging, and repeatability. Models estimated from TDEM data using different transmitter loop sizes suggest that there is a definite geoelectrical boundary between saltwater-saturated and freshwater-saturated sediments, but with different resolutions. Large loop cannot resolve shallow structures in a freshwater-saturated zone, while small loop could resolve it clearly. On the other hand, Large loop data could resolve the saltwater-saturated zone at depth, in which smaller loop could not penetrate to it. Because the different transmitter loops provide complimentary information about the hydrogeological structure at different depths, further analysis using a joint inversion approach of both data sets may improve the estimated geoelectric models of the subsurface. Monitoring or time-lapse TDEM measurements should be carried out at the same time of the year along the same profiles to monitor the movement of the fresh/saline water interface over time.

Acknowledgements

The author would like to thank Sultan Qaboos University for providing the resources to conduct the field survey. The author would like to thank the Ministry of Regional Municipalities and Water Resources for providing the well data and large loop TDEM raw data. The author also would like to thank the students who helped him during the field work: Raja AlRahbi and Ibrahim AlSawafi

References

Al-Garni, M. El-Kaliouby, H. 2011. Delineation of saline groundwater and sea water intrusion zones using transient electromagnetic (TEM) method, Wadi Thuwal area, Saudi Arabia. Arabian Journal of Geosciences 4 (3-4), 655-668

Constable, S.C., Parker, R.L., and Constable, C.G., 1987. Occam's inversion—A practical algorithm for generating smooth models from EM sounding data: Geophysics, 52, p. 289–300.

El-Kaliouby, H. and Abdallah, O. 2015. Application of time-domain electromagnetic method in mapping saltwater intrusion of a coastal alluvial aquifer, North Oman. Journal of applied Geophysics v.115, p 59-64.

McNeill, J.D., 1994. Principles and application of time domain electromagnetic techniques for resistivity sounding: Geonics technical note TN-27, Mississauga, Ontario, p.1-15.

Nenna, V., Herckenrath, D., Knight, R., Odlum, N., and McPhee, D., 2013. Application and evaluation of electromagnetic methods for imaging saltwater intrusion in coastal aquifers: Seaside Groundwater Basin, California. Geophysics, 78, p. B77–B88.

Stanger G., 1986. The hydrogeology of the Oman Mountains. PhD Dissertation, Open University, Milton Keynes, England, 355 pp.

Young, M, Macumber, P , Watts, M and Al Touqy, N. 2004. Electromagnetic detection of freshwater lenses in a hyper-arid terrain. Journal of Applied Geophysics, 57, p. 43-61.

Young, M, de Beuijn, R, and Al-Ismaily, A. 1998. Exploration of an alluvial aquifer in Oman by time-domain electromagnetic sounding. Hydrogeology Journal, v. 6, p. 383-393.

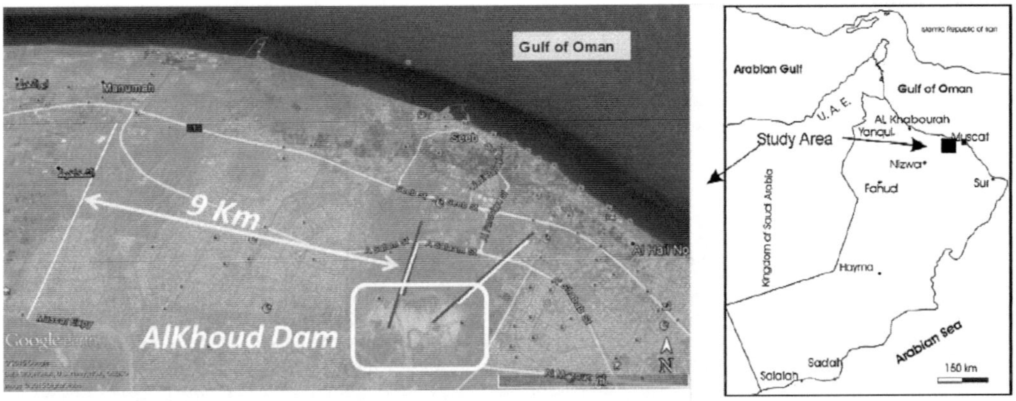

Figure 1. Location of the study area showing the TDEM profiles.

Figure 2. Field photo showing wadi gravel and silt from a road cut in the study area.

Figure 3. TDEM central loop sounding configuration setup using Geonics system.

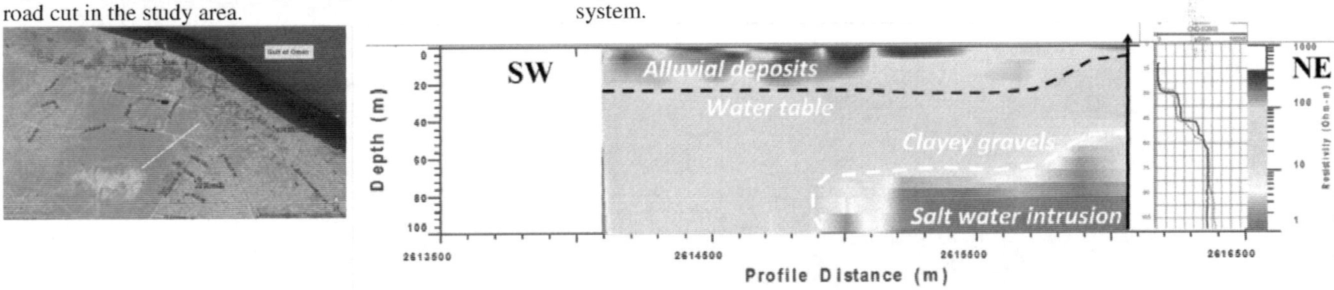

Figure 4. TDEM profile (3 km length) using 20mx20m loop showing the water table and the top of the salt-water intrusion zone.

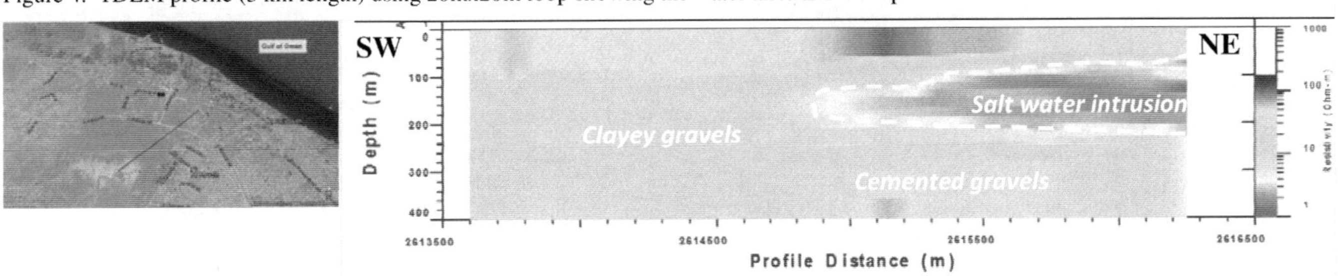

Figure 5. TDEM profile (4 km length) using 300mx300m loop showing the full salt-water intrusion zone.

Geophysical Prospecting of Groundwater in Carbonate Discontinuous Aquifer in Morocco

Mohamed Rouai & Khaoula Qarqori*
Earth Sciences Department, Faculty of Sciences of Meknes, Moulay Ismail University, Morocco.
**mohamed.rouai@gmail.com*

Summary

We present a case study of hydrogeophysical prospecting in discontinuous karstic Lias reservoir from Moroccan Middle Atlas. The geophysical surveys applied mainly Electrical Resistivity Tomography imaging the 2D subsurface karst. This technique appears to be an appropriate geophysical tool, especially by improving understanding of fracture network and reservoir geometries.
Additional investigations using electromagnetics and microgravimetry surveys were also performed for more precision and to characterize favorable hydrological zones.

Introduction

The Tabular Middle Atlas (TMA) reservoir is one of the most important aquifers in northern Morocco. It is mainly a water table fractured and karstified reservoir consisting of Lias limestone and dolomite (Martin, 198; Amraoui, 2005). The Bittit Spring belongs to this karstic system which constitutes an important aquifer lying at the junction between the tabular reservoir and the Sais basin (Fig.1). Bittit spring, with an average annual discharge of about 1600 l/s, contributes largely to water supply of the big city of Meknes (600. 000 inhabitants), despite its overexploitation and drought years. Groundwater circulation is complex due to tectonics.

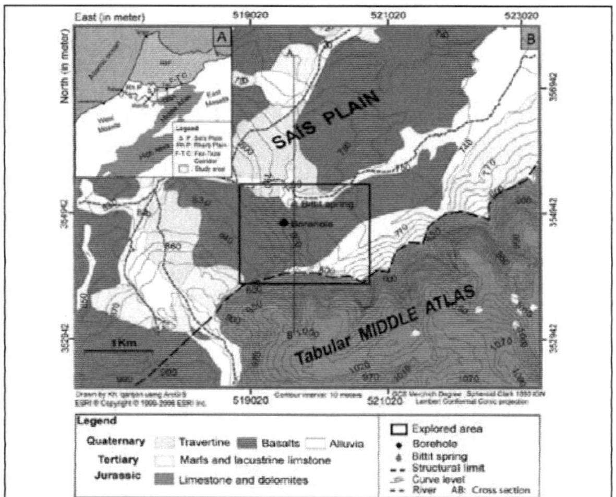

Fig. 1: Location of study area, and geological map of surveyed Bittit area (outlined in black).

Up to now water flow paths and the underground karst organization remain misunderstood, and turbidity affects the water quality after each important rain event. To highlight these issues and in order to protect the resource an integrated geophysical survey was performed in this area consisting mainly by Electrical Resistivity Tomography (ERT).

Methods

Seven ERT profiles were carried out around Bittit spring (Figure 2). Apparent resistivity data were acquired by an ABEM Terrameter SAS1000 and a multi-electrode Lund system imaging using Wenner array configuration of 64 electrodes and 5m spacing. It allows investigating a depth of about 50 m. The lengths of the profiles vary from 315 to 395 m with a roll-up of the cables.
Geological map of Bittit area provides major NE-SW and NW-SE faults direction that helped to define the direction of the profiles. Some profiles have also been crossed to obtained resistivity information in perpendicular directions.

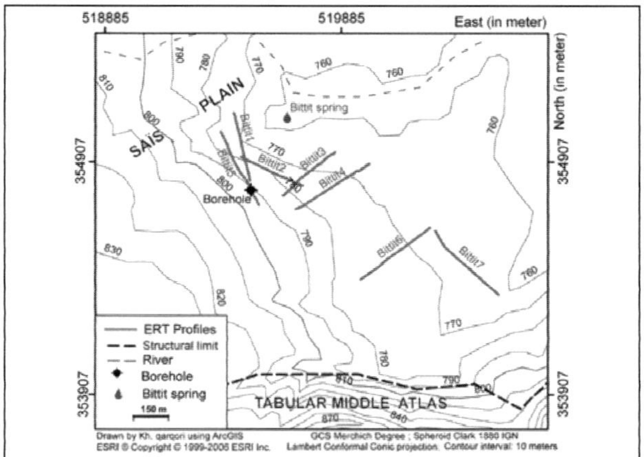

Fig. 2: Positioning plan of electrical tomography profiles

The electrical resistivity tomography (Fig. 3) have been obtained after mathematical inversion using Res2Dinv software (Loke and Baker, 1995).
Topographic corrections were carried out for each electrode position through an inclinometer and a GPS.

Results

The obtained 2D high resolution resistivity images allow to detect, delineate important fractures and also to hydrogeological characterization of the underground karstic reservoir.

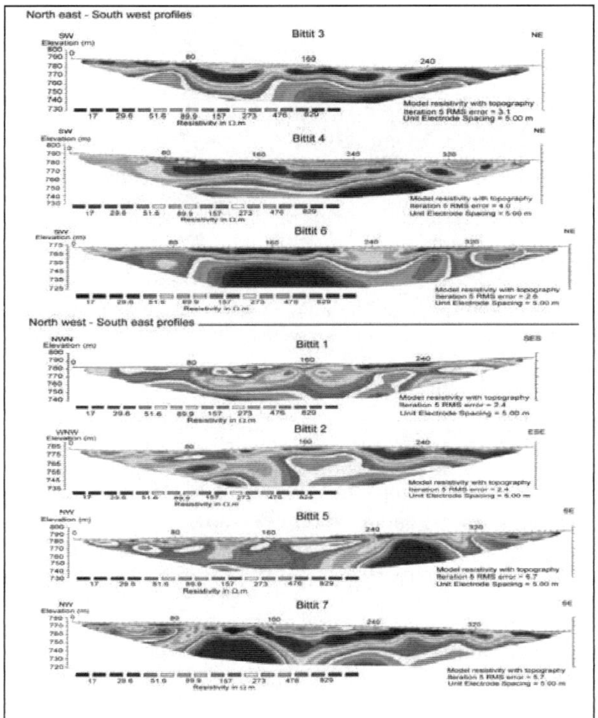

Fig. 3: Inverse 2D resistivity sections.

The resulting sections display various electrical resistivity's from less than 10 Ωm to more than 1000 Ωm. Observing the inversion results, we have classified the resistivity values into three ranges of resistivity: [10 to 70 Ωm] represents the conductive structures, [70 to 215 Ωm] for the intermediate structures and [≥ 215 Ωm] for the resistant structures as Lias limestones..

The electrical sections on NE-SW direction (Bittit 3, Bittit 4 and Bittit 6, respectively from the spring to the TMA) generally show the presence of horizontal structures, with three distinctive features. We observe first a very resistant layer (~ 500 Ωm) on the upper part of Bittit 4 section, secondly, a conducting layer (< 40 Ωm) which become increasingly deep and thick while going from TMA (depth of -10 m on Bittit 6) towards Bittit spring (depth of -35 m on Bittit 3), and thirdly, a very resistant layer (> 215 Ωm) which appears from -15 m depth on Bittit 6 section and deeps while going towards the spring to reach a depth of about -30 m to -40 m on the Bittit 3 section. This resistant block is always limited by sub-vertical edges and is in contact with a layer with intermediate resistivity value.

Contrary to the NE-SW profiles, the electrical sections on NW-SE direction (Bittit 1, Bittit 2, Bittit 5 and Bittit 7) show the presence of vertical structures and heterogeneous and complex aspect. First, the Bittit 7 profile was carried out perpendicularly to the NE-SW profiles in order to verify the presence of the upper conducting layer and its extension towards the TMA. The result shows that this structure seems to be almost continuous from the TMA at the SE to the spring at the NW. We observe also a very resistant block (~ 1000 Ωm) with a length of approximately 80 m, limited by conducting vertical structures on the Bittit 5 and Bittit 7 sections.

We can observe that these resistive blocks are individualized and dislocated by vertical conducting structures on Bittit 2, Bittit 5 and Bittit 7 and also by a heterogeneous distribution of the resistivities, especially on Bittit 1 and Bittit 2 sections.

To sum up, the main electrical structures consist of (1) a shallow sub-horizontal conductive layer, and (2) a deeper resistive layer, discontinuous due to sub-vertical conducting structures and heterogeneous resistivity media.

Two sub-vertical fracture families have been identified with NE-SW and NW-SE directions respectively. These results correlate well with fracture data analysis gathered from remote sensed Spot images and field fracture surveys. Geophysical scans and interpreted fractures were displayed with Gocad GIS software leading to geometric construction of aquifer units and to 3D modeling of the underground.

From a hydrological point of view, water preferential draining zones coming from the tabular Middle Atlas to Bittit spring seem to be essentially localized on a shallow sub-horizontal level of porous calcareous travertines (Figs. 3,4 &5). Furthermore, sub-vertical features seem not to be main water drains.

Other geological observations lead to the same conclusions. One of the faults extracted from Bittit5 profile could supply the Bittit spring by extrapolating the fault plane toward NE. However, the Bittit5 profile is located on a hill, 50 m higher than the spring. If this fault was draining, water should emerge at the intersection between topography and the fault plane on the hill slope, but no source rises at this point. It must be assumed that these sub-vertical faults are not draining features.

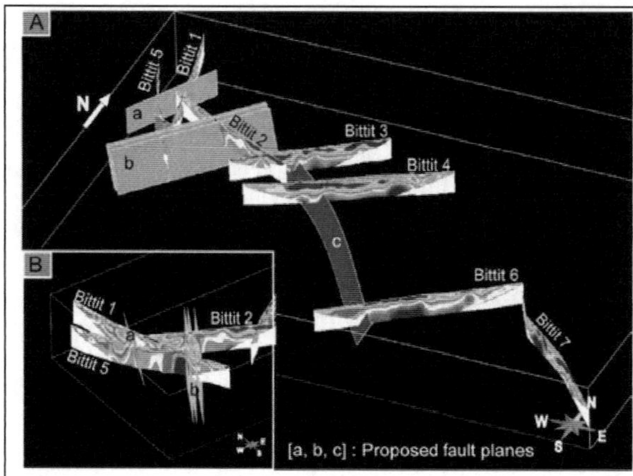

Fig. 4: Gocad 3D visualization of ERT profiles and three proposed fault surfaces (a,b,c).

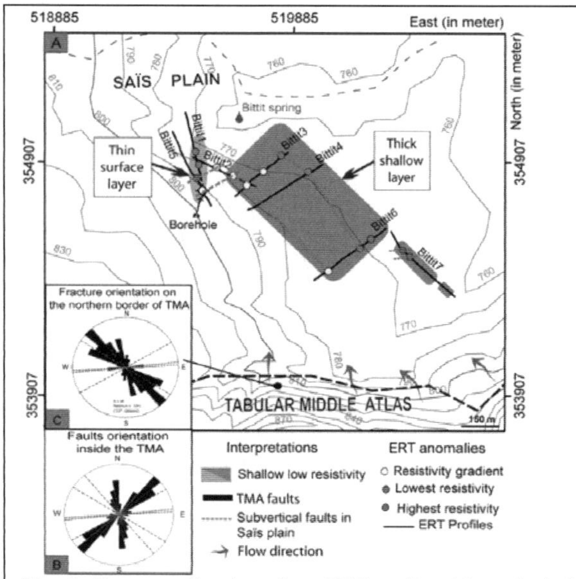

Fig. 5: Fractures directions from ERT profiles (a), geological observations (b and c) and possible flow.

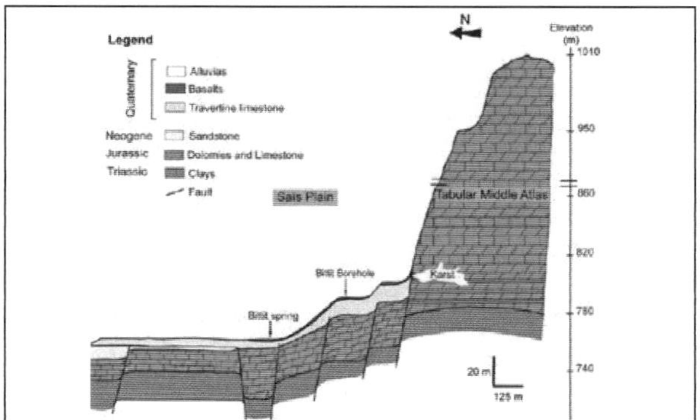

Fig. 6: Interpretative schematic geological cross section (AB in Fig.1)

Figure 6 is a synthetic interpretative section using ERT profiles and some well data, showing an obvious graben-like structure. By contrast, the dislocation of the deeper Liassic blocks should prevent water to flow continuously.

Conclusion

ERT appears to be an appropriate geophysical method in this issue, especially by improving understanding of fracture geometry and reservoir geometries.

Additional investigations using electrical resistivity profiles, and electromagnetic VLF were also performed in order to discriminate between dry and favourable hydrological zones.

The inversion of these pseudosections was performed using Res2DInv software, adding a Vertical to Horizontal flatness filter to discriminate sub-vertical resistivity anomalies that could be associated to geological or hydrological features.

Two sub-vertical fault families with NE-SW and NW-SE trends respectively were identified and correspond to the known geological evolution of the region. This fracture network highlighted dislocation of the blocks of underground substratum. It was shown that the NE-SW faults cannot be related to the mean spring and thus do not constitute draining features. Observations lead to conclude that the main water paths from the TMA to the Bittit spring are located inside a shallow horizontal layer of porous travertine. The preferential flow is meanly on the SE-NW direction.

A spontaneous potential (SP) survey is planned to confirm again this important observation.

Acknowledgments
This work has been performed in the framework of the French-Moroccan scientific projects Volubilis MA/07/169, IRD-CORUS2-Project and the Programme d'Appui à la recherche de l'Université My Ismail (Meknes).
We are very grateful also to Drs G. Saracco, D. Hermitte from CEREGE (France), and to M. Boujamaoui from FST Errachidia for help and fruitful discussions.

REFERENCES

Amraoui F., 2005, Contribution à la connaissance des aquifers karstiques. Cas du Lias de la plaine du Sais et du Causse moyen atlasique tabulaire (Maroc). Doctorat d'Etat ès Sciences Thesis, University of Casablanca, Morocco.

Loke M.H. and Baker R.D., 1995, Least-squares deconvolution of apparent resistivity pseudosections: GEOPHYSICS, **60**, 1682–1690.

Martin J., 1981, Le Moyen Atlas Central: Etude géomorphologique. Notes & Mém. Sér. Geol. Maroc, 258, 445p.

Hydrogeophysical study for the groundwater flooding in Al-Borgeig area, Northern Sudan.

(1) *Mohamed-Ali, M. A.**, (1) *Hassan, E. A* (1) *Ali, A. I.*

(1) *Faculty of Earth Sciences and Mining – University of Dongola, Wadi Halfa, Sudan.*

Summary:

Some people lost their houses because of collapse due to groundwater flooding in Al-Borgeig area in Northern Sudan. The objective of the current study was to evaluate if and how the decreased groundwater pumping and the increased direct use of water from the Nile increase the groundwater flooding, control its patterns and groundwater flow directions. The groundwater level and the estimated groundwater flow directions in the surrounding Quaternary aquifer were determined with electrical resistivity tomography surveying and water level data from monitoring wells. ERT results indicate in some local areas where groundwater flooding negative impacts severe are associated with irrigation in the nearby topographical high farms of sandy deposits. The high transitivity and high permeability of the underneath sandy layers and mud clay soils respectively play vital role in the degree of the groundwater flooding.

Key words: *Al Borgeig, Groundwater, flooding, ERT*

Introduction:

Al-Borgeig area is located in northern Sudan (Figure 1), in a generally low-lying palaeo-drainage depression, much undulated, filled with wind-blown sandy layers as part of Es-Silem basin with thickness of the sediments and ground water diminishing northwards (Figure 2). The bottom sand layers are covered by sand deposits overlain partly by cotton fertile black soil. 'Groundwater flooding' is the result of a rise, either natural or man-made, in the water table (Vainu et al., 2014). The water table is the level of water within rock or soil that makes up the land surface. When the water table reaches ground level, water starts to emerge on the surface and flooding occurs (Pate 1 & Plate 2). Since establishing Al-Borgeig Agricultural Scheme in 1946 and afterwards groundwater level drop and rise are regular due to the balanced rate of discharge and recharge. In the 2010s, the abstraction rate decreased and brought about a partial water-level recovery. Periods of abnormal increase of irrigation from the River Nile resulted in groundwater flooding and resulted the emergence of groundwater at the ground surface, causing damage to property and infrastructure (Plate 2).

* **Corresponding to:** *Dr. Mohamed Mohamed-Ali, Faculty of Earth Sciences and Mining – University of Dongola, Wadi Halfa, Sudan. Email: mohdabdwhab@yahoo.com*

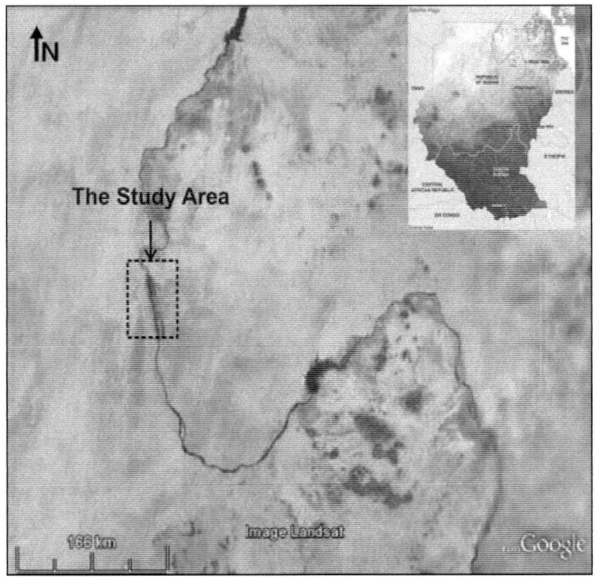

Figure 1: Location map of Al-Borgeig area, northern Sudan.

The objective of the current study was to evaluate how the increased groundwater pumping affected the flooding patterns and groundwater flow directions. The groundwater level and estimated groundwater flow directions in the Quaternary aquifer were determined with electrical resistivity tomography surveying and water level data from the productive and monitoring wells (Plate 2). The previous studies of Adminstration of Groundwater and valleys (Dongola Office) and Dam Authority Unit were very good source of materials to have a general understanding for the problem.

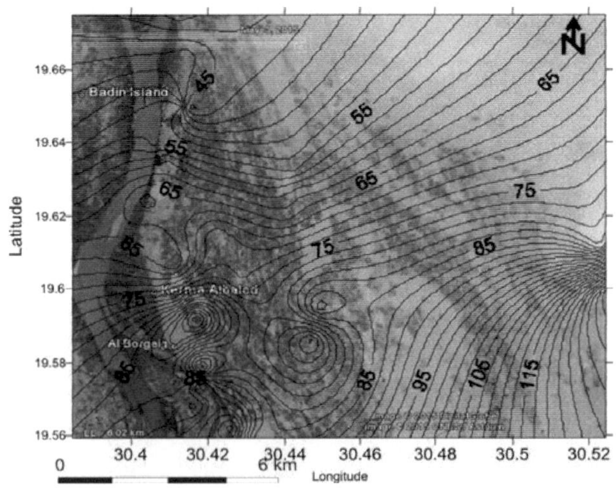

Figure 2: Groundwater contour map of Al-Borgeig Area

Plate 1: Flooding of groundwater in the study area

Figure 3: Location of ERT06 and ERT07, topographically high farm area and low lying settlement area

Theory and Method:

A more accurate model of the subsurface can be produced using a two-dimensional (2-D) electrical resistivity tomography (Figure 4), where the resistivity changes in the vertical direction, as well as in the horizontal direction along the survey line (Loke, 2000). The inversion process applied Res2Dinv code by which forward modeling subroutine was used to calculate the apparent resistivity values, the standard least-squares optimization technique was used for the inversion routine (Loke et al., 2003).

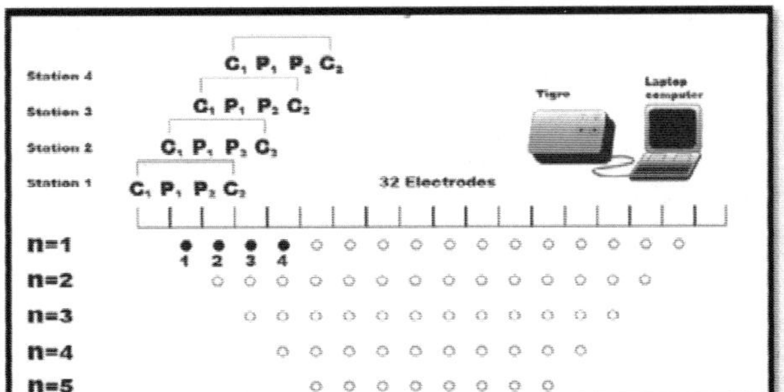

Figure 4: Sketch for layout and the computerized data acquisition of ERT survey.

Plate 2: Groundwater flooding in hand-dug well, most of such wells are not working now.

Plate 3: ERT data acquisition in the area of ERT06, which much affected by flooding.

Metwaly et al. (2012) traced the subsurface oil pollution leakage using 2D electrical resistivity tomography (ERT).

ERT Data & Results:

ERT survey was acquired in the current study using a multi-electrode resistivity imaging system (Figure 4 & Plate 3). The ERT profiles were distributed to assess the degree of the groundwater flooding and its impacts on buildings and infrastructures; the profiles were carried out using Tigre32, Allied, UK with Wenner array and electrode spacing of 5m. The following text will present and discuss some selected examples of ERT lines. ERT data was inverted using Res2Dinv The inversion plan included, data sorting, data filtering, smooth inversion, logarithm of apparent resistivity was chosen, method of least-squars equation, model discretization was used with model refinement of half the unit spacing for better resolution of near-surface variation of resistivity. The presented figures include an upper section showing the measured geoelectric pseudosection and the lower one act for the inverse model produced by inversion.

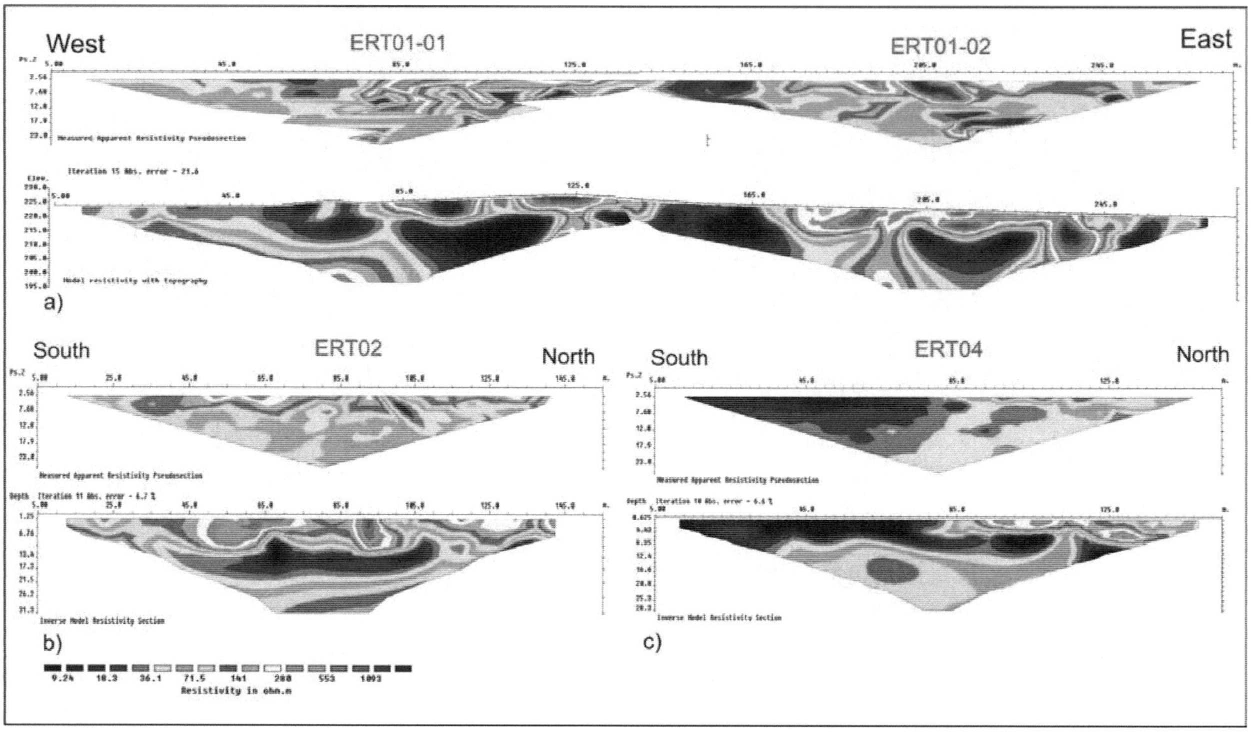

Figure 5: ERT results showing different degrees of flooding, a) ERT01 in an elevated area close to the Nile, b) ERT02 in low lying area suffering flooding, c) ERT03 in farm area much affected (to the degree of collapse). Wenner array was used, spacing, 5m

It was interpreted that a resistive (90-500 Ω-m) surface layer corresponding to the streambed sediments. A relatively conductive (10-30 Ω-m) middle layer corresponding to mud-clay soils. However, it is common for the range of resistivity 50-80 Ω-m to be linked to sandy deposits with contents of water of different salinity. ERT01 and ERT02 were carried out closed together, the resistivity results confirm the actual situation in the site that the area of ERT01 (Fig. 5a) is less affected by flooding than area of ERT02 (Fig. 5b), because ERT01 was located in relatively topographical high area (227), whereas ERT02 with elevation of 222 m. ERT03 (Fig. 5c) coincides with actuality that the area is suffering much groundwater flooding cause collapse of houses in El Dafofa G. Meskit area was studied in details, ERT06 is a long profile composed of three subsections with length 155m for each (Figure 6).

It shows low resistivity range, which indicates the spread of the flooding in this area (Figure 7), ERT07 measured in N-S trend almost perpendicular to ERT06. The measured and inverse sections of ERT07 show (Figure 8) the same situation of low resistivity layers, and give another indicator for severe groundwater flooding in Meskit area. To resolve the resistivity anomalies more clearly a higher resolution line (ERT08) with spacing of 2.5 m was carried out (Figure 6, Plate 3).

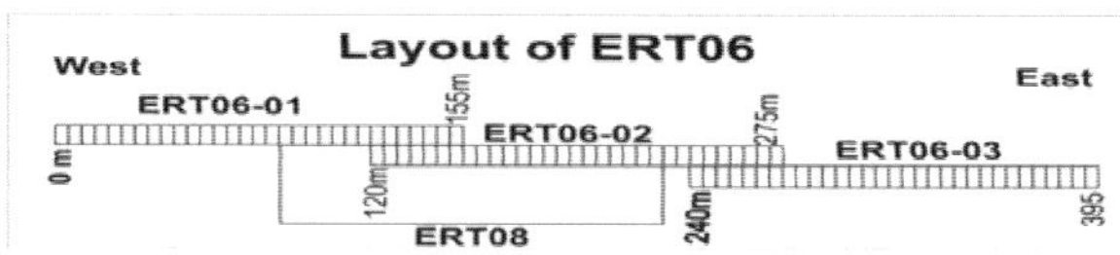

Figure 6: Layout of ERT06 survey showing of three subsections of 5m electrode spacing, ERT08 is 2.5m spacing.

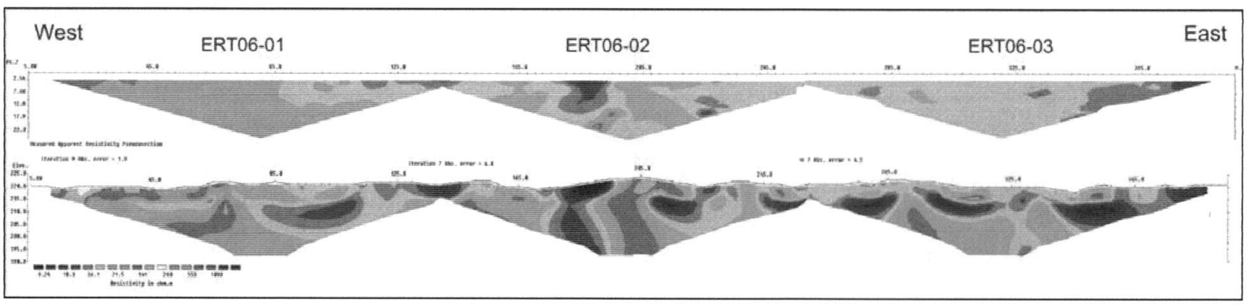

Figure 7: ERT06 results showing different degrees of floodings in the farm area in Figure 2.

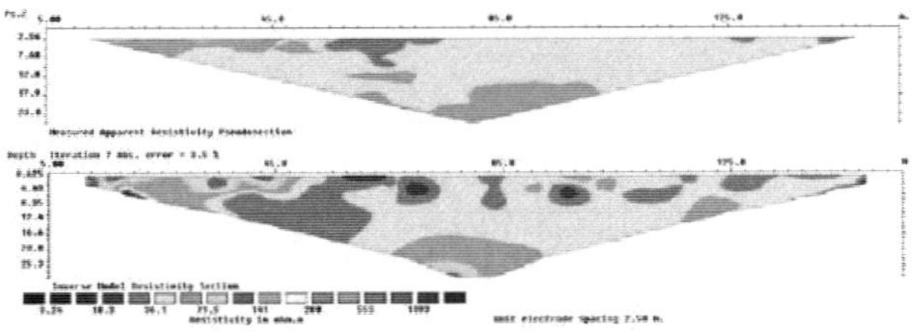

Figure 8: ERT07 results showing different degrees of groundwater flooding in the farm area in plate 3.

The inversion results of ERT08 are shown in Figure 9, these sections show generally topographically controlled higher and lower resistivity anomalies (Figure 9). The blue color zone associated with black-cotton damp soil, the underneath green-brown zone is sand deposit rich with layer is the layer extend laterally and accordingly help groundwater flooding.

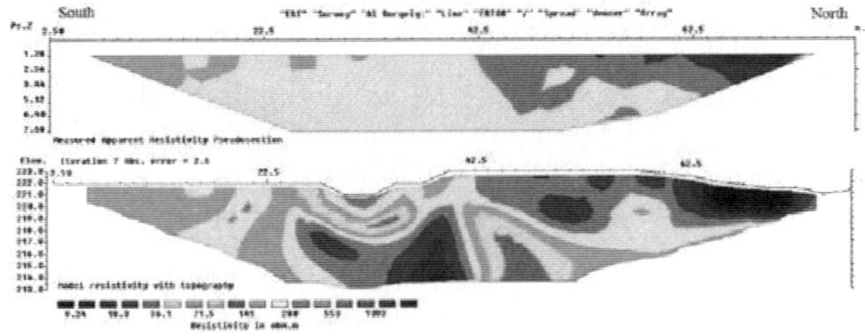

Figure 9: ERT08 results showing different degrees of groundwater flooding in the farm area in Figure 3.

Conclusions:

ERT results show that most of the dark-blue layers and anomalies are associated with groundwater flooding negative impacts. Ground-water floods occurred where the resistivity data showed that the residual clays were thinnest and bedrock was closest to the streambed and sandy layers provide good transitivity for groundwater to seep and percolate to permeable soil, which in a widely spread favor water flooding process. The places where these layers spatially correlated with sandy deposit confirm that the flooding patterns were following the estimated direction of groundwater flow. A severe flooding are associated with irrigation in the nearby topographical high farms of sandy deposits.

It is recommended that groundwater management for balancing the outtake, intake and infiltration is highly required to reduce the negative impacts of the flooding on the settlement and crops.

It is recommended to further study the problem in details to conclude final points for the solution.

Acknowledgements:

The current study was carried out in framework of the projects of the final year of undergraduates of Department of Hydrogeology, Faculty of Earth Sciences and Mining – University of Dongola, Sudan. It is our sincere gratitude to the Government of Al-Borgeig Locality for sponsoring the study, thanks to the Adminstration of Groundwater and valleys (Dongola Office, Eng. Abdalla Omer). Thanks to Mustafa Arman (Al-Borgeig) for valuable helps.

Reference:

Loke, M., 2000, Electrical imaging surveys for environmental and engineering studies. A practical guide to 2-D and 3-D surveys.

Loke, M., Acworth, I., & Dahkin, T., 2003. A comparison of the smooth and blocky inversion methods in 2d electrical imaging surveys. Exploration Geophysics **34**: 182-187.

Salinger Peters Limited, 2014, Groundwater flooding A guide to protecting your home from flooding caused by groundwater, Envirnoment Agency.

Metwaly M., Khalil M., Al-Sayed, E. & El-Kenawy A., 2012, Tracing subsurface oil pollution leakage using 2D electrical resistivity tomography. Arab J Geosci, DOI 10.1007/s12517-012-0600-z.

Vainu M., Terasmaa J., Vaasma T. & Vandel E., 2014, Groundwater Flooding Patterns in a Closed-Basin Lake Before and After an Increase in Groundwater Pumping Rates from an Unconfined Aquifer in The Kurtna Kame Field, Estonia, Proceedings: 2nd International Conference - Water resources and wetlands. Tulcea (Romania), Petre Bretcan; ISSN: 2285-7923; Pages: 189-196.

Imaging the saline/fresh water interface at Bafra, Samsum, Turkey using joint inversion of seismic refraction and direct current electricity data

Hamdan Hamdan[*1], Emin Candansayar[2], Ismail Demirci[2], Nikos Economou[3], Nikos Andronikidis[3], Hakan Arslan[4], Pantelis Soupios[5] and Antonis Vafidis[3]*
[1] Arab Center for Engineering Studies (ACES), Dubai, UAE.
[2] Ankara University, Faculty of Engineering Department of Geophysical Engineering, 06100 Tandoğan, Ankara, Turkey
[3]Laboratory of Applied Geophysics, Mineral Resources Engineering School, Technical University of Crete, Chania, Greece.
[4]Department of Agricultural Structures and Irrigation, Faculty of Agriculture, Ondokuz Mayis Universit, 55139 Samsun, Turkey
[5]Technological Educational Institute of Crete, Department of Environmental and Natural Resources Engineering, Lab. of Geophysics & Seismology, Lab. of GeoInformatics, Chania, Greece.

Summary

A geophysical survey was conducted at Bafra, Samsun, Turkey, in order to support the geological and hydrogeological studies of the existing phenomenon of seawater intrusion. This survey employed seismic refraction tomography (ST), time domain electromagnetic (TDEM) and Electrical Resistivity Tomography (ERT). The objective of this survey was to image the saltwater intrusion at this area and map possible favorable sea water pathways. We present the interpretation of geophysical data obtained under the research project ClearWater project. We applied both individual and joint inversion of electric and seismic data from Bafra. These geophysical methods proved useful as prediction and prevention tools in qualitative characterization of water bearing formations for an optimum groundwater management plan for the Bafra basin.

Introduction

A geophysical survey was conducted at Bafra Plain, Turkey in order to support the geological and hydrogeological studies of the existing phenomenon of seawater intrusion. This survey employed seismic refraction tomography (ST), time domain electromagnetic (TDEM) and geoelectrical methods. The later, consisted of Electrical Resistivity Tomography (ERT). The objective of this survey was to image the saltwater intrusion at this area and map possible favorable sea water pathways. Geophysical data processing consisted of separate processing for each geophysical method, as well as joint inversion of ERT and ST data using the cross-gradients joint inversion algorithm (Hamdan and Vafidis, 2013; Gallardo & Meju, 2004). A calibration with nearby boreholes data and geological mapping data was utilized for the final interpretation results.

The Site

Bafra is a delta plain, which is located on the middle part of the Black Sea coast in Turkey (Figure 1). The plain is within the Samsun province boundaries. Kızılırmak River passes through the Plain in N-S direction. Bafra is the one of two major delta plains near the Black Sea shore. Previous studies have shown that high chlorine concentration is present at boreholes near the seashore. These boreholes clearly show that ground water is affected from the seawater. Thus, Bafra Plain is a suitable site for seawater intrusion studies. Bafra plain consists of slightly sloped alluvium formations. Bafra delta plain is created by the river Kızılırmak. The alluvium around the flooding area consists of sandy and silty gravels. The origin of the gravel basically comes from volcanic rocks and limestones. The alluvium field out of the flooding area is characterized by silty-clay, sandy-clay and clay minerals. Terrace deposits could be partially observed at the East and West side of the River Kızılırmak in the form of firm silty-sandy gravel levels (Arpacı and Yüksel, 1996). Elevation of the plain gradually increases to the South starting from the wide and flat alluvium area. The south side of the delta consists of upper Cretaceous flysch, Eocene flysch and volcanics, Miocene and Pliocene sediments (DSI 1985).

Geologically, Delta Kızılırmak is a young area approximately 2000 years old. There are Neogene sedimentary rocks (marl, clay stone, silt stone and gravel) and Eocene aged flysch units (sand stone, clay stone, marl, and partially limestone) at the hillside area of the plain. Mesozoic formations are located at the east side of the area and in the form of thin and medium layered, grey colored sand stone, clay stone, marl alterations. Eocene formations consist of andesite and basalt conglomerates and dykes inside the volcanic breach and tuffs.

Previous geophysical studies

A geophysical investigation during the DSI project (DSI, 1985) was carried out by employing vertical electrical soundings (VES). Existing geophysical data consist of 71 VES measurements, which were organized in sections indicating three geoelectrical layers. A digital elevation model was produced from a contour elevation map for Bafra (DSI, 1985). In addition, geological data comprise digitized formation boundaries and faults as well as borehole information.

Geophysical survey

The geophysical survey at Bafra, involved electrical methods (ERT), electromagnetic methods and the seismic refraction method (Figure 1). Electrical Resistivity Tomography (ERT) was carried out at the north-eastern part of the Bafra plain, in order to map the saline water intrusion (Figure 2). Seismic Tomography (ST) was conducted at two of the ERT study lines in order to jointly invert the geophysical data. Time Domain Electromagnetic (TDEM) measurements were also acquired.

At the study area, twenty four (24) Electrical Resistivity Tomography (ERT) lines covered 6600 m. The line length was 275 m, and the electrode spacing was set to 5 m (56 electrodes were used). Sting R8/IP AGI system collected apparent resistivity data using both Dipole-dipole (DD) and Inverse Wenner-Shclumberger (IWS) electrode arrays at selected lines (ERT P1-ERT P14), while the rest of the lines (ERT P15 – ERT P24) were surveyed only with the IWS method. The depth levels were 16 and 20, corresponding to maximum depth 55 and 65 for the IWS and DD methods respectively. The DD array enhances lateral variation of the electrical resistivity, while the IWS array is more suitable to describe the vertical variation of the electrical resistivity. ERT resistivity sections indicate that the whole northern area of Bafra is under sea water intrusion (Figure 2). The depth of the water table is almost 3 m. The geoelectrical sections detect the interface between fresh and sea water at almost 10-15 m depth (a change in electrical resistivity from >7 Ohm.m to <5Ohm.m).

Additionally, 3 Seismic Tomography (STP) lines were surveyed whose total length is 370m. Each line consisted of two spreads with a 5 m interval between the spreads and coincides with specific ERT lines. Each line had a 2.5m geophone interval, and 18 shots. A twenty four channel Geometrics StrataView was used together with 14 Hz geophones (Figure 3).

The geoelectrical section ERT P1 depicts three geoelectrical layers (Figures 2 and 3). The intermediate geoelectrical layer exhibits very low resistivity (< 4 Ohm.m). The first layer lacks lateral homogeneity. Both the first and the third geoelectrical layers are characterized by higher resistivity values (>7 Ohm.m). The thickness of the first layer is 15m, while the thickness of the second layer is less than 10 m.

The far shot of seismic line STP1 (Figure 1) is located at the 50th m of line ERT P2 (11th electrode). The geophone spacing was 2.5 m and the number of shots per spread 18. This line consists of two spreads and 24 geophones per spread, while in between the spreads there was a 5 m distance. The seismic velocity model depicts two seismic layers (Figure 3a). The first is a low velocity layer (<500 m/s), while the second is characterized by a higher seismic velocity (>1200 m/s). The thickness of the first seismic layer does not exceed 3 m. Part of ERT P2 is depicted in Figure 3c. We applied both individual and joint inversion of electric and seismic data from Bafra (Hamdan and Vafidis, 2013; Gallardo & Meju, 2004). Figures 3a and 3c are the individual inversion results. Joint inversion increased the lateral resolution of the seismic model (Figures 3a and b), while no significant differences occur between individual and joint inversion regarding the geoelectrical sections (Figures 3c and d).

Conclusion

At Bafra, geophysical methods imaged the saltwater intrusion which is present at the north west of the basin. These geophysical methods are useful as prediction and prevention tools in qualitative characterization of water bearing formations for an optimum groundwater management plan for the Bafra basin.

Acknowledgements

This research has been financed by Greek national funds through the Operational Program "Education and Lifelong Learning" of the National Strategic Reference Framework (NSRF) - Research Funding Program: BS-ERA.NET. Networking on Science and Technology in the Black Sea Region within the framework of the BLACK SEA ERA.NET research project ClearWater.

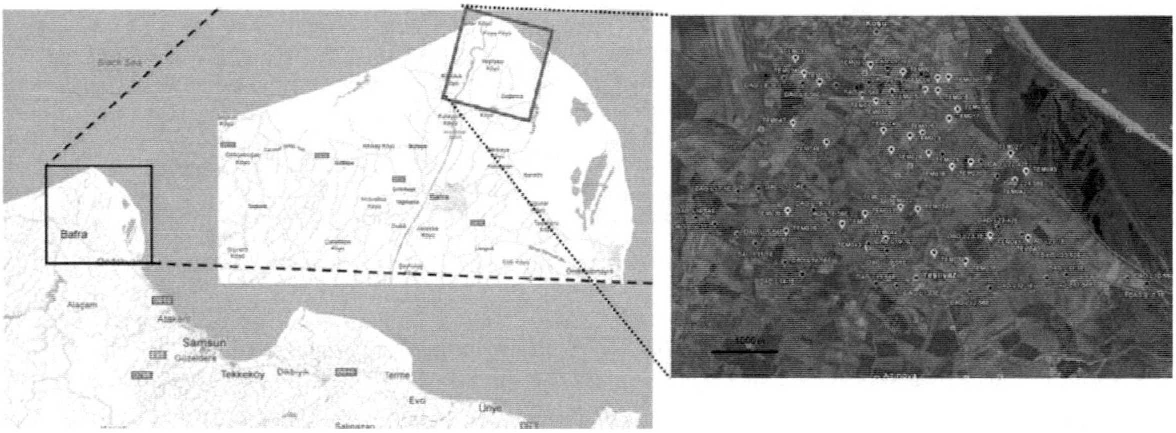

Figure 1: Location map of Bafra plain. The investigated area is shown in red rectangle (left).Study lines and stations of the geophysical survey at Bafra. ERT lines are shown in red color, ST lines, in blue, and TDEM soundings, in yellow.

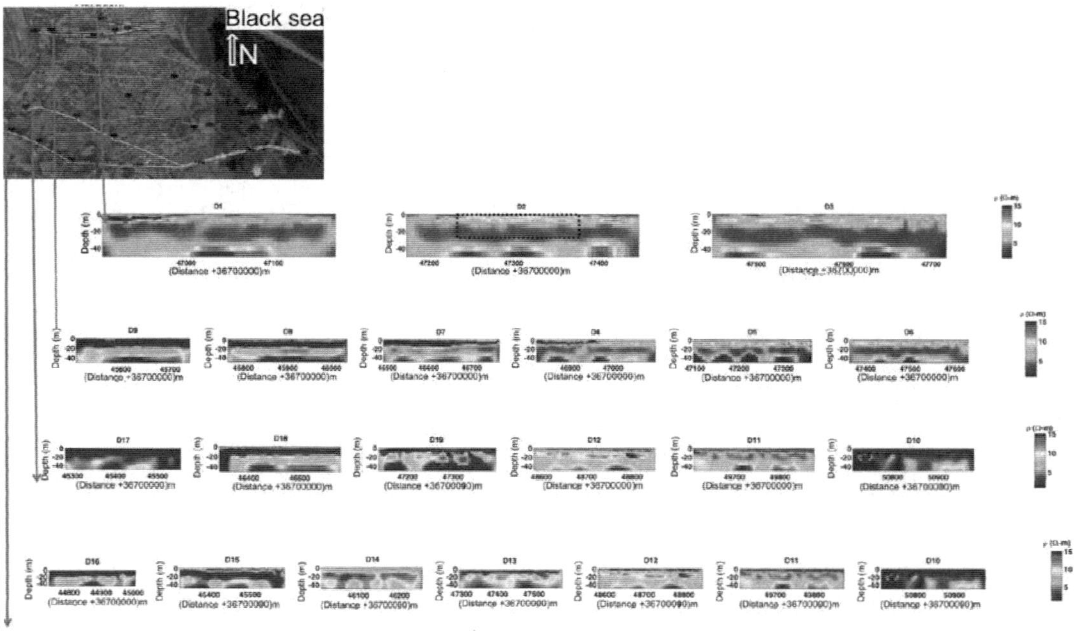

Figure 2 Google map showing the study lines at Bafra and the corresponding ERT sections indicating the northern and eastern increased salt water intrusion. Dotted line indicates the extend of sections in Figure 3.

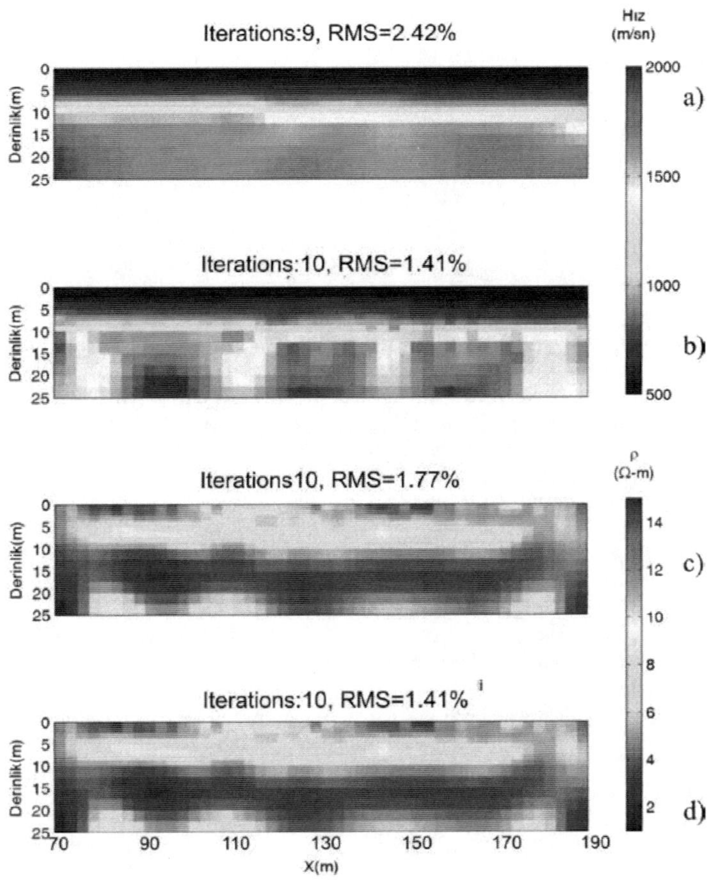

Figure 1: Inversion for line P2, Bafra site : Seismic velocity models (a) from individual inversion and (b) from joint inversion. ERT resistivity models: (c) from individual inversion and (d) from joint inversion. The direction of the study line is WNW-ESE.

References

Arpacı K., and Yüksel M., 1996, Bafra Ovası Sol Sahili Topraklarının Sınıflandırılması: Tarim Bilimleri Dergisi, **2**, 87-93.

Gallardo L.A. and Meju M.A., 2004, Joint two-dimensional DC resistivity and seismic travel time inversion with cross-gradients constrains: Journal of geophysical research, **109**, B03311.

DSI, 1985, Geophysical Report on Bafra plain (in Turskish): Halil Bakan, 1985, Samsun-Bafra Ovasi Detayli Jeofisik Etut Rapopu, Enerji ve tabii kaynaklar bakanligi, Devlet su isleri genel mudurlugu, Samsun..

Hamdan, H., and Vafidis, A., 2013, Joint inversion of 2D resistivity and seismic travel time data to image saltwater intrusion over karstic areas, Environmental Earth Sciences, **68, 7**, 1877-1885.

Utilization of isotopes in hydrogeological studies in UAE: A review

Parimalarenganayaki. S [a], Mohamed Mostafa Mohamed [b], Ahmed A. Murad [c]*

[a,b]*Department of Civil and Environmental Engineering, UAE University, P.O. Box: 15551, Al-Ain, United Arab Emirates*
[c]*Department of Geology, College of Science, UAE University, P.O. Box: 17551, Al-Ain, United Arab Emirates*
email: pari@uaeu.ac.ae; m.mohamed@uaeu.ac.ae; ahmed.murad@uaeu.ac.ae

Abstract

The United Arab Emirates (UAE) is located in the southeastern corner of the Arabian Peninsula. The water demand for domestic, industrial and agricultural activities are met by conventional (rainfall recharge, falajes, springs and flash floods) and non-conventional water resources (desalinated water and treated waste water). The limited availability of fresh water resources and increasing population in the UAE creates more demand on the fresh water resources. Over usage of freshwater resources creates hydrological stress on the aquifer system. This situation alerts the hydrogeologists to plan for better freshwater management systems in UAE. In order to adopt the best water management plan, it is necessary to understand the rainfall pattern, sources of groundwater recharge, movement of groundwater, evaporation process and changes in water quality which are taking place in the hydrological system by means of suitable technology. During past three decades many studies had been carried out using environmental isotopes to understand the hydrological problems associated with arid climate in UAE. The aim of the present study is to review the utilization of isotopes in hydrogeological studies in UAE. This study will help decision makers to understand the past and present conditions of freshwater systems in UAE.

Introduction

Stable and radioactive isotopes are commonly used environmental isotopes for the water resources research. Application of isotope hydrology includes assessment of evaporation rates, hydrograph separation, recharge rate estimation, identification of recharge sources, travel time of water, interaction between surface and groundwater, dating groundwater, groundwater salinization and groundwater pollution (Clark and Fritz, 1997). Review of the use of isotopes in groundwater studies in UAE is discussed in the following section.

Shortages of water resources threaten the sustainability development and classified the UAEas one of the poorest countries in the world in availability of water resources. The arid climate of the study area accompanied with a low amount of recharge to major aquifers is the main challenge to water resources in the region (Murad 2010). UAE is located in arid zone with low amount of rainfall and high rate of evaporation. Based the rainfall data from 2000 to 2006, the highest and lowest rainfall of this region varied from 2.8 mm and 168.2 mm respectively (Metrology Department 2007). The rate of evaporation always exceeds the amount of rainfall which may go up to 2000 mm/year (Jones and Marrei 1982). The water demand for domestic, industrial and agricultural activities are met by conventional (rainfall recharge, falajes, springs and flash floods) and non-conventional water resources (desalinated water and treated waste water). The limited availability of fresh water resources and increasing population in the UAE creates more demand on the fresh water resources. Over usage of fresh water resources in UAE created hydrogeological stress on the fresh water system which results into decline in water level and deterioration of water quality. This situation alerts the hydrologists for proper management of fresh water resources to augment groundwater storage and to protect surface water resources of UAE. In order to adopt the best water management plan, it is necessary to understand the rainfall pattern, sources of groundwater recharge, movement of groundwater, evaporation process and changes in water quality which are taking place in the hydrological system by means of suitable technology. During the past three decades, many researches had been carried out using environmental isotopes to investigate the hydrological problems associated with the arid region like UAE. Thorough review of these researches will help to understand the condition of freshwater resources in various geological conditions of UAE for better water management. Hence, this study was carried out to review the research work carried out using isotopes in the UAE region.

Application of isotope hydrology in UAE

A study on factors controlling groundwater quality in Eastern UAE was carried out by Murad and Krishnamurthy (2004) in two aquifers which are Gravel plain aquifer and inland Ophiolite aquifer. Isotopic and chemical approach was used to identify the possible sources of salinity in the groundwater of the region. The plot between δD–$\delta^{18}O$ values of the groundwater samples and Gulf of Oman samples were made and compared with the local meteoric water line (LMWL). The slope of the δD–$\delta^{18}O$ indicates that the precipitation evaporates even before the recharge process takes place and hence evaporation was not the reason for salinity. This plot also suggested that the narrow distribution of $\delta^{18}O$ values and plot of isotopes of groundwater samples locations away from the value of Gulf of Oman samples also suggested that seawater intrusion is not the reason for the salinity in this region. Therefore, high salinity in this region was attributed to sources such as fertilizers, pesticides and release of ions such as sodium, calcium from the host rocks to the groundwater system which was identified based on the chemical analysis. This study focused on the use of stable isotopes in identifying the source of salinity in the groundwater of the region.

Changes in groundwater from recharge zones into the carbonate aquifer in an arid environment located southeast of Al Ain were assessed by Murad and Mirghni (2012). Plot on isotopic variations of oxygen and hydrogen of groundwater samples in the study area of different zones are plotted to the right of the LMWL indicated that the recharged groundwater undergoes the process of evaporation after infiltration. The degree of evaporation is not uniform for all the zones, it varies based on the lithology and interactions with the other water bodies. Positive values of *d- excess* from the groundwater samples of Mubazarah (Zone 1) indicates that these water undergoes less degree of evaporation. Whereas, the negative *d-excess* from the groundwater samples of Neima (Zone 2) and Ain BuSukhanah (Zone 3) indicates that these water may undergo higher degree of evaporation. This study helped to identify the degree of evaporation which took place after the groundwater recharge was taken place.

Factors controlling stable oxygen, hydrogen and carbon isotope ratios in regional groundwater of the eastern (UAE) were carried out by Murad and Krishnamurthy (2008) in two aquifers which are Gravel plain aquifer and inland Ophiolite aquifer. Plot on isotopic variations of oxygen and hydrogen of groundwater samples in the study area of different zones are plotted above and below the LMWL. Slopes of both groups (plotted above and below the LMWL) are similar and major difference between two groups is the y-intercept, which shows difference in evaporation process. Groundwater samples above the LMWL were affected by primary evaporation prior to infiltration and samples plotted below LMWL were influenced by evaporation of return flow and evapotranspiration in the unsaturated zone resulting from intense agricultural activities. The evaporative loss of groundwater samples above and below LMWL is 21 and 19% respectively. Variation in concentration of $\delta^{13}C$ and dissolved inorganic carbon (DIC) in groundwater samples from these two aquifers are due to the presence of C-3 and C-4 plants present in the aquifer. This study was useful to identify the percentage of evaporation taking place in different hydrogeological conditions using isotopes and the factors controlling presence of DIC and carbonate in the aquifer system.

Murad et al. (2011) carried out study on Isotope investigations of a carbonate aquifer of the Wadi Al Bih area in the northern part of the UAE. The aquifer in this area is mainly composed of fractured limestone and dolomite. The regression line of the plot between the stable isotopes of groundwater samples was compared with the LMWL, Wadi Al Bih local meteoric water line (WLMWL), Mediterranean Local Meteoric Water Line (MLMWL) and the Arabian Sea Meteoric Water Line (ASMWL). As the LMWL and WLMWL were below the MLMWL, the source for the precipitation for this region was identified as Mediterranean Sea. The *d-excess* of local precipitation of the Wadi Al Bih area decreased to +8.9%, indicating that precipitation in the study area has been affected by raindrop evaporation. This isotopic study helped to identify the source of precipitation and factors affecting the precipitation of this region.

Evolution of isotopic compositions in groundwater of the area between the Gulf of Oman and the Arabian Gulf was studied by Murad (2010). The stable isotopes of oxygen and hydrogen have been implemented to assess the recharge mechanisms in an area. The plot of isotopic compositions of groundwater in the study area are generally above the local meteoric water line (LMWL) of UAE and scatter around the Mediterranean meteoric water line (MMWL). This distribution pattern suggests that the source of precipitation in this region is the Mediterranean Sea. The similarity between the slopes of two regression lines for 1996 and 2006 confirmed the fact that the recharged water diluted the groundwater at each event of precipitation, suggesting modern-day recharge. It is noticed that the values of deuterium excess for collected groundwater samples are higher than those of LMWL (15‰), and also these values are close to 20‰ which is the *d-excess* of MMWL. This suggests that the Mediterranean Sea is the main source of moisture for the region. Stable isotopes and *d-excess* helped to find out the source of precipitation and recharge mechanism in the region.

Study on isotopic evaluation of groundwater in Jabal Hafit area, Eastern Abu Dhabi Emirate, UAE was carried out by Murad et al. (2012). The regression equation of δD and $\delta^{18}O$ for groundwater samples were compared with the LMWL of UAE. The lower slope value and intercept values for groundwater line are significantly different from those for LMWL suggested that the groundwater has gone through evaporation process before or during recharge. The absence of correlation between Cl- and δD indicates that there is more than one source of Cl- in groundwater of the study area. This study intended to identify the various sources of water for groundwater recharge and Cl concentration using isotopic signatures. Further research on this study is required to particularly name the source of groundwater recharge for this region.

A detailed study on water resources in the UAE was carried out by Rizk and Alsharhan (2003). This study discussed the source of groundwater recharge and changes in groundwater for two regions of UAE using isotopic techniques. Most of the groundwater samples from Wadi Al Bih aquifer in Ras al Khaimah indicates that most of isotopes of this aquifer plot close to the global meteoric water line with a deuterium excess (*d-excess*) of +20. The groundwater data which plots to the right of LMWL indicates that the recharged groundwater undergoes evaporation and mixing takes place with the different water types or salt water intrusion from Arabian Gulf. The isotopic composition of this region is lighter than the average values of stable isotopic composition of LMWL indicate that the possible source of rainfall in this region is from Mediterranean Sea. The environmental isotopes (^{2}H, ^{3}H and ^{18}O) in Jabal Hafit thermal water indicate that at least part of the water pumped from Jabal Hafit well field comes from recent recharge. Depletion of ^{2}H and ^{18}O relative to standard reflects recharge at high elevation. These isotopes are lighter than other groundwater in the surrounding area, suggesting local recharge at high elevation. The tritium levels, while relatively low, still indicate mixing with recent locally recharge. This study helped to understand the age of groundwater and various sources of mixing which took place in this region.
Tang et al. (2001) carried out a case study in Al Ain, UAE to understand the role of wadis in regional groundwater recharge using environment isotopic methods. Regression line of the groundwater and stream water samples was compared with the meteoric water line of northern Oman. The gradient values of the collected samples are very different than that of the meteoric water. Stable isotopes of groundwaterin the wadi were lighter than that of the stream water in the mountain area, since stream water was affected by evaporation before it infiltrated into the ground. This suggests that the isotopes of recharged water must have isotopically light values. From the above results, groundwater along the wadi was mainly derived not only from water from the mountain area but also from infiltrated water during the flood period. It is understood that groundwater is mixed with the various types of water in this region.

Summary

The review of research work on the use of isotopes to identify the hydrological conditions of the UAE was carried out in the present study. Most of the researchers used environmentally stable oxygen, deuterium, *d-excess* (derived from oxygen and deuterium isotopes) and Tritium as a tracer to understand the hydrological system. Meteoric water lines of UAE, Mediterranean sea and Arabian sea were used to compare the isotopic regression lines of the water samples of the study area. Based on this comparison, the hydrological pattern of the study region is determined. The review shows that, groundwater salinity Gravel plain aquifer and inland Ophiolite aquifer of eastern part of UAE is not due to the seawater intrusion, but it is mainly due to the agricultural activities and release of ions from host rock. The evaporation loss of groundwater samples in Gravel plain aquifer and inland Ophiolite aquifer was identified to be 21% and 19% respectively. Groundwater in the Mubazara zone undergoes less degree of evaporation whereas groundwater in the Neima and Ain Bu Sukhanam undergoes high degree of evaporation. Source of precipitation for the carbonate aquifer of the Wadi Al Bih area in the northern part of the UAE was identified as Mediterranean Sea and this precipitation is affected by raindrop evaporation. Modern day recharge was taking place between Gulf of Oman and Arabian Gulf and again source of precipitation for this region is identified to be Mediterranean Sea. Most of the groundwater samples from Wadi Al Bih aquifer in Ras al Khaimah indicates that the rainfall is not the only source for recharge in this area, and the groundwater is receives recharge from various other sources. This study was helped to understand that isotopic technique alone is not sufficient to understand the hydrological pattern of the region. Isotope techniques combined with other geochemical characteristic of water would give better insight for the situation. These studies revealed only the present status of the water crisis, but it is also necessary to suggest best water management plan for the present situation to preserve aquifer from hydrological stress for the whole part of UAE.

Acknowledgments

The authors would like to thank the National Water Centre (NWC) for funding this research project under the Grant number 31R008.

References

Clark, I. D., and P., Fritz, 1997, Environmental isotopes in hydrogeology, 328 pp., CRC Press/Lewis Publishers, Boca Raton, FL.

Jones, G. P., and S. H., Marrei, 1982, Groundwater resources in the united Arab emirates, Middle East Water Sci., 6, no. 1, 41–45.

Ministry of Economics, 2007, Data sheet for the climate of UAE, Metrology Department, Abu Dhabi, United Arab Emirates.

Murad, A., 2010, Evolution of isotopic compositions in groundwater of the area between the Gulf of Oman and the Arabian Gulf, Chin.J.Geochem., 29,152–156.

Murad, A.A., and R.V., Krishnamurthy, 2008, Factors controlling stable oxygen, hydrogen and carbon isotope ratios in groundwater of the eastern United Arab Emirates (UAE). Hydrol. Process, 22: 1922–1931.

Murad, A.A., and F.M., Mirghni, 2012, Isotopic variations of oxygen and hydrogen in groundwater of carbonate aquifer in an arid environment, Arab J Geosci., 5,1459-1468.

Murad, A.A., Garamoon, H., S., Hussein, Al-Nuaimi, H.S., 2011, Hydrogeochemical characterization and isotope investigations of a carbonate aquifer of the northern part of the United Arab Emirates, Journal of Asian Earth Sciences, 213–225.

Murad, A.A., S.M., Hussein, and Al-Abdouli, A., 2012, Hydrogeochemical and isotopic evaluation of groundwater in Jabal Hafit area, Eastern Abu Dhabi Emirate, United Arab Emirates (UAE). Scientific Research and Essays,7, no. 30, 2731-2743.

Murad, A.A., and R. V., Krishnamurthy, 2004, Factors controlling groundwater quality in Eastern United Arab Emirates: A chemical and isotopic approach. Journal of Hydrology.,286, 227–235.

Rizk., Z.S., and A.S., Alsharhan, 2003, Timing of Recharge, and the Origin, Evolution and Distribution of Solutes in a Hyperarid Aquifer System, *in* Water Resources Perspectives: Evaluation, Management and Policy. Ed. A.S. Alsharhan and W.W. Wood. Elsevier Science Publisher, The Netherlands, 245-264.

Tang, C., I., Machida, Shindo, S., A., Kondoh, and Sakura, Y., 2001, Chemical and isotopic methods for confirming the roles of wadis in regional groundwater recharge in a regional arid environment: A case study in Al Ain, UAE. Hydrol. Process, 15, 2195–2202.

Effect of flood analysis on the foundation investigation using integrated approach

Lukumon Adeoti[1], Shakirudeen Odunuga[2], Rafiu B.Adegbola[3], Sunday Oladele[1], and Saheed Adekunle Raji[2]*
[1]Department of Geosciences, University of Lagos, Akoka, Lagos, Nigeria, [2]Department of Geography, University of Lagos, Akoka, Lagos, Nigeria, [3]Department of Physics, Lagos State University, Ojo, Lagos, Nigeria,

Summary

The effect of flood analysis on the foundation investigation of the proposed Gas-Fired Power Plant in Badagry, Lagos, Nigeria was carried out using geophysical, geotechnical and geographical information system (GIS) methods. The subsurface information was obtained by using two electrode arrays for the resistivity data acquisition. Geotechnical data were obtained through two boreholes via percursion drilling within the study area. The integration of geophysical and geotechnical data reveals that the subsoil materials comprises fibrous peat, organic silty clay, clayey sand, sandy clay and sand. The analysis of the subsoil conditions via geotechnical data shows the feasibility of foundation around 25 m via piling while the safe grade elevation has been worked out to be 6.672 m for 100yrs return period via GIS approach. Thus, in order to avoid adverse effect of flood on the proposed structure, the safe grade elevation value has to be added to the foundation floor.

Introduction

The growing population in Lagos State, Nigeria is on a geometrical rise because it is considered as the economic beehive of the country. This rising population has overstressed the infrastructural facilities which have led to the mass movement of people from the cities to the suburb arising from its development via the provision of infrastructure. The initiative has led to the proposed Gas-Fired Power Plant in Badagry Lagos in order to cope with demand of the people and business environment. The proposed site is a flood plain. More than thirty-five years, the flood's magnitude and frequency are tremendously accelerated particularly triggered by human actions; for example land clearing for urbanisation or farming activities,building of structure such as freeways, roads, and bridges which repeatedly alters the flood behaviour (Rabie et al., 2007;Toriman et al., 2009).

The impacts of floods on any society are primarily dependent on the management of flood reduction. Mitigation is a series of activities and measures that are put in place before the occurrence of any natural hazard event in order to reduce or eliminate the impacts of such hazards on the society and the environment at large (Adiat et al., 2012). The identification of the vulnerable or risk zones will assist in the effective implementation of the other aspects of the preparedness phase which includes early warning, alert system and awareness raising campaign and flood response by emergency and disaster relief (Chapman and Canaan, 2001)

Many studies have been conducted on floor susceptibility mapping using remote sensing data and GIS tools (Adiat et al., 2012). The integration of GIS and 2D imaging was used to investigate the effects of subsurface condition on flood occurrence in Kedah State, Peninsula Malaysia (Adiat et al., 2012).The application of geophysical and geotechnical methods becomes necessary as they are capable of imaging the subsurface targets or providing additional information useful for generating geological models and understanding the engineering properties of the soil in order to ensure stability of structures (Rucker et al ., 2011).

In this study, the geophysical, geotechnical and GIS data were integrated to assess the effect of flood analysis on the foundation investigation.

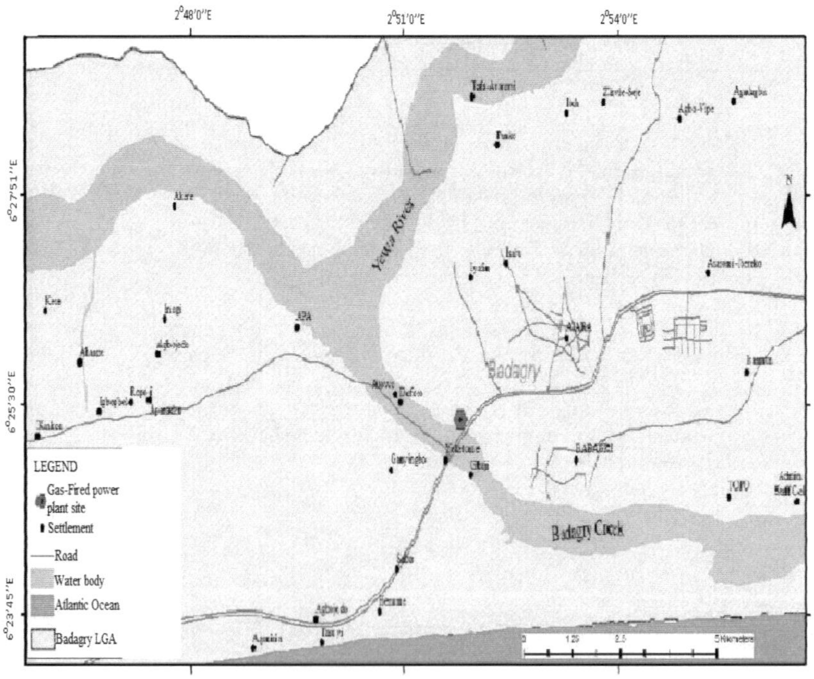

Figure 1. Map of the study area

Site Description & Geology

The site of the proposed project is located near Iyafin village within Badagry town, Badagry local government area of Lagos State. Geographically, it lies approximately within latitudes $6^{0}22'$ N and $6^{0}36'$ N and longitudes 2^{0} 50' E and 2^{0} 54' E. The land area of the proposed project site is about 4.093 Ha (40,930 m^2). It is proximate to the Badagry Creek (Figure1). The proposed project site is located entirely within the floodplain of the Badagry creek. The creek is influenced by tides and floods from the Lagos Lagoon and Cotonou harbour through Lake Nokue and Lake Porto-Novo (Anyanwu and Ezenwa, 1988).The general geology of Lagos and its environs can be described as an integral part of Dahomey basin, which is the eastern part of sedimentary basin that extends from Volta Delta in Ghana to Okitipupa ridge in Nigeria (Jones and Hockey, 1965; Whiteman, 1982). The area investigated is situated in the southern part of Dahomey basin and consist of stratified series of sedimentary sequence made up of silt clay, peat and sand.

Methods

In geophysical survey, twenty (20) Vertical Electrical Sounding (VES) via schlumberger array and three (3) 2D resistivity data using Wenner configuration distributed along three traverses (T1-T3) were acquired by resistivity meter called PASI. The electrode spread (AB) for schlumberger area varied from 2 m to 320 m while the total spread of the 2D array configuration varied from 0 to a maximum of 200 m with the electrode spacing between 10 m and 60 m. The geotechnical data comprised drilling two, shell and Auger boreholes to 30.00 m depth.

Table 1. Data set used

s/n	Data	Data Scale & Resolution	Data Extracted	Data Source
1	Badagry Creek Gauge Height Data (1980-2012)	Point-based	Annual average water level	Ogun Osun River Basin Development Authority (OORBDA)
2	Topographic maps of Badagry Sheet 278A NE 1, 2	1:25,000; graticle interval (7' 30"N by 7' 30"E)	Spot heights and contour	Federal Surveys Department. Data acquired in 1982
3	Shuttle Radar Topographic Mission	30 metre	Digital Elevation Model (DEM)	USGS via EarthExplorer portal. Data acquired in 2014

Boring was executed using a shell and Auger cable percussion boring technique by means of a pilcon wayfarer Rig equipped with the in-situ standard penetration tests (SPT) accessories. During boring, samples were recovered at regular intervals of 0.75m while standard penetration tests (SPT) were carried out at 1.5m intervals in cohesionless strata. The dataset used for determining the safe grade elevation are shown in Table 1. Design flood estimation was based on both probabilistic approach on L-moments methods using rainfall frequency analysis (Gumbel EV-1 distribution) and deterministic approach using unit hydrograph method that incorporated tidal effect and long term ocean surges. The maximum flood level above mean sea level (Safe Grade Elevation) was determined for the various flooding conditions for 2yrs, 5yrs, 25yrs, 50yr, 100yrs and 200yrs return periods. However, the severe flooding condition of 100-year flood with additional freeboard of 1.0 m was considered for this study. Analysis of inundation area for the 100- year return period was conducted using Digital Elevation Model (DEM) derived from USGS via EarthExplorer portal in a GIS environment.

Results & Discussion

Samples of the results of the 2-D inverse resistivity models, geoelectric section and borehole log obtained for the study area are presented in figure 2 (A-C) respectively. The resistivity profile show relevant changes in the vertical direction as well as lateral changes typically occurring along the survey lines. On the pseudo-section, arrows were used to indicate the position of the two VES and a borehole with a view to constraining the 2D inverse resistivity model. The range of resistivity values of 2D structure vary from a low value of 0.45 Ohm-m to a high close to 87 Ohm-m. The low resistivity values are indicative of conductive subsurface soil materials which are representative of peat, organic silty clay, clayey sand, silty sand, and sand. The integrated approach shows that subsoil materials are highly saturated and affirms the susceptibility of the study area to flood. The borehole log reveals the feasibility of the foundation via piling at about 25m.

The statistical distribution of the area and proportional analysis of the spatial extent of inundated and non flooded area within the basin for seven return periods (Table 2) shows flood elevation is not uniform within the study area for flood elevation which depends on surface topography and storm intensities. Hence, a single maximum flood elevation level might not be anticipated in the entire proposed project area. The computed flood frequency elevation by considering the 100-year flood elevation gave 5.672 m. This reveals the maximum flood level during the simulation and computation of the flood level extremes. This was adopted as the maximum expected flood level around within the study area. This now led to the generation of the 2D flood inundation map for modeled return period of 100yrs which reflects areas liable to flood and the non-flooded area as shown in figure 3. Considering this level as base flood elevation or flood submersion line and adopting freeboard of 1.0 m, the safe grade elevation was calculated as 6.672 m.. Furthermore, the maximum flood level is not uniform within the study area, therefore a land development plan may be considered in which the topography of the site is retained while the land is filled to maximum flood level + freeboard. With this elevation (6.672 m), the dimensions of the study area will be at safe grade for developmental purposes.

Conclusion

The low resistivity values are indicative of conductive subsurface soil materials which are representative of peat, organic silty clay, clayey sand, silty sand, and sand. The integrated approach shows that subsoil materials are highly saturated and affirms the susceptibility of the study area to flood. The borehole log reveals the feasibility of the foundation via piling at about 25m. . In consideration of the surface slope and movement of water, with a freeboard of 1.0 m and the lowest flood elevation at par with river surface (at mean sea level), the calculated safe grade level of 6.672 m above mean sea level should be adhered to. Hence, in order to avoid adverse effect of flood on the proposed structure, the safe grade elevation value has to be added to the foundation floor.

Acknowledgement

The authors would like to express their gratitude to World Environmental Systems Nigeria Limited for providing fund for this research work.

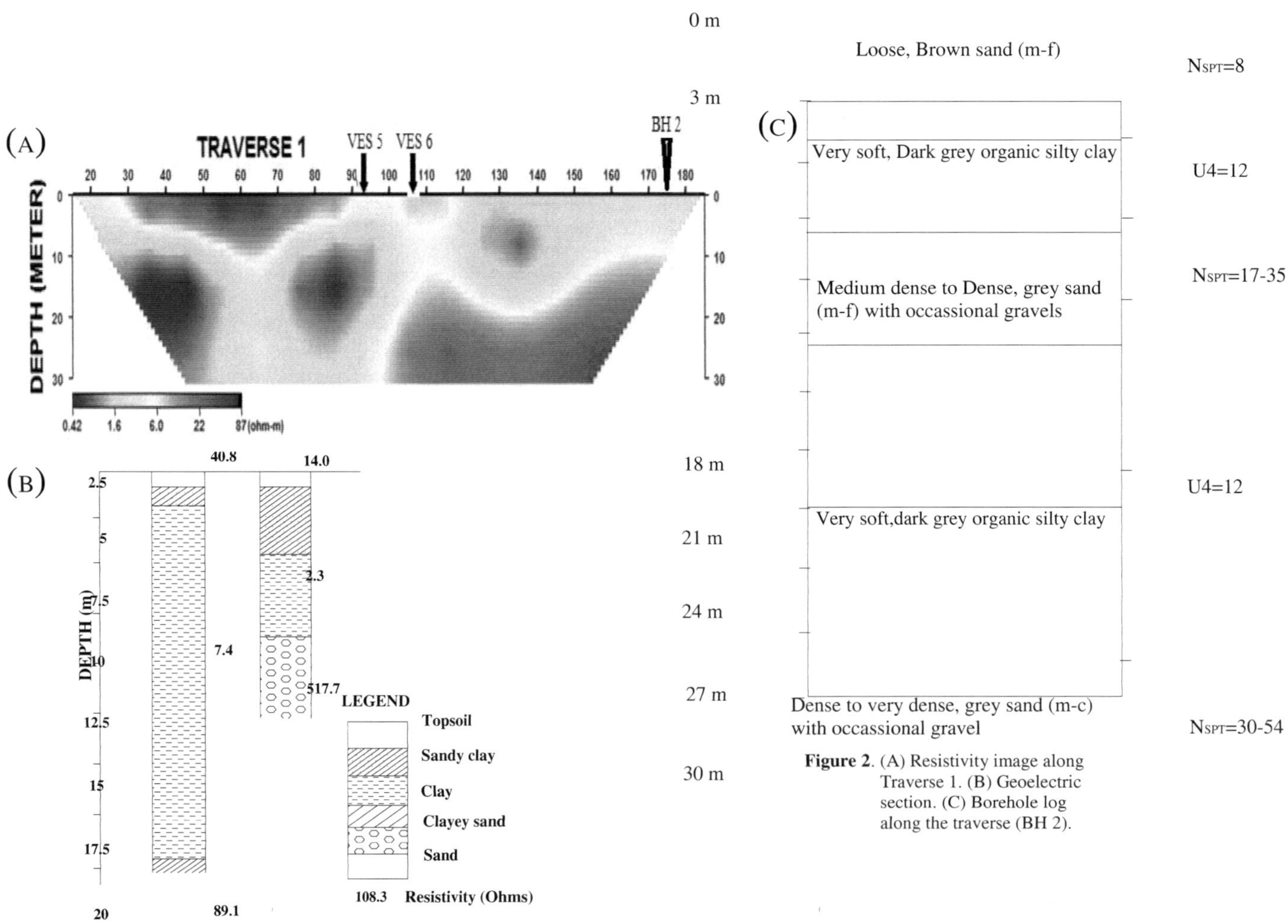

Figure 2. (A) Resistivity image along Traverse 1. (B) Geoelectric section. (C) Borehole log along the traverse (BH 2).

Table 2. Statistical Analysis of the Return Periods based Inundation Scenarios.

Return Period	Modelled Water elevations (metres)
2	2.020
5	2.935
10	3.568
25	4.393
50	5.026
100	5.672
200	6.336

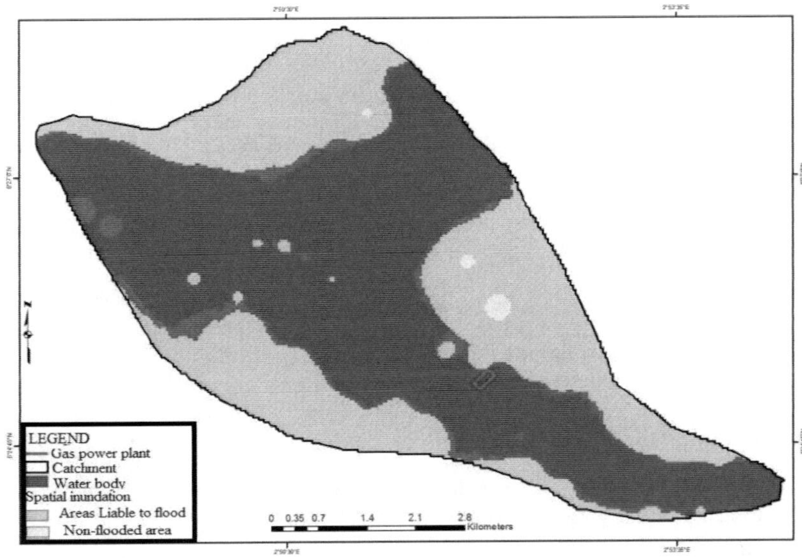

Figure 3.2D flood inundation map for modeled return period of 100yrs

References

Adiat, K.A.N., M.N.M. Nawawi, K. Abdullah, 2012, Integration of geographic system and 2D imaging to investigate the effects of subsurface conditions on flood occurrence. Modern Applied Science, 6, no.3,11-21.

Anyanwu, A. J. And B.I.O. Ezenwa, 1988, Incidence of parasitic infection of pond raised Tilapia spp. and some cultivable fish species from three ecological areas of Lagos state. Nigeria Institute of Oceanography and Marine Research, Technical Paper No. 32.

Chapman, J.B. and W.D. Cannan, 2001, Flood maps are key to better flood damage control. CE News, 58-60.

Jones, H.A., and R.D. Hockey, 1965, Geology of Parts of Southwestern Nigeria :Ibadan University Press.

Rabie, A.H., A.G. Aminuddin, Z. Nor Azizi, S.A. Mohd Sanusi and A.H. Zorkeflee, 2007, Modeling floodplain inundation by integrating hydrogeological with hydraulic model. Case study: Muda River, Kedah.: 2nd International Conference on Managing Rivers in the 21st century: Towards PWTC sustainable River Basins, Kuching.

Rucker, D.F., J. B.Fink and M. H. Loke, 2011, Environmental monitoring of leaks using time lapsed long electrode electrical resistivity: Journal of Applied Geophysics, **74**, 242- 254

Toriman, M.E., A.J. Hassan, M.B. Gazim, M. Mokhtar, S.A. Sharifah, O. Jaafar, O. Karim and N.A. Abdul Aziz, 2009, Integration of 1-d Hydrodynamic Model and GIS Approach in floodplain Management Study in Malaysia. Research Journal of Earth Sciences, 1,22-27.

Whiteman, A.J. 1982, Its Petroleum Geology Resources and Potential, Nigeria: Graham and Trotman.

Uncoupled Mass Geophone, "UMG"

Ibrahim Gwerah, Waha Oil Company.*

Summary

The conventional geophone was modified in order to achieve a wider frequency response starting from 0 to more than 1000Hz with high sensitivity, so, the factors which limiting the performance of the conventional geophone are either neutralized or become controlled.

This paper presents the theoretical and practical applications of the UMG geophone, and a primary evolution of the results obtained in the field using UMG Geophone and SM-24 geophone as well.

Introduction

Seismic acquisition is to emit elastic waves into the earth, these waves propagate in the subsurface and part of their energy is reflected back to the surface when they encounter the interfaces between different layers are then detected by geophones, and converts into electrical signals which are recorded on digital media.

The principle of the conventional geophone is a fixed permanent core magnet surrounded with a moving coil, which suspended by leaf springs to the geophone housing; as illustrated in figure (1);

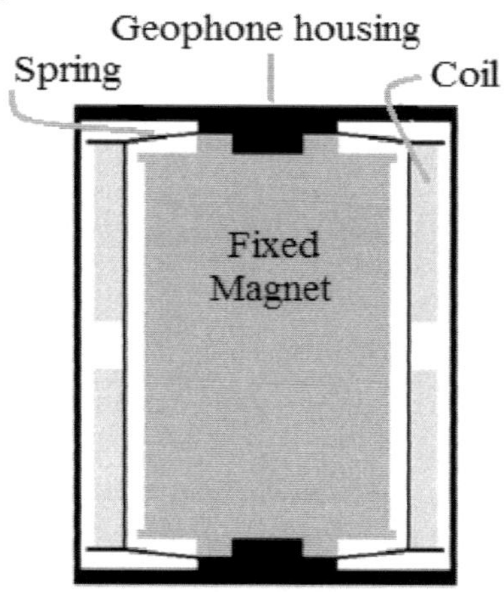

Figure 1: Schematic cross-section of the conventional geophone.

When the conventional geophone is disturbed by elastic waves, both the fixed magnet and the coil are move responding to that disturbances for frequencies below the resonance frequency providing a response falling, and generates an oscillatory output signal at the resonance frequency, but for frequencies above resonance frequency the coil will no longer follow the movement of the fixed magnet and remains settled providing a response proportional to ground displacement, which of only about 10^{-10}m, see figure (3). The resonance frequency is determined by the equation;

$$f_r = \frac{1}{2\pi}\sqrt{\frac{k}{m}} \qquad (1)$$

where k is spring constant and m is the mass of the suspended coil, as the resonance frequency depends on the mass of the moving coil and the stiffness coefficient of the spring, therefore the presence of springs in the conventional geophone limits its performance, where the response bandwidth is restricted between the resonance frequency and the spurious frequency, with constant sensitivity depends on strength of the permanent core magnet, number of turns of wire in the coil and the stiffness coefficient of the spring, hence, to increase the response bandwidth of the conventional geophone is to lower the spring constant (k) and/or increase the suspended mass (m), so many developments have been made to overcome these limitations to improve resolution of the wavelet, but these developments are practically limited because by enlarging the mass or reducing the spring constant would result a larger preload of the spring which well cause in mechanical failure of the spring.

Also to eliminate the oscillatory output signal at resonance frequency a damping of an appropriate value is required, so the maximum amount of damping is obtained when $R = 2\sqrt{km}$, whereas over damping leads to a reduction of the sensitivity, therefore the typical damping is about 0.7 of maximum value.

Therefore to overcome these limitations in performance of the conventional geophone, which are caused by the leaf springs, is to exploit the magnetic repulsion force to uncouple the permanent core magnet instead of using the leaf springs to suspend the moving coil, therefore there is no tension on the leaf springs, so, their function are confined to prevent friction between the permanent core magnet and the coil's cylinder.

Theory

The principle of UMG geophone is to use magnetic repulsion force instead of springs to uncouple the permanent core magnet by fixing a permanent magnet in the bottom of the geophone housing situated in inverse polarity to the core magnet, while the coil is attached to the geophone housing; as illustrated in figure (2);

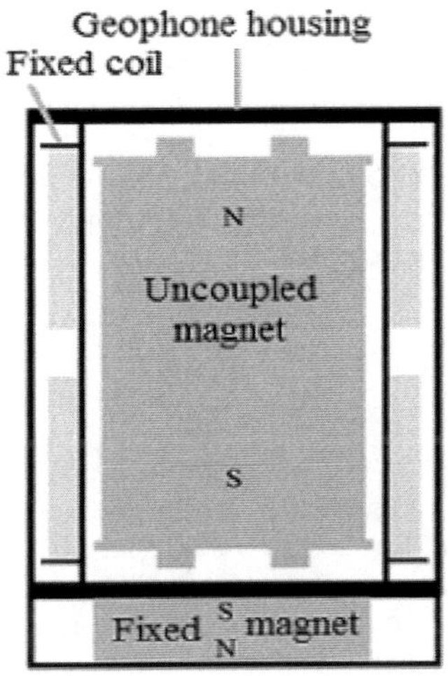

Figure 2: Schematic cross-section of the UMG geophone.

When the UMG geophone is disturbed by elastic waves it shows a response to all frequencies from 0 to more than 1000Hz, as shown in figure (6). As there are no springs in UMG geophone or become neutralized, because no tension on them, therefore there is no resonance frequency neither spurious frequency nor oscillatory output signal at resonance frequency related to them, consequently the sensitivity of the UMG geophone is increased, and allows to increase it more by increasing the mass and/or number of turns in the coil, in addition to that there is no damping is required, which leads to increase the sensitivity as well.

Testing

A set of SM-24 geophones fig(1), were modified to UMG geophones without springs fig(2), then a series of tests has been conducted in the field in order to verify that the behavior of the modified geophones was conforming to what is predicted by the theoretical considerations.

Some of that tests consisted of the following:

- Tap test was conducted to characterize response of the UMG geophone using UMG geophones and SM-24 geophone, all of them without shunt resistance.
- Recording the earth response by geophones of UMG geophone and SM-24 geophone using a sledgehammer as a source, the distance between the source and the geophones is 20meter.
- UMG geophones inserted within a spread of 24 geophones per group of SM-24 geophones in a production 3D seismic survey.

In each of the above mentioned tests, the response of the UMG geophone conforms what is theoretically expected as illustrated by the fallowing figures;

In the tap test, figure (3) shows an amplitude response of tap test obtained using SM-24 geophone, where the response falls below the resonance frequency and an oscillatory output at the resonance frequency, so damping is required, and figure (4) shows an amplitude response of a tap test obtained using UMG geophone without shunt resistance, where the UMG geophone shows a regular response started from 0Hz, and is approximately equal at all frequencies to 250Hz, so no damping is required.

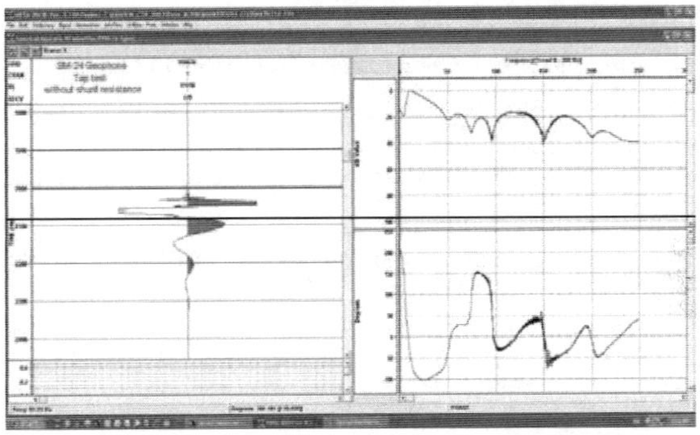

Figure 3: Amplitude response of a tap test on SM-24 geophone without shunt resistance, the horizontal scale is from 0 to 300Hz and the vertical scale is from 0 to -100dB.

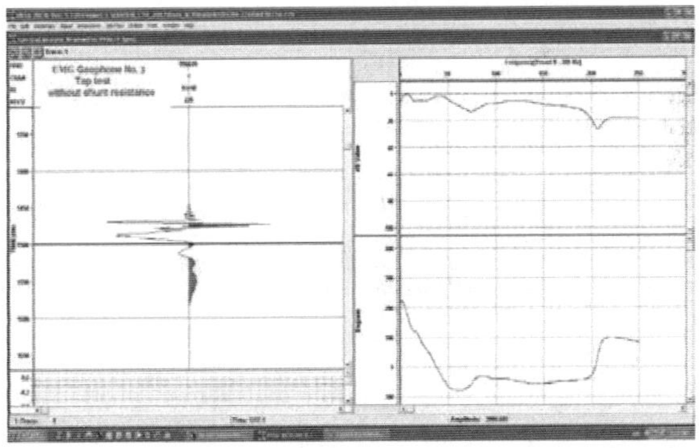

Figure 4: Amplitude response of tap test on UMG geophone without shunt resistance, the horizontal scale is from 0 to 300Hz and the vertical scale is from 0 to -100dB.

In the test using sledgehammer as a source, where the distance between the source and the geophone is 20m, figure (5) shows an amplitude response of a trace recorded by single SM-24 geophone, figure (6) shows an amplitude response of a trace recorded by single UMG geophone without shunt resistance, where the response started from 0 to more than 1000Hz, and figure (7) shows an amplitude response of the all traces, where the response is approximately equal from 0 to 200Hz.

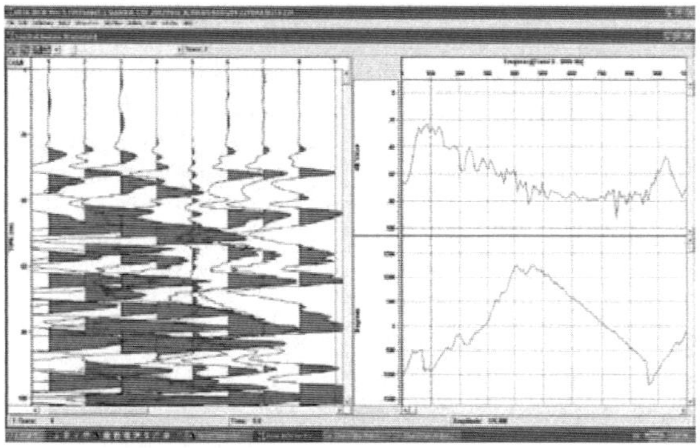

Figure 5: Amplitude response of trace No. 1. which detected by a SM-24 geophone using a sledgehammer as a source. The horizontal scale is from 0 to 1000Hz and the vertical is from 0 to -100dB.

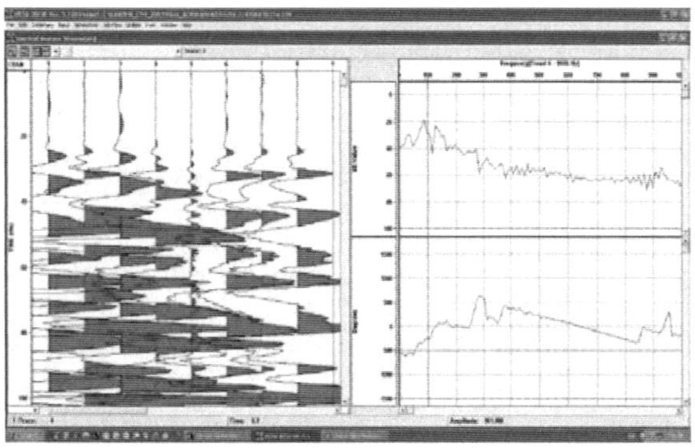

Figure 6: Amplitude response of trace No. 2 which detected by a UMG geophone using a sledgehammer as a source. The horizontal scale is from 0 to 1000Hz and the vertical is from 0 to -100dB.

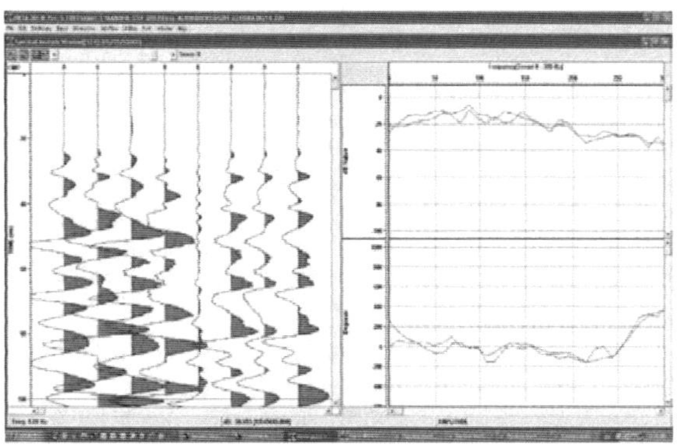

Figure 7: Amplitude response of all traces detected using a sledgehammer as a source, the horizontal scale is from 0 to 300Hz and the vertical is from 0 to -100dB.

In a 3D seismic survey, figure (8) shows a 7-68Hz / 3dB/octave non-linear sweep used in a 3D seismic survey, figure (9) shows the amplitude response of a trace recorded by a group of 24 geophones of SM-24 geophone, where high frequencies are attenuated, while figure (10) shows amplitude response of a trace recorded by single UMG geophone without shunt resistance, where the shape of the sweep is kept, consequently high frequencies are preserved.

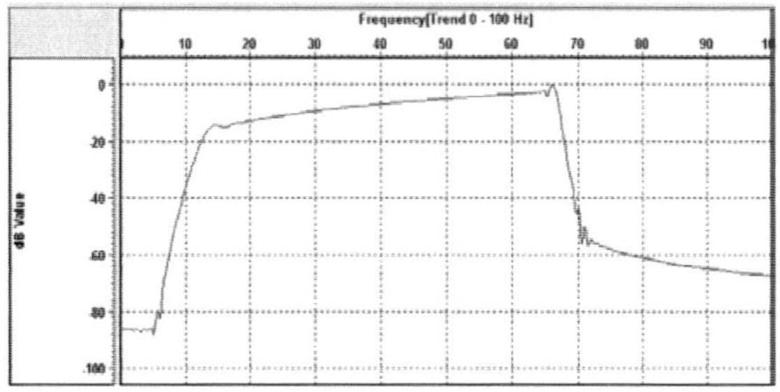

Figure 8: Amplitude response of a 7-68Hz / 3dB/octave non-linear sweep was used in a 3D seismic survey, the horizontal scale is from 0 to 100Hz and the vertical is from 0 to -100dB.

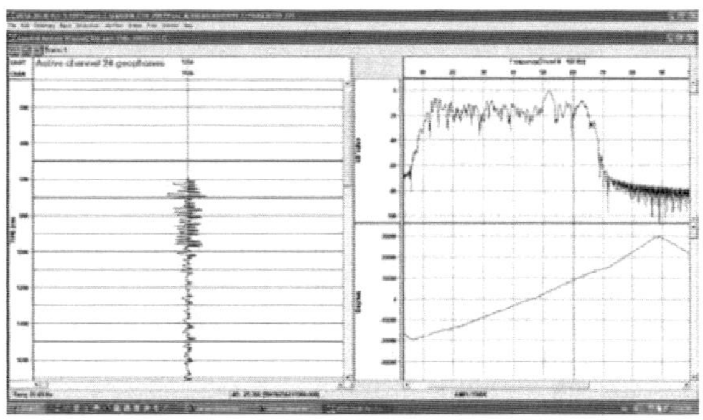

Figure 9: Amplitude response of a trace recorded by a group of 24 geophones of SM-24 geophone, where high frequencies are attenuated.

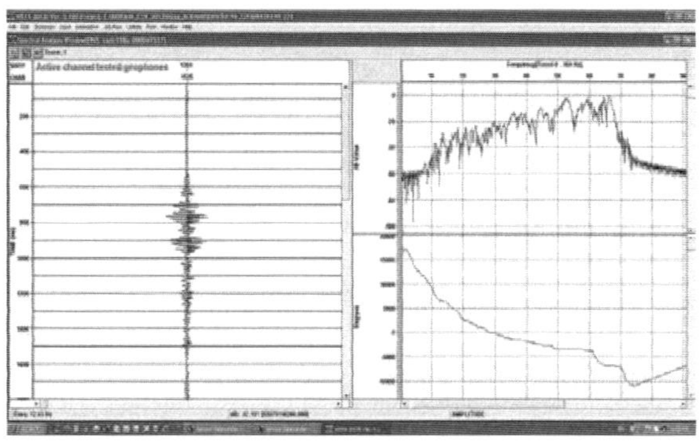

Figure 10: Amplitude response of a trace recorded by UMG geophone, where the shape of the sweep is kept, consequently high frequencies are preserved.

Conclusions

It is apparent from theoretical and practical applications that the performance of the UMG geophone overcomes the limitations of the conventional geophone, where the improvement in bandwidth is extended in both lower and higher sides of the frequency spectrum, as well as no damping is required, and the most significance is that the UMG geophone's design allows to increase the uncoupled mass up to an acceptable weight, and to increase the number of turns in the coil up to an acceptable number, consequently increases the sensitivity, which results in higher seismic resolution to improve the chance of success with geophysical efforts that better resolves the subsurface images and the geological model.

References

Introduction to Geophysical Prospecting-third edition,1981. Milton B. Dobrin.p.58-63

*An Introduction to Applied and Environmental Geophysics-second edition*2011. John M. Reynolds.p.170

An Introduction to Geophysical Exploration-third edition,2002. Philip Kearey, Michael Brook, and Ian Hill.p.39-40

Low-frequency with passive geophone magnetic spring. Johan Dams.

What Receivers will we use for Low Frequencies?
R. Malcolm Lansley and Peter Maxwell.

Testing the quality of geophone plants for 3-C land acquisition.
Henry C. Bland, Malcolm B. Bertram, and Eric V. Gallant.

Acknowledgments

The author would like to thank Ali Kheder, Mohamed Moftah, Carol Heitmann, and Tamer Fares, Waha Oil Company; Ata Elias, Conocophillips Company for their assistance.

Airborne Geophysics applied to Engineering and Environmental Problems

Greg Hodges, Craig Annison, CGG MultiPhysics

Introduction

Airborne geophysics has seen increasing application to environmental and engineering geophysical problems over the last 20 years, driven in part by increased understanding of the application of geophysics, and improved resolution and calibration of airborne systems.

The Special Character of Geophysics for Engineering

The use of airborne geophysics for environmental and engineering problems requires innovation and flexibility in the application, interpretation, and products. Unlike exploration, where clients are often geophysicists or at least familiar with geophysics, engineering clients may be geotechnical or civil engineers with very limited understanding of geophysics and its use.

It is not uncommon to be considering a survey for an application that has never been tried before. This puts pressure on the geophysicist to understand the problem and the geophysical properties that define the target– information that the end-user might not be able to provide.

Geophysics measures resistivity, density, susceptibility and other rock properties. Engineering and environmental users need to know factors like thickness, salinity, friction ratio, rippability, or porosity. It is the geophysicist who must make the translation to material properties the user needs, and advise the user on reliability and ambiguity in the results.

It is vitally important that a geophysicist remain as part of the project team through to the last use of the geophysical data. The risk of misinterpretation or misuse is significant, otherwise.

Airborne Geophysics and Drilling

Drilling is the preferred method of sampling the sub surface for many engineers. The drill hole gives clear results with samples for testing, but they can be fooled by boulders (for example). However, drill holes tell the engineer nothing about the conditions just metres away from the hole. Airborne geophysics, conversely, has less resolution at any point but provides complete coverage over the area surveyed.

The two methods – drilling and geophysics – are highly complementary and when used together return more than the sum of the two. Drilling provides the point samples and precise measurements, while the geophysics provides the full area coverage; mapping between the holes and showing which areas have continuity and which show change.

Collection of airborne geophysical data generally does not depend on detailed knowledge of the ground, and can be collected before drilling. The results have immediate value in indicating where drilling should be concentrated – in areas of high complexity or serious hazards, for example – and where drilling can be sparse because of indicated continuity. Further interpretation of the geophysics can then be conducted constrained by the initial drilling, then used to guide further drilling, and then be re-interpreted with the new knowledge. This iterative, interactive process uses the strengths of both data types to progressively build a superior model of the subsurface in the most efficient manner.

Airborne Geophysics Applications

The airborne geophysical methods most often used for engineering include electromagnetic (measuring resistivity or conductivity), magnetic (to detect ferrous materials and geology), gamma-ray spectrometry (to detect anthropogenic radioactive materials and map the geology by changing potassium concentration) and gravity gradiometry (to map density changes).

Airborne surveys cover large areas very quickly, with no need for ground access or exposure to hazards. Coverage can be continuous across all kinds of terrain, including continuous coverage from land across tidal zones and water. High resolution RESOLVE EM surveys acquire full coverage at a rate of 800 Ha per hour at typical spacing of 100m. Figure 1 shows a conductivity section with continuous coverage (and changing interpretation) from sea (left) across land onto tidal flats.

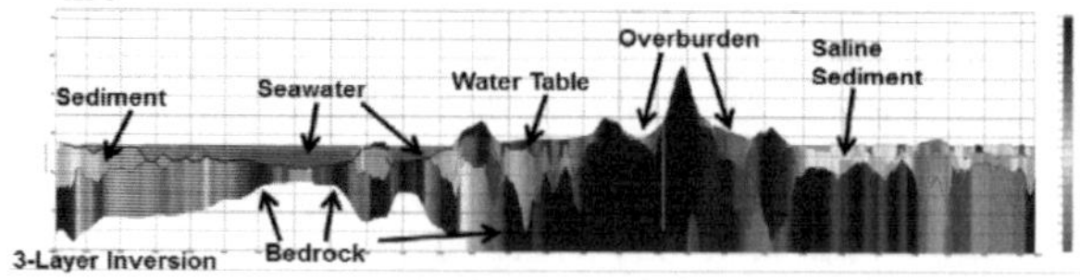

Figure 1: Continuous conductivity section from water (left) to land and tidal zone.

Construction and Infrastructure

Almost all construction projects will require characterization of the ground for one or more of: compaction, stability, rippability, groundwater, salinity, electrical conductivity, and other factors. Airborne electromagnetic surveys measure ground resistivity in three dimensions. A straight-forward application is mapping resistivity (or conductivity) of the ground for design of infrastructure lighting protection grounding (Chisholm et al, 2003). Figure 2 shows a resistivity cross-section from New Brunswick, Canada, from the presentation by Chisholm et al (2003), where warm colours indicate more conductive ground (low resistivity) and cool colours more resistive ground.

The resistivity distribution with depth can be used to design custom grounding arrays to take maximum advantage of the depth distribution of conductive soil and rock. Two layer resistivity models can be generated and used with standard ground electrode resistance equations to design electrode arrays to provide the necessary protection from lightning.

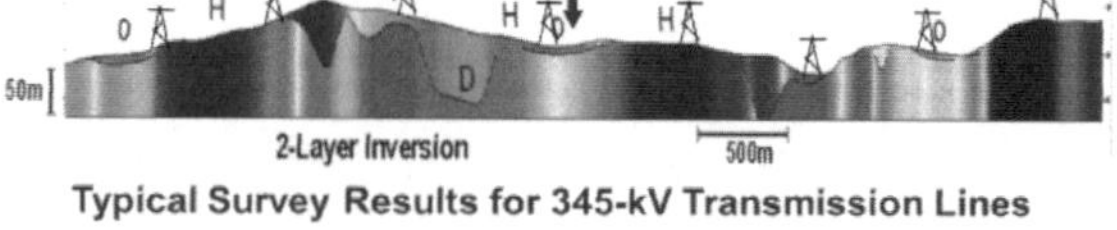

Figure 2: Conductivity section from airborne EM and ground recommendations for 345kV transmission line.

In the example shown in Figure 3 (Brown and Hodges 2006) a planned pipeline corridor in Quebec, Canada, was flown with helicopter-borne EM to map the presence and thickness of overburden (principally glacial till related sediments). Pipeline construction was planned for 2m depth, so knowledge of the depth of overburden allowed for planning where trenching with a backhoe would suffice and where blasting of bedrock would be necessary – at a much higher cost and level of complexity.

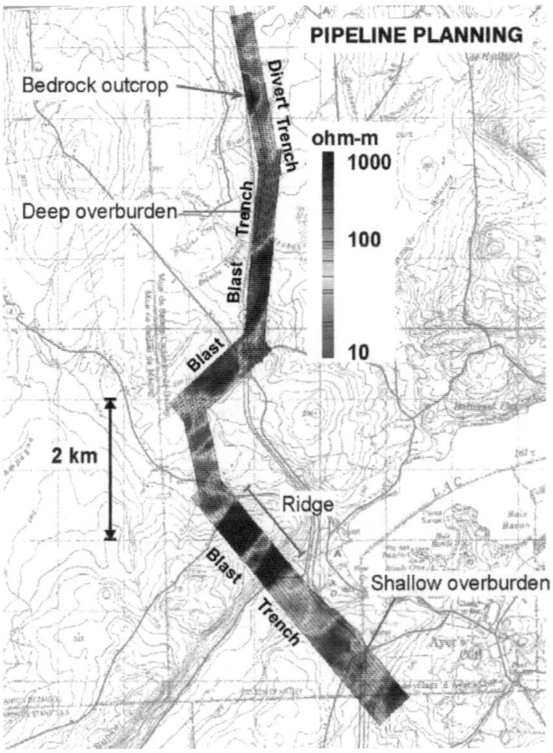

Figure 3: Resistivity map used to interpret overburden thickness to estimate need for blasting for pipeline construction. From Brown and Hodges, 2006

Surveys of a similar nature have been conducted in a wide range of geological environments, looking for soil types, soil depths, rock type, permafrost, paleochannel aquifers, and karst features.

Karstic Geology and other Hazards

Karstic cavities, sinkholes, and zones of weakness can be a hazard to many engineering projects. Airborne geophysical data can be used to detect karst where the geophysical properties of the cavity are distinct from their host. The example shown in Figure 4 is from the Victor Diamond Mine in northern Ontario, Canada. The mine was planned to be close to a known sinkhole, and the survey was conducted to outline the sinkhole and search for other, unknown karstic hazards in the area before construction began. The limestone surface and sinkhole are covered by about 12m of Quaternary glacial till, and no evidence of the sinkhole is apparent on surface. However, the sinkhole is clearly mapped with airborne EM by the change in conductivity between the till filling the hole and the limestone surrounding. The airborne EM data were used in conjunction

Figure 4 Bedrock surface (showing sinkhole) derived from airborne EM survey, constrained with seismic data.

with seismic refraction data over the deepest parts of the sinkhole to construct a complete map of the sinkhole.

Conductivity contrast can also be created by water filling a cavity, and often by the higher porosity (and therefore saturation) of the extensive weak zones on which cavities occur. This was the case in a survey described by Smith et al, 2003, which detected the previously-known Valdina Farm sinkhole in the Edward Aquifer in Texas, USA (Figure 5).

An air-filled cavity is a difficult target for EM, but may be detectable with airborne gravity gradiometry (AGG) by the density contrast depending on the size and depth. Helicopter-borne AGG provides the lowest, slowest flight and therefore the highest sensitivity of an airborne method.

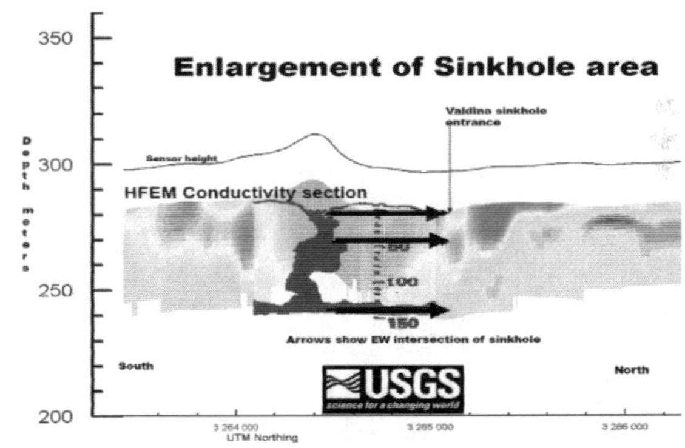

Figure 5: Resistivity section and sinkhole section at Valdina Farms Sinkhole, Texas (Smith, 2003)

Oil Field Applications

There are many demands for engineering and environmental work in the oil fields. The magnetic data in Figure 6, from Quarantine Bay in the Mississippi delta, shows anomalies from surface infrastructure, wellheads and pipelines. Records of wellhead locations are often inaccurate, and old wells were often poorly sealed, if at all (Veloski et al, 2008). If a field is to be pressurized for enhanced oil recovery all the wells ever drilled have to be located and the seal verified.

Figure 6: Magnetic data over Quarantine Bay, Louisiana, showing anomalies of wellheads, surface infrastructure, and pipelines.

Pipeline location surveys are often conducted in oilfield areas to locate them before new seismic surveys are shot, to prevent risk to the pipeline and interference in the data collected.

Oil well produced brine is a major environmental problem, and in the past was often dumped into brine pits on the surface (Paine et al, 2007). It is possible to map the subsequent distribution of this highly conductive saline water using airborne EM surveys. Figure 7 shows a map of 7200Hz (mid-range) resistivity from a DIGHEM survey over the Permian Uplands and Red River basin and including part of the Nocona Oil Field. The brine pits are shown as black triangles. Both the airborne

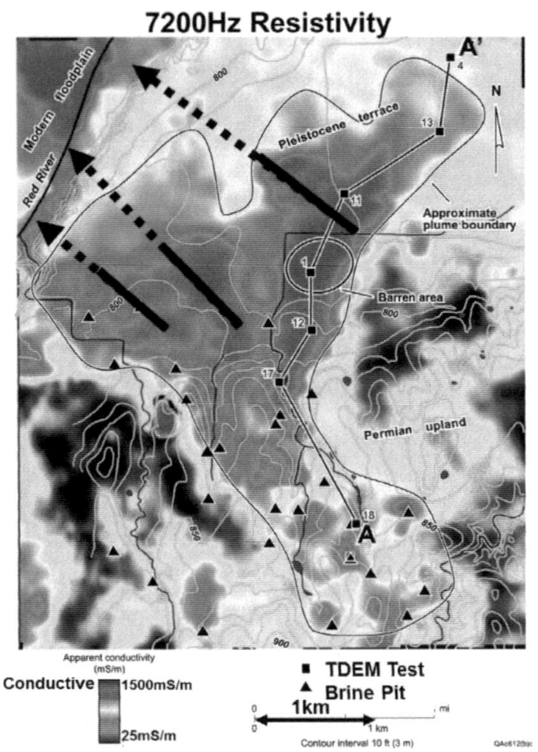

Figure 7 Resistivity map showing near-surface brine produced from oil wells (From Paine. 2003)

EM and ground TDEM surveys show the brine is at the surface near the pits, travels below the surface under the Permian upland (following ground water flow) to re-emerge in the Red River valley.

Contaminants

One could argue whether oil field produced brine (Figure 8) is a "contaminant", but there is no dispute that abandoned coal mines and hard rock mines can produce highly contaminated water, particularly those with high sulfide content.

The Sulphur Bank Mercury mine was worked from late in the 19th century intermittently until it was closed in 1957 (Hammack et al 2001). It was known that the mine was leaking acidic, metal-contaminated water into Clear Lake, but the pathway was not known. The high-frequency 56kHz resistivity map (Figure 8) shows the near-surface resistivity surrounding the water-filled open pit (the Herman Impoundment), and the conductive plume of contaminated water leading into the lake to the west. Lower frequency EM data (not shown) shows several other, possible contaminant pathways along faults in the geology trending northwest to northeast.

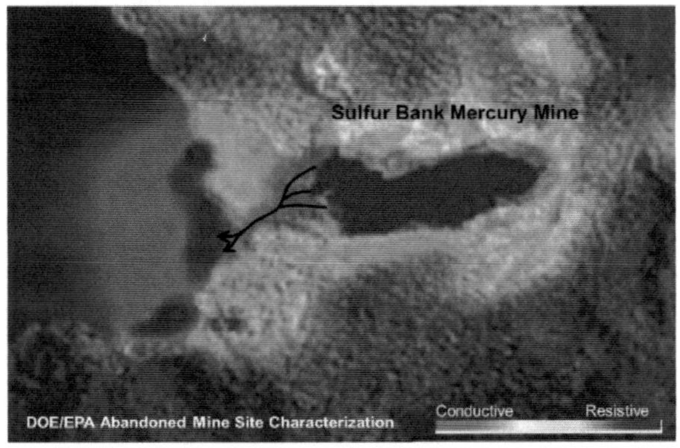

Figure 8: Resistivity map over Sulphur Bank Mercury Mine showing conductive contaminant plume from open pit flowing west to Clear Lake. (Hammack et al, 2001)

Hammack et al, in 2005, showed the results of an airborne EM survey to map contaminated water leaking from a coal mine impoundment (Figure 9). The survey detected the strong conductivity in the decant basin, as well as the leakage pathway through the embankment downhill to come out at surface at the base of the impoundment.

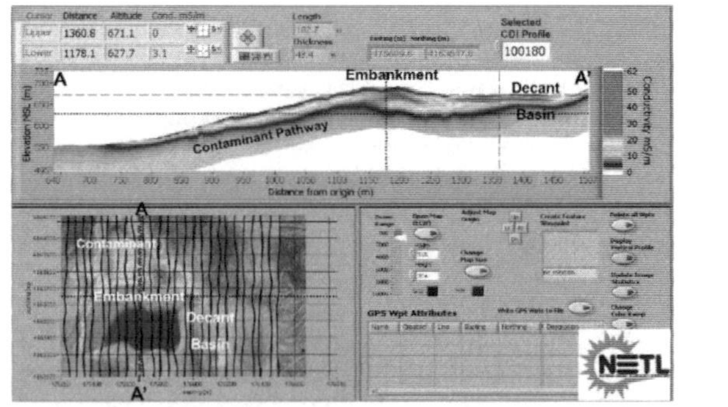

Figure 9 Resistivity section through coal tailings impoundment showing high conductivity of tailings and leakage path for contaminated waters (Hammack et al, 2005)

Figure 10 shows a map of a different kind of contaminant: airborne Cs^{134} and Cs^{137} contamination carried by winds from the Fukushima nuclear power plant leak, mapped using airborne gamma-ray spectrometry. Gamma ray surveys have also been conducted to locate lost military and industrial radioactive sources – one case found in a scrap metal yard.

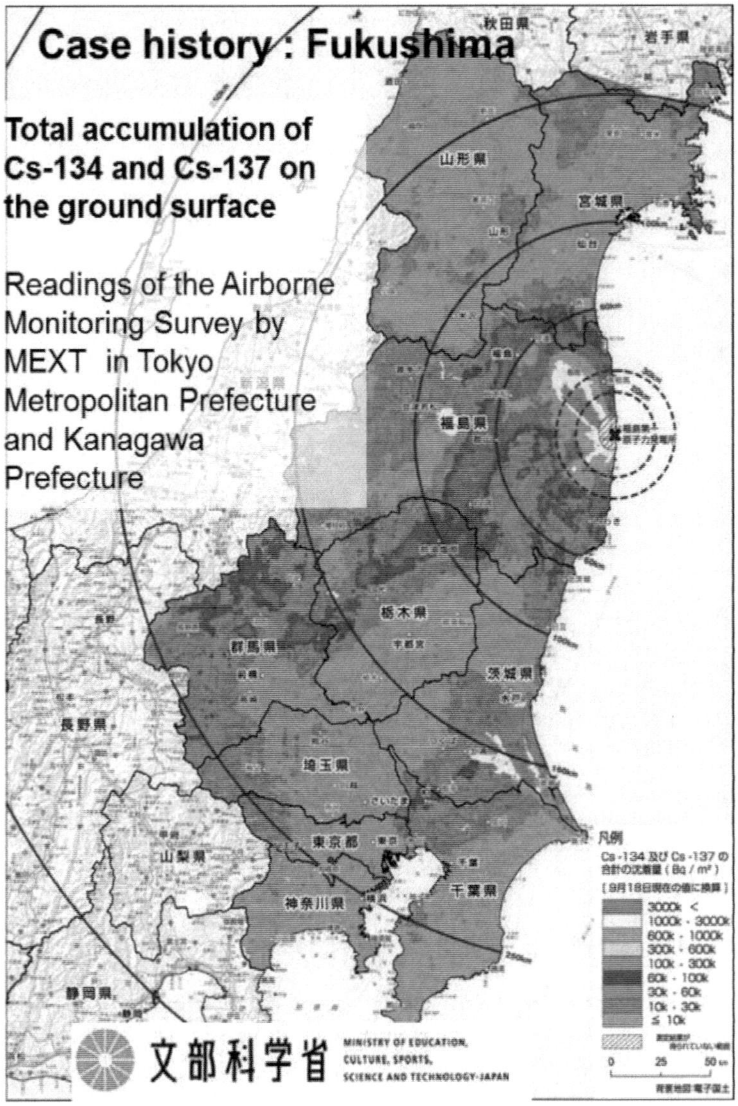

Figure 10 Airborne gamma ray spectrometry map showing Cs distribution from Fukushima nuclear plant disaster

Conclusions

There are many applications of airborne geophysics to engineering and environmental projects. Through the use of one or more of electromagnetic, gravity gradiometry, magnetic, or gamma-ray spectrometry, any target that causes a substantial change in resistivity, or density, or has a magnetic signature or natural or man-made radioelement concentration can be detected and mapped. Airborne surveys offer fast coverage of large areas to provide continuous measurement that provides a highly complementary data set to drilling. The two data sets can and should be interpreted together, using the geophysics to guide the drilling to where it is most needed (and saving money in the process) and the drilling used to refine and constrain the interpretation of the airborne geophysics.

Sensing the Presence and Transport of Engineered Nanoparticles in Saturated Porous Media using Spectral Induced Polarization (SIP) Method

Gamal Z Abdel Aal[1,2]*, Estella E Atekwana[1] and Dale Werkema Jr.[3]
[1]Boone Pickens School of Geology, Oklahoma State University, Stillwater, OK, 74078
[2]Geology Department, Faculty of Science, Assiut University, Assiut, 71516, Egypt
[3]U.S. Environmental Protection Agency, Las Vegas, NV 89119

Summary

Laboratory column experiments were conducted to investigate the sensitivity of the spectral induced polarization (SIP) measurements to the presence and transport of Silver (Ag) and Zero Valent Iron (ZVI) nanoparticles with particle diameters of 90-250 nm in saturated quartz sand packed column. SIP measurements were obtained between 0.1 to 10000 Hz using the dynamic signal analyzer (DSA) National Instruments (NI) – 4461. The impedance magnitude $|\sigma|$ and the phase shift φ of the sample were measured relative to a high-quality resistor. The real (σ') and imaginary (σ'') parts of the sample complex conductivity were calculated. The first experiment measured the SIP response to the mixture of one pore volume of the nanoparticle suspension with the porous medium. The same procedure was repeated with the two nanoparticles using different nanoparticle concentrations (0-20 mg/ml). At frequencies above 1 Hz, the magnitudes of the phase and imaginary conductivity increased with increasing concentration of the nanoparticle concentration. The magnitudes of the phase and imaginary conductivity of the ZVI were two times the magnitude of the Ag. The transport experiment was conducted using a 12 mg/mL concentration nanoparticle suspension. A peristaltic pump with flow at 1 mL/minute was used to introduce the suspension into the sand columns. Effluent samples of the suspensions from the sand columns and SIP measurements were collected every five minutes for 180 minutes. The optical density of the nanoparticles in the effluent samples were determined at 430 nm wavelength using a spectrophotometer and then converted into concentrations using relationships determined from the linear correlation between concentrations of nanoparticle suspensions and their measured optical densities. A breakthrough curve (BTC) was obtained by plotting the normalized outlet nanoparticle concentration as a function of time. A similar procedure was followed for the SIP parameters. The BTC of normalized outlet Ag was higher in magnitude compared to the BTC of the normalized ZVI, suggesting retention of ZVI in the sand column was higher than Ag. Higher retention of ZVI was reflected in higher magnitudes for the normalized phase and imaginary conductivity component. Interestingly, the trend of the BTC of the normalized imaginary conductivity component was similar to the BTC trends of normalized outlet nanoparticle concentration. These results demonstrate the sensitivity of the SIP technique to nanoparticles and their transport in saturated porous media.

Introduction

Nano-materials are emerging into the global marketplace. Engineered Nano-particles, and other throwaway nano-devices may constitute a whole new class of non-biodegradable pollutants of which scientists have very little understanding. Therefore, the production of significant quantities of nano-materials and their accidental introduction into the subsurface may pose a potential risk and hazard to the environment and public health which needs an urgent and exhaustive evaluation (Luque-Garcia et al. 2013). Additionally, NPs are extensively used or tried for environmental remediation (Zhang et al. 2013), where information on their fate and impact is largely limited. Therefore, there is a demand to develop a technique that is sensitive to the presence and transport of these nanoparticles in the subsurface in order to assess the risks associated with these nanoparticles. Geophysical methods, specifically spectral induced polarization (SIP), have shown to be sensitive to the very small physiochemical changes that occur at the grain fluid interface and has been used to locate disseminated metallic mineral deposits (AbdelAal et al., 2014). The main objectives of the present study are to (1) investigate the sensitivity of spectral induced polarization (SIP) measurements to the presence of different concentrations of Ag and ZVI nanoparticles in saturated porous media and (2) assess the potential to use SIP measurements to monitor the transport of Ag and ZVI nanoparticles in the subsurface.

Materials and Method

A schematic representation of the spectral induced polarization (SIP) setup used in this study is shown in Figure 1. The SIP measurements were obtained using a dynamic signal analyzer (DSA) (National Instruments (NI) – 4461) between 0.1 Hz to 10000 Hz at 16 equal logarithmic intervals (sensitivity of 0.1 mrad over this frequency range). The conductivity magnitude $|\sigma|$ and phase shift φ of the samples were measured relative to a high-quality resistor at 1000 Ohms. Once measured, the complex sample conductivity which includes the real ($\sigma' = |\sigma| \cos \varphi$) and imaginary ($\sigma'' = |\sigma| \sin \varphi$) conductivity components was calculated. The porous media (Ottawa sands) used to fill the column consisted of fine silica sand (U.S. Silica Company). The silver nanoparticles (Ag) and the nano Zero Valent Iron (ZVI) were in a powder form (90-250 nm) that had a purity of 99% (Nanostructured and Amorphous Materials Incorporated, Inc). The surface area (m^2/g) ranges from 2.4-4.4 and 3-7 for Ag and ZVI, respectively. The fluid used to prepare the nanoparticles suspensions was artificial groundwater (AGW) similar to that used by Abdel Aal et al., (2009).

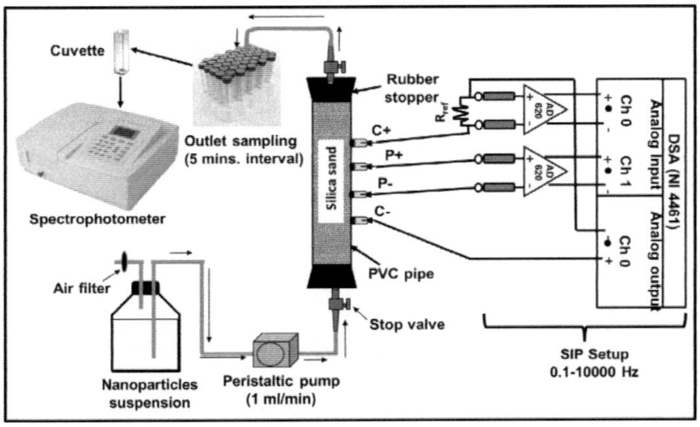

Figure 1. Schematic of column setup and instrumentation used in SIP measurements.

The first experiment measured the SIP response to the mixture of one pore volume of the nanoparticle suspension with the porous medium. The same procedure was repeated with the two nanoparticles using different nanoparticle concentrations (0-20 mg/ml). The transport experiment was conducted using a 12 mg/mL concentration nanoparticle suspension. A peristaltic pump with flow at 1 mL/minute was used to introduce the suspension into the sand columns. Effluent samples of the suspensions from the sand columns and complex conductivity measurements were collected every five minutes for 180 minutes. The optical density of the nanoparticles in the effluent samples were determined at 430 nm wavelength using a spectrophotometer and then converted into concentrations using relationships determined from the linear correlation between concentrations of nanoparticle suspensions and their measured optical densities. A breakthrough curve (BTC) was obtained by plotting the normalized outlet nanoparticle concentration as a function of time. A similar procedure was followed for the SIP parameters.

Results and Discussion

Figure 2 shows the SIP results of different concentrations of Ag and ZVI nanoparticles (NPs) in saturated sand. The magnitude of the phase shift (φ) (figure 2a) and imaginary conductivity (σ'') (figure 2b) increases with increasing nanoparticle concentration as shown in figure 2a and b, respectively. The real conductivity component (σ') shows insignificant changes with increasing nanoparticles concentrations (figure 2c). Conduction and polarization at the grain-fluid interface are a function of surface area, surface-charge density, surface ionic mobility, and interfacial geometry (Revil and Glover, 1998; Lesmes and Frye, 2001). Relationships between the magnitude of the IP effect and the percentages of metallic minerals are well documented (Bigalke and Junge, 1999). The magnitudes of the phase and imaginary conductivity of the ZVI are two times the magnitude of the Ag as shown in figure 2 a and b, respectively. This can be explained by the high surface area of ZVI compared to Ag nanoparticles which considered a major control on the magnitude of SIP response of metallic minerals. These results indicate that both the phase shift and imaginary conductivity are more sensitive to the presence of metallic nanoparticles compared to the real conductivity component.

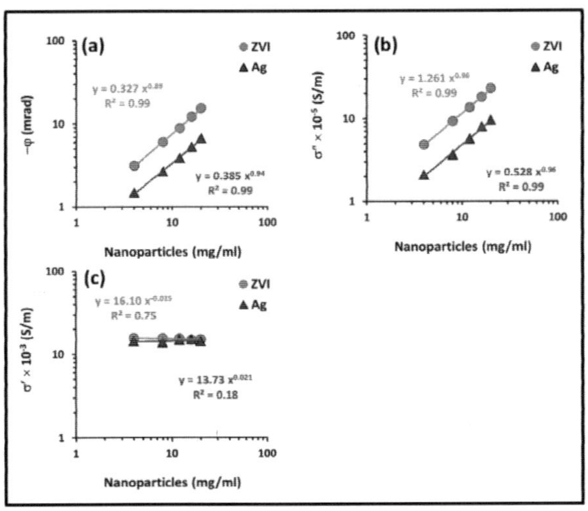

Figure 2. Results of SIP measurements at 1 HZ; (a) phase shift (φ), (b) imaginary conductivity (σ''), and (c) real conductivity (σ') of different concentration (0-20 mg/ml) of silver (Ag) and Zero Valent Iron (ZVI) nanoparticles in saturated sand.

Figure 3 shows the break through curves of the normalized outlet concentrations of the nanoparticles (Ag and ZVI) and the measured SIP parameters. The BTC of normalized outlet Ag is higher in magnitude compared to the BTC of the normalized ZVI which indicates higher retention of ZVI in sand column than AgNPs (figure 3a). The higher retention of ZVI reflected in higher magnitudes for the normalized phase shift (φ) and imaginary conductivity (σ″) as shown in figure 3b and c, respectively. Again, no significant changes are observed for the BTC of the normalized real conductivity (σ′) of Ag and ZVI nanoparticles (figure 3d). Figure 4 shows a comparison between the BTC of the normalized outlet concentrations of Ag and ZVI nanoparticles and their normalized imaginary conductivity (σ″). The trend of the BTC of the normalized imaginary conductivity component is similar to the BTC trend of normalized outlet nanoparticle concentration for both Ag (figure 4a) and ZVI (figure 4b).

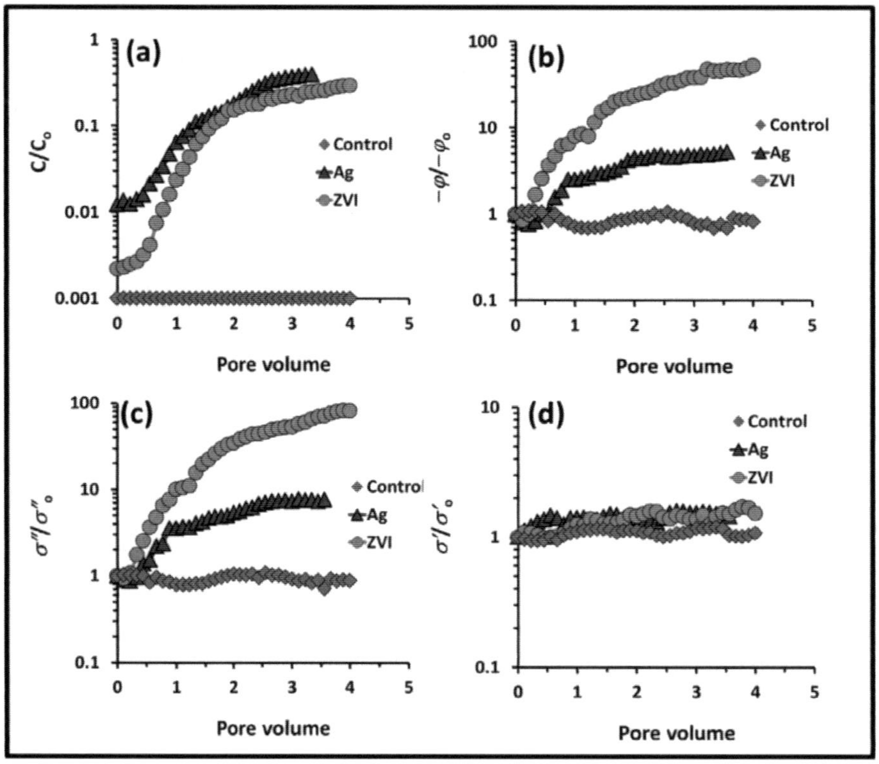

Figure 3. The break through curves (BCTs) of normalized (a) outlet concentrations , (b) phase shift (φ), (c) imaginary conductivity (σ″), and (d) real conductivity (σ′) of control (no nanoparticles), Ag and ZVI nanoparticles (12 mg/ml).

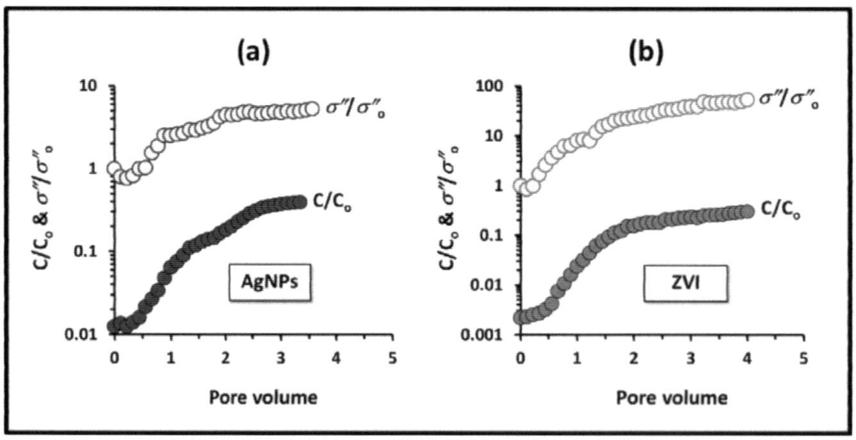

Figure 4. Comparison between the BTCs of the normalized outlet concentrations and imaginary conductivity (σ″) for (a) Ag and (b) ZVI nanoparticles (12 mg/ml).

Conclusions

The results from this study demonstrated the sensitivity of SIP measurements, specifically the imaginary conductivity (σ") component, to the fate and transport of engineered nanoparticles (Ag and ZVI) in in the subsurface saturated porous media. Breakthrough curves obtained from normalized imaginary conductivity component showed a similar trend and shape to those obtained from normalized concentrations of nano-silver dispersions. This suggests that geophysical measurements can be used to monitor the presence and transport of nano-particles in the subsurface. However, further studies are needed to investigate the effect of pH, ionic strength, and surface chemistry of nano-particles and porous media on the transport of nano-particles in the subsurface environment and their associated geophysical signatures.

Acknowledgments

The United States Environmental Protection Agency through its Office of Research and Development partially funded and collaborated in the research described here. It has been subjected to Agency review and approved for publication.

References

Abdel Aal G.Z., E. A. Atekwana, S. Radzikowski, S. Rossbach, 2009, Effect of bacterial adsorption on low frequency electrical properties of clean quartz sands and iron oxide coated sands: GEOPHYSICAL RESEARCH LETTER, **36**, L04403, doi:10.1029/2008GL036196.

Abdel Aal, G. Z., E. A. Atekwana, and A. Revil, 2014, Geophysical signatures of disseminated iron minerals: A proxy for understanding subsurface biophysicochemical processes: JOURNAL OF GEOPHYSICAL RESEARCH-BIOGEOSCIENCE, **119**, 1831–1849,

Bigalke, J., and A. Junge, 1999, Using evidence of non-linear induced polarization for detecting extended ore mineralizations: GEOPHYSICAL JOURNAL INTERNATIONAL, **137**, 516–520.

Lesmes, D. P., and K.M. Frye, 2001, Influence of pore fluid chemistry on the complex conductivity and induced-polarization responses of Berea Sandstone: JOURNAL OF GEOPHYSICAL RESEARCH, **106**, no. B3, 4079–4090.

Luque-Garcia, J. L., Sanchez-Díaz, R., Lopez-Heras, I., Camara, C., Martin, P., 2013, Bioanalytical strategies for in vitro and in vivo evaluation of the toxicity induced by metallic nanoparticles: TRENDS IN ANALYTICAL CHEMISTRY JOURNAL, **43**, 254-268.

Revil, A., and P. W. J. Glover, 1998, Nature of surface electrical conductivity in natural sands, sandstones, and clays: GEOPHYSICAL RESEARCH LETTERS, **25**, 691–694.

Zhang, H., Peng, C., Yang, J.J., Shi, J.Y., 2013, Eco-toxicological effect of metal-based nanoparticles on plants: research progress: CHINESE JOURNAL OF APPLIED ECOLOGY, **24**, 885-892

Landfill characterization using geoelectrical tools

*Abdullahi *Abdulrahman, Mohd Nordin Mohd Nawawi, Rosli Saad*
Geophysics Section, School of Physics, Universiti Sains Malaysia

Summary

A combined application of electrical resistivity and time domain Induced Polarization (IP) methods was used to characterize an old landfill area. A grid of the geoelectrical inverse models classified various subsurface sections of the waste disposal site. The results of the investigations confirmed high chargeability values representing waste deposits and saturated clayey layer while the contaminant plume reflected low resistivity. Moreover, physiochemical analyses confirmed that the diffused leachate plume exhibits a weak chargeability zone. The outline of resistivity and chargeability profiles at the site displayed different zones of the subsurface strata; buried waste, regions containing only leachate, uncontaminated subsoil volumes. Delineation of the contaminated sites in this study supports contaminant plume monitoring and shows promise as a tool for site characterization and management.

Introduction

Open dumping and unregulated landfills are the dominant waste management procedures in the developing countries (Singh et al. 2011). With the rapid growth of population and the subsequent increasing need for shelter, the expanding residential and commercial areas sprawl over old landfill sites. The contaminant leachate plume from these dumpsites seeps into underground and surface water resources, which creates a potential for hazard to the inhabitants and the ecosystem. In order to manage the vulnerability of the water resources to pollution, continuous monitoring and evaluation of the sites are crucial for the environmental risk assessment.

Monitoring wells are usually installed within a landfill and it's neighbourhood to gauge the extent of contamination of the water resources. Cost-effective and non-invasive geophysical tools are ready complementary approaches to the monitoring wells. Resistivity technique is the most common geophysical application in the monitoring of the contaminant plume. The plume has a high concentration of ions, and this lowers the resistivity of the contaminated areas marking out the contaminated zone from its background (Bernstone and Dahlin 1999). Nevertheless, when there is a negligible variation in resistivity between the contaminant plume and the background, the resistivity approach will be insufficient to map the leachate.

The combination of resistivity and IP profiling at landfill sites minimizes the interpretation ambiguity and could demarcate various sections of a municipal solid waste dumpsite. Several formations could exhibit similar resistivity anomaly with waste areas and leachate plumes. Such cases could be distinguished if the chargeability ratings of the sediments and landfill sections are established. Information on leachate plume monitoring and landfill characterization is the basis for waste management practices as it affects the protection of the environment and remediation or reclamation of waste dumpsites.

In the joint application of resistivity and IP surveys, the interpretation of resistivity models is more definite than analysis of chargeability models. There is negligible controversy on resistivity status of various subsurface sections of the waste disposal sites. However, the correct assessment of chargeability status of the diffused leachate is still under discussion. Some authors delineated the contaminant plume as high chargeability zone (Abu-Zeid et al. 2004; Martinho and Almeida 2006). Whereas Gallas et al. (2011) postulate that the leachate plume has low chargeability status.

Additionally, Dahlin et al. (2010) demarcate leachate zones with resistivity values (<10 Ωm) at two landfill sites while chargeability (> 70 ms) values delineate the waste areas. This interpretation suggests that the leachate plume is either susceptible to a resistivity evaluation rather than an IP assessment or exhibits low chargeability signals. The investigation in this work attempts to address the chargeability status of the disseminated contaminant plume and generate a procedure for landfill characterization with minimal ground truthing.

Methodology

This study selected an old landfill on the Mainland of Penang, Malaysia for the characterization. The site is Pulau Burung Landfill (PBL), located at Latitude 5° 12′ N Longitude 100° 25′ E (Fig. 1). Five resistivity/chargeability inverse models were conducted on the landfill. Two lines (400 m) in SE-NW direction and perpendicular to the slope of the landfill and three lines (400 m) conducted SW-NE parallel to the slope of the site (Fig 1). In order to assess the chargeability status of the contaminant, leachate was sampled from aeration ponds for physiochemical analysis.

The measured resistivity/IP data were inverted using RES2DINV program (Loke and Barker 1996). Surfer 8 of the Golden Software was used to improve visualization of the 2D sections. This study adopted the robust model constraint (L_1-norm) for the inversion process.

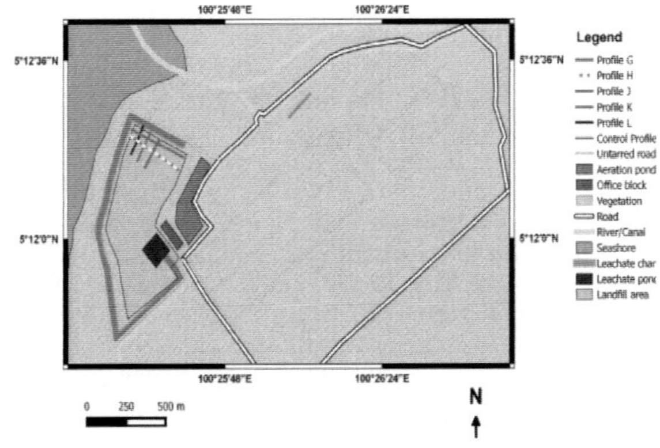

Figure 1: Map of Pulau Burung Landfill area

Parameter	1st Pond (mg/L)	2nd Pond (mg/L)	3rd Pond (mg/L)
TDS	13470	5656	2735
TSS	411	307	22
Heavy Metals	9.5	8.8	4.7

Table 1: Physiochemical analyses of the leachate samples

Subsurface Section	Resistivity	Chargeability
Mixture of leachate and waste	***Low*** *(< 10 Ωm) (Dahlin et al. 2010)*	***High*** *(> 70 ms) (Dahlin et al. 2010)*
Saturated clay (uncontaminated)	***Low*** *(< 10 Ωm) (Guérin et al. 2004)*	***High*** *(> 70 ms) (Gazoty et al. 2012)*
Unsaturated waste	***High*** *(> 1000 Ωm) (Dahlin et al. 2010)*	***High*** *(> 70 ms) (Dahlin et al. 2010)*
Unsaturated soil	***High*** *(> 1000 Ωm)* (Leroux et al. 2007)	***Low*** *(< 20 ms)* (Leroux et al. 2007)
Leachate Plume	***Low*** *(< 10 Ωm) (Guérin et al. 2004)*	***Low*** *(< 20 ms) (Gallas et al. 2011)*
Saturated waste (without leachate)	***Intermediate*** *(30 –150 Ωm) (Dahlin et al. 2010)*	***High*** *(> 70 ms) (Dahlin et al. 2010)*
Saturated soil (uncontaminated)	***Intermediate*** *(30 –150 Ωm) (Guérin et al.2004)*	***Low*** *(< 20 ms) (Dahlin et al. 2010)*

Table 2: Interpretation template

The L_1-norm inversion allows for considerable variation in the inverted models and it is more robust against noise in the data. This process is the most appropriate for the IP data that is sensitive to noise (Dahlin et al. 2010).

Results and Discussion

The probable sections in the subsurface of a municipal landfill are the leachate plume, the waste deposit (saturated or unsaturated) and the uncontaminated soil (saturated or unsaturated), while clay with distinct chargeability profile is used to cover waste intermittently at landfill sites. Among these segments, the interpretation of chargeability anomaly of the disseminated leachate is the most controversial in the analysis of subsurface strata at solid waste disposal sites.

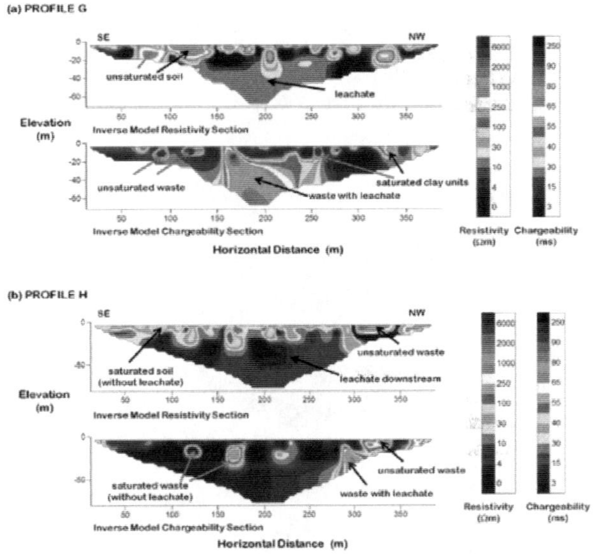

Figure 2: Resistivity and chargeability inverse models (a) Profile G; (b) Profile H

Abu-Zeid et al. (2004) express that heavy metallic loads exist in the leachate plume as suspended materials, and these insoluble particles promote high IP response through the electronic "electrode polarization." On the contrary, Gallas et al. (2011) explain that the ionic "membrane polarization," is the only IP response in the leachate, and it reduces in the progressive accumulation of ions. They emphasize that the membrane polarization diminishes while ions are accumulating near to the membrane in sediments. However, the quantity of TDS (Total Dissolved Solids) in disseminated plume samples is greater than that of TSS

(Total Suspended Solids) and invariably higher than the suspended metallic loads (Table 1).

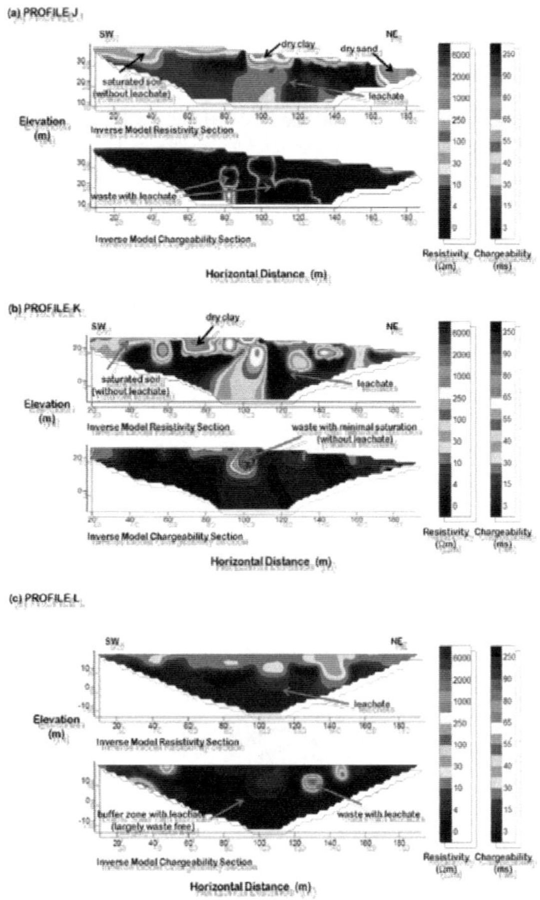

Figure 3: Resistivity and chargeability inverse models (a) Profile J; (b) Profile K; (c) Profile L

The higher proportion of TDS over TSS establishes the dominance of membrane polarization over electrode polarization in the overall potential of the plume's IP responses. The dominant membrane polarization thus determines the plume's chargeability status, although the polarization in turn weakens. This impediment in the vicinity of high concentrations of ions is due to cation adsorption by clay particles (Schön 1996).

In view of the appraisal on contaminant plume, this study affirms low chargeability status to the diffused leachate. On the basis of this assessment and corroboration of the results from the cited literature, resistivity and chargeability ratings were determined for other strata of the landfills' subsurface (Table 2). Subsequently, the various sections of the landfill generated by the inverse models were differentiated (Figs. 2 and 3).

Fig. 2 shows two 400 m models (for Profiles G and H), which are perpendicular to the gradient of the landfill. Fig. 3 has three 200 m sections (for Profiles J, K, and L) parallel to the slope of the dumpsite. The high resistivity (>1000 Ωm) spread on the ground surface from 65 m to 180 m on Profile G likely an unsaturated soil, which is part of the clay cap (Fig. 2a). The dry units extend to a depth of about 20 m, with unsaturated refuse evident within these units based on the high chargeability and high resistivity.

Profile G has a large waste unit at its midpoint. To the right of this deposit are small high IP units at shallow subsurface (Fig. 2a). The near subsurface spread of these high chargeability sections is likely to be part of the saturated and uncontaminated clay cap material. The models of Profile H display a surface with intermediate resistivity and weak chargeability (Fig. 2b), which signifies a saturated and uncontaminated soil (Guérin et al. 2004; Dahlin et al. 2010). The right end of the models depicts an exposed dry waste unit to a depth of about 20 m, which was affirmed by high IP and resistivity values.

Two saturated waste deposits, likely extensions of the massive waste marked out previously in Profile G, are located at 120 m and 170 m in the chargeability section of Profile H (Fig. 2b). These units are leachate free, confirmed by the intermediate resistivity values. However, a waste section saturated with contaminant plume (Dahlin et al. 2010) is present at almost the similar depth coinciding with 300 m.

The middle of Profile J shows a significant waste unit submerged in leachate plume while Profiles K and L show minute quantity of waste units (Fig. 3). Unpolluted saturated soil covers Profile J from 0–90 m, while 90–145 m is unsaturated land, which is most likely the clay cover material (Fig. 3a). Exposed units of dry soil confirmed to be sand based on surface observations lie at the extreme of the model.

Most part of the surface of Profile K (Fig. 3b) is unpolluted saturated soil (intermediate resistivity and low IP response), with a few areas that are unsaturated soil [high resistivity and low IP response (Leroux et al. 2007)]. Fig. 3c presents Profile L, which consisted of a leachate largely waste free and considered to be within the buffer zone between the waste sections and the leachate channel. The chargeability models of Profiles K and L show isolated waste units of similar size. A proportion of the waste deposits in Profile K is saturated and leachate free while those in Profile L are completely immersed in leachate.

Information from the Control Profile (Fig. 4) upstream further confirmed the contrast of resistivity and chargeability results between surveyed landfill sites and the uncontaminated areas, although responses from unpolluted saturated clay zones could upset the signals of waste deposits. However, the non-chargeable character of the leachate plume, which is corroborated in this study, distinguishes the contaminant plume from saturated clayey sections while outside the landfill waste boundaries, both of which display low resistivity signals.

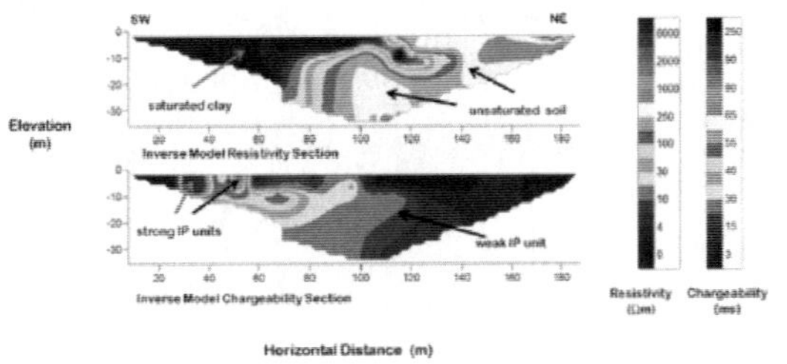

Figure 4: Resistivity and chargeability inverse models for the Control Profile

Conclusions

With inference from the physiochemical evaluation of leachate plume and results of the cited literature, this study adopted an analysis scheme for resistivity and chargeability models generated at the landfill site. The template assigns Low (< 10 Ωm), Intermediate (30 - 150 Ωm) and High (> 1000 Ωm) resistivities. The IP responses have Low (< 20 ms) and High (> 70 ms) chargeability values. The ranges of values missed out in the chart are considered as the transition zones between neighbouring strata.

The result of the assessment of chargeability profile of the waste-free leachate plume is encouraging in the utilization of IP surveys at solid waste dumpsites. However, the limitation of the layouts hinders the desired depth to investigate possible contaminant leakages. A 3D survey with a lower separation between the profiles would provide an enhanced image of the subsurface, but this would be more expensive and time-consuming.

References

Abu-Zeid, N., G. Bianchini, G. Santarato G, and C. Vaccaro, 2004, Geochemical characterisation and geophysical mapping of landfill leachates: the Marozzo canal case study (NE Italy). Environmental Geology, **45**, 439–447.

Bernstone C., and T. Dahlin, 1999, Assessment of two automated electrical resistivity data acquisition systems for landfill location surveys: two case studies. Journal of Environmental and Engineering Geophysics, **4**, 113–121.

Dahlin T, Rosqvist H, Leroux V (2010) Resistivity-IP for landfill applications. First Break, **28,** 101–105.

Gallas, J. D. F., F.Taioli, and W. Malagutti Filho, 2011, Induced polarization, resistivity, and self-potential: a case history of contamination evaluation due to landfill leakage. Environmental Earth Sciences, **63**, 251–261.

Gazoty A, Fiandaca G, Pedersen J, Auken E, Christiansen A, Pedersen J, 2012, Application of time-domain induced polarization to the mapping of lithotypes in a landfill site. Hydrology and Earth System Sciences **16**, 1793-1804.

Guérin, R., M. L. Munoz, C. Aran, C. Laperrelle, M. Hidra, E. Drouart, and S.Grellier,2004, Leachate recirculation: moisture content assessment by means of a geophysical technique. Waste Management, **24**, 785–794

Leroux V, T. Dahlin, and M. Svensson, 2007, Dense resistivity and induced polarization profiling for a landfill restoration project at Härlöv, Southern Sweden. Waste Management & Research, **25**, 49–60.

Loke, M. H., and R. D. Barker,1996, Rapid least-squares inversion of apparent resistivity pseudosections by a quasi-Newton method. Geophysical Prospecting, **44,** 131–152.

Martinho, E., and F. Almeida, 2006, 3D behaviour of contamination in landfill sites using 2D resistivity/IP imaging: case studies in Portugal. Environmental Geology, **49,** 1071–1078.

Schön, J. H., 1996, Physical properties of rocks – fundamentals and principles of petrophysics: Pergamon Press.

Singh, R. P., P. Singh, A. S. F. Araujo, M. H. Ibrahim, O. Sulaiman, 2011, Management of urban solid waste: Vermicomposting a sustainable option. Resources, Conservation and Recycling, **55**, 719–729.

Neotectonic evidences of Soummam active fault by electrical resistivity tomography

Djamal Machane, Rabah Bensalem*

Centre National de Recherche Appliquée en Génie Parasismiuque, CGS, Rue Kaddour Rahim Prolongée, Hussein-Dey, Alger, Algérie, tel : +213 21 49 55 60, fax: +213 21 49 55 36, email : machanedjamel@yahoo.fr

Summary

The Soummam Basin, located in the north-eastern part of Algeria is bounded on its north border by an active fault. This fault is clearly expressed on the surface in Kabylia. Several epicenters of earthquakes are near this fault. At Tazmalt region, the passage of fault is expressed on the ground by tilting vertical formations of the Quaternary. Electrical imaging profile was used to determine the location and geometry of the fault at depth. Accurate mapping of the active fault identified in its passage through several urban centers, while Algerian regulations (Algerian seismic code) prohibits construction on either side of an active fault.

Introduction

The location of active faults in Algeria is one of the most complex aspects to its introduction at the Algerian earthquake regulations. Indeed, it states that it is prohibiting any construction in 50 to 100 m on either side of an active fault. In practice, these days, it is almost impossible to apply these regulations without first defining all sites within which such faults. Specific studies for each active fault should be undertaken to arrive at the possibility of applying this Article in the field of regulation. It is in this context that one is interested in this work on the active Soummam fault by electrical resistivity.

The Soummam valley corresponds to a syncline containing post-plies Miocene formations, alluvium (gravel terraces) of Pliocene age and terraces and cones of Quaternary alluvial resting on a bedrock Cretaceous allochthonous or affected parautochthonous a ductile and brittle tectonics.

The geology of the region (Figure 1) of the Soummam, especially that of Tazmalt is taken from the Ficheur map (1912). Nevertheless, the datings were updated by Courme-Rault and Contelle [1982] and Courme-Rault [1985] for the Miocene formations of Soummam.

The Soummam faults have gradually formed thereafter and in geological time, progressing from north to south (Miocene, Pliocene and finally Quaternary). According to this hypothesis ho is the most likely, the fault that is considered active, is the southernmost and is closest to the plain and corresponds to the Soummam scarp (Fig. 2).

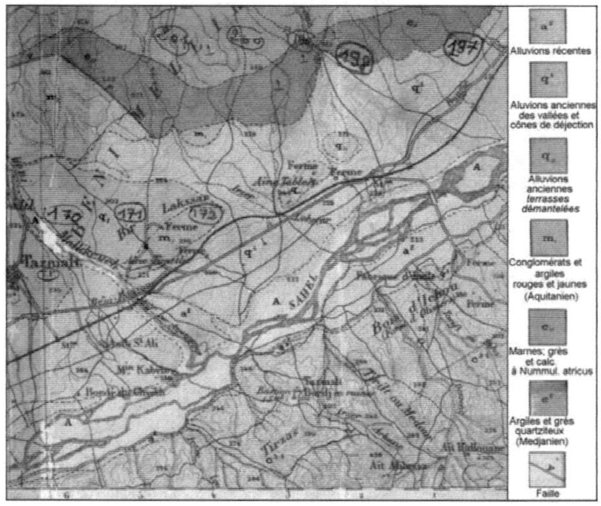

Figure 1: Geological map of Miocene and Quaternary conglomerates formations in Oued Sahel-Soummam (in Ficheur, 1912).

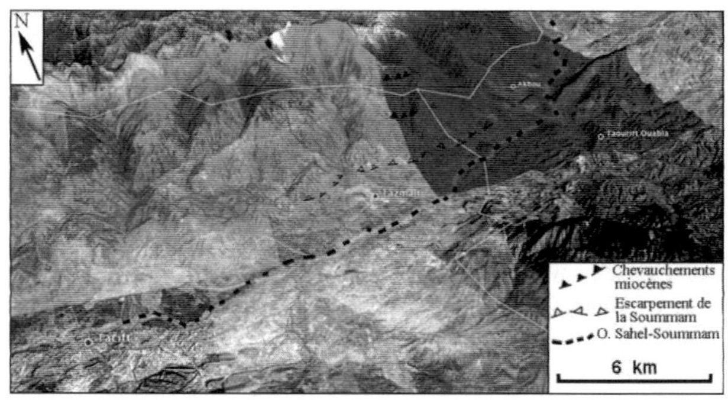

Figure 2: Soummam scarp where we see the high relief with steep slopes and a flat topography south of the scarp. The three overlapping resumed in this image were highlighted on the ground. The fault scarp in the South materialize the biggest quaternary active thrust fault in the Soummam Basin

Geological field work

The geological filed work helped highlight several tectonic faults materializing globally the Soummam active fault in this region of Bejaia.

One of these faults is materialized on the ground by a large overlap in the late Miocene formations. This intra-Miocene accident (Fig. 3) is marked by tectonic poly-phase that several families of striates are displaced by the more recent other. The reverse game this accident is attested by the great fault crochon that is seen at the base of red clay and conglomeratic formations. Microstructures forming ramps and dishes are clearly identified this intra-Miocene accident. This accident affects the training of the upper part of the series of outcropping Miocene at Tazmalt and probably corresponding to the Tortonian continental conglomerate and devoid of identifiable remains

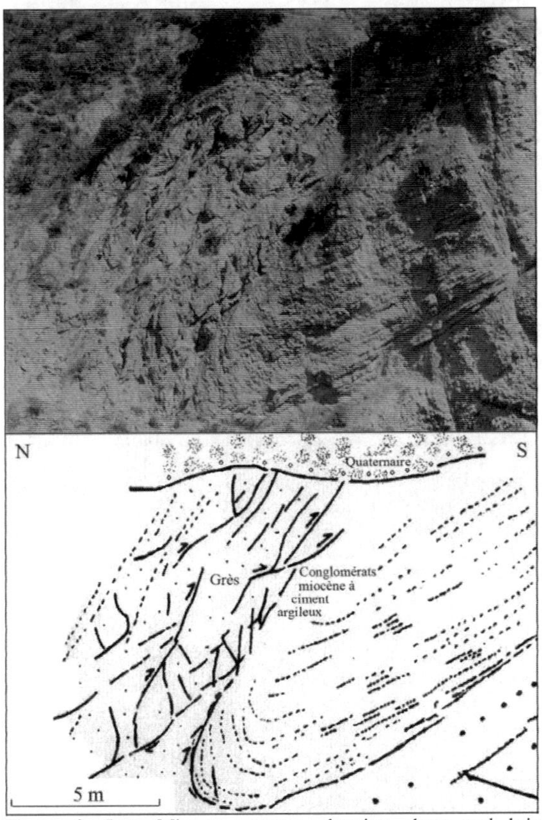

Figure 3: Late Miocene outcrop showing shear and brittle deformation in connection with a regional duplication of alpine formations south on post-Miocene layers of Soummam: reverse thrust faults or sometimes together by forming connections type "shuttle" or sigmoid specifying an overall kinematics of the North to the South

Other tectonic compressive structures (fig. 4) have been identified at the Soummam basin. They reflect the magnitude of expression by compressing the surface in this region. In addition to these fragile structures, training of Soummam are affected by deca-metric folds.

Figure 4: photography showing a reverse fault with a rejection of a foot shifting the clay marl bench overcome by the conglomerates of Miocene formations of the western basin of Soummam (Tazmalt area on the left bank of Oued Beni Mellikech , himself a left bank tributary of the Wadi Soummam-Sahel).

The photo of Figure 5 shows the overlap of regional Cretaceous formations (flysch and Tellian) on Soummam Basin Miocene formations. This accident was mapped by Coutelle (1979), it is located north of the accident intra-Miocene. On the map in Figure 2.36, it corresponds to the "post-Miocene abnormal contact Soummam". At the accident area, the striations showed both games that reverse fault and normal fault. This probably corresponds to the presence of more than one tectonic episode. Indeed, the cracks between the marl and sandstone formations of mudstones probably played in normal faults at the time of filing (as they show a domino expansion stage) and were at times in compression overlapping cretaceous. At this level, one is at the top of the Miocene series flush with Tazmalt.

At this site, the stratification is almost horizontal. Indeed, the dip of the basal conglomerates gives direction N102 ° and a dip of 6 ° to the NNE. This is also confirmed by the Stereogram of Figure 6 taken to another site that gives us a dip of less than 20 °.

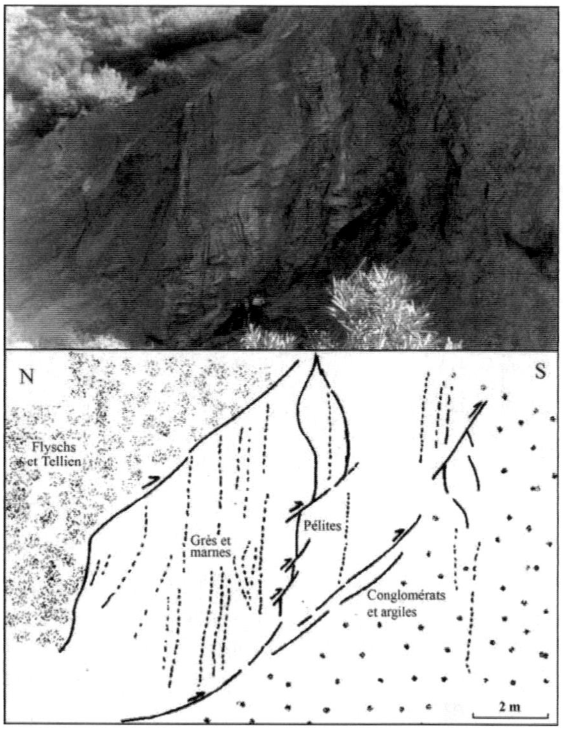

Figure 5: Découpage cassant des grès du Miocène supérieur en sigmoïdes plurimétriques constituant un copeau tectonique à la base du chevauchement régional du domaine tello-massylien sur les formations molassiques miocènes post-nappes du bassin de la Soummam à Tazmalt.

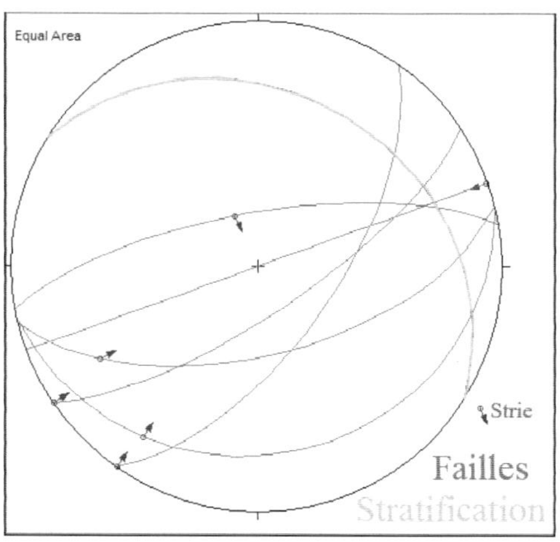

Figure 6: Stereogram of stratification, faults and striates, got close to the overlap of the Tellian Miocene

Geophysical acquisitions and results

An automatic acquisition system was used with the Sting R1 IP resistivity and Swift (box providing the connection between the electrodes 36 and resistivity). The device used in this acquisition is the Wenner and is characterized by good depth of investigation and sensitivity to resistivity variations.

A profile with the number of electrodes is 33 with and inter-electrode spacing of 6 meters was achieved at the Wadi Beni Mellikech (tributary of the Wadi Soummam) to Tazmalt perpendicular to the escarpment of the Soummam accident at the site (fig. 7).

Figure 7: Site of geophysical acquisition along the Soummam fault at Oued Beni Mellikech.

The processing of data obtained from this profile allows us to have a two-dimensional image of the basement. This represents the spatial distribution of resistivity of the basement. The resulting image due to an inversion method is a very reliable and precise treatment to obtain a solution that converges closer to reality. The processing of geophysical data was performed with the program RES2Dinv Geotomo software.

Interpretation of the data in the methods of electrical prospecting is based on resistivity contrasts between the different geological formations. These contrasts can have several causes: facies change, the presence of water in geological formations, heterogeneities existence of a geological accident, there was a stratigraphic contact (lateral facies variation) or the presence of any other conducting body or resistant. The models obtained after inversion highlight contrasts between resistant and conductive formations. These contrasts, according to their size and geometry, can be interpreted geologically.

The resistivity of cut obtained (figure 3) after treatment shows a very large resistivity contrast and a variation ranging from 5 to 300 ohm.m Ohm.m. On this cut at the

Tazmalt area easily can be seen on the profile a strong structure to the north, which is superimposed on a less strong structure in the south. Contact between the two has a dip to the north (according to what is known about the accident of Soummam) and strong enough dip deep values and low surface.

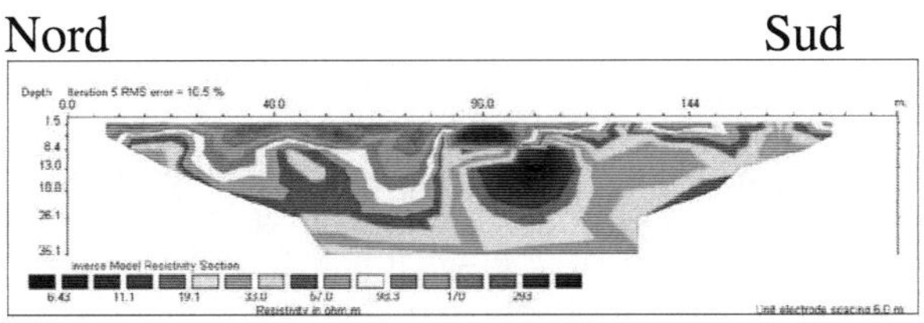

Figure 3: Geoelectric section perpendicular to a segment of the Soummam fault at Oued Beni Mellikech to Tazmalt. We see clearly resistant formations (red) come to rest on the conductive formations (blue).

Conclusions

Satellite imagery showed us the great escarpment affecting Soummam Basin and the mountains located further north, going towards the Djurdjura.

Geological field work have located the evidence of the fault Soummam, particularly the evidence related to the neotectonic activity of this accident. We have seen that this flaw affects most often quaternary training, although it is best expressed in the ancient formations.

These studies reinforce and complement those of Boudiaf *et al.* (1999) on the high seismic potential of the Soummam active fault. Indeed, these tectonic evidences have been obvious for the first time in the field.

The geophysical work conducted at Soummam region in order to materialize the Soummam active fault allowed first to mount his deep geometry, at least the first tens of meters, is not constant, Indeed, his dip increases with depth. They also showed the ability to locate the fault in places where it was still unknown, and therefore be able to make precision mapping.

Study methodology applied for the asset accident should be a prelude to the study of all Algerian active faults, leading to precise mapping in order to implement Algerian code building Article [RPA, 2003] on non urban sites through which the active faults, article that prohibits construction of 50 100 m of both sides of an active fault.

References

Boudiaf H., Philip H., Coutelle A. Ritz J.F., 1999, Découverte d'un chevauchement d'âge Quaternaire au sud de la Grande Kabylie (Algérie). Geodinamica Acta. 71-80. DOI:10.1080/09853111.1999.11105332

Bouillin J.-P., 1986, Le bassin maghrébin: une ancienne limite entre l'Europe et l'Afrique à l'Ouest des Alpes. Bull. Soc. géol. France. 2 (4), série 8, 547-558.

Courme-Rault M.D., 1985, Stratigraphie du Miocène et chronologie comparée des déformations suivant deux transversales des Atlasides orientales (Algérie, Sicile). Doc. thesis, université of Orléans.

Courme-Rault M.D. and A. Coutelle, 1982. Le Miocène de la Soummam (Algérie) : précisions sur l'âge de ses niveaux de base, existence d'un cycle inférieur anté-nappes sud-telliennes, corrélations. Géologie méditerranéene. Tome IX, n°2. 99-107.

Ficheur E., 1912, Carte géologique au 1/50 000 de Tazmalt (feuille n° 66). Publication du Service Géologique d'Algérie.

Frizon de Lamotte D., B. Saint Bézar, R. Bracène and Mercier E. 2000. The two main steps of the Atlas building and geodynamics of the western Mediterranean. Tectonics 19, 740–761.

Geodynamic Framework and Petroleum Potential of the Cap Bon- Gulf of Hammamet Province –Tunisia

O. ABIDI [1], M H.INOUBLI [1], K. SEBEI [1], H. BOUSSIGA[1], A. AMIRI [1], I. HAMDI [1], A H. BEN SALEM [2]*
1:Faculty of Sciences of Tunis, University Tunis-El Manar, Tunisia, Tunis, Tunisia.
2: Entreprise Tunisienne D'Activités Pétolières

Summary

Subsurface analysis of the Cap Bon-Gulf of Hammamet province CGP of Tunisia, based on well logs, seismic data and outcrops reveals the dynamic basin evolution and portrays its hydrocarbon accumulation. Thousands of kilometers of seismic profiles, dozens of well logs as well as gravity measurements together with surface geology were fully examined and integrated in depth.
Seismic-based mapping of Key-horizons of specific sequences illustrate the partition of the dynamic sedimentation all over the strike-slip faulted basins. Selected time slices spanning upper Cretaceous to Tertiary illustrate the spatial environmental and depositional differentiation and allow retracing its geodynamic evolution and its integration into the Mediterranean dynamic evolution.
3D analyses indicate episodes of major regional change in paleogeographic and tectonic setting. These occurred in the late Maastrichtian to lower Paleocene, middle to late Eocene, late Oligocene, early Middle Miocene, late Miocene and late Pliocene to Villafranchian. Episodes are probably induced by the relative motions of the African and Eurasian plates. Plate motions induced discrete steps in the regional kinematics and geodynamics that governed the palaeogeographic evolution of the CGP.
Several of oil and gas fields have been discovered; a great number are accumulated in the Middle Miocene Birsa formation sands: Tazarka, Birsa, Oudna, Baraka fields. Other hydrocarbon reservoirs were recognized. These are drawn by: the Maamoura field in the Langhian Ain Grab Formation and the Senonian Fractured chalky limestone, the Yasmin oil field producing from the Langhian Ain Grab Formation, the Halk El Menzel oil field reservoired in the Oligocene limestones, the Belli oil field in the intensively fractured Ypresian Bou Dabbous carbonate formation, the Zinia and the Dougga fields reservoired in the Campanian–Maastrichtian chalky limestone of the Abiod Formation, the Lotus gas field accumulation in the Cretaceous Serdj Formation, the most ancient Cap Bon gas field producing from the Hauterivian-Valanginian M'cherga Formation sands,etc.

The paper intends a regional review of the area through a comprehensive and consistent interpretation and modeling. It illustrates that significant exploration potential still exists in the studied province with quite some sizeable undrilled structures. The dominant play remains within the Neogene. The study is directed toward unraveling and understanding the Complexity of the CGP.

Introduction

The Cap Bon (fig.1) Peninsula is known by the Sidi Abderrhaman Gas field discovered in the fifties of the previous century which produced from Lower Cretaceous sands and located on the anticline having with the same name.
The Gulf of Hammamet (fig.1) is a complex zone with salt structures, living anticlines, deep Mio-Pliocene depocentersand grabens. Cretaceous and Tertiary reservoirs are interesting with the presence of the main Mesozoic and Cenozoic source rock intervals in the area. Several discoveries have been made during the last decades such as Tazerka, Birsa, Oudna, Yasmin, Cosmos, Zelfa and Baraka oil fields (producing from Mid-Miocene Ain Grab limestones and Birsa sands) and Maamoura (tested oil from the Campanian-Maastrichtian Abiod fractured chalky limestones).

Lithostratigraphy

Outcrops are characterized by significant thickness and facies variations. Sedimentary records vary from Eocene to Quaternary age within the study area (Fig. 2). Older series are described from neighboring outcrops exposed further to the South-West in central Tunisia. Surface geology reveals non-depositional or erosional hiatuses witnessing the complexity of the tectono-sedimentary events that controlled the structuring of the region.

Structural Evolution

The integrated compilation of seismic and borehole data show that the area of study undergoes great lateral and vertical variations which are induced by the global geodynamic evolution resulting from the Eurasian-African plate convergence. Borehole calibrated seismic sections lead to the following episodes of major tectonic change.
The geodynamic evolution started from the Tethyan openings by the migration of the African plate to the north east (Biju et al, 1977;. Olivet et al., 1982; Dercourt et al., 1985; Bouaziz, 2002; Guiraud, 2005). The Trias is characterizes by a distension phase, resulting the horst and graben structure (Bedir, 1995).

Petroleum system and Play

The Tunisian stratigraphic column presents many source rock intervals with good potential; these are spread from Paleozoic to Cenozoic. A number of source rock intervals were evidenced in the area and its neighboring through studies and analyses based on outcrops and borehole data samples. The main proven source rocks here are the Albian Fahdene formation,the Cenomanian to Turonian Bahloul formation and the Ypresian BouDabbous formation.

The Albian shales and limestones cover northeastern Tunisia. It is an organic rich interval with good TOC values varying from 0.65 % to 4 %, reaching 400m thickness in the CGP area. The Late–Upper Cenomanian Bahloul Horizon is one of the main source rocks in Tunisia; it consists of laminated and marly limestone with large Globigerinids of Late Cenomanian age (Robaszynsky et al. 1990a and 1990b). The Bahloul fm. has been recently documented by Zagrarni et al. (2008). Surfaces sections offered TOC values varying from 1.1% to 3.6% in central Tunisia. Borehole samples within the organic rich part of the Bahloul Horizon lead to a TOC ranging from 0.8 to 6.4%. The hydrocarbon potential may reach 33g/kg at Oued Bahloul.

The BouDabbous Fm. is made up of a well bedded mudstone with abundant planktonic foraminifers. It covers the Gulf of Hammamet and parts of the Pelagian Sea. With a thickness varying between 50m to 350m, the TOC content of the Bou Dabbous ranges between 0.5 and 2.5% (Fournié, 1978; Burollet et Oudin, 1980). Due to its proximity, the BouDabbous sourced the Nummulite packstone of the Sidi El Itayem and Ashtart oil fields.

Most hydrocarbon accumulations are reservoired within the Mid-Miocene Birsa Formation sandstones, such as Dougga, Tazerka, Oudna, Birsa, Cosmos and Yasmin fields. These sands exhibit excellent reservoir quality. The Ain Grab limestonesfmflowed some 3,500bopd on test on Maamoura. Eocene reservoirs include the Halk el Menzel Formation and the Lower Eocene BouDabbous Formation. Fractured BouDabbous Formation is the significant reservoir onshore Cap Bon province such as at Belli, Al Manzah and Beni Khalled fields. The Abiod Formation which possess some reasonable matrix porosity a typically low permeability. Fracture porosity and connected pore volume is required for the Abiod Formation to be a viable reservoir.

The critical moment of the CGP petroleum systems is the late Miocene to early Pliocene where peak hydrocarbon migration has occurred. Trap retention since this time appears not to be an issue in the studied area.

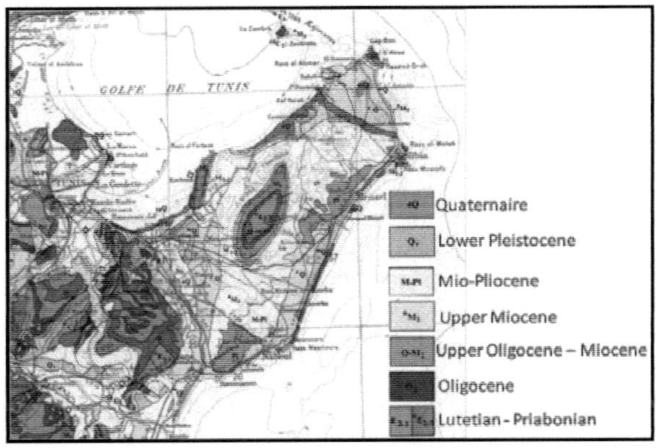

Figure 1: Geological map of the Cap Bon area (map 1/500, 000)

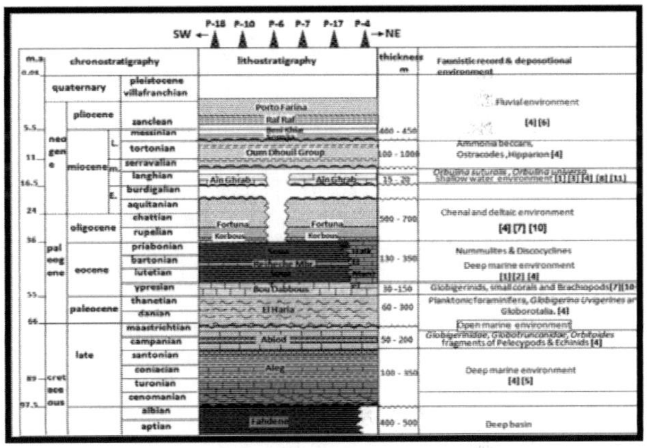

Figure 2: Synthetic lithostratigraphic Chart for the study area

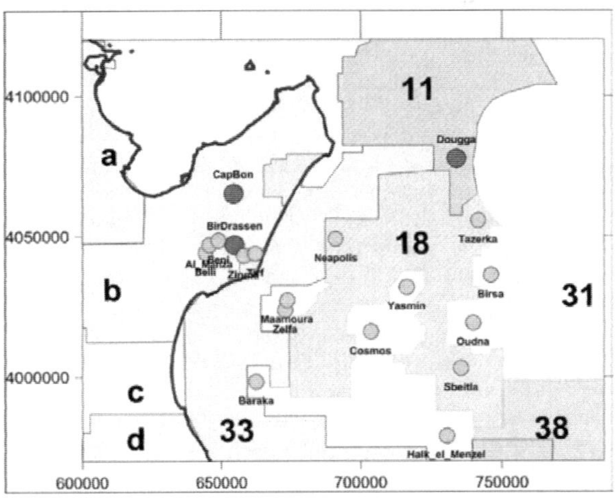

Figure 3: Distribution of Research blocks and producing concessions all over the are. (origine: E.T.A.P. 1999).

Conclusions

This is a compilation based on surface geology, borehole data (well logs and subsurface geology), and surface seismic data covering the CGP area. The region expresses imbricate extensive and compressive structures with complicated arrangements.Across-calibrated structural and stratigraphic consistent interpretation and assessment show that significant exploration potential still exists within the Gulf of Hammamet. The dominant exploration plays within the CGP area remain within the upper cretaceous tertiary interval while focusing on the lower to middle Miocene series. These results quality are well supported with the abundance of source rocks. Finally, due to the structural complexity of the area, detailed geological modeling using 3D seismic survey is well recommended.

Reference

BEDIR M. (1995). Mécanismes géodynamiques des bassins associés aux couloirs de coulissement de la marge atlasique de la Tunisie. Sismo-tectonique et implications pétrolières. Thèse Sci., Université Tunis II. 417p.

Biju-Duval B., Dercourt J., Le Pichon X., (1977). From Tethys ocean to the Mediterranean Sea, In: B. Biju-Duval and L. Montadert (Eds.), Structuralhistory of the Mediterranean basin, Split 1976, 1977, pp. 143–164.

Bouaziz, S., Barrier, E., Soussi, M., Turki, M.M., Zouari, H., (2002): Tectonic evolution of the northern African margin in Tunisia from paleostress data and sedimentary record. Tectonophysics 357, 227–253

BUROLLET P. F. and OUDIN J. L. (1980) : Paléocène et Eocène en Tunisie : Pétrole et Phosphates - Doc. BRGM N° 24, Orléans, pp. 203-216, fig3.

Dercourt J., Zonenshain L.P., Ricou L.E., Kazmin V.G., Le Pichon X., Knipper A.M., Grandjacquet C., Sborshikov J.M.,Boulin J., Sorokhtin O., Geyssaut J., Lepvrier C., Biju-Duval B., Sibuet J.M., Savostin L.A., Westphal M., Lauer J.P. (1985) : Présentation de 9 cartes paléogéographiques au 1/20.000.000 s'étendant de l'Atlantique au parmir pour la periode du Lias à l'Actuel. Bull. Soc. Géol. Fr., (8), I, 637-65.

Entreprise Tunisienne d'Activites Petrolieres, C.1999, Information packet. Tunis, Tunisia, Entreprise Tunisienne d'Activites Petrolieres (ETAP).

FOURNIE D. (1978) : Nomenclature lithostratigraphique des séries du Crétacé supérieur au Tertiaire de Tunisie, Bull. Centre Rech. Explor. Prod. Elf Aquitaine, Pau, vol. 2 n°1, pp. 97-148.

Guiraud R., Bosworth W., Thierry J., Delplanque, A. (2005): Phanerozoicgeological evolution of Northern and Central Africa: Anoverview. Journal of African Earth Sciences. 43, 83–143.

OlivertJ.L., Bonnin J., Beuzart P., Auzende J.M., (1982) : Cinématique des plaques et paleogeographie. Bull. Soc. Geol. France 24, 875–892.

ROBASZYNSKI F., CARON M. , DUPUIS G., AMEDRO F., CALADRA F., DELOFFRE R.,GARTNER S.,GONZALEZ DENOSO J.M., HAEDENBOL J. and LINARES D. (1990a) : Litho-biostratigraphie, microfaciès et séquences stratigraphiques dans le Turonien d'un secteur de laTunisie Centrale. Bull. Soc. Géol. France, 8, VI, pp 1011-1024.

ROBASZYNSKI F., CARON M. , DUPUIS G., AMEDRO F., CALADRA F., DELOFFRE R.,GARTNER S., GONZALEZ DENOSO J.M., HAEDENBOL J. and LINARES D. (1990b): A tentative integrated stratigraphy in the Turonian of Central Tunisia : Formation zones and sequential stratigraphy in the Kalaat Senan area. Bull.C.R. Expl. Prod. Elf Aquitaine, Pau, 14, 1, pp 213-384

Acknowledgments

The authors express their appreciation to 'Entreprise Tunisienne d'Activités Pétrolières" of Tunisia for permission to use the information on discussed areas, and also for their support and access to the various seismic data sets.

MAPPING FLOOD PRONE AREAS IN EKPOMA, EDO STATE, NIGERIA USING ELECTROMAGNETIC PROFILING AND ELECTRICAL RESISTIVITY SOUNDINGS

Aigbedion Isaac

Email: isaacaigbedion @yahoo.com

ABSTRACT

This study exploits the integrated approach of using the electromagnetic profiling and electrical resistivity soundings in flood management with the goal of mapping areas vulnerable to flood in Ekpoma- Edo State-Nigeria. Profiles length of 900m, 200m, 140m and 500m were covered using the Abem Wadi (West-East Orientation) at a measuring interval of 20m. The electromagnetic data were plotted against the station position, and the plots were found to exhibit positive peak and negative trough. The Schlumberger vertical electrical sounding was conducted in the conductive zones prone to flooding. Resistivity values were obtained by taking readings using the Ohmaga resistivity meter. The readings were further processed accordingly using Resist software meter. The study reveals that subsurface geophysical method is very useful for flood control.

Characterization by electrical resistivity imaging 3D and GPR of dissolutions in Quaternary carbonate sandstone formations in Algiers-West

F. Khaldaoui[1,], Y. Djediat[2], M. Djeddi[1], A. Ydri[1]*
[1] Laboratoire de Géophysique, [2]Laboratoire de Géodynamique.
FSTGAT-Université des Sciences et de la Technologie Houari Boumediene

Summary

In recent years Algeria has experienced significant development in urbanization. Open areas often do not meet the objective of a planning. The presence of cavities/fractures affects the entire geodynamic instability processes, which is acting on the soil surface. Problems of this type may cause time delays and cost increase of the project and in certain cases may stop the project. For this purpose, a geophysical study of three-dimensional (3D) electrical imaging and ground carried out penetrating radar (GPR) on a site located in the Northwest of Algiers, to map cavities. This site reveals the presence of any cavities when the excavation work for the foundation platform started.

Introduction

The sources of hazards in civil engineering disciplines result essentially from undetected near-surface structures, such as cavities, fracture zones (natural and anthropic). The presence of this structure is a real threat to infrastructures. The implementation of civil works requires a detailed survey of the selected areas. In the north-west of Algiers (fig.1), important dimensions of voids were discovered during excavations of the transaction for the location of residential buildings foundations.

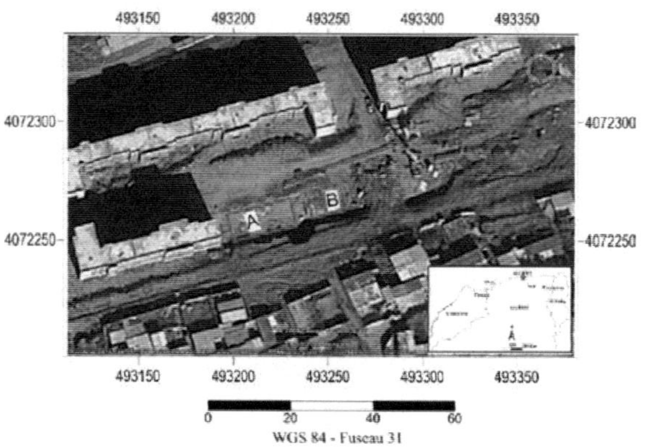

Figure 1: Satellite image of the study area

In this context, a geophysics study by three-Dimensional (3D) electrical imaging (Chambers and al., 1999; Dahlin and al., 2002; Tejero and al, 2002; Ortega et al., 2010 Gambetta et al., 2011) and ground penetrating radar (Benson, 1995; Hruska and Hubatka, 2000; Ansón, 2008; Carbonel 2015) have been applied at the site, to map cavities. These surveys were aimed at characterizing the shallow subsurface in order to delineate features that may cause structural instability.

Geological background

At this site, we identified the pseudokarsts in Quaternary Tyrrhenian formations (Aymé, 1952, Betrouni, 1983). These pseudokarsts result from a form of erosion by dissolving limestone cements which are the primary and secondary connections dune sandstone. The dune formations are partially cemented because in these materials, carbonate cement dissolution has undergone several phases (fig.2) and recrystallization during the early diagenesis in the Tyrrhenian. The fracturing network promoted the drainage of seepage and therefore the dissolution of carbonate inter-particle bonds (fig.3).

Figure 2: Karstic Environment at Northwest of Algiers: A and B: The entrance of the cave, C) pseudokarst cavity opens in the sandy limestones, D) Karst fissure with infill of red sandy clays.

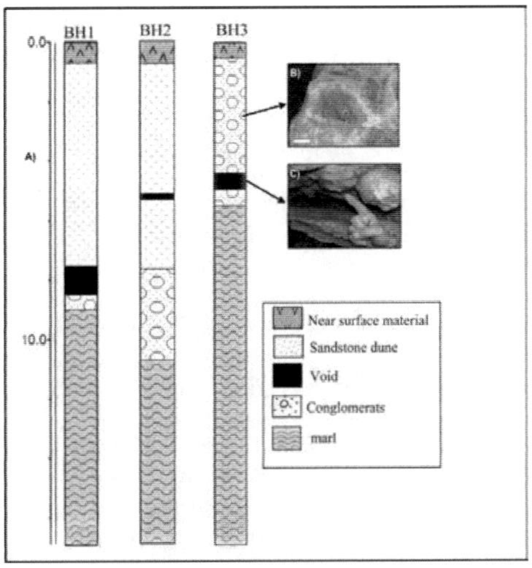

Figure3: A) Example of a Lithology Columns for three boreholes located near the areas study, B) Fill in empty geodic calcite dissolution of pseudokarstic, C) Crystallization of calcite dissolution on the walls of the cavity

Methodology and tools

Geophysical field measurements were conducted using electrical resistivity imaging (ERI) and ground penetration radar (GPR). Two areas (A and B) were chosen to be studied (fig. 4). The Resistivity data were acquired using the dipole–dipole electrode array. The instrument used was a Syscal Junior multi-electrode system from IRIS Instruments. The resistivity data were inverted using 3D algorithms based on smoothness constrained least-squared algorithm (Loke and Barker, 1995, 1996).

GPR measurements have also been made for locating shallow dissolution zones round Penetrating Radar (GPR) data were collected using a GSSI model SIR-20 system with 200 MHz Antenna. Its principle is based on the propagation of electromagnetic high-frequency waves. The electromagnetic signal is emitted by the transmitter into the surface of investigation; the receiver captures their reflections from interfaces where there is dielectric contrast. The acquisition time range was 100 ns, the number of scans was set at 40 scans per meter and 512 samples per scan. The data were collected in continuous mode along parallel and perpendicular profiles.
This survey consisted of 18 profiles in two terraced areas A and B.

Figure 4: Picture of the two areas A and B for survey

Discussion and conclusions

The geophysical surveys in Algiers-West area reveal a complex geology. They have detected a number of karst cavities and fractures in the calcareous sandstone associated with carbonate dissolution by groundwater.
3D electrical imaging (fig.5) shows degree heterogeneity in the spatial distribution of electrical resistivity. A karst system was highlighted. It is characterized by a high resistivity gradient (>3000 Ohm.m) that can be probably the extension of cavities discovered during the excavation operation.

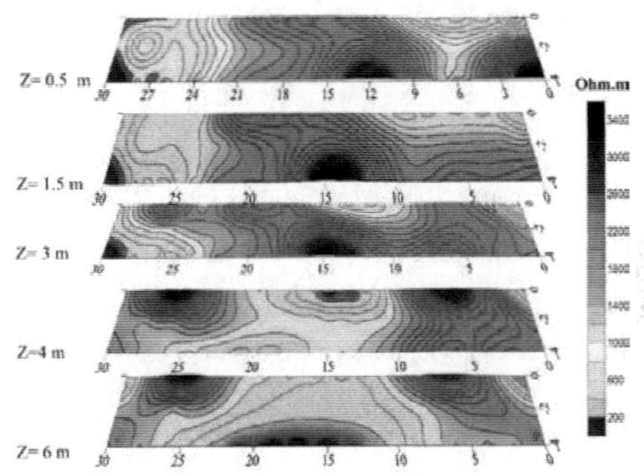

Figure 5: A 3D resistivity survey (area A) can be used to assist in the detection of dissolution zones

The application of the GPR confirmed the presence of these anomalous structures. Note that these structures are manifested by multiple reflections on GPR section and associated with the presence of cavities (fig.6a, b). They are located between 2 and 6 m at a depth.
Finally, The ERI and GPR methods prove the efficacy for detection and characterize areas of dissolution.

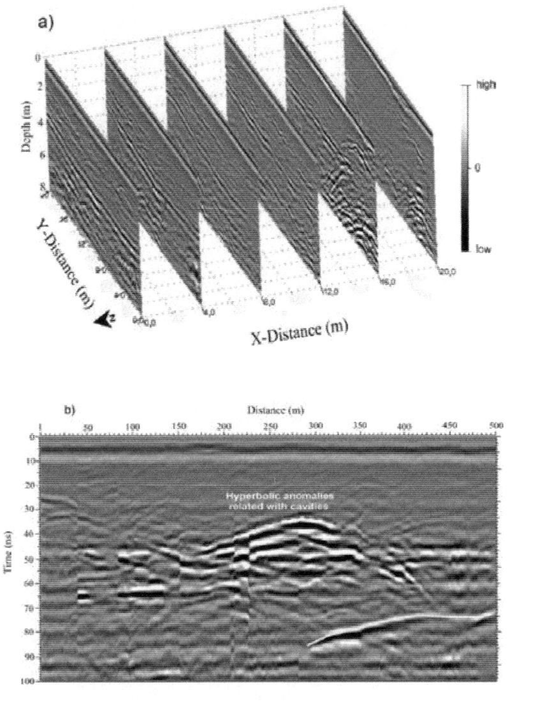

Figure 6: a) A series of five radargrams performed in area B, b) Example of GPR section with 200 MHz in study area, hyperbolic anomaly associated with the presence of cavity

References

Ansón, D., 2008. Detection of underground cavities by combining gravity, magnetic and ground penetrating radar surveys: a case study from the Zaragoza area, NE Spain. Environmental Geology. **53** 1067-1077.

Aymé, A. 1952. Le Quaternaire littoral des environs d'Alger. IIe Congrès panafricain de Préhistoire et d'études du Quaternaire, Alger.

Benson, A.K., 1995, Applications of ground-penetrating radar in assessing some geologic hazards examples of groundwater contamination, faults, and cavities: JOURNAL OF APPLIED GEOPHYSICS, **33**, 177-193.

Betrouni, M. 1983. Le Pleistocène supérieur du littoral ouest Algérois. Thèse université d'Aix-Marseille.

Carbonel, D., Rodríguez-Tribaldos V, Gutiérrez F, Galve JP, Guerrero J, Zarroca M, Roqué C, Linares R, McCalpin JP, Acosta E, 2015. Investigating a damaging buried sinkhole cluster in an urban area (Zaragoza city, NE Spain) integrating multiple techniques: Geomorphological surveys, DInSAR, DEMs, GPR, ERT, and trenching. Geomorphology, **229**, 3-16.

Chambers, J., Ogilvy, R., Meldrum, P., and Nissen, J., 1999, 3D Resistivity imaging of buried oil- and tar-contaminated waste deposits:European. JOURNAL OF ENVIRONMENTAL AND ENGINEERING GEOPHYSICS, **4,** 3–15.

Dahlin, T., Bernstone, C., and Loke, M. H., 2002, A 3-D resistivity investigation of a contaminated site at Lernacken, Sweden: GEOPHYSICS, **67**, 1692–1700.

Hruska, J., Hubatka, F., 2000: Landslides investigation and monitoring by a high performance and cavities in limestone: GPR 2000. 8th International Conference on Ground-Penetrating Radar, Ausralia.

Gambetta, M., Armadillo, E., Carmisciano, C., Stefanelli, P., Cocchi, L., and Tontini, F.C., 2011, Determining geophysical properties of a near-surface cave through integrated microgravity vertical gradient and electrical resistivity tomography measurements: Journal of cave and karst studies, **73**, 11–15.

Loke, M.H., Barker, R.D., 1996. Rapid least-squares inversion of apparent resistivity pseudosections by a quasi-newton method. GEOPHYSICAL PROSPECTING **44**, 131–152.

Loke, M. H., Barker R. D. 1995. Least-squares deconvolution of apparent resistivity. GEOPHYSICS, **60**: 1682-1690.

Ortega, A.I., Benito-Calvo, A., Pérez-González, A., Porres, A., Martín, M.A., 2010. Applyingelectrical resistivity tomography to the identification of endokarstic geometries in the Pleistocene sites of the Sierra de Atapuerca (Burgos, Spain). Archaeological Prospection,**17**, 233–245.

Tejero A., Chávez R.E., Urbieta J., Flores-Márquez E.L., 2002, Cavity detection in the south-western hilly portion of Mexico City by resistivity imaging: JOURNAL OF ENGINEERING AND ENVIRONMENTAL GEOPHYSICS, **7-3**, 130-139

Natural radioactivity of carbonate rocks as materials for construction and cement industry

*D. Alshamsi[a], A. Murad[a], A. Aldahan[a], X. L. Hou[b,c]

[a,*]Department of Geology, United Arab Emirates University, Al Ain, UAE

[b]Technical University of Denmark, Roskilde, Center for Nuclear technologies, 4000, Roskilde, Denmark

[c]Xi'an AMS centre and SKLLQG Institute of Earth Environment, CAS, China

Summary

In this report, we present an investigation on the naturally occurring radioactivity in carbonate rocks and possible radiological hazards. Carbonate rocks represent a vital source of materials used for construction as well as in the cement industry in the United Arab Emirates (UAE). For this pilot study samples of carbonate rocks from two regions (Ras Al Khaimah and Al Ain) were analyzed for ^{238}U, ^{232}Th and ^{40}K. The rocks are mainly Upper Triassic (230- 215 million years ago) to Lower Cretaceous (145- 140 million years ago) and vary in lithology from dolomitic to argillaceous carbonates. Chemical decomposition of powdered rock samples, using $HF+HNO_3$, was performed to prepare the samples for ^{238}U and ^{232}Th measurement which was done using ICP-MS. Potassium was measured using ICP-SFMS after fusion of the powdered rock sample with $LiBO_2$. The results indicate ^{238}U, ^{232}Th and ^{40}K activity at (4.1 – 39.1) Bq kg^{-1}, (<0.04 – 1.3) Bq kg^{-1} and (<0.2 – 0.7) Bq kg^{-1}, respectively. Also activity of the radioisotopes is generally higher in the geologically older carbonate rocks in Ras Al Khaimah than the younger ones in Al Ain. The external radioactivity hazard index is below 0.11 for all the samples which is negligible in terms of impact on health according to the guidelines recommended by United Nations Scientific Committee on the Effects of Atomic Radiation.

Introduction

Carbonate rocks are widely used in the United Arab Emirates (UAE) as construction materials and in the cement industry. In both bulk rocks and as industrial products and waste, such as dust from the cement factories, the undesirable chemical components are spread out without monitoring. Radioactive elements may form hazardous components in the rocks and knowledge about the quantities of these is needed. Generally, the major sources of radioactivity in carbonate rocks are uranium, thorium and their decay products such as radium and radon, and to a lesser extent potassium-40. In the piolt investigation presented here, samples of carbonate rocks from two selected areas, namely Ras Al Khaimah and Al Ain, are used for evaluation of naturally occurring radioactivity. The carbonate rocks in Ras Al Khaimah are mainly Upper Triassic (230- 215 million years ago) to Lower Cretaceous (145- 140 million years ago) which vary in lithology from dolomitic to argillaceous carbonates (Rizk et al., 2007; Breesch et al., 2010). Al Ain carbonate rocks are younger than Ras Al Khaimah having age ranges extending from Paleogene to Neogene (younger than 40 million years). These rocks are composed mainly of nodular and partly dolomitic limestones with interlayers of marls, anhydrites and some shale (El-Saiy and Jordan, 2007).

Materials and methodology

Seven carbonate samples were collected from Wadi Bih in Ras Al Khaimah and Jabel Hafit in Al Ain (Fig. 1). These carbonate rocks are generally composed of calcite ($CaCO_3$) and dolomite ($CaMg(CO_3)_2$) with occurrence of chert (SiO_2) and minor iron oxides. For ^{238}U and ^{232}Th measurements, powdered rock samples were digested with HF and HNO_3, the decomposed sample solution was evaporated to dryness, and the residue was dossolved in 3% HNO_3, and the measurements were performed using ICP-MS. After fusion of the powdered rock sample with $LiBO_2$, the concentration of potassium was measured using ICP-SFMS.

Results and discussion

The activity concentration of ^{238}U, ^{232}Th and ^{40}K for the carbonate samples were found to range at (4.1 – 39.1) Bq kg^{-1}, (<0.04 – 1.3) Bq kg^{-1} and (<0.2 – 0.7) Bq kg^{-1}, respectively (Table 1). The older carbonate rocks in Ras Al Khaimah contain more radioisotopes than the younger ones in Al Ain. This probably is due to the higher organic content in Ras Al Khaimah rocks. The ^{238}U, ^{232}Th and ^{40}K activities of the carbonate rocks investigated here are comparable to other carbonate rocks in the Bahamas platform, Indiana State., Brazil and India (Kenneth, 1963; Chakrapani, 2002; Conceição and Daniel, 2003; Romaniello et al., 2013).

Generally, evaluating radioactive hazards from solid natural substance involves calculating both external and internal doses. External hazard occurs when the source of radiation is outside the body, whereas internal hazard occurs if the source of radiation is inhaled or ingested. The annual effective dose of radiation includes the sum of the internal and external doses. The external hazard index (H_{ex}) is considered here to include ^{238}U and ^{232}Th and ^{40}K activities.. In some studies ^{238}U concentration is substituted by ^{226}Ra. The internal hazard index gives the internal exposure to carcinogenic or other diseases causing radioactivity exposure within the human body. Radon-222 which is progeny of ^{238}U decay may become concentrated and released from the carbonate host rocks to groundwater and into the human body through inhalation. This might cause health hazards such as breathing diseases or lung cancer.

In this study we focused on the external hazard index that may be delivered when dealing with carbonate rocks (Table 1). The index is described by the following equation according to United Nations Scientific Committee on the Effects of Atomic Radiation reports from 1988 to 2008 (UNSCEAR, 2010):

$$H_{ex} = A_U / 370 + A_{Th} / 259 + A_K / 4810 \qquad (1)$$

Where A_u, A_{Th} and A_K are the activity concentrations of ^{238}U, ^{232}Th and ^{40}K in Bq kg^{-1} and the division constant values are also in Bq kg^{-1}. The calculation indicates H_{ex} of the samples that range between 0.025 and 0.106 with the older carbonate rocks in Ras Al Khaimah showing relatively higher radiation external hazard index than the carbonates from Al Ain. The lowest calculated H_{ex} occurs in a sample form the younger age rocks (Al Ain) and was collected from a fault plane of pure calcite-dolomite minerals precipitated from thermal fluids. The highest calculated H_{ex} was in a sample from the older age rocks (Ras Al Khaimah) which is composed of chertified limestone interbedded with shale. The high H_{ex} value is probably related to adsorption of uranium on coating around chert grains as well as high organic matters in the rocks.

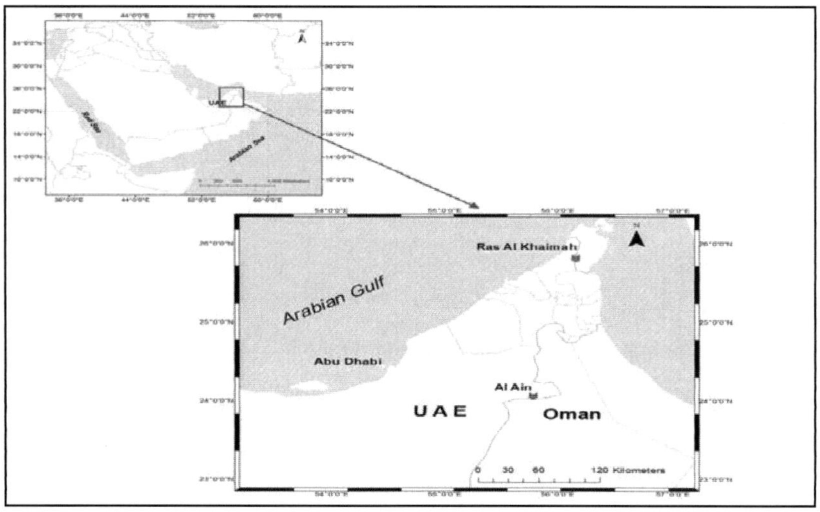

Fig. 1 Sampling location map from two the sampled regions (Ras Al Khaimah and Al Ain) in the UAE.

Table 1 Activity concentrations of U-238, Th-232 and K-40 obtained from different carbonate rock formations and their external hazard index.

Sample ID	Rock description	Age and Formation	^{238}U (Bq kg^{-1})	^{232}Th (Bq kg^{-1})	^{40}K (Bq kg^{-1})	External hazard index (H_{ex})
r-3	Grayish brown muddy limestone	Lower Jurassic, Musandam Formation	26.9	0.15	<0.2	0.073
r-8	Grayish brown chertified limestone interbedded with shale	Middle Triassic, Milaha Formation	39.1	0.18	0.34	0.106
r-15	Gray lime-mudstone	Upper Permian, Hagil Formation	7.4	1.3	0.7	0.025
r-20	Gray argillaceous limestone	Lower Triassic, Ghail Formation	16.6	<0.04	<0.2	0.044
r-30	Gray laminated marl	Middle Permain, Bih Formation	15.5	0.33	0.61	0.043
JH-1	White colored mixture of calcite and dolomite at a fault plane	Lower Eocene, Rus Formation	4.1	<0.04	<0.2	0.011
JH-2	Grayish brown chertified granular limestone	Lower Eocene, Rus Formation	14.6	0.32	0.58	0.041

Conclusion

In this pilot study samples of carbonate rocks from two regions (Ras Al Khaimah and Al Ain) were analyzed for ^{238}U, ^{232}Th and ^{40}K in order to evaluate the possible external radioactivity hazards. The rocks are mainly Upper Triassic (230- 215 million years ago) to Lower Cretaceous (145- 140 million years ago) and vary in lithology from dolomitic to argillaceous carbonates. The relatively higher ^{238}U, ^{232}Th and ^{40}K activity in the geologically older carbonate rocks of Ras Al Khaimah than the younger ones of Al Ain is related to varying organic matter content and adsorption on mineral surfaces. The external radioactivity hazard index of all the samples is < 0.11 which is negligible in terms of impact on health according to the guidelines recommended by the United Nations Scientific Committee on the Effects of Atomic Radiation.

References

Bell, Kenneth G. *Uranium in carbonate rocks*. No. 474-A. 1963.

Breesch, Liesbeth, Rudy Swennen, B. Vincent, R. Ellison, and Ben Dewever. "Dolomite cementation and recrystallisation of sedimentary breccias along the Musandam Platform margin (United Arab Emirates)." *Journal of Geochemical Exploration* 106, no. 1 (2010): 34-43.

Chakrapani, G. "Water and sediment geochemistry of major Kumaun Himalayan lakes, India." *Environmental Geology* 43, no. 1-2 (2002): 99-107.

Conceição, F. T., and Daniel Marcos Bonotto. "Use of U-isotope disequilibrium to evaluate the weathering rate and fertilizer-derived uranium in São Paulo state, Brazil." *Environmental Geology* 44, no. 4 (2003): 408-418.

El-Saiy, A. K., and B. R. Jordan. "Diagenetic aspects of tertiary carbonates west of the Northern Oman Mountains, United Arab Emirates." *Journal of Asian Earth Sciences* 31, no. 1 (2007): 35-43.

Kuroda, Paul K. "The Oklo phenomenon." In *The Origin of the Chemical Elements and the Oklo Phenomenon*, pp. 31-55. Springer Berlin Heidelberg, 1982.

Rizk, Zeinelabidin S., Abdulrahman S. Alsharhan, and Warren W. Wood. "Sources of dissolved solids and water in Wadi Al Bih aquifer, Ras Al Khaimah Emirate, United Arab Emirates." *Hydrogeology Journal* 15, no. 8 (2007): 1553-1563.

Romaniello, Stephen J., Achim D. Herrmann, and Ariel D. Anbar. "Uranium concentrations and 238 U/235 U isotope ratios in modern carbonates from the Bahamas: Assessing a novel paleoredox proxy." *Chemical Geology* 362 (2013): 305-316.

United Nations Scientific Committee on the Effects of Atomic Radiation. "UNSCEAR 2008 report, vol. I: sources of ionizing radiation. Annex A: medical radiation exposures." (2010).

Evaluation and treatment of the swelling behaviour of expansive soil derived from Enugu Shale, South Eastern Nigeria.

Author name: Ogochukwu Ozotta and Osita Okeke*
Affiliation: Federal University of Technology Owerri, Imo State, Nigeria

Summary

The focus of this project is to evaluate the swelling behaviour of expansive soil and its treatment with lime and cement. A representative soil sample from Enugu area was selected for the investigation. Their geotechnical prospective were measured and assessed as a means of evaluating their swelling behaviour before and response to treatment with lime and cement.
The results of the geotechnical investigation indicate medium liquid limit (38%) and medium plasticity index (18%) which implies medium swelling potential. This soil was stabilized using lime and cement which reduced the liquid limit and the plasticity index. Hence the result of treatment with lime and cement indicates low liquid limit (29.8% and 29.5% respectively) and low plasticity index (13.5% and 14.25% respectively) which implies low potential swell after stabilization. Soil stabilization of expansive soil derived from Enugu Shale would be one of the ways to unravel the challenge.
The constructive outcomes of this research have the potential to improve the safety of the communities by assisting homeowners in promoting proper design, helpful constructions and maintenance attitudes.

Introduction

Expansive soil exhibit volume change when subjected to moisture variation. Alternative names for these soils include swelling or clayey soils. They possess the ability to swell and shrink in response to fluctuation in moisture content [4]. The shrinkage and swelling may cause damages to lightweight structures supported on expansive soils such as distortion and cracking of pavements and on-grade floor slabs, cracks in grade beams, wall etc. Expansive soils contain clay-sized particles usually less than two microns (<0.002mm) in diameter. [5] Defined clays as natural plastic earth materials comprised of hydrous aluminium silicates. The direct consequence of its shrink-swell behaviour is associated with soils containing the clay mineral smectite.

Content and theory

The aims and objectives of this research project work include;

a. To determine and evaluate the geotechnical properties of expansive soils, to investigate the swelling potential through comprehensive and detailed laboratory analyses.
b. To stabilize the expansive soils using lime and cement.
c. Investigate the possibility of having plausible relationship between the lime stabilization and cement stabilization of expansive soils.
d. To use the results to characterize the soils and their implications on construction project, which gives clue on problems encountered during engineering construction: thereby advising either to abandon the site or apply certain engineering methods to improve the strength of the soil before erecting structures on such area.

The study area is located in Garki in Enugu south, Enugu state, south eastern Nigeria. Garki is located between latitude 7°24′ to 7°41′ and longitude 6°19′ to 6°31′. The location in which sample was obtained is a town within the Awkuke- Enugu geographical area in south eastern Nigeria, it occupies much of the high lands of Awgu, Udi and Nsukka. The topographic and geologic map of the study area is shown in figure 2 and 3. The study area is accessible by a network of roads. There is also provision for minor access by footpaths and tracks.[3][2]

Laboratory Analysis

Different laboratory analyses were carried out on the soil sample collected. These analyses were done to determine the geotechnical index properties as well as its swelling potential and stabilization.
The geotechnical properties, strength of material and stabilization that were analyzed included the following;

I. Atterberg limits (Plastic limit and Liquid limit)
II. Linear shrinkage
III. Grain size distribution
IV. California Bearing Ratio
V. Specific gravity
VI. Natural moisture content
VII. Density
VIII. Lime stabilization
IX. Cement stabilization

Result

Swelling potential

The swelling potential for expansive soil from Enugu Shale is medium having averaged linear shrinkage of 9.2%, averaged plasticity index of 18%, when stabilized with lime; the swelling potential will then be low, as well as when stabilized with cement with averaged linear shrinkage of 4.44% and 4.3% respectively and averaged plasticity index of 13.5% and 14.25 respectively.

Table 1: Evaluation of effectiveness of Cement stabilization on some geotechnical index properties of an expansive soil from Enugu Shale, South Eastern Nigeria,

C=Cement

R=Reduction,

LL=Liquid Limit

PI=Plascity Index

LS=Linear Shrinkage

%C	LL (%)	% R	PI (%)	% R	LS (%)	% R
0	38	-	18	-	9.2	-
2	36	5.26	15.5	13.89	8.0	13.04
4	30	21.05	12.7	29.44	4.4	52.17
6	25	34.21	11.1	38.33	3.8	58.70
8	23	39.47	10.5	41.67	2.9	68.48
10	30	21.05	12.0	33.33	2.4	73.91

Table 2: Evaluation of effectiveness of Cement stabilization on California Bearing Ratio (CBR) of an expansive soil from Enugu shale, South Eastern, Nigeria

C=Cement

I=Increase

% C	Unsoaked CBR (%)	% I	Soaked CBR (%)	% I
0	22.0	-	3	-
2	26	18.18	9	200.00
4	33	50.00	10	233.33
6	35	59.09	15	400.00
8	39	77.27	17	466.67
10	34	54.55	8	166.67

Swelling potential

The swelling potential for expansive soil from Enugu Shale is medium having averaged linear shrinkage of 9.2%, averaged plasticity index of 18%, when stabilized with lime; the swelling potential will then be low, as well as when stabilized with cement with averaged linear shrinkage of 4.44% and 4.3% respectively and averaged plasticity index of 13.5% and 14.25 respectively.

Table 3: The results of the test conducted on the soil sample obtained from Enugu Shale are summarized above. From the analyses, it is observed that the sample obtained has the following (averaged values) in the table below:

LL=Liquid Limit

PL=Plascity Limit

PI=Plascity Index

LS=Linear Shrinkage

	LL (%)	PL (%)	PI (%)	LS (%)
Natural	38.0	20.00	18.00	9.20
Lime	29.8	16.65	13.50	4.44
Cement	29.5	16.50	14.25	4.30

According to the British soil classification, clay is classified in 5 categories:

i. Clay of low plasticity (CL), having a liquid limit less than 35%

ii. Clay of medium plasticity (CM), having a liquid limit from 35% to 50%

iii. Clays of high plasticity (CH), having a liquid limit from 50% to 70%

iv. Clays of very high plasticity (CV), having liquid limit from 70% to 90%

v. Clays of extreme high plasticity (CE), having a liquid limit exceeding 90%

Based on the classification, the clay of interest has medium plasticity, but when treated or stabilized with lime and cement it has low plasticity.

The use of Atterberg limits as a tool for recognition of problem soils for construction purposes and in estimating the amount of swell and shrinkage is now widely acknowledged. This can be explained from its simplicity and ease in the analysis of the results obtained. The most useful classification data for identifying the relative swell potential are liquid limits and plasticity index.

The results indicated that the soil when stabilized with lime had a liquid limit (LL) from 20% to 38% averaging 29%. Also the average plasticity index (PI) was found to be 13.5% ranging from 9% to 18%. Soil from Enugu Shale can be classified as having low swelling potential based on the model proposed by [1]. The results indicated that the soil when stabilized with cement had a liquid limit (LL) from 23% to 38% averaging 30.5%. Also the average plasticity index (PI) was found to be 14.25% ranging from 10.5% to 18%. Therefore expansive soil from Enugu shale can be classified as having low swelling potential based on the model proposed by [1] when stabilized with cement.

Grain size results

Table 4: Mechanical analysis of expansive soil derived from Enugu Shale

SS=Sieve Size
WR=Weight retained
WA=Weight actual
WF=Weight Factor
PP=Percent of whole soil passing

SS (mm)	WR (g)	WA (g)	W F(g)	PP (%)
3				
2.5				100
2	18	9.8	9.8	90.2
1.5	10	5.4	15.2	84.8
1	6	3.3	18.5	81.5
0.75	2	1	19.5	80.5
0.5	4	2.2	21.7	78.3
0.375	2	1.1	22.8	77.2
0.25	2	1.1	23.9	76.1
0.19	2	1.1	25	75

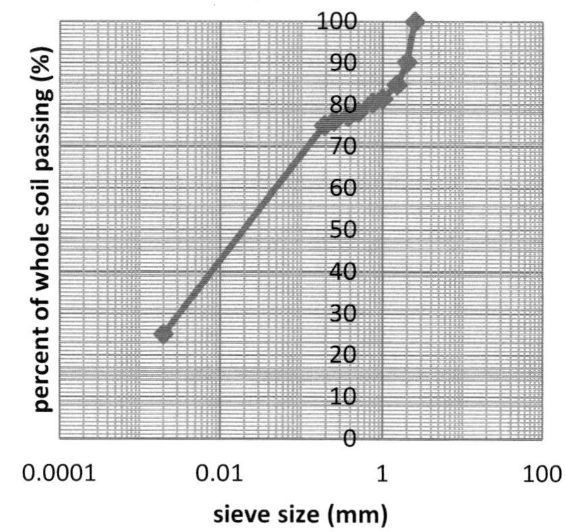

Figure 1: particle size distribution of expansive soil derived from Enugu Shale in Enugu Area

Conclusions

Clays could be used as construction material, but owing to their low to medium strength, they require individual site attention and careful designing appropriate stabilization if they must bear safely very heavy super structure. Expansive soils have been investigated in this study. This research project has helped identify the propensity of soil to shrink and swell and the associated problems in the area. The more positive outcome of the research will be to sensitize the implementation of the proposed mitigation measures to preclude structural damages originating from the behaviour of expansive soils.

This awareness is a very positive development in terms of ensuring the durability of the properties in the area. The constructive outcomes of this research have the potential to improve the safety of the communities by assisting homeowners in promoting proper design, helpful constructions and maintenance attitudes.

Based on the actual study results, the following general conclusions can be drawn;

a) The sample of interest consists mainly of combined silt and clay sized particles comprises 75%

b) The sample of interest was characterised as clay of 18% plasticity.
c) The sample of interest was characterised as clay of medium swelling potential
d) The sample of interest was characterised as clay of 13.5% plasticity when stabilized with lime.
e) The sample of interest was characterised as clay of low swelling potential when stabilized with lime.
f) The sample of interest was characterised as clay of 14.25% plasticity when stabilized with cement.
g) The sample of interest was characterised as clay of low swelling potential when stabilized with cement.

Recommendation

For the study in the case at hand, the recommendations summarized here below have been singled out based on the results of visual observations and field and laboratory investigations.

a) The expansive behaviour of the soil can be controlled by stabilizing it through the addition of chemicals like cement, lime or a combination of these. This would alter the physical and chemical properties by increasing shrinkage limit, plastic limit and decreasing moisture content, plasticity index, liquid limit and swelling potential.
b) High-rise buildings are recommended to be built in the Garki, Enugu South region so as to avoid cracks caused by swelling of the soil when it absorbs moisture.
c) For constructional purposes, the original top expansive soil should be excavated and replaced with non-expansive soil.
d) On an extreme note, since the soils are very good agricultural soils, the area can be used for agricultural purposes to feed the growing populace. However, the inhabitants of the land should be compensated by the government and resettled in another area with lesser expansive soil problems.
e) I recommend lime stabilization using 6% of lime, because it had its lowest plasticity at this point.
f) I recommend cement stabilization using 8% of cement, because from the result it shows low plasticity at this point.

REFERENCE

[1]. HOLTZ, W. G. AND GIBBS, H. J. (1956), Engineering Properties of Expansive Clays. Transactions, American Society of Civil Engineers, vol. 121, pp. 641-677.

[2]. OKEKE, O.C. AND AKAOLISA, C.C.Z. (2004), Geotechnics in Environmental Impact Studies; Journal for Geosciences and Professionals. Vol.2 pp4-5.

[3]. OKEKE, O.C. AND OKOGBUE, C.O. (2010), Effects Of Lime-Stabilization on Geotechnical Properties Of An Expansive Soil From Abakaliki Basin (Lower Benue Trough), Southeastern Nigeria. International Journal Of Environmental Science. Vol. 6, No. 1 2010, Pp 139-145.

[4]. OKOGBUE, C. O AND ENE, G.E. (2008), Geochemical and Geotechnical characteristics and the potential for use in drilling mud, of some clay bodies in Southeastern Nigeria. Journal of Mining and Geology vol 44, No2

[5]. PETTIJOHN, F. J. (1965), Sedimentary Rocks. New York: Harper.

[6]. SELLY, R.C. (1976), An introduction to Sedimentology. Academic press London.

[7]. VELDE, B. (1992), Introduction to Clay minerals, Chapman and Hall, New York.

Importance of Sampling and Sample Preparation in Slake Durability Test

*Abdulkarim Ahmed Owain Alali**
Undergraduate Student at United Arab Emirates University, College of Science, Geology Department, Al Ain, United Arab Emirates

Summary

The slake durability test is intended to assess the resistance offered by a rock sample and to estimate qualitatively the durability of rocks in the service environment. Four type's rock samples, belonging to the Early Oligocene epoch Asmari Formation, were collected from Jabal Hafit, Al-Ain city, United Arab Emirates (UAE). The objective of this study is to emphasis the importance of sampling and sample preparation.

Introduction

The importance of geotechnical examination of weak rocks such as shale, clay bearing sandstones and mudstones for slake durability to evaluate the engineering parameter has been clearly revealed by Dhakal et al., (2002). It analyses the decadence of rocks due to weathering and erosion (Gupta, 2007). Many studies have been carried out to understand the effect of environmental conditions on the mechanical properties of rocks.

The most commonly used of these is the slake durability index test, which has been developed by Franklin and Chandra in 1972 (Dhakal et al., 2002). Then, the American Society for Testing and Materials (ASTM) was standardized the test in 1990.

For rock analysis, the most important things are the collection of correct samples from the field, which means it must be representative and the sample preparation for tests following the available standard.

Field study

Samples of various weak rocks (four types) from the Asmari Formation, which belongs to the Early Oligocene epoch (~30 ma), were collected from Jabal Hafit. The study area is located in the south east of Al-Ain City, Abu Dhabi Emirate, UAE (Figure 1 a) on the western limb of Jabal Hafit, southeast of Al-Ain city. The rock samples consist mainly of greenish friable marl, marly limestone, hard limestone and chalky limestone (Figures 2 a, b and c). The Asmari Formation unconformably overlies the Upper Eocene Dammam Formation (35 ma) and underlies the Lower Fars Formation of Early Miocene epoch (18 ma) and Quaternary deposits. According to Arman et al., 2014, the Asmari Formation can be subdivided into three units. The lower unit is about 10 m thick and is made of greenish grey mudstone interbedded with yellowish brown marl. The middle unit is about 55m thick and consists of thick bedded to massive dolomitic and chalky limestone which in turn, interbedded with greenish to brownish mudstone and marl. This part is highly fossiliferous and characterized by the richness of fractures, and paleokarstic cavities. The thickness of upper unit has 25 m and is consist of chalky limestone, that is massive and very rich in containing different types of coral heads and also *Nummulites* spp. that dominates at all intervals (Figure 1 b).

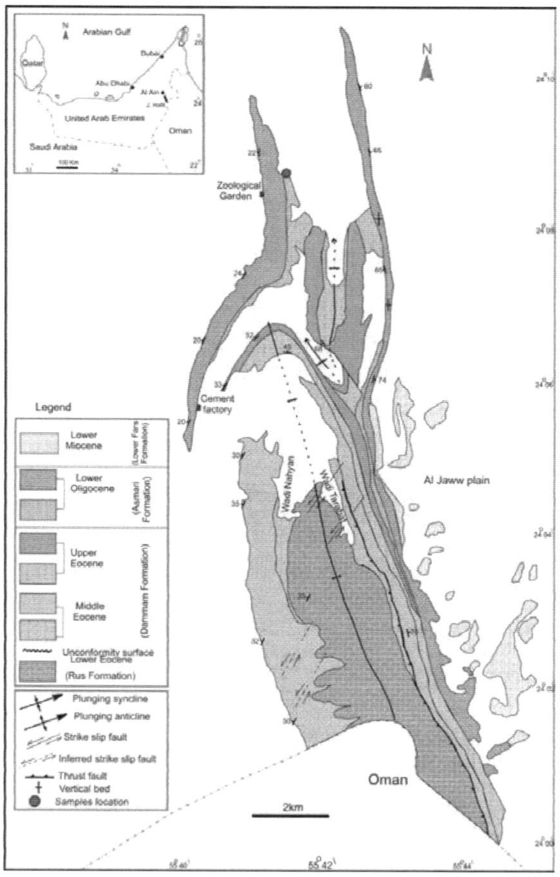

Figure 1 a Location map of the study area and a selected sections shown on geologic map of Jabal Hafit.

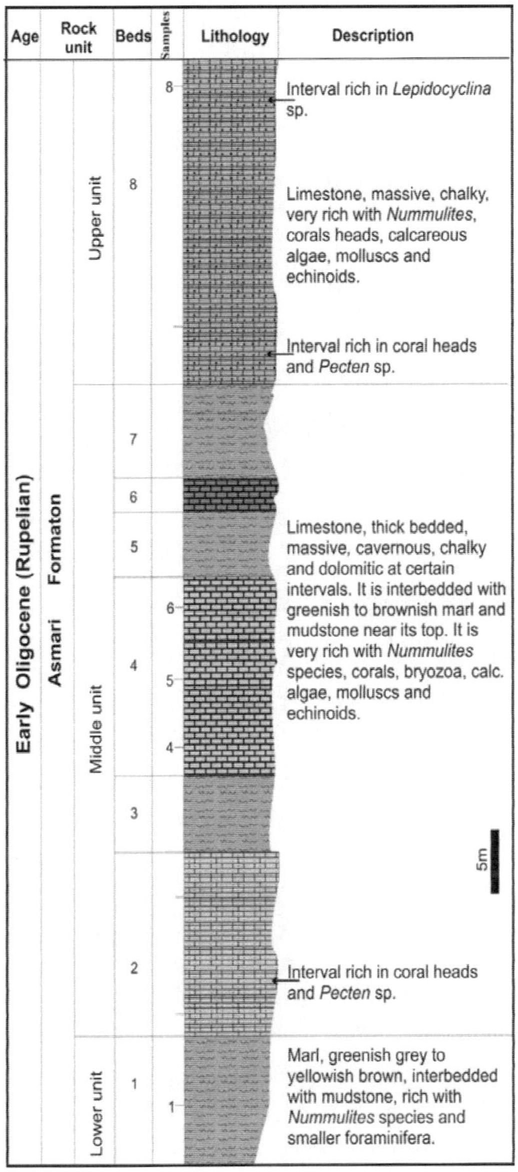

Figure 1 b Lithostratigraphic column of the Asmari Formation.

Figure 2 Sample collection location from the Asmari Formation (a- c).

Laboratory Study

Apparatus

According to standard (ASTM 1990), the following apparatus will be used for slake durability test:

1. Slake durability device, which contains a test drum, made of 2.00 mm square-mesh, woven-wire cloth and a trough to contain the test drum (Figure 3).
2. Drying oven, able to withstand 105 C° to within 5 C°
3. Balance capable of weighing both the drum and the sample.

4. Miscellaneous Apparatus, in addition to a hammer and brush.
5. Distilled water.
6. Digital Camera.

Figure 3 Slake durability device.

Sample Preparation

Test samples were prepared by following ASTM (1990) standard in order to assess the durability of the rocks. Eighty test sample were prepared for the slake durability test Sample preparation process is shown in Figures 4 a and b.

Figure 1 a and b. Sample preparation

Samples should be roughly spherical in shape (10 pieces of about 40 to 50 gm each) and were prepared by breaking with a hammer (Figures 5 a and b), and making sure that each piece had smooth corners to prevent them from breaking during rotation (Gokceoglu et al., 2000). If not, test results would be affected and contain errors. To eliminate this, it is crucial to avoid having sharp corners on each rock sample pieces. Therefore, sample preparation is an essential part of the test. In addition, any dust was removed by brushing the fragment just prior to weighing.
After smoothing the rock pieces and weighting them with the sensitive balance (Figure 6), a photograph of the specimen is taken prior to placement in the drum. Then, the sample is placed into a clean drum and is dried to a constant mass for 2 to 6 hours in the oven. The mass of the drum plus sample is recorded.

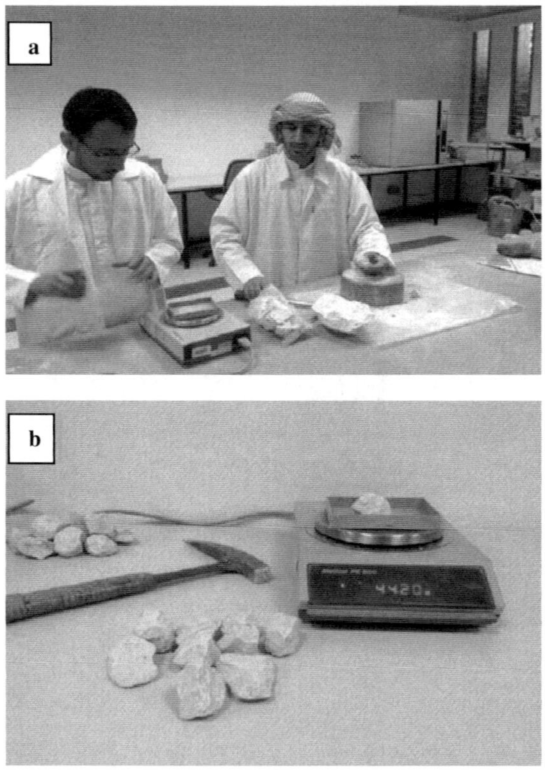

Figure 2. Smoothing sample using hammer and scaling (a and b)

Sample Testing

After cooling, the sample is tested. The lid is replaced, and the drum mounted in the trough and connected to the motor. The trough is filled with distilled water at room temperature 20 C° to a level of 20 mm below the drum axis.

The drum is rotated at 20 rpm for a period of 10 minutes. Then, the drum is removed from the trough, and the drum plus the retained portion of the sample dried in the oven at 105 C°. After cooling, this is recorded again. The process is repeated to obtain the final oven dried mass for the 2nd cycle. A photograph of the specimen is taken after drying.

Figure 3. Prepared samples.

Calculation

The slake durability index (second cycle) is calculated as the percentage ratio of the final to initial dry sample masses as follows:

$$\text{Slake durability index } Id(2) = \frac{(WF-C)}{(B-C)} \times 100$$

Where:

I_d (2)= slack durability index (second cycle),

B = mass of drum plus oven dried sample (before 1st cycle),

W_F = mass of drum plus oven dried sample (after 2nd cycle),

C = mass of drum.

Discussion and Recommendation

The present study was carried out on greenish friable marl, marly limestone, hard limestone and chalky limestone to assess their durability under weathering and erosion and to take an attention to the importance of sampling and sample preparation in slake durability test. The sample preparation stage still in progress. The testing program has not been initiated yet. Hence, the next step will be to test all samples according to suggested standard.

As strongly emphasized in the suggested standard sample preparation is a crucial step in slake durability test. This was gradely observed during sample preparation process.

It may possible to say that the slake durability test might be greatly influenced by the sampling and sample preparation processes. For that samples should be smoothed to prevent it from breaking during rotation.

Acknowledgments

We like to extend our special thanks to our supervisors (Arman, H., Abdelghany, O., El-Tokhi, M., Mahmoud, B. and Abu Saima, M.) for their valuable guidance and help during this study. In addition, we'd like to express our heartfelt thanks for UAE University-Research Affairs for providing financial aid for the study (COS/IRG-14/14). Special thanks to our department for letting us to use all available facilities in the Geology Department Laboratories to carry out this work.

REFERENCE:

ASTM, 1990. Standard test method for slake durability of shales and similar weak rocks (D4644- 08).

Candan Gokceoglu, Resat Ulusay, Harun Sonmez, 2000. Factors affecting the durability of selected weak and clay-bearing rocks from Turkey, with particular emphasis on the influence of the number of drying and wetting cycles. Engineering Geology (57), 215-237.

Ganesh Dhakal, Tetsuro Yoneda, Masaji Kato, Katsuhiko Kaneko, 2002. Slake durability and mineralogical properties of some pyroclastic and sedimentary rocks. Engineering Geology (65), 31-45.

Hasan Arman, Waheed Hashem, Mohammed El Tokhi, Osman Abdelghany, Ayman El Saiy, 2014. Petrographical and Geomechanical Properties of the Lower Oligocene Limestone from Al Ain City, United Arab Emirates. Arab J Sci Eng. 39:261-271.

Vikram Gupta, Iqrar Ahmed, 2007. The Effect of PH of water and mineralogical properties on the slake durability (degradability) of different rocks from the Lesser Himalaya, India. Engineering Geology (95), 79-87.

INTEGRATION OF ELECTRICAL RESISTIVITY AND GROUND PENETRATING RADAR METHODS IN ENGINEERING SITE CHARACTERIZATION: CASE STUDY AROUND AJIBODE, IBADAN, SOUTHWESTERN NIGERIA

*Adeyemi O. Adelekan * and Michael A. Oladunjoye, Dept. of Geology, University of Ibadan, Nigeria*

Summary

Integrated geophysical investigation involving Electrical Resistivity(ER) and Ground Penetrating Radar (GPR) methods were carried out around Ajibode area, Ibadan underlain by Basement Complex rocks of the southwestern Nigeria. The study was aimed at imaging the subsurface lithological setting and delineating shallow geologic structures for the purpose of characterizing the area for engineering construction suitability. A total of twenty five (25) Vertical Electrical Sounding (VES) data using Schlumberger array, ten (10) traverses of resistivity imaging using Wenner array and Ground Penetrating Radar (GPR) surveys were carried out along the established traverse lines (5 parallel traverse lines each in NW-SE and NE-SW directions with 100 m length) within the area. The VES data were quantitatively interpreted using partial curve matching technique and computer inversion using IPI2Win software to obtain the layer geoelectric parameters. The 2-D data were inverted into 2-D and 3-D resistivity image of the subsurface using Res2Dinv and Res3Dinv inversion software's respectively and GPR data were processed into radar section using RadExplorer software. Vertical electrical sounding results delineated a typical of three to four geologic layers which are the topsoil/lateritic hardpan, weathered basement (clayey and sandy clay) and fractured/fresh basement with layer resistivity value range of 10 – 2684 Ωm, 12 – 242 Ωm and 229 – 3213 Ωm and thickness value range of 0.5 - 2.1 m, 4.0 - 14.1 m respectively. 2-D inverted resistivity results also delineated three major geologic layers which are the topsoil, weathered basement and fresh basement and correlates well with the results obtained from the VES result. Layers 1 to 3 of 3-D inverted resistivity sliced result show high degree of variation in resistivity distribution at shallow depth, consisting of highly resistive material towards the eastern part with less resistive material concentrating at the south-western part. Results of the GPR survey delineated four geologic layers which include the topsoil, lateritic hardpan, weathered basement and fractured/fresh basement. The study area is categorized to have semi-competent to competent basement rock but the nature of overlying material and depression into the bedrock at some location could pose threat of differential settlement to engineering construction works. It should be ensured therefore, that engineering foundation is designed to sit comfortably on a competent material or by employing suitable foundation, such as piling to ensure stability. Thus, electrical resistivity and ground penetrating radar are versatile tools in engineering site characterization.

Introduction

The spate of failures of structures such as road, buildings, dam and bridges has assumed an alarming dimension. The need for pre-foundation studies has therefore become imperative before any structure is laid so as to prevent loss of lives and also goods and properties worth millions of naira. Some earth materials such as sands and fresh rock provide firm support for solid foundation while other host material due to their nature cannot support solid and rigid structures (Coker et al., 2013). Such structures that are improperly sited have failed due to lateral and vertical inhomogeneity in the subsurface earth materials (Carlsten et al., 1995; Adewumi and Olorunfemi, 2005 and Oladapo et al., 2008). Such fallout can however be prevented when necessary information aimed at better understanding the subsurface geometry and the structural setting in-situ is available. Geophysical methods that have been found useful in pre-and post-construction geotechnical investigations include Electrical Resistivity, Seismic Refraction and Electromagnetic methods including the Ground Penetrating Radar (GPR) among others (Roth et al., 2002 and Elawadi et al., 2006. The main objective of this work was to investigate subsurface stratigraphical and structural setting of the study area and to define its competency for engineering construction purposes using Electrical Resistivity and Ground Penerating Radar methods.

Methodology

A total of ten (10) traverse lines were obtained with five (5) lines each in approximately NW-SE and NE-SW directions along which Electrical Resistivity and Ground Penetrating Radar data were acquired *(Fig. 3).* Vertical Electrical Sounding (VES) data was acquired using Schlumberger electrode array. A total of twenty five (25) VES stations were occupied with each located at the grid points of the traverse line within the study area with current electrode half-spacing (AB/2) ranging from 1.0 up to 50 m in successive steps. Wenner array was employed for Electrical Resistivity Imaging (ERI) with spacing between adjacent" electrodes represented by 'a', all the possible measurement made with Wenner array is of electrode spacing of "na' where n = 1, 2....5 and 'a' = 5m. GPR data was acquired using "Mala RAMAC/GPR" geophysical survey equipment. The equipment consists of radar antennae housed in a box attached to the survey wheel which rolled along the traverse lines. The box was connected to the computer monitor unit which automatically recorded and store the data as the survey wheel is being moved across the study area. The survey was carried out using two different antenna frequencies (250 MHz and 500 MHz) with antenna separation of 0.36 m and 0.18 m respectively. The GPR data positioning was calibrated using survey wheel. Vertical Electrical Sounding data were interpreted by partial curve matching technique (Keller and Frischknecht, 1966) with the result of the layer resistivity parameters derived from the partial curve matching technique being used as the initial interpretation of the field curves for the computer inversion using IPI2Win software. Forward inversion modeling technique using regularized least-square optimization method (Sasaki, 1989 and Loke et al., 2003) of Res2Dinv and Res3Dinv inversion software's were used in the interpretation of the 2-D and 3-D data. GPR data was processed using "RadExplorer V1.4 (2005) software" package which is commercially available. The processing parameters applied, as made available by the software" and that which gives acceptable result on the radargram includes: DC removal, Time zero adjustment, Bandpass filtering and Amplitude correction.

Results

Vertical Electrical Sounding Results

Quantitative interpretation of the VES curves resulted in determination of geoelectric layer parameters (layer resistivity and thickness) for the subsurface characterization. Based on the approximate range of resistivity values of common rock types (Kearey et al., 2002) and correlation with the pit information obtained within the study area, three major geologic layers were delineated from the geoelectric layer parameters obtained from the study area *(Fig. 1)* and these includes the topsoil/lateritic hardpan, weathered basement (clayey/sandy clay) and fractured/fresh basement.

Electrical Resistivity Imaging Results

2-D Wenner data results were presented as an inverted model resistivity section representing the image of the subsurface along each established traverse line. This shows both lateral and vertical variations in resistivity of the subsurface material. The 2-D inverted resistivity sections generated along NW-SE and NE-SW *(Fig. 2)* respectively within the study area delineated three major subsurface layers which are; the topsoil, the weathered layer and the bedrock. The inverted resistivity model section delineated topsoil with high degree of variations in resistivity value ranging between 10 and 1000 Ωm and layer thickness of approximately 3.5 m. Some areas of the topsoil with higher resistivity value are dominated by lateritic clay pan (T2, T8 and T9). The topsoil is underlined by a conductive material characterized by resistivity of <100 Ωm and is typical of clay-rich weathered material which is similar to the range established for similar geological provinces by Kearey et al (2002). The thickness of the second layer ranges from 2 - 8 m and cannot be ascertain along some traverse lines as it forms depression into the basement. Such depression was noticed beneath traverse line T3, T7, T8, T9 and T10. It occurs at horizontal distance 35 - 45 m beneath traverse line T3 and T7, between horizontal distance 35 and 60 m beneath T8 and T9 and at horizontal distance 40 and 55 m beneath T10. The weathered layer is underlain by basement rock with characteristic resistivity value of about 200 - 9036 Ωm. The depth to competent bedrock delineated by 2-D inverted resistivity results within the area is about 11 m, but could not be ascertained beneath regions where the weathered basement forms depression into the basement. Inverted resistivity section for traverse lines T3, T7, T8, T9 and T10 generally show a continuous trend of depression into the basement and cut across all the traverse lines along the NE-SW direction.

The data acquired along the ten gridded traverse lines with 2-D Wenner profile were combined to produce the slice depth and a collective 3-D view of the inverted resistivity models *(Fig. 3)* for all the traverse lines (T1 to T10). X and Y coordinates represent the NW-SE and NE-SW directions respectively with slice been made along Z-direction to a maximum depth of 21.9 m. Layers 1 to 3 shows high degree of variations in resistivity distribution at shallow depth revealing formation consisting of highly resistive topsoil material (lateritic hardpan) at the eastern part with low resistivity (highly conductive) material concentrating at the south-western part of the study area.

Ground Penetrating Radar Results

Radar section results for 250 MHz antenna frequency *(Fig. 4)* was able to image up to 10 m depth within the study area. Four radar facies were recognized and correlated to their lithological equivalence in geolectric section and pit information within the study area. The upper layer (above the red dotted lines) shows high amplitude, parallel and horizontal reflections with good continuity and is interpreted as topsoil. The thickness is approximately 0.7 m across the entire 250 MHz antenna frequency profiles. This is underlain by high amplitude, sub-parallel and a chaotic reflection with moderate continuity reflection characterized to be lateritic hardpan (between red and yellow dotted lines) across the 250 MHz antenna frequency profiles with thickness range between 0.3 m and 2 m except beneath traverse line 5 where the layer is absent. The third layer (between yellow and green dotted lines) depicts the weathered unit (clay/sandy clay) with low to moderate reflection amplitude, parallel and horizontal reflection with poor continuity and ranges in thickness between 4 m and about 8 m along the entire traverse lines. The weathered layer is underlain by fresh basement (from green lines to infinity) consisting of low reflections amplitude with poor continuity and depth to fresh bedrock ranging from 6 m to 8.5 m.

The radar section results for 500 MHz antenna frequency *(Fig. 4)* delineated a maximum depth of 5 m. Three radar facies were recognized and correlated to the lithological equivalence from geoelectric section and pit information obtained within the area. The high amplitude, parallel, horizontal and continuous reflections characterized by the upper layer (above red dotted lines) is interpreted as the topsoil with approximate depth of < 0.5 m across the traverse lines. The second layer (between red and yellow dotted lines) was interpreted as lateritic hardpan with high amplitude, chaotic reflectivity and moderate continuity with approximate thickness of 1.5 m except beneath traverse 5 where lateritic hardpan was not delineated. The last layer (beneath yellow dotted lines) is characterized by low to moderate amplitude, parallel and horizontal reflection with moderate continuity and is interpreted as weathered basement with depth at infinity.

Conclusions

Integrated geophysical techniques involving Vertical Electrical Sounding, Electrical Resistivity Imaging and Ground Penetrating Radar have been adopted in this study around Ajibode, southwestern Nigeria. The result of the Electrical Resistivity (ER) investigation has effectively delineated the subsurface geological information and gives detailed description of the lithological setting of the area. It was observed from ER surveys that the depth to competent basement varies around the study area with shallow depth at some location but can go up to 22 m at locations where presumed depression are delineated as a result of the heterogeneity in the weathering nature of basement rock. Thus, the area under study can be categorized to have semi-competent to competent basement rock considering its resistivity distribution. The nature of the overlying material (clay-rich weathered material) and depression into the bedrock at some location could pose threat of differential settlement to engineering construction works. Therefore, it should be ensured that subsurface region where depression was noticed is avoided when sitting mega structures in engineering projects. Moreover in engineering construction within the area, foundation should be designed to sit comfortably on a competent material or by employing suitable foundation work (such as piling) to ensure the stability of such proposed structure. Reliable information about the lithological characteristics of the study area has been deduced from the results of this research work. Thus, electrical resistivity imaging and ground penetrating radar are versatile tools in shallow engineering site characterization. It is therefore advised that the methods should be incorporated in any engineering foundation design for better understanding of the subsurface.

Acknowledgements

We appreciate the CEO of Geoterrain Nigeria Limited and all the staff of the company for their assistance during this work by making the GPR equipment and processing software available for use and their contributions toward data acquisition and processing. Our sincere thanks goes to the management of Geo-sore Multi-transaction Limited for making the DC resistivity meter available for use in the course of this research work.

References

Adewumi I. and Olorunfemi, M. O. (2005). Using geoinformation in construction management. Journal of Applied Science, Vol. 5, No. 4, pp. 761-767.

Carlsten S., Johansson S. and Worman A. (1995). Radar techniques for indicating internal erosion in embankment dams. Journal of Applied Geophysics. Vol 33, pp. 143-156.

Coker, J. O., Makinde, V., Mustapha, A. O. and Adesokan, J. K., (2013). Electrical resistivity imaging for foundation failure investigation at Remo Secondary School, Sagamu, Southwestern Nigeria. International Science and Investigation Journal, Vol. 2, No. 4, pp. 40-50.

Elawadi E., El-Qady G., Nigm A., Shaaban F. and Ushijima K. (2006). Integrated Geophysical Survey for site investigation at a new dwelling area, Egypt. Journal of Environmental and Engineering Geophysics, Vol. 11, No. 4, pp. 249-259.

Keary P., Brooks M. and Hil I. (2002). An Introduction to Geophysical Exploration, 3rd Edition,Blackwell Science Limited.

Keller, G. V., and Frischknecht, F. C. (1966). Electrical Methods in Geophysical Prospecting, Oxford: Pergamon, 283 pp.

Loke, M. H., Acworth, I. and Dahlin, T. (2003). A comparison of smooth and blooky inversion methods in 2D electrical imaging surveys. Exploration Geophysics. 34:182-187.

Oladapo, M. I., Olorunfemi, M. O. and Ojo, J. S. (2008). Geophysical investigation of road failures in the basement complex areas of southwestern Nigeria. Journal of Applied Geophysics, Vol. 3, No. 2, pp. 103-112.

Roth, J. S., Mackey, J. R., Mackey C., Nyquist, J. E. (2002). A case study of the reliability of multielectrode earth resistivity testing for geotechnical investigations in Karst terrains. Eng. Geol. Vol 65, pp. 225-232.

Sasaki, Y. (1989). Two-dimensional joint inversion of magnetotelluric and dipole-dipole resistivity data. Geophy, 54: 254-262.

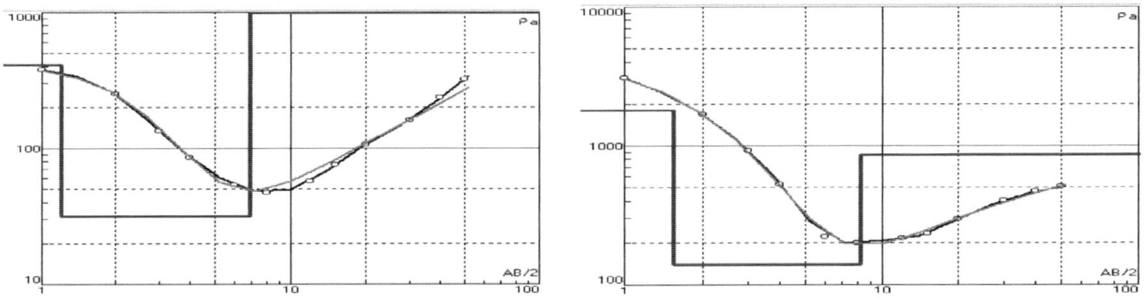

Figure 1: Representative VES curves obtained in the area

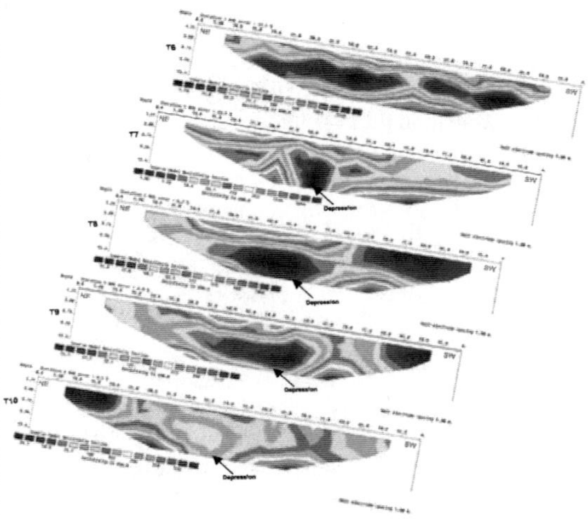

Figure 2: Panel view of inverted resistivity model section along NE-SW direction

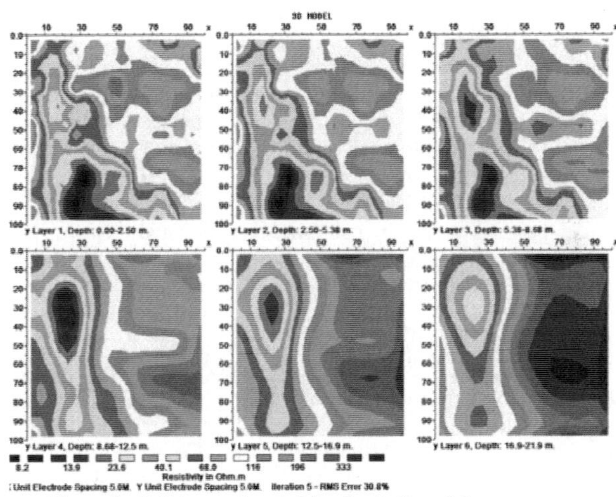

Figure 3: 3-D inverted resistivity depth slice of the area

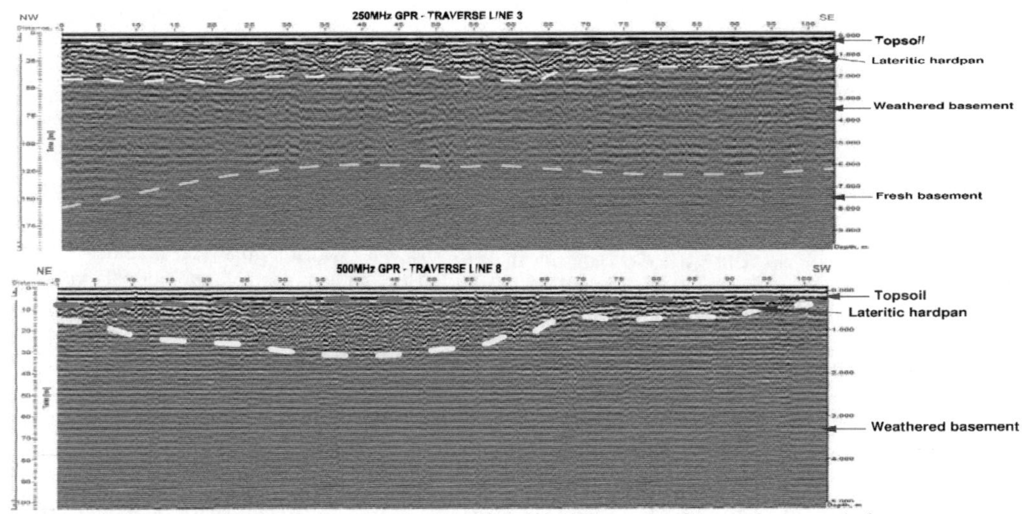

Figure 4: Radar section along traverses 3 & 8 using 250 & 500 MHz antenna frequencies respectively

Geochemical behavior of elements of marine sediments of Abu Dhabi area, United Arab Emirates

M. El Tokhi, B. M. Amin & S. Alaabed
Geology Department
UAE University
United Arab Emirates
meltokhi@uaeu.ac.ae

Summary

The geochemical behavior of specific REEs makes them useful tracers in marine environment pollution where its distributions influenced by geogenic or anthropogenic sources.

Abstract
Concentrations of rare earth elements (REE) in the bottom sediments of Abu Dhabi area were recently measured on bottom sediments collected from seventy nine offshore stations by using marine divers and grab sampler and analyzed for rare earth elements (REEs) using inductively coupled plasma mass spectrometry (ICP-MS). Results showed that the REEs patterns in sediments reflected the source rock patterns with an overall order of abundance such as: light rare earth elements (LREE) > high rare earth elements (HREE. The chondrite normalized patterns of REEs showed enrichment of LREEs over HREEs with La/Yb of 8.7 while the chondrite normalized value showed low $(Gd/Yb)_N$ ratios (1.55) which was illustrated in the flat HREE pattern. A semi positive Eu anomaly and a slightly negative Ce anomaly, which is an indicator of marine depositional environments. The negative Ce anomaly was also partially due to early stage weathering which led to less pronounced REE fractionation. This might be due to the location of the sampling area that was completely covered with sands and shell fragments where the abundance of quartz and calcite as natural carbonate sediments at this site would tend to dilute the REE abundance. In the study area, the REEs concentration not related to anthropogenic sources.

Introduction

The rare earth elements (REE), or lanthanide elements, are a group of 14 elements that have similar chemical properties principally due to the same electronic configuration. They are mainly similar crystal ionic radii and valance state. They have a decreasing crystal ionic radius from La (1.016 Å) to Lu (0.85 Å) and a single 3+ valence (with the exception of Ce and Eu which are stable in the tetra and divalent oxidation states respectively) within the siliciclastic fraction of sediments and sedimentary rocks. The geochemical behavior of specific REEs makes them useful tracers in marine environments. For example, light REEs (LREEs: La, Ce, Pr, Nd, Pm, Sm, and Eu) are preferentially scavenged on the surface of particles such as Mn–Fe oxyhydroxides and clay minerals, or are precipitated as REE phosphates (Byrne and Kim, 1990). In most natural waters, the REEs form strong complexes with carbonate ions (Millero, 1992), while the heavy REEs (HREEs: Gd, Tb, Dy, Ho, Er, Tm, Yb, and Lu) can form dissolved complexes which remain in solution. Certain minerals such as aluminosilicates or apatite tend to have high concentrations of REEs while other minerals such as quartz have very low concentrations (Graf, 1978).The relative abundances of HREEs to LREEs have successfully been used in aquatic and solid-phase geochemistry (Duncan and Shaw 2003; Caccia and Millero 2007). For instance, REEs are used to evaluate the mixing among terrestrial and marine sources under natural (Santos et al. 2007; Kim and Kim 2011) or anthropogenically altered conditions (Xu et al. 2010; Silva-Filho et al. 2011). This similarity allows for only limited fractionation during their passage through the ocean. Thus, although their absolute concentrations may vary between phases, the concentration of any one REE relative to the other will show only slight variations from one marine phase to another and from source rock to sediment. Thus, they are an ideal group to identify the sources of the lithogenous fraction of marine and non-marine sedimentary deposits, e.g.mafic- and felsic-igneous and sedimentary rocks.

This paper is the first to report on the REE composition of offshore marine sediments from Abu Dhabi area. The aim of this work was to study the REE distribution in bottom sediments from the coastal area of Abu Dhabi area and to assess industrial effluents as sources of anomalous REE concentration. In this area, the geochemistry of recent sediments is poorly investigated; in particular, data of REE composition in sediments from Abu Dabhi area not have been reported in literature until now. At light of these data lacking, this work should represent a baseline dataset in the study of REE in marine sediments from this area of the Arabian Gulf and in the environmental pollution assessment.

Nature of Abu Dhabi coastline

The United Arab Emirates is located on the eastern side of the Arabian Peninsula, between latitudes 220 40' and 260 00', and longitudes 510 00' and 560 00', where the interior platform of the Arabian shelf occurs in the subsurface. It is bounded on the northwest by the Qatar–South Fars Arch, and on the northeast and east by the foreland basin of the Arabian Gulf, and adjacent foreland fold and thrust belt of Oman (Fig. 1).
The study area located in the United Arab Emirates (UAE) on the offshore for Abu Dhabi area of the Arabian Gulf approximately 120km from Gantoout area in the north east to Abu Al Abyad island in the south west. We found that concentrations of REE in the bottom sediments varied between different locations in the studied areas.

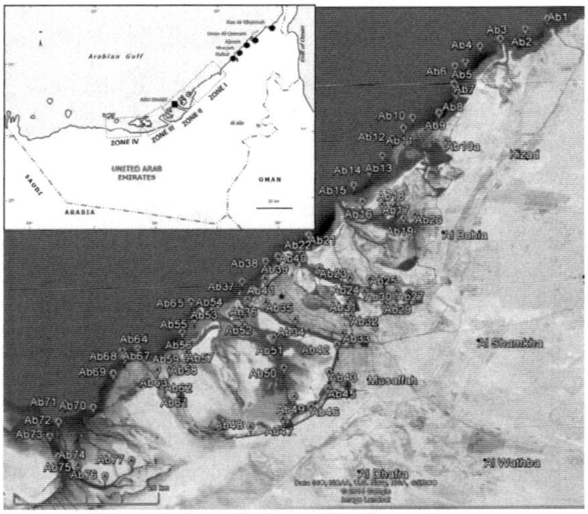

Figure 1: Location map for the study area

Therefore. We identified four zones (Zone I: from Gantoout area to Khor al Baghi, Zone II: from to Khor al Baghi to Khor al Batin, Zone III: from Khor al Batin to end of al Dhabaiya area and Zone IV: from al Dhabaiya area to Abu Al Abyad island (Fig.1). It includes a nearly closed lagoon connected to the Arabian Gulf by narrow channels, and is being enveloped by a large sabkha constituting the supratidal part of the lagoon (Baltzer et al., 1994). Southwest along the coast of the United Arab Emirates, barrier islands become numerous and widespread and include Abu Dhabi, Al Sadiyat, Al Qanatir, Abu Al Abyad, and Marawwah (Fig. 1). These islands and the peninsula of Al Dhabaiya are nucleated around Pleistocene aeolian calcareous sands and marine sediment, that is locally elevated some 2 to 4 m above sea level (Kirkham, 1998b).

Anthropogenic influence

The Arabian Gulf coastline of the UAE (Fig. 1) is undergoing massive infrastructure development on an unprecedented scale. The shoreline is increasingly dissected by the construction of oil and gas pipelines, road construction and shipping canals. Terrestrial- and marine-derived sediments are excavated and transported over considerable distances for use in land reclamation projects. Large-scale dredging projects deepen and widen channels whilst simultaneously constructing offshore islands, causeways and bridges. These activities are significantly modifying the dynamics of the coastline at an unparalleled rate. The impending loss of much of the remaining natural coastal system (Lokier and Steuber, 2008) gives added impetus to the need to accurately record these natural sedimentary systems before they are irrecoverably obliterated by this anthropogenic overprint.

Samples and Experimental methods

Bottom sediment samples were collected using marine divers and a 0.05 m2 Van Veen grab sampler at 79 stations across the Abu Dhabi shore line from Gantoout area in the north east to Abu Al Abyad island in the south west in April to June 2014 (Fig. 1). A Teflon spatula was used to extract the sediment samples from the center of the grab and the samples were stored in pre-cleaned 30 ml HDPE (high density polyethylene) bottles. They were then frozen until analyzed. Precautions were taken to minimize possible metal contamination throughout all sampling and analytical procedures. All the bottles and material used for the collection and analysis were cleaned, dried in a clean laminar reverse flow bench (Class 100) and stored in Zip-lip low-density polyethylene bags, before use. For the chemical analysis, bulk sediments samples were digested and the obtained solutions were analyzed in Acmalabs of Canada by using inductively coupled plasma mass spectrometry (ICP-MS).

Results and Discussion

The average mean values of detected concentrations of REE in the bottom sediments are presented in Table1. Results show that the concentrations of REEs in the bottom sediments varied between 9.77 and 25.53 with an average value of 17.65 μg/g.

Table1: REEs concentrations in the four zones of study area (μg/g)

Zones	La	Ce	Pr	Nd	Sm	Eu	Gd	Tb
I	1.62	3.14	0.32	1.5	0.29	0.11	0.29	0.1
II	3.74	6.45	0.81	3.06	0.69	0.19	0.53	0.1
III	4.73	7.87	1.13	3.87	0.84	0.27	0.64	0.13
IV	3.26	6.38	0.71	2.88	0.58	0.16	0.53	0.1
Average	3.34	5.96	0.74	2.83	0.60	0.18	0.50	0.11
Oceanic Crust	16	28	4.7	21	5.3	1.4	5.9	0.9
NASC value	230.4	521.1	56.1	228.9	37.9	7.9	33.1	5.4
Chondrite value	0.367	0.957	0.137	0.711	0.23	0.087	0.31	0.06
Zones	Dy	Ho	Er	Tm	Yb	Lu	Y	
I	0.29	0.2	0.19	0.1	0.16	0.1	1.43	
II	0.51	0.13	0.31	0.1	0.28	0.1	3.05	
III	0.73	0.16	0.47	0.1	0.39	0.1	4.06	
IV	0.39	0.13	0.21	0.1	0.24	0.1	2.69	
Average	0.48	0.16	0.30	0.10	0.27	0.10	2.81	
Oceanic Crust	4.9	-	3.1	-	2.7	-		
NASC value	30.8	8.5	20.3	3.4	17.9	2.9	N/A	
Chondrite value	0.381	0.085	0.249	0.036	0.26	0.038	2.25	

The concentrations of the 14 RREs in sediments followed the order of Ce>La>Nd>Pr>Sm>Gd>Dy>Er> Yb>Eu> Ho>Tb>Tm=Lu which is less than mean REE values in oceanic crust (Ronov et.al 1974). The distribution patterns of REEs in the study area are shown in Fig. 2. From the concentration map, it is clear that samples in zone 3 from

from Khor al Batin to end of al Dhabaiya have the highest concentrations rather than the other areas. These sampling stations are located near heavy industrial activity and lie inside marine traffic area. However, the concentrations of REEs are still low compared to the oceanic crust values.

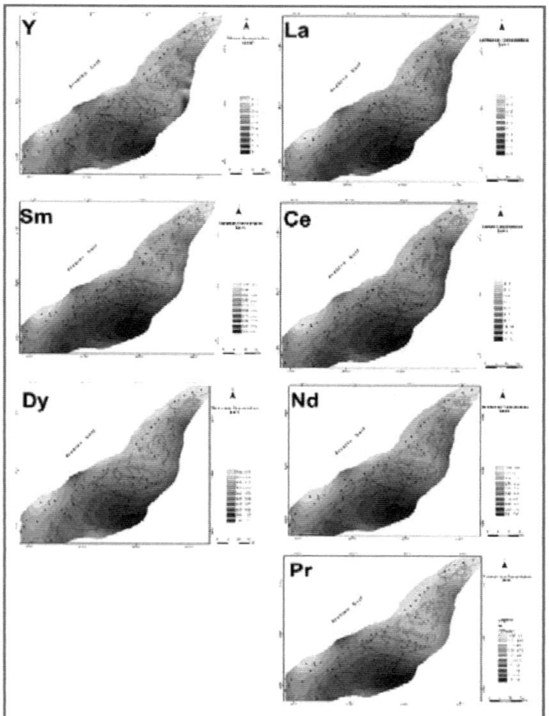

Figure 2: Distribution patterns of REE of study area (μg/g)

It is very important to discuss what material is suitable for normalization of geochemical samples. The chondrite normalized REE patterns to the value of Taylor and McClennan (1985) in sediment are shown in Figure 3. This fractionation is characterized by a significant enrichment in LREE over HREE. The chondrite-normalized REE pattern of marine sediments collected in the study area present in (Table 2).

Table 2: Chonderite Normalization of the studied zones (μg/g) after Taylor and McClennan (1985)

Zone	La	Ce	Pr	Nd	Sm	Eu	Gd	Tb	Dy	Ho	Er	Tm	Yb	Lu	Y
I	4.42	3.28	2.34	2.11	1.25	1.31	0.94	1.72	0.77	2.35	0.77	2.81	0.64	2.62	0.64
II	10.2	6.74	5.92	4.31	2.97	2.15	1.74	1.72	1.34	1.57	1.22	2.81	1.1	2.62	1.36
III	12.9	8.23	8.25	5.45	3.63	3.13	2.1	2.3	1.9	1.94	1.87	2.81	1.5	2.62	1.81
IV	8.89	6.67	5.2	4.04	2.49	1.84	1.72	1.72	1.02	1.57	0.85	2.81	0.94	2.62	1.19

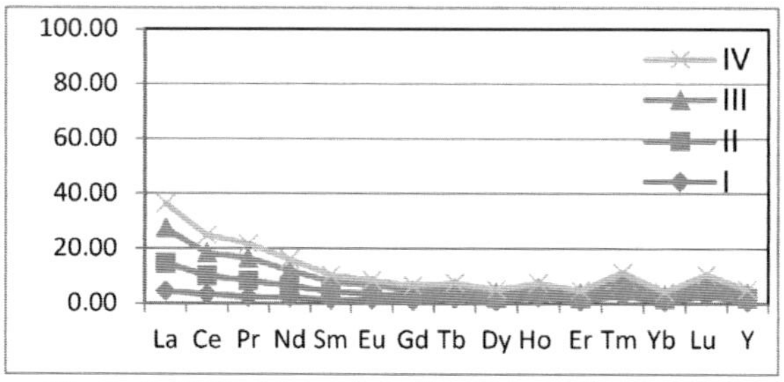

Figure 3: REE patterns in sediment normalized to the concentrations in chondrite in the study area

This normalization allows one to identify within the typical sedimentary REE patterns, an enrichment or deficiency of a single element or group of elements (Henderson, 1984). In general, clays contain over one and a half times more REEs than the sandstones, and the LREE/HREE ratios of sediments decrease in the order of clays, sandstones, carbonates (Balashov et al., 1964; Henderson, 1984). In the study area the sediments mainly coarse to fine sands and shell fragments (calcium carbonate contents). Caccia and Millero (2007) found that there are relation between the increase of calcium carbonate and decrease of REEs concentration. In general the studied area characterized by lower REEs concentrations.

The average of REEs concentration was observed to 17.65 (μg/g). The Fig.3 showed REEs have higher LREE concentrations, a slightly positive Eu anomaly and lower in HREE concentrations. The REEs concentrations showed a slightly negative Ce anomaly which is an indicator of marine depositional environments after Murray el al. 1990. The negative Ce anomaly was also partially due to early stage weathering, which led to less pronounced REE fractionation.

Chondrite normalization showed high (La/Yb)N ratios (8.5) which reflect highly fractioned REE compositions. This value concordant with values of La/Yb ratio in the marine sediments of south China Sea which ranging from 7.4 to 16.7 (Khadijeh Saraee et al. 2009). (Gd/Yb)N ratio was calculated to investigate the relative depletion of HREE. The chondrite normalized value shows low (Gd/Yb)N ratio (1.55) which is illustrated by the flate HREE pattern (Fig.3). The (Ce/La) N ratios in the study area is low value (0.68) and this might be due to an increasing amount of carbonate, biogenic and terrigenous components in the sediments sample.

The proportion enrichment of REE in sediments remained similar with increasing atomic number. HREE are more soluble and more complex than light REE and are more strongly absorbed on most substrates (Byrne and Kim 1993). This might explain the enrichment patterns of LREE in Abu Dhabi sediments.

Reference:

Balashov, Y.A., Ronoc, A.B., Migdisov, A.A., Turanskaya, N.V., 1964, Geochemistry 10, 951.

Baltzer, F., Kenig, F., Boichard, R., Plaziat, J.-C., Purser, B.H., 1994, Organic matter distribution, water circulation and dolomitization beneath the Abu Dhabi Sabkha (United Arab Emirates). In: Purser,

Byrne, R.H., Kim, K.H., 1990, Rare earth element scavenging in seawater. Geochimica et Cosmochimica Acta 54, 2645–2656.

Byrne, R.H., Kim, K.H., 1993, Rare earth precipitation and coprecipitation behavior: the limiting role of PO4 − 3 on dissolved rare earth concentrations in seawater. Geochimica et Cosmochimica Acta 57, 519–526.

Caccia, V. G., & Millero, F., 2007, Distribution of yttrium and rare earths in Florida Bay sediments. Marine Chemistry, 104, 171–185.

Duncan, T., & Shaw, T. J. 2003, The mobility of rare earth elements and redox sensitive elements in the groundwater/seawater mixing zone of a shallow coastal aquifer. Aquatic Geochemistry, 9, 233–255.

Evans, G., Kinsman, D.J.J., Shearman, D.J., 1964b. A reconnaissance survey of the environment of recent carbonate sedimentation along the Trucial Coast, Persian Gulf. In Van Straaten, L.M.J.U. ed, Deltaic and Shallow Marine Deposits. Elsevier, Amsterdam, 129–135.

Graf Jr., J.L., 1978, Rare earth elements, iron formations and seawater. Geochimica et Cosmochimica Acta 42, 1845–1850.

Henderson, P., 1984, Rare Earth Element Geochemistry. Elsevier, Oxford, p. 510.

Khadijeh S.R.I., Elias, S. B., Wood A. K and Reza, A. M, 2009, Rare Earth Elements Distribution in Marine Sediments of Malaysia Coasts, Journal of Rare Earths, 27, 1066-1077.

Kim, I., & Kim, G. (2011). Large fluxes of rare earth elements through submarine groundwater discharge (SGD) from a volcanic island, Jeju, Korea. Marine Chemistry, 127, 12–19.

Kinsman, D.J.J., 1964b, Reef coral tolerance of high temperatures and salinities. Nature, 1280– 1282.

Kirkham, A., 1998b, Pleistocene seif dunes and their role in the development of complex past and present coastlines of the U.A.E. GeoArabia 3, 19–31.

Lokier, S., and Steuber, T., 2008, Quantification of carbonate ramp sedimentation and progradation rates for the late Holocene Abu Dhabi shoreline: Journal of Sedimentary Research, 78, 423-443.

Millero, F.J., 1992, Stability constants for the formation of rare earth inorganic complexes as a function of ionic strength. Geochimica et Cosmochimica Acta 56, 3123–3132.

Murray, R. W., Buchholtz, T., Brink, M. R., Jones, D. L., Gerlach, D. C and Russ, G. P., 1990, Rare Earth Elements as Indicators of Different Marine Depositional Environments in Chert and Shale," Geology, 18, 268-272.

Ronov, A. B. , Balashov Y. A. and Girin, Y. P., 1974, Regulari- ties of Rare Earth Element Distribution in the Sedimen- tary Shell and in the Crust of the Earth," Sedimentology. 21, 171-193.

Santos, I. R., Fávaro, D. I. T., Schaefer, C. E. G. R., & Silva-Filho, E. V. 2007, Sediment geochemistry in coastal maritime Antarctica (Admiralty Bay, King George Island): evidence from rare earths and other elements. Marine Chemistry, 107, 464–474.

Silva-Filho, E. V., Sanders, C. J., Bernat, M., Figueiredo, A. M. G., Sella, S. M., & Wasserman, J. 2011, Origin of rare earth element anomalies in mangrove sediments, Sepetiba Bay, SE Brazil: used as geochemical tracers of sediment sources. Environmental Earth Science, 64, 1257–1267.

Taylor, S. R., & McLennan, S. M. 1985, The continental crust:Its composition and evolution. Oxford: Blackwell.

Xu, Y. Y., Song, J. M., Duan, L. Q., Li, X. G., Zhang, Y., & Sun, P. Y. 2010, Environmental geochemistry reflected by rare earth elements in Bohai Bay (North China) core sediments. Journal of Environmental Monitoring and Assessment, 12, 1547–1555.

Subsurface Geologic Structures Investigation using Magnetic Method at Dibba area, United Arab Emirates (UAE)

A. Gabr, S. Darwish, S. Khosravi and R. Al Buraiki*

Geology Department, United Arab Emirates University, Al Ain, UAE

Summary

The main objective of this research is to investigate the subsurface geologic structures in a highly faulted and fractured area which is Dibba area, using magnetic survey. To achieve the goal of the study, Total Field magnetic intensity map (TF), Residual magneic map (Res), upward continuation (100 to 700 m), as well as derivative, and tilt filtering and many other filtering techniques have been applied using Geosoft (Oasis Montaj) software to enhance the magnetic anomalies for interpretation. Qualitative and Quantitative interpretation of magnetic maps showing that, the main Dibba fault zone extension to the west is well recognized and traced and the estimated depth to the fault is about 600 to 700 m. The importance of this study may be leads to the association of many of the economic resources within the area of interest like underground water, mineralization and ore deposits.

Introduction

The aim of magnetic survey is to investigate the subsurface structure. Magnetic survey is used to explore the highly structured area at Dibba zone ,located at 25° 30' 43.98" N 56° 18' 18.04" E (Figure 1), Fujairah, North East of the United Arab Emirates. The magnetic method is a very popular and inexpensive approach for detecting the anomaly within the earth's subsurface. The anomaly was analyzed to get the favorable area of water well.

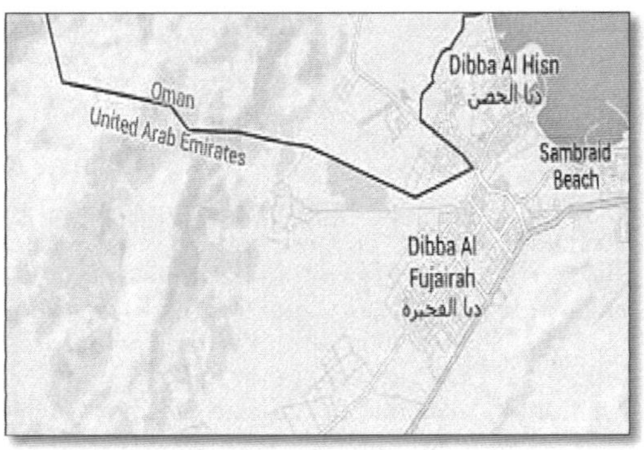

Figure (1): Location map of Dibba zone

Geology of Dibba Zone

The Dibba zone is a structurally complex, NE-SW trending part of the Oman mountains thrust front and it's dominated by polydeformed sedimentary and igneous rocks which are exposed in: Wadi Fai, Wadi Batha Mahani, Jabal Al Mahayiah and the town of Al Tawyen. The northern part of Dibba zone formed by a system of high angle brittle faults which are knows as Wadi Batha Mahani fault (Figure 2) the sedimentary and volcanic rocks of Dibba zone classified into five units and two major mélange units (Manhal, M. , 2008), Dibba City Guide. (n.d.) , (2014) AlBuloshi K., 2010 and others.

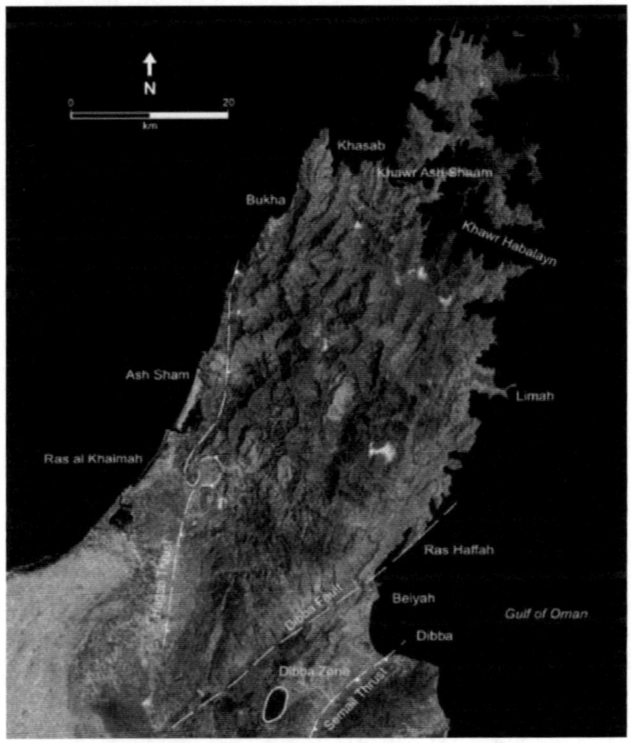

Figure (2): Satellite Image showing location of Dibba

Dibba zone

Methodology

Magnetic survey is acquired to investigate the subsurface geology on the basis of the anomalies in the earth's magnetic field resulting from the magnetic properties of the underlying rocks. In general, the magnetic content (Susceptibility) of rocks is extremely variable depending on the type of rock and the environment. Common causes of magnetic anomalies include dykes, faults and lava flows. A magnetometer is the instrument which measures both the orientation and strength of a magnetic field. When the magnetic field of a rock sample is measured, the result is actually a measure of the field as it is being affected by the earth's magnetic field, as well as any other large bodies of

Data acquisition:

The most important thing for data acquisition in magnetic survey include: Instruments, Points and the Base Station. The important instrument is Magnetometer which measures the scalar intensity of the local magnetic field (Photo1) Also we GPS coordinates of each station have been taken. The interested surveyed area (Dibba) is about 35x5 Km. Base station using magnetometer reader have been taken every one hour. Constant interval with 500 m is applied. Measurements of the magnetometer at each observation point are taken three times, the total acquired points is approximately 904 points. After the points completed and covered all the area the base station readings have been used to correct data for the diurnal variation correction to import the data to the software for processing.

Photo (1): The magnetometer at the base station

Processing:

Oasis Montaj V.8.2 is used to process the acquired magnetic data to remove the noise and to get accurate residual magnetic anomalies. In processing many formulas and techniques have been used to filter the data and enhance the magnetic anomalies for better interpretation.

Total Magnetic Intensity Field

The first gridded map of the survey is the total magnetic intensity field map (TF). The total field show important anomaly in blue color which may represent the extension of the Dibba fault zone to the west. Which is oriented NE-SW. While, the red color may reflect the positive or high magmatic anomalies which may be related to the ophiolite outcrop extension effect to the west.

Residual Magnetic Anomaly (RES)

This magnetic map (Figure 4) shows the result of removal the regional magnetic component (IGRF) from the observed total magnetic field.

total magnetic field map.

The IGRF map has been calculated by considering the coordinates of the area and the time period of the field measurements. In overall there are resemble in the shapes of the anomalies and observed on the total magnetic field map but with different field intensities.

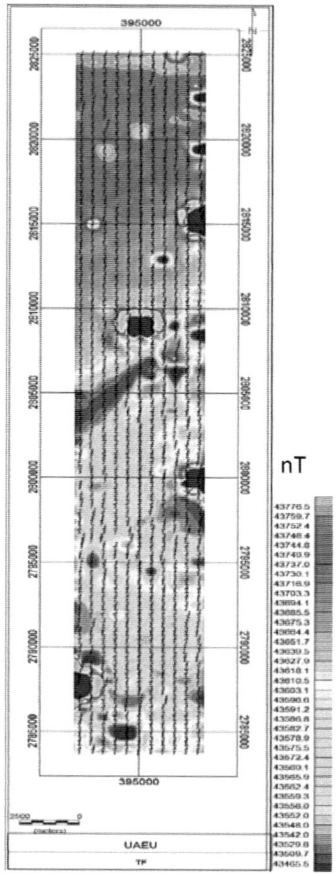

Figure (3): TF magnetic map

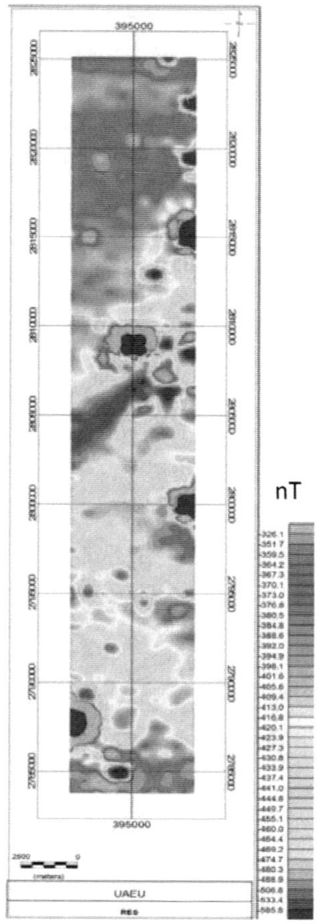

Figure (4): RES Magnetic map

Upward Continuation

The upward continuation filter was applied to the magnetic data to determine whether or not the magnetic sources associated with the mapped anomalies are deep seated. The upward continuation results from the Dibba area with 100 m until 600 and 700 m every 100 m respectively (Figure 5a, b & c) suggest that the linear anomaly shown in (Figure 5 a) persists in the upward continued data to a height of 700m.

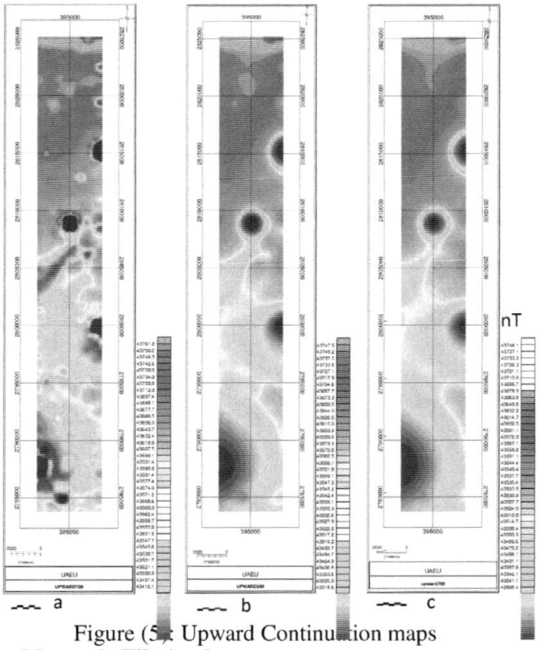

Figure (5): Upward Continuation maps

Magnetic Tilt Angle

This is a new method, known as the tilt depth method which provides an intuitive means of understanding the variation in depth of magnetic source bodies also shows a number of magnetic sources which have not been observed on the other basic magnetic maps. This map shows that the magnetic anomalies are delimited and vertically situated on top of the geologic source (Figure 6) present the eastern structures and the presence of a possible magnetic structure situated mainly in the western and eastern parts of the area and observed also in the central part of the region.

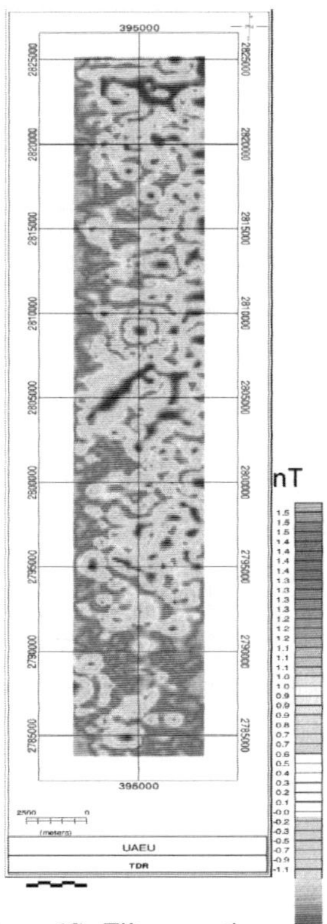

Figure (6): Tilt magnetic map

Structural Analysis

Interpretation of the resulted magnetic maps shows that, the Dibba fault zone is represented by a magnetic lineament oriented NE-SW along the study area. Small and localized lineaments are also observed across the area. However, there are some clear intersections between the localized structures,. Note that the lineaments intersection areas may represent the most promising locations for drilling for groundwater investigation. Also the black solid lines show lineaments in the area. Additionally horst and graben are terms referring to regions that lie between normal faults and are either above or lower than the area beyond the faults. A horst represents a block pushed upward by the faulting, and a graben is a block that has dropped due to the faulting.

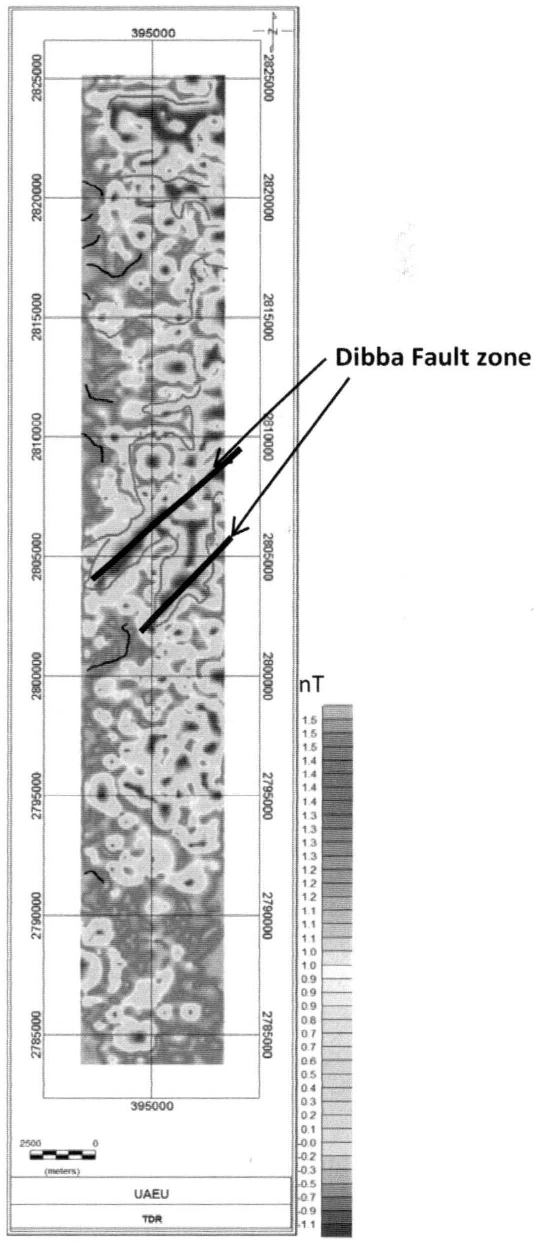

Figure (7): Interpreted Dibba fault on tilt magnetic map

References:

Manhal, M. (2008, February 13). Geophysical investigation of Al Jaww Plain, eastern Abu Dhabi: Implications for structure and evolution of the frontal fold belts of the Oman Mountains. Retrieved May 30, 2015, from http://www.pi.ac.ae/PI_ACA/pge/faculty_staff/ma_pdfs/Ali_Al-Jaww.pdf

A.S., A. (1989, July 12). Petroleum Geology of the Unite Arab Emirates. Retrieved May 30, 2015, from http://www.megeuae.com/SharhanPDFpubs/Petroleum-Geology-UAE/Petrol-Geol-UAE.pdf

AlBuloshi K., 2010, "Investigation the source of thermal anomalies in the northern United Arab Emirates (UAE) desert using geophysical methods" PhD Thesis, 147p

Dibba City Guide. (n.d.)_2014. Retrieved May 30, 2015, from http://www.dibbacityguide.com/

THE COUNTRY & THE ENVIRONMENT - GEOLOGY. (n.d.). Retrieved May 30, 2015, from http://www.uaeinteract.com/news/default3.asp?ID=51

Subsurface Structures and The Effect On Flexible Pavement Performance: The Shagamu-Benin Expressway, Southwestern Nigeria, As Case Study

J. O. Fatoba[1], M. O. Olorunfemi[2] and A. B. Eluwole[3]

1 Dept. of Earth Sciences, Olabisi Onabanjo University, Ago-Iwoye, Ogun-State, Nigeria

2 Dept. of Geology, Obafemi Awolowo University, Ile-Ife, Osun-State, Nigeria

3 Dept. of Geology. Ekiti State University, Ado – Ekiti, Ekiti – State, Nigeria

ABSTRACT

Geoelectric and electromagnetic methods were conducted on failed highway pavement segments at Ilisan and Ore localities along Shagamu-Benin expressway. The objective was to determine the cause(s) of perpetual pavement failure along segment of the highway which have always defied rehabilitation.

Twenty schlumberger Vertical Electrical Soundings (VES) were carried out with the electrode spacing (AB/2) varying from 1 m to 75 m with the total spread of 150 m. The VES data were presented as sounding curves and interpreted by partial curve matching technique and computer assisted 1-D forward modeling. The interpretation results of the VES were presented as geoelectric sections. 2-D resistivity imaging using Dipole-dipole array with electrode spacing (a) = 5 m and expansion factor n = 1 – 5 was conducted along the road segments and along the same profile with the VES. The 2-D resistivity imaging data were inverted using the Dipro software. VLF-EM inline profiling with station interval of 5 m was carried out to confirm the geologic features depicted by the VES and 2-D resistivity imaging. The VLF-EM results were presented as profiles.

The geoelectric sections identified the lateritic sand of resistivity values of 1230 ohm-m and an average thickness of about 2.2 m as the subgrade materials at Ilisan locality and partly weathered basement with resistivity values ranging from 283 ohm-m to 7848 ohm-m as the subgrade materials at Ore locality. The subgrade materials at both localities are adjudged competent to sustain the stability of the highway pavement. However, the 2-D resistivity imaging and VLF-EM identified low resistivity zones (<100 ohm-m), which diagnostic the presence of buried stream channel and shear zone suspected to be a fault beneath the subgrade materials at Ilisian and Ore localities respectively. These features may have been responsible for the flexible pavement failure always experienced at the investigated highway segments.

Keywords: Subsurface Structure, Subgrade Materials, VES, Resistivity and VLF-EM.

4-D Seismic over 3-D

Abstract

Neeharika Shukla
Indian Institue of Technology
Banaras Hindu University, Varanasi, India
neeshukla24@gmail.com

Ever increasing maturity of the reservoirs as we are moving ahead in the Geological Time Scale, a need arose to perform the survey over the same field to examine changes in various parameters characterizing the reservoir. It is because the existing fields are depleting and there is a low probability of obtaining new reservoirs.

The 3-dimensional surveys are performed to get a real picture of the sub-surface of the 3-Dimensional kind. The image will provide us information regarding the sub-surface geological structures for demarcating the regions of hydrocarbons presence. But, we superimpose the images of the same place taken over different times; we get an idea of the changes in the parameters of the reservoir as well as the quality of the image gets improved.

The surveys are performed by assuming certain parameters. They are bin size, group interval, fold, spread length and stretch effects, AGC, filters, line spacing, shot interval, line direction, swath design, line length and roll along. The most important parameter, including the bin, is the area under which each CMP must be stacked to obtain a clear image of the reflector.

However, 3-D seismic is performed to map the traps of oil or, gas or, map the faults that may be the barriers to oil flow in the producing field. It would be good to have a map of the quality of the reservoir e.g. its thickness and porosity, partly to estimate the volume of oil that may be present in a given trap and, partly to plan best to get the oil or, gas out of the ground. So, for reducing the risks of drilling an unsuccessful well, we perform 4-D survey to get the proper estimates of oil contained in it.

4-D seismic identifies the ways in which production well could be drilled and it detects the opportunities for well intervention. It helps in locating the reserves which are not yet exploited and also monitors the progress in the field development. Understanding the effects of changes in fluid saturation over the seismic response in a reservoir is an important part of 4D monitoring. Relating seismic responses to fluid changes may dictate the fluid distribution and the movement of fluids within the reservoir.

Therefore, it is appropriate to impart changes in the existing trend of 3-D surveys and perform a 4-D survey to study the changes in the reservoir in order to drill a successful well. This would help in getting knowledge of the existing field and delineate the places where hydrocarbons are yet to be explored.

References:

- Chaouch, A., and Mari,J. L. ; 2006, "3-D Land Seismic Surveys: Definition of Geophysical Parameters,Oil & Gas Science and Technology", Rev. IVP, 61(5), pp. 611-630.
- Mitra, P. P. et. al., 2007: "Post Stack 4D Seismic Data Conditioning, Matching and Validationto Map the Changes in Fluid Phase", Geohorizons, SPG, Vol. 91, pp. 5–12.
- Sheriff, R. E. and Geldart, L. P.;1995, "Exploration Seismology", Second Edition, pp. 451-459.
- Yilmaz, O.; 2001, "Seismic Data Analysis", Vol. 2, pp. 1213-1242.

Meteorological conditions influence on quantification of site effects at low frequency using the seismic ambient noise

Rabah Bensalem[1], Djamal Machane[1], Jean-Luc Chatelain[2], Mohamed Djeddi[3], Hakim Moulouel[1], El Hadi Oubaiche[1], Mustapha Hellel[3].*

1-Centre National de Recherche Appliquée en Génie Parasismique, CGS, Rue Kaddour Rahim Prolongée, Hussein-Dey, Alger, Algérie, tel : +213 21 49 55 60, fax: +213 21 49 55 36, email : rabah63@gmail.com
2-Institut de Recherche pour le Développement (IRD), Institut des Sciences de la Terre (ISTerre) Maison des Géosciences, Université Joseph Fourier (UJF), Grenoble, France
3- Faculté des Sciences de la Terre, de la Géographie et de l'Aménagement du Territoire, Université des Sciences et de la Technologie Houari Boumedienne (USTHB), Alger, Algére.

Summary

One of the objectives of the studies seismic microzoning or local hazard is to identify possible site effects, which are at the origin of the amplification of seismic motion incident. These motions are amplified in the resonant frequency of a soft layer at low velocity based on a rigid substratum. Evaluation of site effects by the seismic ambient noise method, in the case of deep sedimentary basin, can be induce us into error if precautions are not taken into consideration because the weather conditions and oceanic they also active in the field of low frequency.

Introduction

The sources of ambient noise are natural and anthropogenic origins. The sources of natural origin are related to meteorological, oceanic, etc., and sources of anthropogenic origins to human activities. In literature we denote by microseismic the noise of natural origin, concerning the low frequency domain (<1 Hz), and by the microtremors the anthropogenic noise. Studies of a number of researchers, concerning the evolution of the noise, have shown that the behavior of spectral amplitudes of microseismic and microtremors are different (Kanai and Tanaka 1961; Haubrich et al., 1963;. Akamatsu et al., 1992; Yamanaka et al., 1993; Seo 1997; Bonnefoy 2004). To perform this study, the ambient noise in the low frequency domain was registered for eight weeks on the site of Dar El Beida (Algiers), in order to study the temporal variability of the spectral amplitude, according to the different periods of the day, different days of the week and depending on weather conditions.

Data acquisition and processing

We conducted the ambient vibration experiment in the Dar El Beida city by using a CityShark a single station (Chatelain et al., 2000) coupled to three-component five-second Lennartz seismometers. Such combination is suitable for this type of experiment Guillier et al., (2008). Continuous recordings are acquired at a 200 sps sample rate. Data acquisition was made following SESAME recommendations (SESAME 2005). All data involving spectral analysis have been processed with the freely distributed Geopsy software (http://www.geopsy.org). The spectral amplitude was computed on stable window of 50 seconds to ensure a good representation of the frequencies to 0.2 Hz.

Temporal variation in the spectral amplitude of ambient noise in low frequency domain (microseismic)

The spectral amplitudes showed no correlation between the different days of the week in the low-frequency domain (Figure 1). On the recording period of eight weeks, the amplitudes of the three components are in general relatively weak, but with quite pronounced amplitudes some days. In the latter cases, the amplitudes are generally stronger on the horizontal components on the vertical component, and the North component is slightly greater than the East component.

Temporal variation in the spectral amplitude of ambient noise in high Frequency Domain (microtremors)

Unlike to the results obtained in the domain of microseismic, those achieved in the field of microtremors show a very good correlation between the day and night activities and between different days of the week (Figure 2).
During the period of eight weeks, the spectral amplitudes vary relatively little and are depending on the intensity of human activity, as shown by the recordings made on Fridays (day off in Algeria), characterized by low amplitudes, the amplitude levels on the two horizontal

components are comparable and are greater than the vertical component.

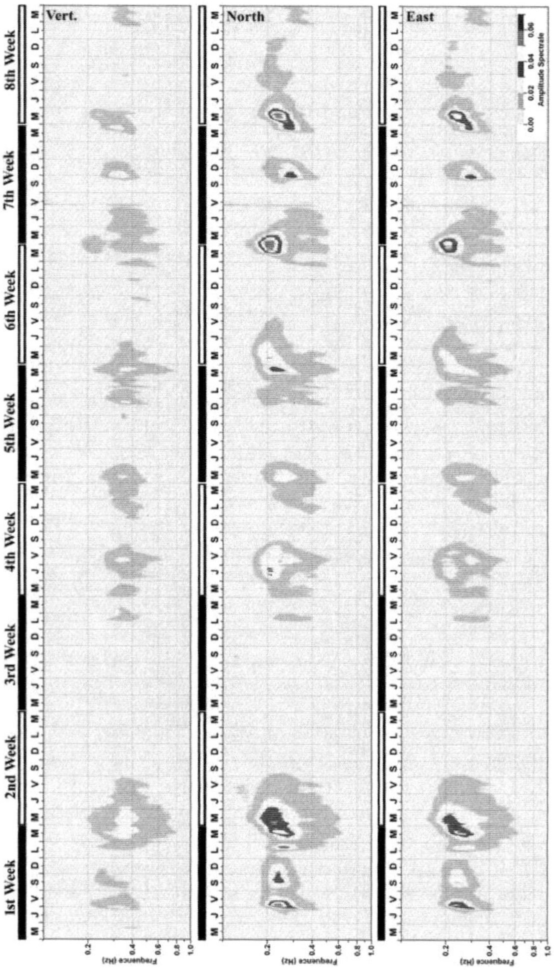

Figure 1 Temporal variation in the spectral amplitude of ambient noise at low frequency (0.1–1.0 Hz) (microseismic) during the eight weeks (26/01/2011–22/03/2011) of Dar El Beida site recording. Vertical components (left), North-South (center) and East-West (right).

Frequencies highlighted on the horizontal components are different from those obtained on the vertical. On this vertical component, two distinct peaks are observed at 2.5 Hz and 5.0 Hz, with slightly higher amplitude for the first peak. The both horizontal components exhibit a peak at 3.5 Hz, but the northern component includes a second peak, of low amplitude at 6.0 Hz. These peaks are always observed, but their amplitudes fluctuate according to the noise level generated by anthropogenic activities.

The records also show that in our study region, the activity starts between 5:00 and 6:00 AM and reaches its maximum between 07:30 and 18:00, then is attenuated beyond, reflecting daily human activities.

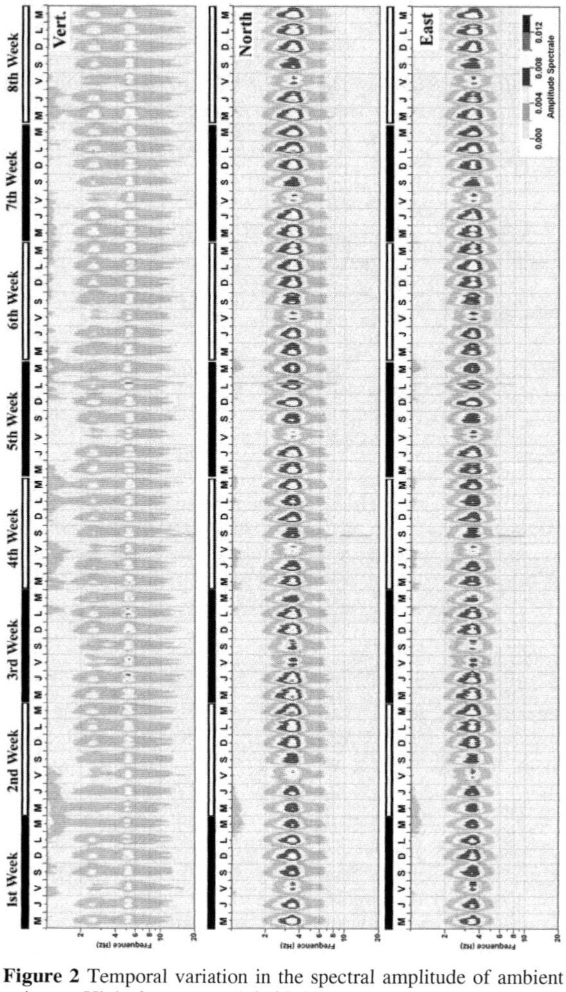

Figure 2 Temporal variation in the spectral amplitude of ambient noise at High frequency (1.0–20 Hz) (microtremor) during the eight weeks (26/01/2011–22/03/2011) of Dar El Beida site recording. Vertical components (left), North-South (center) and East-West (right).

Discussion

At low frequency domain (microseismic), strongest peaks observed for the horizontal components appear between 0.2 and 0.3 Hz (Figure 3).

We have not been able to compare these observations with the real weather conditions data, otherwise, we have exploited the forecast data for qualitative interpretation.

The frequency peaks are well corroborated with a strongly agitated sea, except in two cases (Figure 3): the first peak (**A**) is obtained with a agitated sea, and the peak (**B**) does not match any particular weather forecasting, probably because this event is of short duration.

The forecast of the state of the sea to the days of the 3rd and 4th week were agitated at strongly agitated while the peaks obtained for these days are of low amplitude (Figure 3).

Still according to the forecast data, the wind velocity, which has logically also a directly impact on the state of the sea, has an impact on the amplitude of peaks for the velocity higher than or equal to 40 km/h.

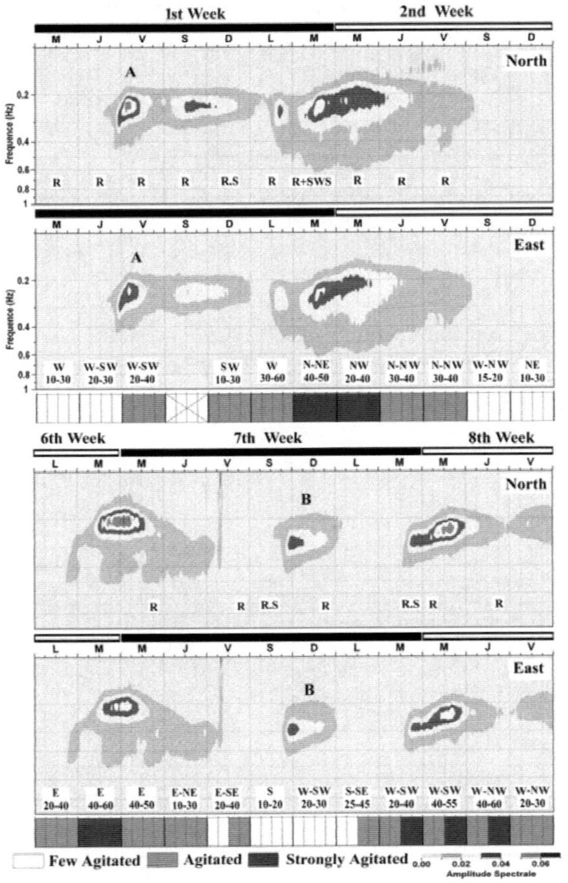

Figure 3 Zoom of low frequency spectral amplitudes for the periods during which the puffs of energy are maximal, Rainfed days (R) and Stormy (R.S) are indicated at the bottom of the northern component diagram. The direction and wind velocity are listed at the bottom of the Eastern component diagram. The sea state is shown at the bottom of the figure. Weather data is provisional data.

The wind direction has no influence on the appearance of high amplifications peaks because they appear with different wind directions (W-N, N-NE, NW, E).

AT high Frequency Domain (microtremors), by superimposing the weather forecast data with the curves of spectral amplitudes, no correlation was observed, whether the state of the sea or rainy and / or stormy days.

By comparing the periods of rain with the curves of the spectral amplitudes, at low frequency, no influence is observed. However, we note that the stormy rains that were held during the 7th week (Figure 3), coincides well with a peak of high amplitude whose frequency is shifted to 0.3 Hz. This frequency shift expresses a decrease in the wavelength, which could find an answer in the foyer of the storm, which would have shifted the origin of the source of these low frequencies.

To explain these variations in frequency we determined the direction of the sources that generate them to determine their spatial distribution. To do this, we studied the particle motion versus time on the two horizontal components between 0.2 and 0.4 Hz (Figure 4).

All the diagrams show highly energetic particle motion in a predominant direction NE-SW with a relatively low component NW-SE.

On days when the low frequencies are attenuated peaks, are always the predominant directions NE-SW

The wind direction has no influence on the occurrence of amplification peaks or on the direction of the source in the event of high wind.

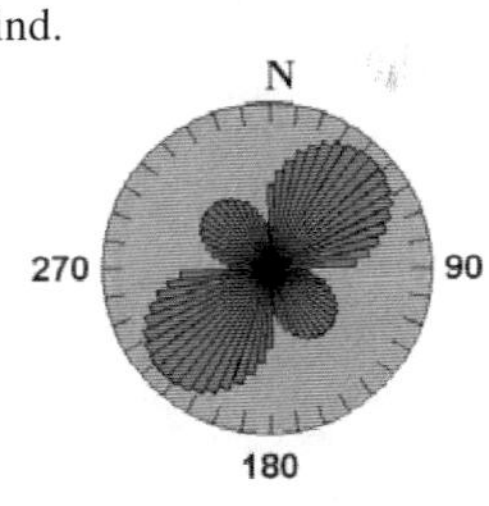

Figure 4 Rosace directions of particle motion between 0.2 -. 0.4 Hz The radius of the rosace represents the amplitude of the normalized resulting particulate movement.

The NE-SW direction coincides with the maximum aperture, in this direction the Mediterranean Sea, over a distance of 970 km (Figure 5), while the NW-SE direction coincides with most southern part of the Mediterranean sea, confined between the Tunisian and Libyan coast in the form of "U" over a distance of 950 km. The waves are more energetic in the NE-SW direction as to reach the recording station, they propagate through the Earth's crust over a very short distance, on the order of a few kilometers, while in the direction NW -SE the waves are attenuated because they travel a great distance from the Tunisian coast to the recording station.

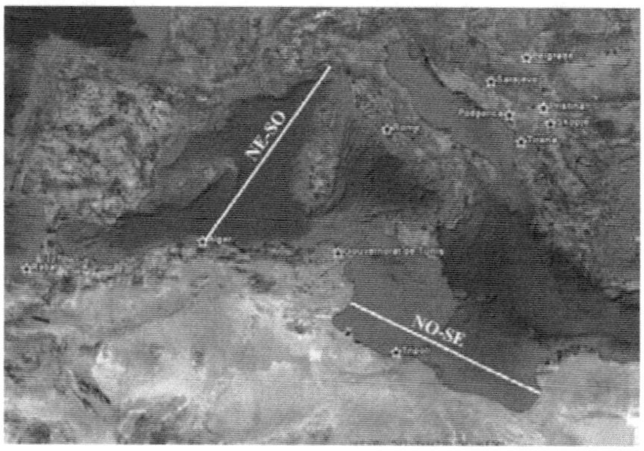

Figure 5 Projection of two directions obtained by the analysis of particle motion, on background map of the Mediterranean Sea (Google earth).

Conclusion

In the low-frequency domain, amplitude peaks occur depending on weather conditions. So for site response studies in the low-frequency domain, as the case of deep sedimentary basins, we must be vigilant and carefully choose the recording moments, when the climatic conditions are favorable, to ensure that results represent the real response of the site and not induced by climate conditions.

References

Akamatsu J., M. Fujita et K. Nishimura (1992). Vibrational characteristics of microseisms and their applicability to microzoning. Journal of Physics of the Earth, 40, pp.137-150.

Bonnefoy-Claudet S. (2004). Nature du bruit de fond sismique : implications pour les études des effets de site. Laboratoire de géophysique interne et tectonophysique. Thèse de doctorat, Université Joseph Fourier-grenoble-France. 241pp.

Chatelain, J.-L., P. Guéguen, B. Guillier, J. Fréchet, F. Bondoux, J. Sarrault, P. Sulpice, and J.-M. Neuville (2000). CityShark: A user-friendly instrument dedicated to ambient noise (microtremor) recording for site and building response studies. Seismological Research Letters 71-6, 698–703.

Guillier, B., K. Atakan, J. L. Chatelain, J. Havskov, M. Ohrnberger, F. Cara, A. M. Duval, S. Zacharopoulos, P. Teves-Costa, , and the SESAME (2008). Influence of instruments on the H/V spectral ratios of ambient vibrations, Bull. Earthquake Eng. 6, 3–31, doi 10.1007/s10518-007-9039-0.

Haubrich R. A., W. H. Munk and F. E. Snodgrass (1963). Comparative spectra of microseisms and swell. Bulletin of the Seismological Society of America, 53 (1), pp.27-37.

Kanai K. et T. Tanaka (1961). On microtremors. VIII. Bulletin of the Earthquake Research Institute, 39, pp.97-114. Tokyo University.

Seo K., (1997). Comparison of measured microtremors with damage distribution. In JICA, Research and Development Project on Earthquake Disaster Prevention.

SESAME European research project, (2005). Guidelines for the implementation of the H/V spectral ratio technique on ambient vibrations measurements, processing and interpretation. Project No. EVG1-CT-2000-00026 SESAME. http://sesame-fp5.obs.ujf-grenoble.fr

Yamanaka H., M. Dravinski and H. Kagami (1993). Continuous measurements of microtremors on sediments and basement in Los Angeles, California. Bulletin of the Seismological Society of America, 83-5, 1595-1609.

Application of Time-Domain Electromagnetic method in Mapping Sea water intrusion in Al-Batinah Coastal plain, Al Batinah, Sultant of Oman

Name:Hiba Abdullah Al-Salmi

Supervisor name: Dr. Hesham El-Kaliouby

Abstract :

Geophysical methods are considered to be the most popular tools for the hydrological applications and TDEM method is one of method that was used widely in mapping potential aquifers and sea water intrusion. Also it was applying in term of its ability to map subsurface for high depth and this method quick. Therefore, this study explores the groundwater and its properties within Al Batinah Coastal plain area, northern of Oman. Al Batinah Coastal plain is contain a large number of wells however because of Increased water demand more than supply for agricultural , industrial and other human requirements a hydrological problems appear which is pump the water frequently which lead to seawater intrusion from that ,the objective of my project become to investigate the saline water intrusion whiten this area .

The work was done on 11 profiles in different orientation N-S and E-W profiles to cover a large area as possible using a large loop of size 100X100 which Increase capacity to mapping depth of saline-intrusion. Then data was processing using IX1D program to generate a modeling and inversion that describe the amount of salinity intrusion Al Batinah Coastal plain. Also, lithological, resistivity log and gamma log were used as ground truth for interpretation.

The TDEM study reveals that area near the coastal had low resistivity (1-4) Ohm-m the due to increase in sea water intrusion the things which drop resistivity whiten this layers. However the region inland is of higher resistivity region (20-100 Ohm-m) correspond to fresh water .Also, because of presence of Gravel ,Marle and limestone .

Gravity interpretation using multi-scale analysis in the Algiers area (Algeria)

L. Benhenni[1], Z. Benaissa[1], S. Samai[1], M. Djeddi[1], H. Boukerbout[2] and A. Boudella[1]*

[1]*Geophysics Laboratory, Faculty of Earth Sciences, USTHB, Algeria. lbenhenni@usthb.dz; zabendz@yahoo.fr*
[2]*Centre de Recherche en Astronomie, Astrophysique and Géophysique – Bouzaréah, Algiers, Algeria*

Summary

The aim of this study is to examine the contribution of multi-scale analysis to the interpretation of gravity map of the Algiers area. The method used is based on the continuous wavelet transform of the first horizontal derivative of gravity anomaly field signal with the first derivative of a Gaussian, considered analyzing wavelet. Software, based on the calculation of the maximum values of Shannon entropy, allowed us to localize the sources.
Overall, the results obtained on synthetic and real data show that multi-scale analysis is more accurate than Euler deconvolution. Its application to the data, extracted from the gravity map of residual anomalies of the Algiers area, highlighted practically all faults and contacts observed at the surface.

Introduction

The explosion of the modern theory of wavelets is not as old, it dates back to the 80s (Morlet, 1980; Grossmann and Morlet, 1984; Meyer, 1987; Mallat, 1989; etc.). The use of wavelets has been increasing since then in various fields such as geophysics.
The continuous wavelet transform (CWT) provides a detailed representation of the signal at different scales, through the passage of large scales to fine scales. This makes it interesting for the potential fields analysis (Moreau et al., 1999; Hornby et al., 1999).
The purpose of this work is to outline the geological structures of the Algiers area and its surroundings using a method based on the CWT of the first horizontal derivative of gravity anomaly field signal with the first derivative of a Gaussian, considered analyzing wavelet as recommended by Cooper (2005).
For the localization of the sources, we used software developed by Boukerbout et al. (2003), based on the calculation of the maximum values of Shannon entropy. Four profiles approximately directed north to south, extracted from the Algiers residual anomaly map, are considered and the obtained results are compared with those of Euler deconvolution method (Thompson, 1982).

Method

The continuous wavelet transform
The CWT of a function $f(x)$, $\in R^n$, is defined as a convolution product:

$$W(\psi, f)(b,a) = \int f(x)\frac{1}{a^n}\psi\left(\frac{b-x}{a}\right)dx = (f * \psi_a)(b)$$

where a and b are the dilation and translation parameters, respectively, and $\psi_a(x)$ is the analyzing wavelet defined by:

$$\psi_a(x) = \frac{1}{a^n}\psi(\frac{x}{a})$$

The covariance of the CWT relative to the dilation $\grave{a} > 0$ is given by the following formula:

$$W(\psi, \psi_{\acute{a}} f)(b,a) = \frac{1}{\acute{a}^n} W(\psi, f)(\frac{b}{\acute{a}}, \frac{a}{\acute{a}})$$

A function is homogeneous of degree α if it satisfies the following equation:

$$f(\acute{a}x) = \acute{a}^\alpha f(x), \forall \acute{a} > 0, \alpha \epsilon R$$

The condition of homogeneity simplifies the covariance formula as follows:

$$W(\psi, f)(\acute{a}b, \acute{a}a) = \acute{a}^\alpha W(\psi, f)(b,a)$$

This can be generalized by the following formula:

$$W(\psi, f)(b, \acute{a}) = \left(\frac{\acute{a}}{a}\right)^{n+\alpha} \psi_{\acute{a}/a} W(\psi, f)(b,a)$$

The extrema of this function are geometrically represented by the union of lines forming a cone. These lines correspond to:

$$\frac{\partial^m W(\psi, f)(b,a)}{\partial b^m} = 0$$

This latter formula represents the edges of the wavelet transform (Holschneider, 1995; Moreau et al., 1999).

Gravity sources location

Moreau et al. (1999), Mallat (2000) and Mallat and Hwang (1992) estimated the location of the sources by extending the lines of maxima in the negative dilation $-a = z$ (Figure 1).

Boukerbout (2003) estimated it by extending the lines of maxima in the negative dilation plane (x,a) to the lower half plane depths, through the maximum entropy method. In this study, for the localization of gravity sources, the corresponding software developed by Boukerbout is used.

Figure 2 shows an application of this type of wavelet analysis to the gravity anomaly (Figure 2A) and its first horizontal derivative (Figure 2B) from two vertical cylinders,

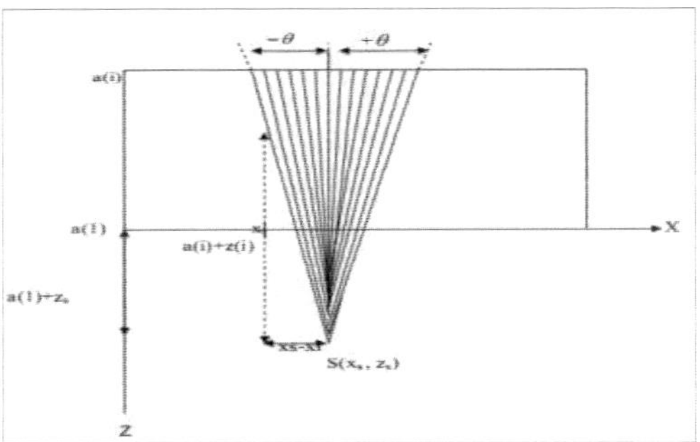

Figure 1: Schematic representation of the maxima lines converging to the source area S (Boukerbout, 2003).

with the first derivative of a Gaussian, considered analyzing wavelet. The best results are those obtained on the first derivative with obvious correlation between the body locations and the maxima of the entropy plot, due to the good correlation between the first derivative and the wavelet (Figures 2A.c and B.c). Euler deconvolution solutions, overlain on the entropy plots for comparison, delineate the bodies but the correlation is lower.

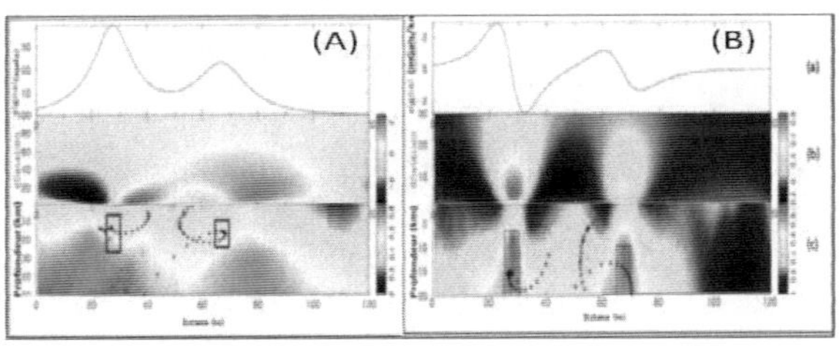

Figure 2: A- (a) Gravity anomaly from two vertical cylinders; (b) CWT analysis; (c) Entropy values.
B- (a) First horizontal derivative of signal A(a); (b) CWT analysis; (c) Entropy values.

The circles represent the solutions of the Euler deconvolution.

Geological setting and used data

The study area is located in the north central region of Algeria. It includes the sedimentary basins of Mitidja, Cheliff, Médéa and Bouira, bounded by several deep faults (Figure 3).

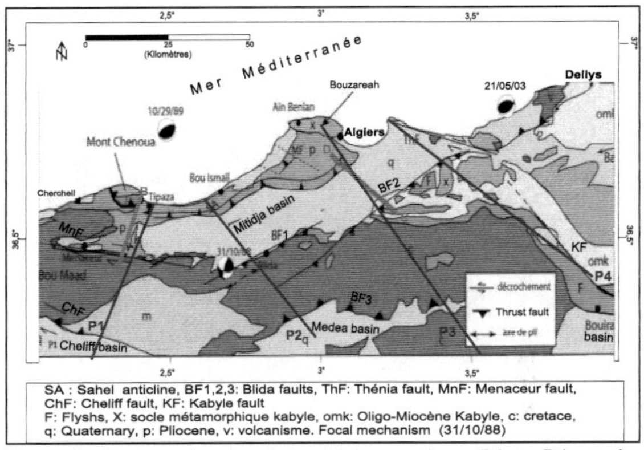

Figure 3: Structural map of the Algiers region. (P1 to P4: study profiles) (Domzig, 2006).

Gravity data used in this study come from the Bureau Gravimétrique International (BGI) database. The corresponding map of residual anomaly field (from -15 to 19 mGal) (Figure 4) shows the four sedimentary basins mentioned previously.

Furthermore, it reveals the existence of four large positive anomalies related to the four massifs of Tipasa, Algiers, Thénia and Dellys.

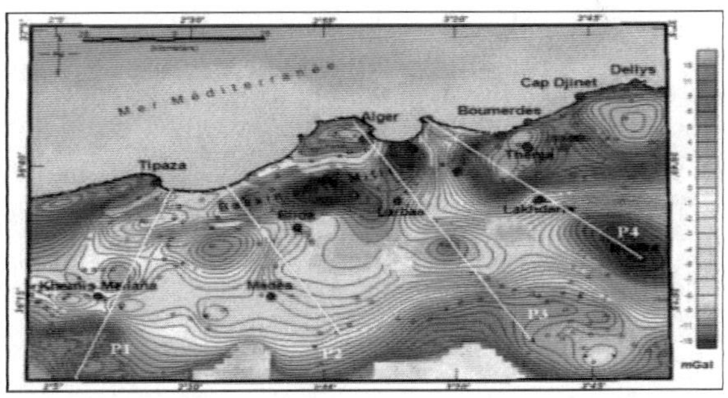

Figure 4: Map of the residual anomaly field of the study area. (P1 to P4: study profiles).

Results and discussion

Profile P1

The analysis of the first horizontal derivative of residual gravity anomaly over profile P1 (Figure 5) has maxima over the eastern extension of the Cheliff fault (10 km, depth

between 8 to 19 km), the Menaceur fault (55 km, at 12 km depth) and, in the North, the geological fault separating Algiers Sahel from Mitidja basin SA (66 km, at 17 km depth).
Euler deconvolution solutions highlight only the Ménaceur and Cheliff faults.

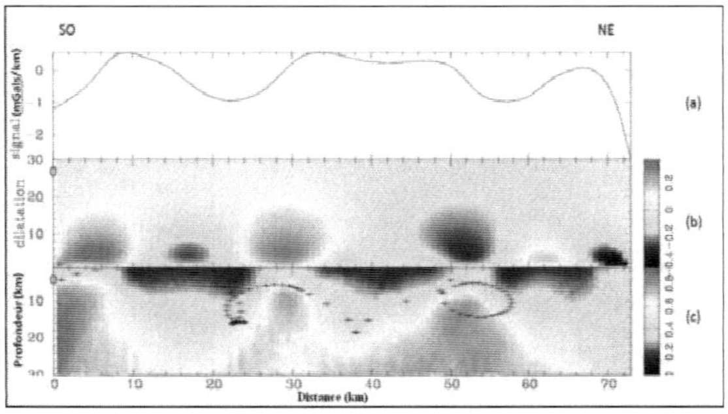

Figure 5: (a) Residual gravity anomaly over profile P1; (b) CWT analysis; (c) Entropy values. The crosses correspond to Euler deconvolution solutions.

Profile P2

Figure 6 shows the analysis of the first derivative of residual gravity anomaly over profile P2. It highlights four contacts which would correspond geologically to the South-East fault of Blida BF3 (0 to 18 km, at 5 km depth), the fault of Blida BF2 (40 to 44 km, at 4 km depth), the fault of Blida BF1 (50 to 57 km, at 6 km depth) and, to the North, the contact SA at a depth of 10 km.
These contacts are somewhat surrounded by Euler solutions.

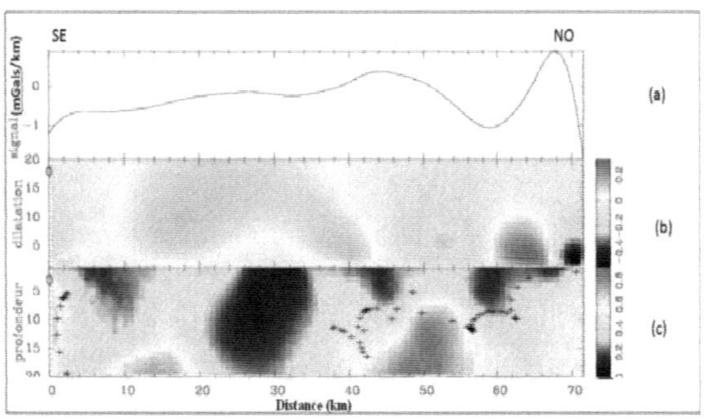

Figure 6: (a) Residual gravity anomaly over profile P2; (b) CWT analysis; (c) Entropy values. The crosses correspond to Euler deconvolution solutions.

Profile P3

The analysis of the first horizontal derivative of residual gravity anomaly over profile P3 (Figure 7) indicate three different discontinuities, geologically corresponding to the inverse fault of Blida BF3 (0 to 20 km, at 10 km depth), the fault of Blida BF2 (38 to 45 km, at 13 km depth) and, at the end of the profile, the contact SA at a depth of 8 to 12 km.
These discontinuities are surrounded by Euler deconvolution solutions.

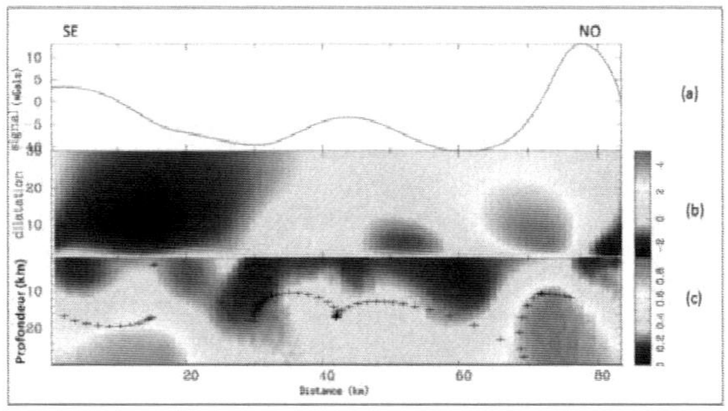

Figure 7: (a) Gravity anomaly over profile P3; (b) CWT analysis; (c) Entropy values. The crosses correspond to Euler deconvolution solutions.

Profile P4

Figure 8 presents the results of the first horizontal derivative of gravity anomaly over profile P4. Three contacts are detected and would be associated to the fault of Bouira, to the South (at 6 km depth), and the faults of Thénia ThF and Blida BF2 to the end of the profile, in the North (at 20 km depth).
These three contacts are also underlined by the Euler deconvolution with, however, much less precision.

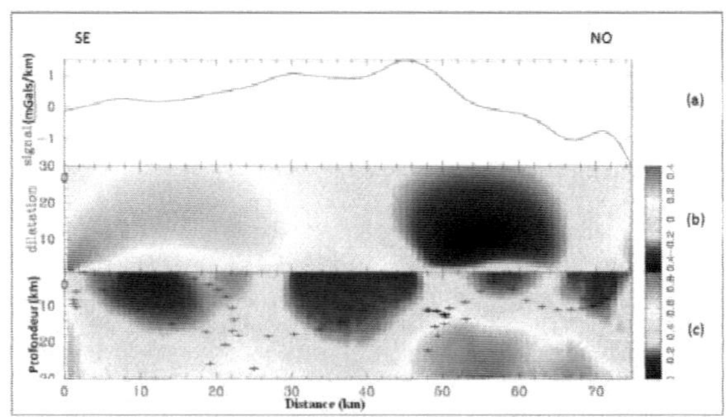

Figure 8: (a) Residual gravity anomaly over profile P4; (b) CWT analysis; (c) Entropy values. The crosses correspond to Euler deconvolution solutions.

Conclusion

In this work, the wavelet analysis, using the first derivatives of the signal and the Gaussian, highlighted almost all of the contacts crossed by the profiles P1 to P4, with more precision than Euler deconvolution. This is due to the signal to noise ratio which is improved when applying the CWT on the first derivative: maximum entropy is due to the good correlation between the first derivatives of the signal and the Gaussian. These results are consistent with the earlier studies of Cooper (2005).

References

Boukerbout, H., D. Gibert, and P. Sailhac, 2003, Identification of sources of potential fields with the continuous wavelet transform: Application to VLF data: GEOPHYS. RES. LETT, **30**, 1427-1429.

Cooper, G.R.J., 2005, Interpreting potential field data using continuous wavelet transforms of their horizontal derivatives: COMPUTERS & GEOSCIENCES, **32**, 984-992.

Domzig, A., 2006, Déformation active et récente, et structuration tectono-sédimentaire de la marge sous-marine algérienne : Doctorat thesis, University of Western Brittany.

Grossmann, A., and J. Morlet, 1984, Decomposition of hardy function into square integrable wavelets of constant shape: SIAM J. MATH.

Holschneider, M., 1995, Wavelets : An analysis Tool: CLARENDON OXFORD, ENGLAND.

Hornby, P. F. Boschetti, and F.G. Horovitz, 1999, Analysis of potential field data in the wavelet domain: GEOPHYS. J. INT., **137**, 175-196.

Mallat, S., 1989, Multiresolution approximations and wavalets orthonormal bases of L2(R): TRANS AMER. MATH. SOC., **315**, 69-87.

Mallat, S., and W. L. Hwang, 1992, Singularity detection and processing with wavelets: IEEE TRANSLATIONS ON INFORMATION THEORY.

Mallat, S., 2000, Foveal orthonormal wavelets for singularities: APPL. COMPUT. HARMON. ANAL., **14**, 133–180.

Meyer, Y., S. Jaffard, and O. Rioul, 1987, L'analyse par ondelettes: POUR LA SCIENCE, **119**.

Moreau, F., D. Gibert, M. Holschneider, and G. Saracco, 1999, Identification of sources of potential fields with the continuous wavelet transform: Basic theory: JOURNAL OF GEOPHYSICAL RESEARCH, **104**, 5003–5013.

Morlet J., G. Arens, I. Fourgeau, and D. Giard, 1982, Wave propagation and sampling theory: GEOPHYSICS, **47**, 203-236.

Thompson, D. T., 1982, EUDPH, A new technique for making computer-assisted depth estimates from magnetic data: GEOPHYSICS, **47**, 31-37.

The integrated application of natural radon-radioactivity and electrical resistivity tomography to investigate karst cave system: a case study from Al-Dahr Cave, N-Jordan

Rasheed Jaradat,[1,2], Eyad Na'amneh[1], Barakat Al-Bataina[1]*
[1]Yarmouk University, Dept. of Earth & Environmental Sciences, Irbid, Jordan. E-mail: rjaradat@yu.edu.jo
[2]Gulf Laboratory for Site Investigation & Material Testing, Abu Dhabi, UAE. E-mail: rasheed.jaradat@gulflab.ae

Summary

In this study, radon-radioactivity and electrical resistivity tomography (ERT) methods were used to investigate an existing karst cave system of northern Jordan.

Initial field reconnaissance and testing led to the selection of appropriate field parameters for the used methods. Various electrode configurations were tested (i.e. Wenner, Schlumberger and Dipole-Dipole). Correlating the ERT results with radon profiles showed that the Dipole-Dipole array has the highest resolution limits and that two meters of spacing was enough to maintain the required level of accuracy. Nine profiles; radon of 2-meter intervals and 0.5-meter depths were collected. The location of each profile was subsequently subjected to ERT surveying using the Dipole-Dipole array with 2-meter inter electrode spacing.

The average value of Radon concentrations across all detectors was equal to 6.18kBq/m^3 which is higher than that reported as background values from Jordan. The maximum radon concentration equals to 25.17kBq/m^3 was collected across the main hall limits of the cave structure. However, the radon concentrations tend to increase across known fault and fracture zone trends. Additionally, elevated concentrations were encountered across the center of the main hall. Some of the fault structure showed possible sealing characteristics, thus hindering the passage of radon gas through specific trend or across thin cave roofs. Several collapse features were seen across the resistivity sections of the subsurface. These localities provide good seepage pathways for the radon gas into the ground surface. Additionally, ERT results showed potential isolated cave extensions that can add to the value of the site.

The results of this study indicated the successful integrated interpretation of radon natural radioactivity and electrical resistivity tomography (ERT) in characterizing karst terrain structures

Introduction

Various direct investigation methods, such as drilling and excavation, are typically used to identify the components of buried subsurface structures. However, these methods have many disadvantages including their high cost, their destructive possibility and the inability to generalize from the obtained results across wide areas. On the other hand, geophysical investigation methods are more cost effective, noninvasive and non-destructive surveying techniques. Moreover, they are able to provide fast and accurate results about the subsurface over a wide area. Geophysical techniques are used in engineering site investigations for the detection of subsurface cavities, abandoned tunnels, sinkholes, buried structural elements, such as faults and fracture zones, and in the exploration for groundwater, fossil energy and mineral resources.

Karst is a distinctive topography in which the landscape is largely shaped by the dissolution of soluble rocks, including limestone (CaCO3) and dolomite (Ca,Mg)(CO3)2. This geological process, occurring over many thousands of years, resulted in unusual surface and subsurface features ranging from sinkholes, vertical shafts, disappearing streams, and springs, to complex underground drainage systems and cave systems.

A geological section is typically consisting of several types of rocks, minerals and pore fluids. Hence, each of these constituents has its own physical properties; therefore, subsequent electrical resistivity would vary in accordance to existing constituents, and their porosity, permeability and availability of a conductive material (i.e. pore fluid) that is able to pass electrical current. However, the possibility of the existence of buried cavities within the subsurface will hinder the ability to transmit injected electrical current and will result in very high resistivity values across the investigated profile.

Electrical resistivity tomography (ERT) is one of the main geophysical methods that is typically used to image and characterize the vertical and horizontal distribution of electrical resistivity within the subsurface. Electrical resistivity surveying methods are based on detecting differences in electrical resistivity from one place to another, and it is one of the best ways to detect cavities and identify their dimensions. Thus, within a karstic topography, zones of large resistivity anomalies strongly suggest the location of possible cavities (Gautam et al., 2000; Batayneh et al., 2000).

Integrated application of radioactivity and ERT to investigate karst cave system

Radon is a natural radioactive gas and it is one of the products of the radioactive decay of Uranium, an element which is usually present in minute amounts in limestone or the adjacent rocks. However, Earth's crust, soil and rocks are the main source of radon gas, where radon concentration varies depending on the type of soil and physical characteristics and the type of rocks and their characteristics and depths (Gautam et al., 2000). Being a gas, radon seeps from the rocks into the surrounding air. Normally, air currents disperse radon to relatively low concentrations, but in a confined space such as a cave, the concentration can build up. So, it is possible to use the changes of radon concentration in the detection of existing ground structures such as faults and caves, because these discontinuities tend to enhance radon gas mobility to the surface and may result in its concentration in the topsoil (Albergi et al., 2005).

The radon (Rn222) is the most important isotope to study because of its long half-life, resulting in very large distributions/concentrations. It is mainly attributed to the decay of lead and bismuth which takes place in solid state form which releases the alpha particles that can affect the human respiratory tract system. These particles can stick on the lung walls from the inside during the process of respiration (Agency for Toxic Substance and Disease Registry, 1990). These elements tend to decay quickly because of their short half-life resulting in danger of radon decay products of alpha particles, beta and gamma.

In this study an attempt was made to investigate the potential use of an integrated Electrical Resistivity Tomography (ERT) and Natural Radon surveying methods to map subsurface karst topography; a case study at Al-Dahr Cave, N. Jordan (Figure 1).

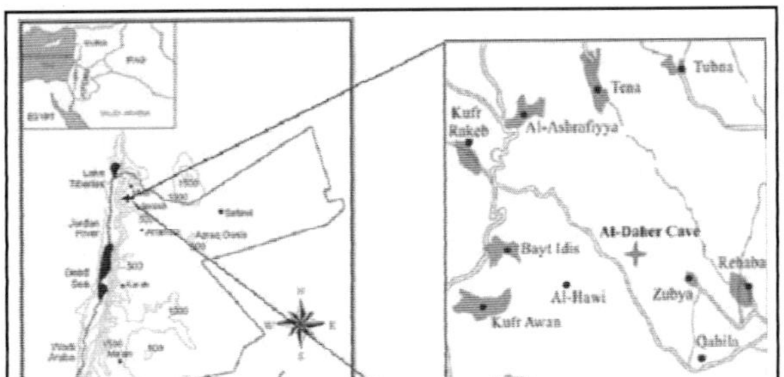

Figure 1: Location map of the Study Area.

Methods

Initial field reconnaissance study was carried out in order to obtain accurate survey design for ERT and radiometric survey parameters. Two profiles (48m) were conducted above the known location of the Dahr Cave location. Natural Radon concentration was determined using calibrated Passive Radon Diffusion Dosimeters consisting of a piece of CR-39; 250μm thickness; of dimension (1.5cm×1.5cm), fixed inside the bottom of a plastic cup (7cm of diameter x 4.6cm of depth), with a double face plastic tape. The cup is covered tightly with plastic lid which has an opening of diameter about 1.5cm, covered using a circular pieces of sponge of a thickness of 0.5cm allowing the passage of Radon gas only and to prevent the entry of impurities and dust to the inner cup. These dosimeters were buried inside 50 cm deep holes with specific horizontal spacing (1m), and recovered after 21 days. The recovered dosimeters were etched with NaOH at a temperature of 70° C for 8 hours in a heat bath and the tracks were counted manually under the microscope (Al-Bataina et al., 1997).

A multi-electrode resistivity meter system (ARES) was employed using different electrode configurations (i.e. Wenner, Schlumberger and Dipole-Dipole) with electrode spacing of 1m. Data inversion was carried out using RES2DINV (Loke, 2004).

Correlating the ERT result with radon profiles showed that the Dipole-Dipole array has the highest correlation trend with Radon collected concentrations and that the 2m electrode spacing was enough to maintain the required level of accuracy. Figure 2 shows the correlation between ERT and Radon concentrations of the first profile.

Results

Nine profiles of radon dosimeters, buried at a depth of 0.5m and at an interval of 2m, were carried out, such that each profile was subsequently subjected to ERT surveying using the Dipole-Dipole array with an inter-electrode spacing of 2m, aiming to carry out an integrated radiometric and resistivity tomography approach to characterize cave structure. Profiles are set such that they are straight and nearly parallel with 4-6 meters apart. The ERT profiles range between 80-94 m. Profile 7 was discarded due the difficult field conditions Figure (3).

Figure 4 shows an example of the integrated interpretation of ERT and Radon concentration profiles (Profile 1), indicating several anomalous radon concentrations. Anomalies B and C were associated with the open space located within the northeastern corner of the main hall (Figure 3). According to Melhem (2004) the height across this location is around 5m, which was confirmed by the inverted resistivity anomaly shown in Figure 4. On the other hand, D, E, and F are located within the northern corner of the main hall, which is truncated by a number of falling rock blocks from the roof of the cave. A pronounced fracture zone is seen across this northwestern boundary of the cave which spatially correlates with the E-anomaly having radon concentration of more than 15 kBq/m^3.

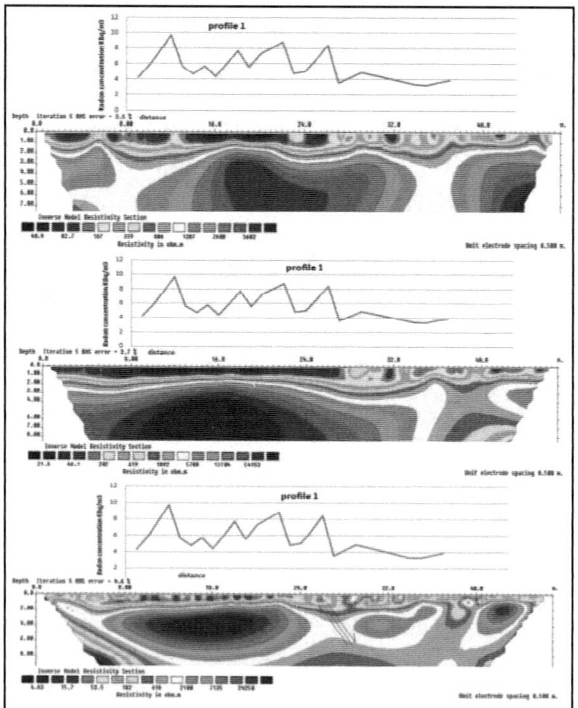

Figure 2: Comparison of ERT results and Radon Concentrations of profile TEST-1.

A similar behavior was noticed across other profiles, suggesting a clear correlation between the location of the walls of the main halls of the known cave structure and/or roof-crossing faults, and the cave's main halls with radon elevated concentrations across measured profiles. Results led to the identification of adjacent unknown cave structures.

The average value of Radon concentrations for all dosimeters was equal to 6.18kBq/m3 which is higher than that reported as background values from Jordan. However, the maximum radon concentration was 25.17kBq/m^3 across profile 5 along the main hall limits of the cave structure.

Conclusions

A successful integration between ERT, using the Dipole-Dipole configuration, and the natural radon radioactivity survey approach was achieved aiming to characterize cave karst structure. Elevated radon and resistivity measurements were seen across known collapse features, cave structures, faults, and fracture zones, that led to the exploration of potential unknown cave structures within the region of interest.

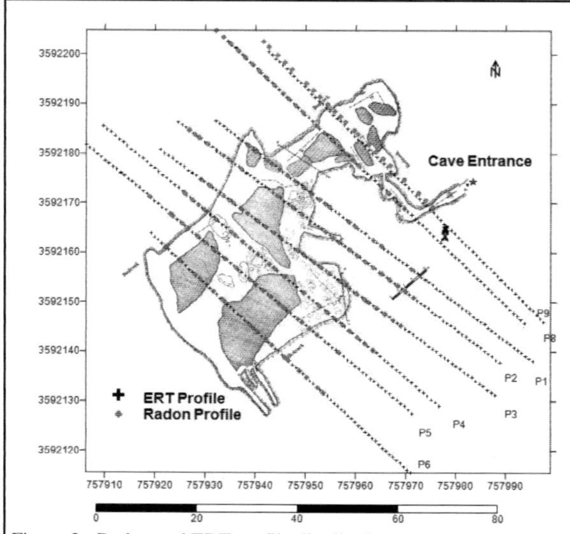

Figure 3: Radon and ERT profile distribution.

Radon concentrations tend to increase across known fault and fracture zone trends. Additionally, elevated concentrations were encountered across the center of the main hall of the known cave structure. Some of the fault structure showed possible sealing characteristics, thus hindering the passage of radon gas. Several collapse features were seen across the resistivity sections of the subsurface. These localities provide good seepage pathways for the radon gas into the ground surface. The results showed potential isolated cave extensions that can add to the value of the site. The results of this study indicated the plausible integrated interpretation of radon natural radioactivity and electrical resistivity imaging in mapping karst terrain structures.

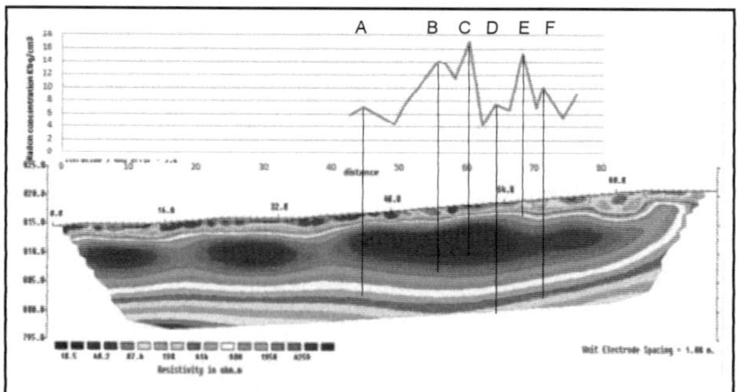

Figure 4: Comparison of radon concetration and ERT section of profile 1.

Reference

Abumurad, K., Atallah, M., Kullab, M., Ismail, A., 1997, Determination of radon soil concentration levels in the Governorate of Irbid, Jordan: Radiat. Meas., **28,** 585-588.

Albergi, S., Pecequilo B. S., and Campos M. P., 2005, Radon concentrations in caves of Parque Estadual do Alto Ribeira (PETAR), SP, Brazil: preliminary results: International Congress Series, **1276,** 403– 404.

Al-Bataina, B., and Abu-Kharma, H., 2004, Study of Radon Concentration in Soil of Samma Town, Jordan: Abhath AL-Yarmouk: Basic Sci. and Eng., **13**, no. 2, 177-190.

Al-Bataina, B. A., Ismail, A. M., Kullab, M. K., Abumurad, K. M., and Mustafa, H., 1997, Radon Measurements in different types of Natural Waters in Jordan: Radiat. Meas., **28,** no. 1, 591-594.

Batayneh, A., Al-Zoubi, A., and Abdallah, S., 2000, Detection of a solution cavity adjacent to a highway in southwest Jordan using electrical resistivity methods: Journal of Environmental Engineering, **5**, 25-30.

Ball,T. K., Cameron, D. G., and Colman, P. D., 1991, Behavior of radon in the geological environment: a review: Quarterly Journal of Engineering Geology, **24** , no. 2, 169-182.

Cigna, A. A., 2005, Radon in Caves: International Journal of Speleology, **34**, no. 1-2, 1-18.

DeGroot-Hedlin, C., and Constable, S., 1990, Occam's inversion to generate smooth, two-dimensional models form magneto telluric data: Geophysics **55**, 1613–1624

Griffiths, D., and Barker, R., 1993, Two-dimensional resistivity imaging and modelling in areas of complex geology: Journal of Applied Geophysics, **29**, 211-226.

Gautam, P., Raj,S., and Ando, H., 2000, Mapping of subsurface karst structure with gamma ray and electrical resistivity profiles: a case study from Pokhara valley, central Nepal, Journal of Applied Geophysics, **45**, 97-110.

Kempe, S., Al-Malabeh, A., Al-Shreideh, A., and Henschel, H., 2006, Al-Dahr Cave (Bergish), Jordan, the first extensive Jordanian limestone cave: a convective Carlsbad-type cave: Journal of Cave and Karst Studies, **68**, no. 3, 107–114.

Knoll, G. F., 2000, Radiation Detection and Measurements: John Wiley & Sons Inc.

Kullab, M., 2005, Assessment of radon-222 concentrations in buildings, building materials, water and soil in Jordan: Applied Radiation and Isotopes, **62**, no. 765-773.

Leucci, G., and De Giorgi, L., 2005, Integrated geophysical surveys to assess the structural conditions of a karstic cave of archaeological importance: Natural Hazards and Earth System Sciences, **5**, no. 1, 17-22.

Loke, M. H., and Barker, R. D., 1995, Least-squares deconvolution of apparent resistivity pseudosections. Geophysics Prospecting, **60**, 1682–1690.

Loke, M. H., and Barker, R. D., 1996a, Rapid least-squares inversion of apparent resistivity pseudosections by a quasi-Newton method, Geophysical Prospecting, **44**, no. 1, 131-152.

Loke, M. H., and Barker, R. D., 1996b, Practical techniques for 3D resistivity surveys and data inversion. Geophysical Prospecting, **44**, no. 3, 499-523.

Melhem, D., 2004, Geology of Bargash Cave system. M. S. thesis, Yarmouk University.

AUTHOR INDEX

AUTHOR INDEX

AUTHOR INDEX

Society of Exploration Geophysicists
8801 South Yale, Suite 500
Tulsa, Oklahoma 74137-3575
USA

ISBN 978-1-5108-9926-1

9 781510 899261

CURRAN ASSOCIATES INC.
proceedings.com